Annals of Mathematics Studies

Number 161

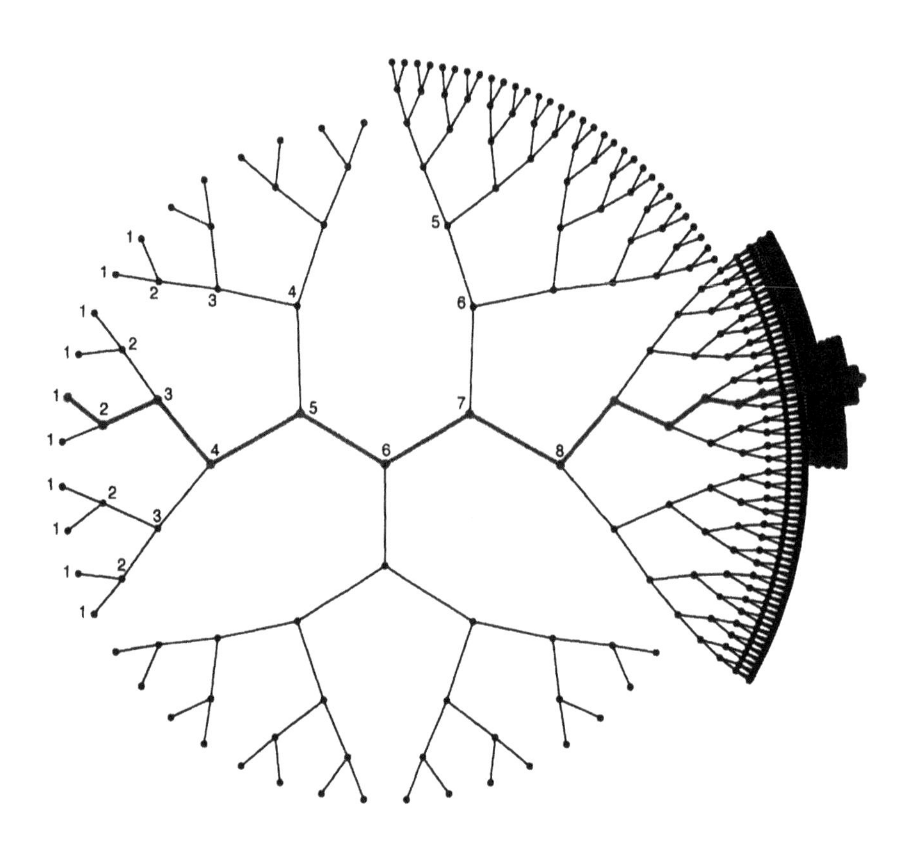

A vertical cycle $Z(j)$ in the case $p = 2$

The vertical cycle $Z(j)$ in the case $p = 2$ for the endomorphism j given by (6A.5.2) in Chapter 6: Appendix. Here the half apartment $\{[\Lambda] = [\Lambda_r] = [\,[e_1, 2^r e_2]\,] \mid \mathrm{mult}_{[\Lambda]}(j) > 0\}$ has been marked, and the multiplicities of components have been indicated.

Modular Forms and Special Cycles on Shimura Curves

Stephen S. Kudla
Michael Rapoport
Tonghai Yang

PRINCETON UNIVERSITY PRESS
PRINCETON AND OXFORD
2006

Published by Princeton University Press, 41 William Street, Princeton, New Jersey 08540

In the United Kingdom: Princeton University Press, 3 Market Place, Woodstock, Oxfordshire OX20 1SY

Library of Congress Cataloging-in-Publication Data

Kudla, Stephen S., 1950–
Modular forms and special cycles on Shimura curves / Stephen S. Kudla, Michael Rapoport, Tonghai Yang.
p. cm. — (Annals of mathematics studies ; 161)
Includes bibliographical references and index.
ISBN-13: 978-0-691-12550-3 (acid-free paper)
ISBN-10: 0-691-12550-3 (acid-free paper)
ISBN-13: 978-0-691-12551-0 (pbk. : acid-free paper)
ISBN-10: 0-691-12551-1 (pbk. : acid-free paper)
1. Arithmetic algebraic geometry. 2. Shimura varieties. I. Rapoport, M., 1948- II. Yang, Tonghai, 1963- III. Title. IV. Annals of mathematics studies ; no. 161.

QA242.5.K83 2006
516.3′5—dc22 2005054621

British Library Cataloging-in-Publication Data is available

This book has been composed in Times Roman in LaTeX

The publisher would like to acknowledge the authors of this volume for providing the camera-ready copy from which this book was printed.

Printed on acid-free paper. ∞

pup.princeton.edu

Printed in the United States of America

10 9 8 7 6 5 4 3 2 1

Contents

Acknowledgments

The authors would like to thank the many people and institutions who provided support over the long period of time during which this project was realized.

SK would like to thank J.-B. Bost, J. Bruinier, J. Burgos, M. Harris, J. Kramer, and U. Kühn for stimulating mathematical discussions. He would also like to thank the University of Cologne and Bonn University for their hospitality during many visits. The extended research time in Germany, which was essential to the completion of this work, was made possible through the support of a Max-Planck Research Prize from the Max-Planck Society and Alexander von Humboldt Stiftung. In addition, SK profited from stays at the Morningside Center, Beijing, the Fields Institute, Toronto, the University of Wisconsin, Madison, the University of Paris XI, Orsay, and the Newton Institute, Cambridge. He was supported by NSF Grants: DMS-9970506, DMS-0200292 and DMS-0354382.

MR acknowledges helpful discussions with T. Zink and with the members of the ARGOS seminar, especially I. Bouw, U. Görtz, I. Vollaard, and S. Wewers. He also wishes to express his gratitude to the mathematics department of the University of Maryland for its hospitality on several occasions.

TY would like to thank the Max-Planck Institute at Bonn, the Hong Kong University of Science and Technology, The Morningside Center of Mathematics in Beijing, and The National Center for Theoretical Research in Taiwan for providing a wonderful working environment when he visited these institutes at various times. He was supported by NSF Grants DMS-0354353 and DMS-0302043, as well as by a grant from the NSA Mathematical Sciences Program.

We thank U. Görtz for producing the frontispiece and B. Wehmeyer for her assistance in TeXing several chapters.

Chapter One

Introduction

In this monograph we study the arithmetic geometry of cycles on an arithmetic surface $\mathcal{M}$ associated to a Shimura curve over the field of rational numbers and the modularity of certain generating series constructed from them. We consider two types of generating series, one for divisors and one for 0-cycles, valued in $\widehat{\mathrm{CH}}^1(\mathcal{M})$ and $\widehat{\mathrm{CH}}^2(\mathcal{M})$, the first and second arithmetic Chow groups of $\mathcal{M}$, respectively. We prove that the first type is a nonholomorphic elliptic modular form of weight $\frac{3}{2}$ and that the second type is a nonholomorphic Siegel modular form of genus two and weight $\frac{3}{2}$. In fact we identify the second type of series with the central derivative of an incoherent Siegel-Eisenstein series. We also relate the height pairing of a pair of $\widehat{\mathrm{CH}}^1(\mathcal{M})$-valued generating series to the $\widehat{\mathrm{CH}}^2(\mathcal{M})$-valued series by an inner product identity. As an application of these results we define an arithmetic theta lift from modular forms of weight $\frac{3}{2}$ to the Mordell-Weil space of $\mathcal{M}$ and prove a nonvanishing criterion analogous to that of Waldspurger for the classical theta lift, involving the central derivative of the L-function.

We now give some background and a more detailed description of these results.

The modular curve $\Gamma \setminus \mathfrak{H}$, where $\mathfrak{H} = \{z \in \mathbb{C} \mid \mathrm{Im}\, z > 0\}$ is the upper half plane and $\Gamma = \mathrm{SL}_2(\mathbb{Z})$, is the first nontrivial example of a locally symmetric variety, and of a Shimura variety. It is also the host of the space of modular forms and is the moduli space of elliptic curves. Starting from this last interpretation, we see that the modular curve comes equipped with a set of special divisors, which, like the classical Heegner divisors, are the loci of elliptic curves with extra endomorphisms. More precisely, for $t \in \mathbb{Z}_{>0}$ let

$$(1.0.1)\quad Z(t) = \{(E, x) \mid x \in \mathrm{End}(E) \text{ with } \mathrm{tr}(x) = 0,\ x^2 = -t \cdot \mathrm{id}_E\},$$

where E denotes an elliptic curve. The resulting divisor on the modular curve, which we also denote by $Z(t)$, is the set of points where the corresponding elliptic curve E admits an action of the order $\mathbb{Z}[x] = \mathbb{Z}[\sqrt{-t}]$ in the imaginary quadratic field $k_t = \mathbb{Q}(\sqrt{-t})$, i.e., E admits complex multiplication by this order. One may also interpret $Z(t)$ as the set of Γ-orbits in $\mathfrak{H}$ which contain a fixed point of an element $\gamma \in \mathrm{M}_2(\mathbb{Z})$ with $\mathrm{tr}(\gamma) = 0$ and

$\det(\gamma) = t$.

It is a classical fact that the degree of $Z(t)$ is given by $\deg Z(t) = H(4t)$, where $H(n)$ is the Hurwitz class number. It is also known that the generating series

$$\sum_{t>0} \deg Z(t)\, q^t = \sum_{t>0} H(4t)\, q^t \tag{1.0.2}$$

is *nearly* the q-expansion of a modular form. In fact, Zagier [58] showed that the complete series, for $\tau = u + iv$,

$$\mathcal{E}(\tau, \tfrac{1}{2}) = -\frac{1}{12} + \sum_{t>0} H(4t)\, q^t + \sum_{n\in\mathbb{Z}} \frac{1}{8\pi}\, v^{-\frac{1}{2}} \int_1^\infty e^{-4\pi n^2 vr}\, r^{-\frac{3}{2}}\, dr \cdot q^{-n^2}, \tag{1.0.3}$$

is the q-expansion of the value at $s = \frac{1}{2}$ of a nonholomorphic Eisenstein series $\mathcal{E}(\tau, s)$ of weight $\frac{3}{2}$, and hence is a modular form.

Generating series of this kind have a long and rich history. They are all modeled on the classical theta series. Recall that if (L, Q) is a positive definite quadratic $\mathbb{Z}$-module of rank n, one associates to it the generating series

$$\theta_L(\tau) = \sum_{x\in L} q^{Q(x)} = 1 + \sum_{t=1}^{\infty} r_L(t)\, q^t. \tag{1.0.4}$$

Here

$$r_L(t) = |\{x \in L \mid Q(x) = t\}|, \tag{1.0.5}$$

and we have set, as elsewhere in this book, $q = e(\tau) = e^{2\pi i\tau}$. It is a classical result going back to the 19th century that θ_L is the q-expansion of a holomorphic modular form of weight $\frac{n}{2}$ for some congruence subgroup of $\mathrm{SL}_2(\mathbb{Z})$. Similarly, Siegel considered generating series of the form

$$\theta_r(\tau, L) = \sum_{x\in L^r} q^{Q(x)} = \sum_{T\in \mathrm{Sym}_r(\mathbb{Z})^\vee} r_L(T)\, q^T, \tag{1.0.6}$$

where $\tau \in \mathfrak{H}_r$, and $q^T = e(\mathrm{tr}(T\tau))$, and

$$r_L(\tau) = |\{\, x \in L^r \mid Q(x) = \frac{1}{2}((x_i, x_j)) = T\,\}|. \tag{1.0.7}$$

He showed that they define Siegel modular forms of genus r and weight $\frac{n}{2}$. Generalizations to indefinite quadratic forms were considered by Hecke and Siegel, and the resulting generating series can be nonholomorphic modular forms. Hirzebruch and Zagier [20] constructed generating series whose coefficients are given by cohomology classes of special curves on Hilbert-Blumenthal surfaces. They prove that the image under any linear functional of this generating series is an elliptic modular form. For example, they identify the modular form arising via the cup product with the Kähler class as an explicitly given Eisenstein series. One can also define special 0-cycles on Hilbert-Blumenthal surfaces and make generating functions for their degrees [25] . These can be shown to be Siegel modular forms of genus two and weight 2.

We now turn to the generating series associated to arithmetic cycles on Shimura curves. We exclude the modular curve to avoid problems caused by its noncompactness. It should be pointed out, however, that all our results should have suitable analogues for the modular curve, cf. [57]. We pay, however, a price for assuming compactness. New difficulties arise due to bad reduction and to the absence of natural modular forms.

Let B be an indefinite quaternion division algebra over $\mathbb{Q}$, so that

$$B \otimes_{\mathbb{Q}} \mathbb{R} \simeq M_2(\mathbb{R}) \qquad \text{and} \qquad D(B) \underset{\text{def.}}{=} \prod_{\substack{B\otimes_{\mathbb{Q}}\mathbb{Q}_p \\ \text{division}}} p > 1. \tag{1.0.8}$$

Let

$$V = \{x \in B \mid \mathrm{tr}(x) = 0\}, \tag{1.0.9}$$

with quadratic form $Q(x) = \mathrm{Nm}(x) = -x^2$, where tr and Nm denote the reduced trace and norm on B respectively. Then V is a quadratic space over $\mathbb{Q}$ of signature type $(1, 2)$. Let

$$D = \{w \in V(\mathbb{C}) \mid (w, w) = 0,\ (w, \bar{w}) < 0\}/\mathbb{C}^\times, \tag{1.0.10}$$

where $(x, y) = Q(x+y) - Q(x) - Q(y)$ is the bilinear form associated to the quadratic form Q. Then D is an open subset of a quadric in $\mathbb{P}(V(\mathbb{C})) \simeq \mathbb{P}^2$, and $(B \otimes_{\mathbb{Q}} \mathbb{R})^\times$ acts on $V(\mathbb{R})$ and D by conjugation. We fix a maximal order O_B in B. Since all these maximal orders are conjugate, this is not really an additional datum. Set $\Gamma = O_B^\times$. The Shimura curve associated to B is the quotient

$$[\Gamma \backslash D]. \tag{1.0.11}$$

Since Γ does not act freely, the quotient here is to be interpreted as an orbifold.

Let us fix an isomorphism $B \otimes_{\mathbb{Q}} \mathbb{R} = \mathrm{M}_2(\mathbb{R})$. Then we can also identify $(B \otimes_{\mathbb{Q}} \mathbb{R})^\times = \mathrm{GL}_2(\mathbb{R})$ and $(B \otimes_{\mathbb{Q}} \mathbb{R})^\times$ acts on $\mathfrak{H}_\pm = \mathbb{C} \setminus \mathbb{R}$ by fractional linear transformations. We obtain an identification

$$D = \mathbb{C} \setminus \mathbb{R}, \ \text{ via } \begin{pmatrix} z & -z^2 \\ 1 & -z \end{pmatrix} \longmapsto z, \tag{1.0.12}$$

equivariant for the action of $(B \otimes_{\mathbb{Q}} \mathbb{R})^\times = \mathrm{GL}_2(\mathbb{R})$.

The Shimura curve associated to B has a modular interpretation. Namely, consider the moduli problem $\mathcal{M}$ which associates to a scheme S over Spec $\mathbb{Z}$ the category of pairs (A, ι) where

- A is an abelian scheme over S
- $\iota : O_B \to \mathrm{End}(A)$ is an action of O_B on A with characteristic polynomial

$$\mathrm{charpol}(x \mid \mathrm{Lie}\ A) = (T - x)(T - x^\iota) \in \mathcal{O}_S[T],$$

  for the induced O_B-action on the Lie algebra.

Here $x \mapsto x^\iota$ denotes the main involution on B. If S is a scheme in characteristic zero, then the last condition simply says that A has dimension 2, i.e., that (A, ι) is a *fake elliptic curve* in the sense of Serre. This moduli problem is representable by an algebraic stack in the sense of Deligne-Mumford, and we denote the representing stack by the same symbol $\mathcal{M}$. We therefore have an isomorphism of orbifolds,

$$\mathcal{M}(\mathbb{C}) = [\Gamma \setminus D]. \tag{1.0.13}$$

Since B is a division quaternion algebra, $\mathcal{M}$ is proper over Spec $\mathbb{Z}$ and $\mathcal{M}(\mathbb{C})$ is a compact Riemann surface (when we neglect the orbifold aspect). By its very definition, the stack $\mathcal{M}$ is an integral model of the orbifold $[\Gamma \setminus D]$. It turns out that $\mathcal{M}$ is smooth over Spec $\mathbb{Z}[D(B)^{-1}]$ but has bad reduction at the prime divisors of $D(B)$. At the primes p with $p \mid D(B)$, the stack $\mathcal{M}$ has semistable reduction and, in fact, admits a p-adic uniformization by the Drinfeld upper half plane $\hat{\Omega}$. In particular, the special fiber $\mathcal{M}_p$ is connected but in general not irreducible.

In analogy with the case of the modular curve, we can define special divisors on the Shimura curve by considering complex multiplication points. More precisely, let $t \in \mathbb{Z}_{>0}$ and introduce a relative DM-stack $\mathcal{Z}(t)$ over $\mathcal{M}$ by posing the following moduli problem. To a scheme S the moduli problem $\mathcal{Z}(t)$ associates the category of triples (A, ι, x), where

- (A, ι) is an object of $\mathcal{M}(S)$
- $x \in \mathrm{End}(A, \iota)$ is an endomorphism such that $\mathrm{tr}(x) = 0$, $x^2 = -t \cdot \mathrm{id}_A$.

An endomorphism as above is called a *special endomorphism* of (A, ι). The space $V(A, \iota)$ of special endomorphisms is equipped with the degree form $Q(x) = x^\iota x$. Note that for $x \in V(A, \iota)$ we have $Q(x) = -x^2$. We denote by the same symbol the image of $\mathcal{Z}(t)$ as a cycle in $\mathcal{M}$ and use the notation $Z(t) = \mathcal{Z}(t)_{\mathbb{C}}$ for its complex fiber. Note that $Z(t)$ is a finite set of points on the Shimura curve, corresponding to those fake elliptic curves which admit complex multiplication by the order $\mathbb{Z}[\sqrt{-t}]$. We form the generating series

$$\phi_1(\tau) = -\mathrm{vol}(\mathcal{M}(\mathbb{C})) + \sum_{t>0} \deg(Z(t))\, q^t \in \mathbb{C}[[q]]. \tag{1.0.14}$$

Here the motivation for the constant term is as follows. Purely formally $\mathcal{Z}(0)$ is equal to $\mathcal{M}$ with associated cohomology class in degree zero; to obtain a cohomology class in the correct degree, one forms the cup product with the natural Kähler class — which comes down to taking (up to sign) the volume of $\mathcal{M}(\mathbb{C})$ with respect to the hyperbolic volume element.

Proposition 1.0.1. *The series $\phi_1(\tau)$ is the q-expansion of a holomorphic modular form of weight $3/2$ and level $\Gamma_0(4D(B)_o)$, where $D(B)_o = D(B)$ if $D(B)$ is odd and $D(B)_o = D(B)/2$ if $D(B)$ is even.*

Just as with the theorem of Hirzebruch and Zagier, this is not proved by checking the functional equations that a modular form has to satisfy. Rather, the theorem is proved by identifying the series $\phi_1(\tau)$ with a specific Eisenstein series[1]. More precisely, for $\tau = u + iv \in \mathfrak{H}$, set

$$\mathcal{E}_1(\tau, s, B) = v^{\frac{1}{2}(s-\frac{1}{2})} \cdot \sum_{\gamma \in \Gamma'_\infty \backslash \Gamma'} (c\tau + d)^{-\frac{3}{2}} |c\tau + d|^{-(s-\frac{1}{2})} \Phi^B(\gamma, s), \tag{1.0.15}$$

where $\gamma = \left(\begin{smallmatrix} a & b \\ c & d \end{smallmatrix}\right) \in \Gamma' = \mathrm{SL}_2(\mathbb{Z})$, and $\Phi^B(\gamma, s)$ is a certain function depending on B. The Eisenstein series $\mathcal{E}_1(\tau, s, B)$ is the analogue for the Shimura curve of Zagier's Eisenstein series (1.0.3). It has a functional equation of the form

$$\mathcal{E}_1(\tau, s, B) = \mathcal{E}_1(\tau, -s, B). \tag{1.0.16}$$

Its value at $s = \frac{1}{2}$ is a modular form of weight $\frac{3}{2}$ and we may consider its q-expansion. Proposition 1.0.1 now follows from the following more precise result.

Proposition 1.0.2.

$$\phi_1(\tau) = \mathcal{E}_1(\tau, \frac{1}{2}, B),$$

i.e., ϕ_1 is the q-expansion of $\mathcal{E}_1(\tau, \frac{1}{2}, B)$.

---

[1]Alternatively, $\phi_1(\tau)$ can be obtained by calculating the integral over $\mathcal{M}(\mathbb{C})$ of a theta function valued in $(1,1)$ forms; this amounts to a very special case of the results of [33]. The analogous computation in the case of modular curves was done by Funke [11].

Proposition 1.0.2 is proved in [38] by calculating the coefficients of both power series explicitly and comparing them term by term. These coefficients turn out to be generalized class numbers. More precisely, for $t > 0$, the coefficient of q^t on either side is equal to

$$\deg\, Z(t) = 2\delta(d; D(B)) H_0(t; D(B)), \tag{1.0.17}$$

where

$$\delta(d; D) = \prod_{\ell | D} (1 - \chi_d(\ell)) \tag{1.0.18}$$

and

$$H_0(t; D) = \frac{h(d)}{w(d)} \sum_{\substack{c|n \\ (c,D)=1}} c \prod_{\ell | c} (1 - \chi_d(\ell)\ell^{-1}). \tag{1.0.19}$$

Here d denotes the fundamental discriminant of the imaginary quadratic field $\mathbf{k}_t = \mathbb{Q}(\sqrt{-t})$ and we have written $4t = n^2 d$; also, $h(d)$ denotes the class number of $\mathbf{k}_t$ and $w(d)$ the number of roots of unity contained in $\mathbf{k}_t$. By χ_d we denote the quadratic residue character mod d. For $t = 0$, the identity in Proposition 1.0.2 reduces to the well-known formula for the volume

$$\mathrm{vol}(\mathcal{M}(\mathbb{C})) = \zeta_{D(B)}(-1), \tag{1.0.20}$$

where in $\zeta_{D(B)}(s)$ the index means that the Euler factors for $p \mid D(B)$ have been omitted in the Riemann zeta function. Note that the fact that the generating series $\phi_1(\tau)$ is a modular form reveals some surprising and highly nonobvious coherence among the degrees of the various special cycles $Z(t)$.

In this book we will establish arithmetic analogues of Propositions 1.0.1 and 1.0.2. In contrast to the above propositions, which are statements about generating series valued in cohomology (just as was the case with the results of Hirzebruch-Zagier), our generating series will have coefficients in the arithmetic Chow groups of Gillet-Soulé [14], [48], (see also [3]). Let us recall briefly their definition in our case.

A divisor on $\mathcal{M}$ is an element of the free abelian group generated by the closed irreducible reduced substacks which are, locally for the étale topology, Cartier divisors. A Green function for the divisor $\mathcal{Z}$ is a function g on $\mathcal{M}(\mathbb{C})$ with logarithmic growth along the complex points of $Z = \mathcal{Z}_{\mathbb{C}}$ and which satisfies the Green equation of currents on $\mathcal{M}(\mathbb{C})$,

$$dd^c g + \delta_Z = [\eta], \tag{1.0.21}$$

where η is a smooth $(1,1)$-form. Let $\hat{Z}_{\mathbb{Z}}(\mathcal{M})$ be the group of pairs $(\mathcal{Z}, g)$, where g is a Green function for the divisor $\mathcal{Z}$. The first arithmetic Chow group $\widehat{\mathrm{CH}}^1_{\mathbb{Z}}(\mathcal{M})$ is the factor group of $\hat{Z}_{\mathbb{Z}}(\mathcal{M})$ by the subgroup generated by the Arakelov principal divisors $\widehat{\mathrm{div}}\, f$ associated to rational functions on $\mathcal{M}$. For us it will be more convenient to work instead with the $\mathbb{R}$-linear version $\widehat{\mathrm{CH}}^1(\mathcal{M})$. In its definition one replaces $\mathbb{Z}$-linear combinations of divisors by $\mathbb{R}$-linear combinations and divides out by the $\mathbb{R}$-subspace generated by the Arakelov principal divisors. Such groups were introduced by Gillet-Soulé [15]; for the case relevant to us, see [3]. Note that restriction to the generic fiber defines the degree map

$$\deg_{\mathbb{Q}} : \widehat{\mathrm{CH}}^1(\mathcal{M}) \longrightarrow \mathrm{CH}^1(\mathcal{M}_{\mathbb{C}}) \otimes \mathbb{R} \xrightarrow{\sim} \mathbb{R}. \tag{1.0.22}$$

The group $\widehat{\mathrm{CH}}^2(\mathcal{M})$ is defined in an analogous way, starting with 0-cycles on $\mathcal{M}$. Since the fibers of $\mathcal{M}$ over Spec $\mathbb{Z}$ are geometrically connected of dimension 1, the arithmetic degree map yields an isomorphism

$$\widehat{\deg} : \widehat{\mathrm{CH}}^2(\mathcal{M}) \xrightarrow{\sim} \mathbb{R}. \tag{1.0.23}$$

Finally we mention the Gillet-Soulé arithmetic intersection pairing,

$$\langle\, ,\, \rangle : \widehat{\mathrm{CH}}^1(\mathcal{M}) \times \widehat{\mathrm{CH}}^1(\mathcal{M}) \longrightarrow \widehat{\mathrm{CH}}^2(\mathcal{M}) = \mathbb{R}. \tag{1.0.24}$$

It will play the role of the cup product in cohomology in this context.

We now define a generating series with coefficients in $\widehat{\mathrm{CH}}^1(\mathcal{M})$ using the divisors $\mathcal{Z}(t)$. For $t > 0$, we equip the divisor $\mathcal{Z}(t)$ with the Green function $\Xi(t, v)$ depending on a parameter $v \in \mathbb{R}_{>0}$, constructed in [24]. Let $\hat{\mathcal{Z}}(t, v)$ be the corresponding class in $\widehat{\mathrm{CH}}^1(\mathcal{M})$. For $t < 0$ note that $\mathcal{Z}(t) = \emptyset$. However, the function $\Xi(t, v)$ is still defined and is smooth for $t < 0$, hence it is a Green function for the trivial divisor, and we may define again $\hat{\mathcal{Z}}(t, v)$ to be the class of $(\mathcal{Z}(t), \Xi(t, v)) = (0, \Xi(t, v))$. To define $\hat{\mathcal{Z}}(0, v)$, we take our lead from the justification of the absolute term in the generating series (1.0.14).

Let ω be the Hodge line bundle on $\mathcal{M}$, i.e., the determinant bundle of the dual of the relative Lie algebra of the universal family $(\mathcal{A}, \iota)$ over $\mathcal{M}$,

$$\omega = \wedge^2(\mathrm{Lie}\, \mathcal{A})^*. \tag{1.0.25}$$

The complex fiber of this line bundle comes equipped with a natural metric. This metric is well defined up to scaling.[2] We denote by $\hat{\omega}$ the class of this

[2] The normalization of the metric we use differs from the standard normalization.

metrized line bundle under the natural map from $\widehat{\text{Pic}}(\mathcal{M})$ to $\widehat{\text{CH}}^1(\mathcal{M})$ and set

$$\hat{\mathcal{Z}}(0, v) = -\hat{\omega} - (0, \log(v) + \mathbf{c}), \tag{1.0.26}$$

where $\mathbf{c}$ is a suitable constant.

The DM-stack $\mathcal{Z}(t)$ is finite and unramified over $\mathcal{M}$. It is finite and flat, i.e., a relative divisor, over Spec $\mathbb{Z}[D(B)^{-1}]$ but may contain irreducible components of the special fiber $\mathcal{M}_p$ when $p \mid D(B)$. This integral extension of the 0-cycles $Z(t)$ is therefore sometimes different from the extension obtained by flat closure in $\mathcal{M}$. Its nonflatness depends in a subtle way on the p-adic valuation of t. Our definition of $Z(t)$ is a consequence of our insistence on a thoroughly modular treatment of our special cycles, which is essential to our method. We strongly suspect that in fact the closure definition does not lead to (variants of) our main theorems and that therefore our definition is the 'right one'. We do not know this for sure since the closure definition is hard to work with.

We form the generating series,

$$\hat{\phi}_1 = \sum_{t \in \mathbb{Z}} \hat{\mathcal{Z}}(t, v)\, q^t \in \widehat{\text{CH}}^1(\mathcal{M})[[q^{\pm 1}]], \tag{1.0.27}$$

where the coefficients depend on the parameter $v \in \mathbb{R}_{>0}$ via the Green function $\Xi(t, v)$. The first main result of this book, proved in Chapter 4, may now be formulated as follows:

Theorem A. *For $\tau = u + iv$, $\hat{\phi}_1(\tau)$ is a (nonholomorphic) modular form of weight $\frac{3}{2}$ and level $\Gamma_0(4D(B)_o)$ with values in $\widehat{\text{CH}}^1(\mathcal{M})$.*

To explain the meaning of the statement of the theorem, recall that the $\mathbb{R}$-version $\widehat{\text{CH}}^1(\mathcal{M})$ of the arithmetic Chow group splits canonically into a direct sum of a finite-dimensional $\mathbb{C}$-vector space $\widehat{\text{CH}}^1(\mathcal{M}, \mu)$, the classical Arakelov Chow group with respect to the hyperbolic metric, and the vector space $C^\infty(\mathcal{M}(\mathbb{C}))_0$ of smooth functions on $\mathcal{M}(\mathbb{C})$ orthogonal to the constant functions. Correspondingly, the series $\hat{\phi}_1$ is the sum of a series $\hat{\phi}_1^0$ in q with coefficients in $\widehat{\text{CH}}^1(\mathcal{M}, \mu)$ and a series $\hat{\phi}_1^\infty$ in q with coefficients in $C^\infty(\mathcal{M}(\mathbb{C}))_0$. The assertion of the theorem should be interpreted as follows. There is a smooth function on $\mathfrak{H}$ with values in the finite-dimensional vector space $\widehat{\text{CH}}^1(\mathcal{M}, \mu)$ which satisfies the usual transformation law for a modular form of weight $\frac{3}{2}$ and of level $\Gamma_0(4D(B)_o)$ whose q-expansion is equal to $\hat{\phi}_1^0$, and there is a smooth function on $\mathfrak{H} \times \mathcal{M}(\mathbb{C})$ which satisfies the usual transformation law for a modular form of weight $\frac{3}{2}$ and of level

$\Gamma_0(4D(B)_o)$ in the first variable and whose q-expansion in the first variable is equal to $\hat{\phi}_1^\infty$. Obviously, the series $\hat{\phi}_1^0$ satisfies the above condition if for any linear form $\ell : \widehat{\mathrm{CH}}^1(\mathcal{M},\mu) \to \mathbb{C}$ the series $\ell(\hat{\phi}_1)$ with coefficients in $\mathbb{C}$ is a nonholomorphic modular form of weight $\frac{3}{2}$ and level $\Gamma_0(4D(B)_o)$ in the usual sense.

Let us explain briefly what is involved in the proof of Theorem A. The structure of $\widehat{\mathrm{CH}}^1(\mathcal{M},\mu)$ is encapsulated in the following direct sum decomposition

$$\widehat{\mathrm{CH}}^1(\mathcal{M},\mu) = \widetilde{\mathrm{MW}} \oplus \mathbb{R}\,\hat{\omega} \oplus \mathrm{Vert}. \tag{1.0.28}$$

Here

$$\widetilde{\mathrm{MW}} \simeq \mathrm{MW}(\mathcal{M}_{\mathbb{Q}}) := \mathrm{Pic}^0(\mathcal{M}_{\mathbb{Q}})(\mathbb{Q}) \otimes \mathbb{R} \tag{1.0.29}$$

is the orthogonal complement to $(\mathbb{R}\,\hat{\omega} \oplus \mathrm{Vert})$, and the subspace Vert is spanned by the elements $(Y,0)$, where Y is an irreducible component of a fiber $\mathcal{M}_p$ for some p. Also, $\mathrm{MW}(\mathcal{M}_{\mathbb{Q}})$ is the Mordell-Weil group of $\mathcal{M}_{\mathbb{Q}}$, tensored with $\mathbb{R}$. By the above remark, we have to prove the modularity of $\ell(\hat{\phi}_1^0)$ for linear functionals ℓ on each of the summands of (1.0.28).

For the summand $\widetilde{\mathrm{MW}}$, this is done by comparing the restriction to the generic fiber of our generating series $\hat{\phi}_1$ with the generating series considered by Borcherds [2], for which he proved modularity. Proposition 1.0.1 is used to produce divisors of degree 0 in the generic fiber from our special divisors.

For the summand $\mathbb{R}\,\hat{\omega}$, the modularity follows from the following theorem which is the main result of [38]. Note that this theorem not only gives modularity but even identifies the modular form explicitly. We form the generating series with coefficients in $\mathbb{C}$ obtained by cupping with $\hat{\omega}$,

$$\langle\,\hat{\omega},\phi_1\,\rangle = \sum_t \langle\,\hat{\omega},\hat{\mathcal{Z}}(t,v)\,\rangle\, q^t. \tag{1.0.30}$$

Theorem 1.0.3. *The series above coincides with the q-expansion of the* derivative *at $s=\frac{1}{2}$ of the Eisenstein series (1.0.15),*

$$\langle\,\hat{\omega},\hat{\phi}_1\,\rangle = \mathcal{E}_1'(\tau,\frac{1}{2},B).$$

Next, consider the pairings of the generating series $\hat{\phi}_1$ with the classes $(Y,0) \in \mathrm{Vert}$, where Y is an irreducible component of a fiber with bad reduction $\mathcal{M}_p$, i.e., $p \mid D(B)$. The corresponding series can be identified with classical theta functions for the positive definite ternary lattice associated to the definite quaternion algebra $B^{(p)}$ with $D(B^{(p)}) = D(B)/p$. This is based on the theory of p-adic uniformization and uses the analysis of the special cycles at primes of bad reduction [36].

Finally, for the series $\hat{\phi}_1^\infty$, we show that the coefficients of the spectral expansion of $\hat{\phi}_1$ are Maass forms. More precisely, if f_λ is an eigenfunction of the Laplacian with eigenvalue λ, then the coefficient of f_λ in $\hat{\phi}_1$ is up to an explicit scalar the classical theta lift $\theta(f_\lambda)$ to a Maass form of weight $\frac{3}{2}$ and level $\Gamma_0(4D(B)_o)$.

To formulate the second main result of this book, Theorem B, we form a generating series for 0-cycles on $\mathcal{M}$ instead of divisors on $\mathcal{M}$. The idea is to impose a *pair* of special endomorphisms, i.e., 'twice as much CM'. Let $\mathrm{Sym}_2(\mathbb{Z})^\vee$ denote the set of half-integral symmetric matrices of size 2, and let $T \in \mathrm{Sym}_2(\mathbb{Z})^\vee$. We define a relative DM-stack $\mathcal{Z}(T)$ over $\mathcal{M}$ by posing the following moduli problem. To a scheme S the moduli problem $\mathcal{Z}(T)$ associates the category of triples $(A, \iota, \mathbf{x})$ where

- (A, ι) is an object of $\mathcal{M}(S)$
- $\mathbf{x} = [x_1, x_2] \in \mathrm{End}(A, \iota)^2$ is a pair of endomorphisms with $\mathrm{tr}(x_1) = \mathrm{tr}(x_2) = 0$, and $\frac{1}{2}(\mathbf{x}, \mathbf{x}) = T$.

Here $(\mathbf{x}, \mathbf{x}) = ((x_i, x_j))_{i,j}$. It is then clear that $\mathcal{Z}(T)$ has empty generic fiber when T is positive definite, since in characteristic 0 a fake elliptic curve cannot support linearly independent complex multiplications. However, perhaps somewhat surprisingly, $\mathcal{Z}(T)$ is not always a 0-divisor on $\mathcal{M}$.

To explain the situation, recall from [24] that any $T \in \mathrm{Sym}_2(\mathbb{Z})^\vee$ with $\det(T) \neq 0$ determines a set of primes $\mathrm{Diff}(T, B)$ of odd cardinality. More precisely, let $\mathcal{C} = (\mathcal{C}_p)$ be the (*incoherent*) collection of local quadratic spaces where $\mathcal{C}_p = V_p$ for $p < \infty$ and where $\mathcal{C}_\infty$ is the positive definite quadratic space of dimension 3. If $T \in \mathrm{Sym}_2(\mathbb{Q})$ is nonsingular, we let V_T be the unique ternary quadratic space over $\mathbb{Q}$ with discriminant $-1 = \mathrm{discr}(V)$ which represents T. We denote by B_T the unique quaternion algebra over $\mathbb{Q}$ such that its trace zero subspace is isometric to V_T and define

$$\mathrm{Diff}(T, B) = \{\, p \leq \infty \mid \mathrm{inv}_p(B_T) \neq \mathrm{inv}(\mathcal{C}_p) \,\}. \tag{1.0.31}$$

Note that $\infty \in \mathrm{Diff}(T, B)$ if and only if T is not positive definite.

If $|(\mathrm{Diff}(T, B))| > 1$ or $\mathrm{Diff}(T, B) = \{\infty\}$, then $\mathcal{Z}(T) = \emptyset$. Assume now that $\mathrm{Diff}(T, B) = \{p\}$ with $p < \infty$. If $p \nmid D(B)$, then $\mathcal{Z}(T)$ is a 0-cycle on $\mathcal{M}$ with support in the fiber $\mathcal{M}_p$, as desired. In fact, the cycle is concentrated in the supersingular locus of $\mathcal{M}_p$. If, however, $p \mid D(B)$, then $\mathcal{Z}(T)$ is (almost always) a vertical *divisor* concentrated in $\mathcal{M}_p$.

Our goal now is to form a generating series with coefficients in $\widehat{\mathrm{CH}}^2(\mathcal{M})$,

$$\hat{\phi}_2 = \sum_{T \in \mathrm{Sym}_2(\mathbb{Z})^\vee} \hat{\mathcal{Z}}(T, v) q^T. \tag{1.0.32}$$

Here the coefficients $\hat{\mathcal{Z}}(T, v) \in \widehat{\mathrm{CH}}^2(\mathcal{M})$ will in general depend on $v \in \mathrm{Sym}_2(\mathbb{R})_{>0}$. How to define them is evident from the above only in the case when T is positive definite and $\mathrm{Diff}(T, B) = \{p\}$ with $p \nmid D(B)$. In this case we set

$$\hat{\mathcal{Z}}(T, v) = (\mathcal{Z}(T), 0) \in \widehat{\mathrm{CH}}^2(\mathcal{M}), \tag{1.0.33}$$

independent of v. Then $\hat{\mathcal{Z}}(T, v)$ has image $\log|\mathcal{Z}(T)| \in \mathbb{R}$ under the arithmetic degree map (1.0.23). If $T \in \mathrm{Sym}_2(\mathbb{Z})^\vee$ is nonsingular with $|\mathrm{Diff}(T, B)| > 1$, we set $\hat{\mathcal{Z}}(T, v) = 0$. In the remaining cases, the definition we give of the coefficients of (1.0.32) is more subtle. If $\mathrm{Diff}(T, B) = \{\infty\}$, then $\hat{\mathcal{Z}}(T, v)$ does depend on v; its definition is purely archimedean and depends on the rotational invariance of the $*$-product of two of the Green functions in [24], one of the main results of that paper. If $\mathrm{Diff}(T, B) = \{p\}$ with $p \mid D(B)$, then the definition of $\hat{\mathcal{Z}}(T, v)$ (which is independent of v) relies on the $\mathrm{GL}_2(\mathbb{Z}_p)$-invariance of the degenerate intersection numbers on the Drinfeld upper half plane, one of the main results of [36]. Finally, for singular matrices $T \in \mathrm{Sym}_2(\mathbb{Z})^\vee_{\geq 0}$ we are, in effect, imposing only a 'single CM', and the naive cycle is a divisor, so that its class lies in the wrong degree; we again use the heuristic principle that was used in the definition of the constant term of (1.0.14) and in the definition of $\hat{\mathcal{Z}}(0, v)$ in (1.0.26). In these cases we are guided in our definitions by the desire to give a construction that is on the one hand as natural as possible, and on the other hand to obtain the modularity of the generating series. We refer to Chapter 6 for the details.

Our second main theorem identifies the generating series (1.0.32) with an explicit (nonholomorphic) Siegel modular form of genus two. Recall that such a modular form admits a q-expansion as a Laurent series in

$$q^T = e(\mathrm{tr}(T\tau)), \quad T \in \mathrm{Sym}_2(\mathbb{Z})^\vee, \tag{1.0.34}$$

and that the coefficients may depend on the imaginary part $v \in \mathrm{Sym}_2(\mathbb{R})_{>0}$ of $\tau = u + iv \in \mathfrak{H}_2$. We introduce a Siegel Eisenstein series $\mathcal{E}_2(\tau, s, B)$ which is *incoherent* in the sense of [24]. In particular, 0 is the center of symmetry for the functional equation, and $\mathcal{E}_2(\tau, 0, B) = 0$. The derivative at $s = 0$ is a nonholomorphic Siegel modular form of weight $\frac{3}{2}$.

Theorem B. *The generating function $\hat{\phi}_2$ is a Siegel modular form of genus two and weight $\frac{3}{2}$ of level $\Gamma_0(4D(B)_o) \subset \mathrm{Sp}_2(\mathbb{Z})$. More precisely,*

$$\hat{\phi}_2(\tau) = \mathcal{E}'_2(\tau, 0, B),$$

i.e., the q-expansion of the Siegel modular form on the right-hand side coincides with the generating series $\hat{\phi}_2$.

Here we are identifying implicitly $\widehat{\mathrm{CH}}^2(\mathcal{M})$ with $\mathbb{R}$ via $\widehat{\deg}$, cf. (1.0.23). Theorem B is proved in Chapter 6 by explicitly comparing the coefficients of the q-expansion of $\mathcal{E}'_2(\tau, 0, B)$ with the coefficients $\hat{\mathcal{Z}}(T, v)$. This amounts to a series of highly nontrivial identities, one for each T in $\mathrm{Sym}_2(\mathbb{Z})^\vee$. Let us explain what is involved.

First let T be positive definite with $\mathrm{Diff}(T, B) = \{p\}$ for $p \nmid D(B)$. The calculation of the coefficient of $\mathcal{E}'_2(\tau, 0, B)$ corresponding to T comes down to the determination of derivatives of Whittaker functions or of certain representation densities. This determination is based on the explicit formulas for such densities due to Kitaoka [22] for $p \neq 2$. For $p = 2$, corresponding results are given in [55]. The determination of the arithmetic degree of $\mathcal{Z}(T)$ boils down to the problem of determining the length of the formal deformation ring of a 1-dimensional formal group of height 2 with two special endomorphisms. This is a special case of the theorem of Gross and Keating [17]. We point out that for both sides the prime number 2 ('the number theorist's nightmare') complicates matters considerably.

Next let T be positive definite with $\mathrm{Diff}(T, B) = \{p\}$ for $p \mid D(B)$. In this case, the corresponding derivatives of representation densities are determined in [54] for $p \neq 2$ and in [55] for $p = 2$. The determination of the corresponding coefficient of $\hat{\phi}_2$ depends on the calculation of the intersection product of special cycles on the Drinfeld upper half space. This is done in [36] for $p \neq 2$. These calculations are completed here for $p = 2$.

Now let T be nonsingular with $\mathrm{Diff}(T, B) = \infty$. Then the calculation of the corresponding coefficients of $\mathcal{E}'_2(\tau, 0, B)$ and of $\hat{\phi}_2$ is given in [24] in the case where the signature of T is $(1, 1)$. The remaining case, where the signature is $(0, 2)$, is given here, using the method of [24].

Next, we consider the coefficients corresponding to singular matrices T of rank 1. For such a matrix

$$T = \begin{pmatrix} t_1 & m \\ m & t_2 \end{pmatrix} \in \mathrm{Sym}_2(\mathbb{Z})^\vee, \tag{1.0.35}$$

with $\det(T) = 0$ and $T \neq 0$, we may write $t_1 = n_1^2 t$, $t_2 = n_2^2 t$, and $m = n_1 n_2 t$ for the relatively prime integers n_1 and n_2 and $t \in \mathbb{Z}_{\neq 0}$. The pair n_1, n_2 is unique up to simultaneous change in sign, and t is uniquely determined. Also, note that, if $t_1 = 0$, then $n_1 = 0$, $n_2 = 1$, and $t = t_2$, while if $t_2 = 0$, then $n_1 = 1$, $n_2 = 0$, and $t = t_1$. Then the comparison between the corresponding singular coefficients of $\hat{\phi}_2$ and $\mathcal{E}_2'(\tau, 0, B)$ in this case is based on the following result, proved in Chapter 5. It relates the singular Fourier coefficients of the derivative of the genus two Eisenstein series occurring in Theorem B with the Fourier coefficients of the genus one Eisenstein series occurring in Theorem A.

Theorem 1.0.4. *(i) Let $T \in \mathrm{Sym}_2(\mathbb{Z})^\vee$, with associated $t \in \mathbb{Z}_{\neq 0}$ as above. Then*

$$\mathcal{E}_{2,T}'(\tau, 0, B) = -\mathcal{E}_{1,t}'(t^{-1}\mathrm{tr}(T\tau), \frac{1}{2}, B)$$
$$- \frac{1}{2} \cdot \mathcal{E}_{1,t}(t^{-1}\mathrm{tr}(T\tau), \frac{1}{2}, B) \cdot \left(\log(\frac{\det\ v}{t^{-1}\mathrm{tr}(Tv)}) + \log(D(B)) \right).$$

(ii) For the constant term

$$\mathcal{E}_{2,0}'(\tau, 0, B) = -\mathcal{E}_{1,0}'(i \det\ v, \frac{1}{2}, B) - \frac{1}{2}\mathcal{E}_{1,0}(i \det(v), \frac{1}{2}, B) \cdot \log D(B).$$

It is this theorem that motivated our definition of the singular coefficients of the generating series $\hat{\phi}_2$. Just as for Proposition 1.0.1, we see that the modularity of the generating function $\hat{\phi}_2$ is not proved directly but rather by identifying it with an explicit modular form.

The coherence in our definitions of the generating series $\hat{\phi}_1$ and $\hat{\phi}_2$ is displayed by the following arithmetic inner product formula, which relates the inner product of the generating series $\hat{\phi}_1$ with itself under the Gillet-Soulé pairing with the generating series $\hat{\phi}_2$. Let

$$\mathfrak{H} \times \mathfrak{H} \longrightarrow \mathfrak{H}_2 \qquad (\tau_1, \tau_2) \longmapsto \mathrm{diag}(\tau_1, \tau_2) = \begin{pmatrix} \tau_1 & 0 \\ 0 & \tau_2 \end{pmatrix} \tag{1.0.36}$$

be the natural embedding into the Siegel space of genus two.

Theorem C. *For $\tau_1, \tau_2 \in \mathfrak{H}$*

$$\langle \hat{\phi}_1(\tau_1), \hat{\phi}_1(\tau_2) \rangle = \hat{\phi}_2(\mathrm{diag}(\tau_1, \tau_2)).$$

Explicitly, for any $t_1, t_2 \in \mathbb{Z}$ and $v_1, v_2 \in \mathbb{R}_{>0}$,

$$\langle \hat{\mathcal{Z}}(t_1, v_1), \hat{\mathcal{Z}}(t_2, v_2) \rangle = \sum_{\substack{T \in \mathrm{Sym}_2(\mathbb{Z})^\vee \\ \mathrm{diag}(T) = (t_1, t_2)}} \hat{\mathcal{Z}}(T, \mathrm{diag}(v_1, v_2)).$$

Theorem C, which is proved in Chapter 7, is the third main result of this book and provides the arithmetic analogue of Theorem 6.2 in [23], which relates to the cup product of two generating series with values in cohomology. Let us explain what is involved here, first assuming that $t_1t_2 \neq 0$.

The proof distinguishes two cases. In the first case $t_1t_2 \notin \mathbb{Q}^{\times,2}$. In this case all matrices T occurring in the sum on the right-hand side are automatically nonsingular; at the same time the divisors $\mathcal{Z}(t_1)$ and $\mathcal{Z}(t_2)$ have empty intersection in the generic fiber, so that the Gillet-Soulé pairing decomposes into a sum of local pairings, one for each prime of $\mathbb{Q}$. Consider the case when $t_i > 0$ for $i = 1, 2$. Then the key to the formula above is the decomposition of the intersection (fiber product) of the special cycles $\mathcal{Z}(t_i)$ according to 'fundamental matrices',

$$\mathcal{Z}(t_1) \times_{\mathcal{M}} \mathcal{Z}(t_2) = \coprod_{\substack{T \\ \operatorname{diag}(T)=(t_1,t_2)}} \mathcal{Z}(T). \tag{1.0.37}$$

Here $\mathcal{Z}(T)$ appears as the locus of objects $((A, \iota), x_1, x_2)$ in the fiber product where $\mathbf{x} = [x_1, x_2]$ satisfies $\frac{1}{2}(\mathbf{x}, \mathbf{x}) = T$. Note that, by the remarks preceding the statement of Theorem B, the intersection of the $\mathcal{Z}(t_i)$ need not be proper since these divisors can have common components in the fibers of bad reduction $\mathcal{M}_p$ for $p \mid D(B)$. Of course, all matrices T occurring in the disjoint sum in (1.0.37) are positive definite. The occurrence in the sum of Theorem C of summands corresponding to matrices T which are not positive definite is due to the Green functions component of the $\hat{\mathcal{Z}}(t, v)$. Similar archimedean contributions occur in the cases where one of the t_i is negative.

In the second case $t_1t_2 \in \mathbb{Q}^{\times,2}$. In this case, $\mathcal{Z}(t_1)$ and $\mathcal{Z}(t_2)$ intersect in the generic fiber. In addition to the contribution of the nonsingular T to the sum in Theorem C, there is also a contribution of the two singular matrices T, where T is given by (1.0.35) with $m = \pm\sqrt{t_1t_2}$. In this case the Gillet-Soulé pairing does not localize. Instead we use the arithmetic adjunction formula from Arakelov theory [10], [40]. To calculate the various terms in this formula we must, among other things, go back to the proof of the Gross-Keating formula and use the fine structure of the deformation locus of a special endomorphism of a p-divisible group of dimension 1 and height 2.

We stress that the proof of Theorem C sketched so far has nothing to do with Eisenstein series. However, the modularity of both sides of the identity in Theorem C allows us to deduce from the truth of the statement for all $t_1t_2 \neq 0$ first the value of the constant $\mathbf{c}$ in (1.0.26) and then the truth of the statement for all (t_1, t_2). In this way we can also prove our conjecture [38] on the self-intersection of the Hodge line bundle.

Theorem 1.0.5. *Let $\hat{\omega}_0$ be the Hodge line bundle on $\mathcal{M}$ metrized with the normalization of Bost* [3]. *Then*

$$\langle \hat{\omega}_0, \hat{\omega}_0 \rangle = 2 \cdot \zeta_{D(B)}(-1) \left[\frac{\zeta'(-1)}{\zeta(-1)} + \frac{1}{2} - \frac{1}{4} \sum_{p|D(B)} \frac{p+1}{p-1} \log p \right].$$

Formally, this result specializes for $D(B) = 1$ to the formula of Bost [4] and Kühn [39] in the case of the modular curve (note that due to the stacks aspect our quantity is half of theirs). In their case they use the section Δ of $\omega^{\otimes 6}$ to compute the self-intersection of $\hat{\omega}_0$ explicitly from its definition. For Shimura curves there is no such natural modular form and our result comes about only indirectly. We note that the general form of this formula is related to formulas given by Maillot and Roessler [42].

The above three theorems are the main results in this book. As an application of these results, we introduce an arithmetic version of the Shimura-Waldspurger correspondence and obtain analogues of results of Waldspurger [53] and of Gross-Kohnen-Zagier [18].

If f is a cusp form of weight $\frac{3}{2}$ for $\Gamma_0(4D(B)_o)$, we can define the arithmetic theta lift of f by

$$\text{(1.0.38)} \quad \widehat{\theta}(f) := C \cdot \int_{\Gamma_0(4D(B)_o)\backslash\mathfrak{H}} f(\tau)\, \overline{\widehat{\phi}_1(\tau)}\, v^{-\frac{1}{2}}\, du\, dv \;\in\; \widehat{\mathrm{CH}}^1(\mathcal{M}^B),$$

for a constant C given in section 3 of Chapter 9. Of course, this is the analogue of the classical theta lift from modular forms of weight $\frac{3}{2}$ to modular forms of weight 2, but with $\widehat{\phi}_1(\tau)$ replacing the classical theta kernel of Niwa [43] and Shintani [47]. By the results discussed above, it follows that

$$\text{(1.0.39)} \qquad \langle\, \widehat{\theta}(f), 1\!\!1 \,\rangle = \langle\, f, \mathcal{E}_1(\tau, \tfrac{1}{2}; B)\, \rangle_{\mathrm{Pet}} = 0,$$

$$\text{(1.0.40)} \qquad \langle\, \widehat{\theta}(f), \hat{\omega} \,\rangle = \langle\, f, \mathcal{E}_1'(\tau, \tfrac{1}{2}; B)\, \rangle_{\mathrm{Pet}} = 0,$$

and

$$\text{(1.0.41)} \quad \langle\, \widehat{\theta}(f), a(\phi)\, \rangle = \langle\, f, \theta(\phi)\, \rangle_{\mathrm{Pet}} = 0, \quad \text{for all } \phi \in C^\infty(\mathcal{M}(\mathbb{C}))_0,$$

since f is a holomorphic cusp form. Here, for $\phi \in C^\infty(\mathcal{M}(\mathbb{C}))_0$, we denote by $a(\phi)$ the corresponding class in $\widehat{\mathrm{CH}}^1(\mathcal{M})$ and by $\theta(\phi)$ the corresponding Maass cusp form of weight $\frac{3}{2}$. Thus $\widehat{\theta}(f)$ lies in the space of $\widetilde{\mathrm{MW}} \oplus \mathrm{Vert}^0$, where Vert^0 is the subspace of Vert orthogonal to $\hat{\omega}$.

In order to obtain information about the nonvanishing of $\widehat{\theta}(f)$, we consider the height pairing $\langle\, \widehat{\phi}_1(\tau_1), \widehat{\theta}(f)\,\rangle$. Using Theorems B and C, we obtain

$$
\begin{aligned}
\langle\, \widehat{\phi}_1(\tau_1), \widehat{\theta}(f)\,\rangle &= \langle\, f, \langle\, \widehat{\phi}_1(\tau_1), \widehat{\phi}_1\,\rangle\,\rangle \\
&= \langle\, f, \widehat{\phi}_2(\mathrm{diag}(\tau_1, \cdot))\,\rangle \\
&= \langle\, f, \mathcal{E}_2'(\mathrm{diag}(\tau_1, \cdot), 0; B)\,\rangle \\
&= \frac{\partial}{\partial s}\Big\{\langle\, f, \mathcal{E}_2(\mathrm{diag}(\tau_1, \cdot), s; B)\,\rangle\Big\}\Big|_{s=0}.
\end{aligned}
\tag{1.0.42}
$$

We then consider the integral $\langle\, f, \mathcal{E}_2(\mathrm{diag}(\tau_1, \cdot), s; B)\,\rangle$ occurring in the last expression. This integral is essentially the doubling integral of Piatetski-Shapiro and Rallis [45] (see also [41]), except that we only integrate against one cusp form.

Theorem 1.0.6. *Let F be a normalized newform of weight 2 on $\Gamma_0(D(B))$ and let f be the good newvector, in the sense defined in section 3 of Chapter 8, corresponding to F under the Shimura-Waldspurger correspondence. Then*

$$
\langle\, f, \mathcal{E}_2(\mathrm{diag}(\tau_1, \cdot), s; B)\,\rangle = \mathcal{C}(s)\cdot L(s+1, F)\cdot f(\tau_1),
$$

where

$$
\mathcal{C}(s) = \frac{3}{2\pi^2} \prod_{p \mid D(B)} (p+1)^{-1} \cdot \left(\frac{D(B)}{2\pi}\right)^{s} \Gamma(s+1)\cdot \prod_{p\mid D(B)} \mathcal{C}_p(s),
$$

with

$$
\mathcal{C}_p(s) = (1 - \epsilon_p(F)\, p^{-s}) - \frac{p-1}{p+1}\,(1 + \epsilon_p(F)\, p^{-s})\, B_p(s).
$$

Here $L(s, F)$ is the standard Hecke L-function of F, $\epsilon_p(F)$ is the Atkin-Lehner sign of F,

$$
F|W_p = \epsilon_p(F)\, F,
$$

and $B_p(s)$ is a rational function of p^{-s} with

$$
B_p(0) = 0 \qquad \text{and} \qquad B_p'(0) = \frac{1}{2}\cdot\frac{p+1}{p-1}\log(p).
$$

Note that $\mathcal{C}_p(0) = 2$ if $\epsilon_p(F) = -1$ and $\mathcal{C}_p(0) = \mathcal{C}_p'(0) = 0$ if $\epsilon_p(F) = 1$. As a consequence, we have the following analogue of Rallis's inner product formula [46], which characterizes the nonvanishing of the arithmetic theta lift.

Corollary 1.0.7. *For F with associated f as in Theorem 1.0.6,*

$$\langle\, \widehat{\phi}_1(\tau_1), \widehat{\theta}(f)\,\rangle = \mathcal{C}(0)\cdot L'(1,F)\cdot f(\tau_1).$$

In particular,

$$\langle\, \widehat{\theta}(f), \widehat{\theta}(f)\,\rangle = \mathcal{C}(0)\cdot L'(1,F)\cdot \langle\, f, f\,\rangle,$$

and hence

$$\widehat{\theta}(f)\neq 0 \qquad \Longleftrightarrow \qquad \begin{cases} \epsilon_p(F) = -1 & \text{for all } p \mid D(B), \text{ and} \\ L'(1,F)\neq 0. \end{cases}$$

Let $S_2^{\text{new}}(D(B))^{(-)}$ be the space of normalized newforms of weight 2 for $\Gamma_0(D(B))$ for which all Atkin-Lehner signs are -1. Note that, for $F \in S_2^{\text{new}}(D(B))^{(-)}$, the root number of $L(s,F)$ is given by

$$\epsilon(1,F) = -\prod_{p\mid D(B)} \epsilon_p(F) = -1.$$

Since the vertical part of $\widehat{\phi}_1(\tau)$ is a linear combination of theta functions for the anisotropic ternary spaces $V^{(p)}$, for $p \mid D(B)$, and since the classical theta lift of a form F with $\epsilon(1,F) = -1$ to such a space vanishes by Waldspurger's result [50], [53], it follows that $\widehat{\theta}(f) \in \widetilde{\text{MW}}$. Recall from (1.0.29) that this space is isomorphic to $\text{MW}(\mathcal{M}_{\mathbb{Q}})$ via the restriction map $\text{res}_{\mathbb{Q}}$.

Corollary 1.0.8. *For each $F \in S_2^{\text{new}}(D(B))^{(-)}$, let f be the corresponding good newvector of weight $\frac{3}{2}$. Then*

$$\text{res}_{\mathbb{Q}}(\,\widehat{\phi}_1^B(\tau)\,) = \mathcal{E}_1(\tau,\frac{1}{2};B)\cdot\frac{\omega_{\mathbb{Q}}}{\deg\omega_{\mathbb{Q}}} + \sum_{\substack{F\in S_2^{\text{new}}(D(B))^{(-)} \\ L'(1,F)\neq 0}} \frac{f(\tau)\cdot \text{res}_{\mathbb{Q}}\,\widehat{\theta}(f)}{\langle\, f, f\,\rangle},$$

where $\omega_{\mathbb{Q}}$ is the restriction of the Hodge bundle to $\mathcal{M}_{\mathbb{Q}}$.

Next, for each $t \in \mathbb{Z}_{>0}$, write $Z(t)(F)$ for the component[3] of the cycle $Z(t) = \mathcal{Z}(t)_{\mathbb{Q}}$ in the F-isotypic part $\text{CH}^1(\mathcal{M}_{\mathbb{Q}})(F)$ of the Chow group $\text{CH}^1(\mathcal{M}_{\mathbb{Q}})$. Note that $Z(t)(F)$ has zero image in $H^2(\mathcal{M}_{\mathbb{C}})$ and hence defines a class in $\text{MW}(\mathcal{M}_{\mathbb{Q}})$.

[3]Here we transfer F to a system of Hecke eigenvalues for the quaternion algebra B via the Jacquet-Langlands correspondence.

Theorem 1.0.9. *The F-isotypic component of the generating function*

$$\mathrm{res}_{\mathbb{Q}}(\,\widehat{\phi}_1^B(\tau)\,) = \sum_{t\geq 0} Z(t)\, q^t,$$

is

$$\mathrm{res}_{\mathbb{Q}}(\,\widehat{\phi}_1^B(\tau)\,)(F) = \sum_{t\geq 0} Z(t)(F)\, q^t = \frac{f(\tau)\cdot \mathrm{res}_{\mathbb{Q}}\widehat{\theta}(f)}{\langle\, f, f\,\rangle}.$$

In particular,

$$Z(t)(F) = \frac{a_t(f)\cdot \mathrm{res}_{\mathbb{Q}}\,\widehat{\theta}(f)}{\langle\, f, f\,\rangle},$$

where

$$f(\tau) = \sum_{t>0} a_t(f)\, q^t$$

is the Fourier expansion of f. Moreover, for t_1 and $t_2 \in \mathbb{Z}_{>0}$, the height pairing of the F-components of $Z(t_1)$ and $Z(t_2)$ is given by

$$\langle\, Z(t_1)(F), Z(t_2)(F)\,\rangle = C(0)\cdot L'(1,F)\cdot \frac{a_{t_1}(f)\cdot a_{t_2}(f)}{\langle\, f, f\,\rangle}.$$

This result is the analogue in our case of the result of Gross-Kohnen-Zagier [18], Theorem C, p.503. The restriction to newforms in $S_2^{\mathrm{new}}(D(B))$ with all Atkin-Lehner signs equal to -1 is due to the fact that our cycles are invariant under all Atkin-Lehner involutions. To remove this restriction, one should use 'weighted' cycles, see section 4 of Chapter 3.

In fact, we construct an arithmetic theta lift of automorphic representations σ in the space $\mathcal{A}_{00}(G')$ on the metaplectic extension $G'_{\mathbb{A}}$ of $\mathrm{SL}_2(\mathbb{A})$. This theta lift, which is only defined for representations corresponding to holomorphic cusp forms of weight $\frac{3}{2}$, is the analogue of the classical theta lift considered by Waldspurger [50], [51], [53]. We formulate a conjectural analogue of Waldspurger's nonvanishing criterion and prove it in certain cases as an application of Theorem 1.0.6 and Corollary 1.0.7. For forms F with $\epsilon(1,F) = +1$, Waldspurger proved that the classical theta lift is nonzero if and only if (i) certain local conditions (theta dichotomy) are satisfied at every place, and (ii) $L(1,F)\neq 0$. In the arithmetic case, we show that for (certain) forms F of weight 2 with $\epsilon(1,F) = -1$, the arithmetic theta lift is nonzero if and only if (i) the local theta dichotomy conditions are satisfied, and (ii) $L'(1,F)\neq 0$. A more detailed discussion can be found in section 1 of Chapter 9 as well as in [29]. Our construction is similar in spirit to that of [16], where Gross formulates an arithmetic analogue of another result of Waldspurger [52] and shows that, in certain cases, this analogue can be proved using the results of Gross-Zagier [19] and their extension by Zhang [60].

We now mention some previous work on such geometric and arithmetic-geometric generating functions. The classic work of Hirzebruch-Zagier mentioned above inspired much work on modular generating functions valued in cohomology. Kudla and Millson considered modular generating functions for totally geodesic cycles in Riemannian locally symmetric spaces for the classical groups $\mathrm{O}(p,q)$, $\mathrm{U}(p,q)$, and $\mathrm{Sp}(p,q)$ [31], [32], [33]. Such cycles were also considered by Oda [44] and Tong-Wang [49]. In the case of symmetric spaces for $\mathrm{O}(n,2)$, the generating function of Kudla-Millson [33] and Kudla [23] for the cohomology classes of algebraic cycles of codimension r is a Siegel modular form of weight $\frac{n}{2}+1$ and genus r. In the case $r=n$, i.e., for 0-cycles, the generating function was identified in [23] as a special value of an Eisenstein series via the Siegel-Weil formula. A similar relation to Eisenstein series occurs in the work of Gross and Keating [17] for the generating series associated to the graphs of modular correspondences in a product of two modular curves. Borcherds [2] used Borcherds products to construct modular generating series with coefficients in CH^1 for divisors on locally symmetric varieties associated to $\mathrm{O}(n,2)$ and proved that they are holomorphic modular forms. We also mention recent related work of Bruinier [5], [6], Bruinier-Funke [8], Funke [11], and Funke-Millson [12], [13].

The results in the arithmetic context are all inspired by the theorem of Gross and Zagier [19]. Part of a generating series for triple arithmetic intersections of curves on the product of two modular curves was implicitly considered in the paper by Gross and Keating [17], where the 'good nonsingular' coefficients are determined explicitly, cf. also [1]. For Shimura curves, Kudla [24] considered the generating series obtained from the Gillet-Soulé height pairing of special divisors. It was proved that this generating series coincided for 'good' nonsingular coefficients with the diagonal pullback of the central derivative of a Siegel Eisenstein series of genus two. The 'bad' nonsingular coefficients were determined in [36]. However, the singular coefficients were left out of this comparison. In [37] we considered the 0-dimensional case, where the ambient space is the moduli space of elliptic curves with complex multiplication. In this case we were able to determine the generating series completely and to identify it with the derivative of a special value of an Eisenstein series. Another generating series is obtained in [38] by pairing special divisors on arithmetic models of Shimura curves, equipped with Green functions, with the metrized dualizing line bundle. Again this can be determined completely and identified with a special value of a derivative of an Eisenstein series. A generating series in a higher-dimensional case is constructed by Bruinier, Burgos, and Kühn [7]. They consider special divisors on arithmetic models of Hilbert-Blumenthal surfaces whose generic fibers are Hirzebruch-Zagier curves, equip them with (generalized) Green functions [9], and obtain a generating series by taking the pairing with *the square* of the metrized dualizing line bundle. They identify this series with a special value of an Eisenstein series. Finally we mention partial results in higher-dimensional cases (Hilbert-Blumenthal surfaces, Siegel threefolds) in [34], [35].

This monograph is not self-contained. Rather, we make essential use of our previous papers. We especially need the results in [24] about the particular Green functions we use, as well as the results on Eisenstein series developed there. We also use the results on representation densities from [54], [55]. Furthermore, for the analysis of the situation at the fibers of bad reduction we use the results contained in [36]. These are completed in [38], which is also essential for our arguments in other ways. Finally, we need some facts from [27] in order to apply the results of Borcherds. These papers are not reproduced here. Still, we have given here all the definitions necessary for following our development and have made an effort to direct the reader to the precise reference where he can find the proof of the statement in question. We also have filled in some details in the proof of other results in the literature. Most notable here are our exposition in section 6 of Chapter 3 of the special case of the theorem of Gross and Keating [17] that we use, and the exposition in Chapter 8 of the doubling method of Piatetski-Shapiro and Rallis [45] in the special case relevant to us. In the first instance, we were aided by a project with a similar objective, namely to give an exposition of the general result of Gross and Keating, undertaken by the ARGOS seminar in Bonn [1]. In the second instance, we use precise results about nonarchimedean local Howe duality for the dual pair $(\widetilde{\mathrm{SL}}_2, \mathrm{O}(3))$ from [30].

We have structured this monograph in the following manner. In Chapter 2 we provide the necessary background from Arakelov geometry. The key point here is to show that the theory of Gillet-Soulé [14], [3] continues to hold for the DM-stacks of the kind we encounter. We also give a version of the arithmetic adjunction formula. It turns out that among the various versions of it the most naive form, as presented in Lang's book [40], is just what we need for our application of it in Chapter 7. In Chapter 3 we define the special cycles on Shimura curves and review the known facts about them. Here we also give a proof of the special case of the Gross-Keating formula which we need. In Chapter 4 we prove Theorem A, along the lines sketched above. In Chapter 5 we introduce the Eisenstein series of genus one and two which are relevant to us and calculate their Fourier expansion. In particular, we prove Theorem 1.0.4. In Chapter 6 we define the generating series $\hat{\phi}_2$ and prove Theorem B by comparing term by term this series with the Fourier coefficients of the Siegel Eisenstein series of genus two determined in the previous chapter. For the 'bad nonsingular' coefficients of $\hat{\phi}_2$, the calculation in the case $p = 2$ had been left out in [36]. In the appendix to Chapter 6 we complete the calculations for $p = 2$. Chapter 7 is devoted to the proof of the inner product formula, Theorem C. In Chapter 8 we give an exposition of the doubling method in our case. The point is to determine explicitly all local zeta integrals for the kind of good test functions that we use. The case $p = 2$ again requires additional efforts. In Chapter 9 we give applications of our results to the arithmetic theta lift and to L-functions and prove Theorems 1.0.6 and 1.0.9 and Corollaries 1.0.7 and 1.0.8 above.

This book is the result of a collaboration over many years. The general idea of forming the arithmetic generating series and relating them to modular forms arising from derivatives of Eisenstein series is due to the first author. The other two authors joined the project, each one contributing a different expertise to the undertaking. In the end, we can honestly say that no proper subset of this set of authors would have been able to bring this project to fruition. While the book is thus the product of a joint enterprise, some chapters have a set of *principal authors* which are as follows:

Chapter 2: SK, MR
Chapter 4: SK
Chapter 5: SK, TY
Appendix to Chapter 6: SK, MR
Chapter 7: SK, MR
Chapter 8: SK, TY

The material of this book, as well as its background, has been the subject of several survey papers by us individually: [25], [26], [28], [29], [56], [57]. It should be pointed out, however, that in the intervening time we made progress and that quite a number of question marks which still decorate the announcements of our results in these papers have been removed.

Bibliography

[1] ARGOS (Arithmetische Geometrie Oberseminar), Proceedings of the Bonn seminar 2003/04, forthcoming.

[2] R. Borcherds, *The Gross-Kohnen-Zagier theorem in higher dimensions*, Duke Math. J., **97** (1999), 219–233.

[3] J.-B. Bost, *Potential theory and Lefschetz theorems for arithmetic surfaces*, Ann. Sci. École Norm. Sup., **32** (1999), 241–312.

[4] ______, Lecture, Univ. of Maryland, Nov. 11, 1998.

[5] J. H. Bruinier, *Borcherds products and Chern classes of Hirzebruch–Zagier divisors*, Invent. Math., **138** (1999), 51–83.

[6] ______, Borcherds Products on $O(2, l)$ and Chern Classes of Heegner Divisors, Lecture Notes in Math., **1780**, Springer-Verlag, New York, 2002.

[7] J. H. Bruinier, J. I. Burgos Gil, and U. Kühn, *Borcherds products in the arithmetic intersection theory of Hilbert modular surfaces*, preprint, 2003.

[8] J. H. Bruinier and J. Funke, *On two geometric theta lifts*, preprint, 2003.

[9] J. I. Burgos Gil, J. Kramer, and U. Kühn, *Cohomological arithmetic Chow rings*, preprint, 2003.

[10] G. Faltings, *Calculus on arithmetic surfaces*, Annals of Math., **119** (1984), 387–424.

[11] J. Funke, *Heegner divisors and nonholomorphic modular forms*, Compositio Math., **133** (2002), 289–321.

[12] J. Funke and J. Millson, *Cycles in hyperbolic manifolds of non-compact type and Fourier coefficients of Siegel modular forms*, Manuscripta Math., **107** (2003), 409–444.

[13] ______, *Cycles with local coefficients for orthogonal groups and vector valued Siegel modular forms*, preprint, 2004.

[14] H. Gillet and C. Soulé, *Arithmetic intersection theory*, Publ. Math. IHES, **72** (1990), 93–174.

[15] ______, *Arithmetic analogues of standard conjectures*, in Proc. Symp. Pure Math., **55**, Part 1, 129–140, AMS, Providence, R.I., 1994.

[16] B. H. Gross, *Heegner points and representation theory*, in Heegner points and Rankin L-Series, Math. Sci. Res. Inst. Publ., **49**, Cambridge Univ. Press, Cambridge, 2004.

[17] B. Gross and K. Keating, *On the intersection of modular correspondences*, Invent. math., **112** (1993), 225–245.

[18] B. H. Gross, W. Kohnen, and D. Zagier, *Heegner points and derivatives of L-functions. II*, Math. Annalen, **278** (1987), 497–562.

[19] B. H. Gross and D. Zagier, *Heegner points and the derivatives of L-series*, Invent. math., **84** (1986), 225–320.

[20] F. Hirzebruch and D. Zagier, *Intersection numbers of curves on Hilbert modular surfaces and modular forms of Nebentypus*, Invent. math., **36** (1976), 57–113.

[21] Y. Kitaoka, *A note on local densities of quadratic forms*, Nagoya Math. J., **92** (1983), 145–152.

[22] ______, Arithmetic of Quadratic Forms, Cambridge Tracts in Mathematics, **106**, Cambridge Univ. Press, 1993.

[23] S. Kudla, *Algebraic cycles on Shimura varieties of orthogonal type*, Duke Math. J., **86** (1997), 39–78.

[24] ______, *Central derivatives of Eisenstein series and height pairings*, Annals of Math., **146** (1997), 545–646.

[25] ______, *Derivatives of Eisenstein series and generating functions for arithmetic cycles*, Séminaire Bourbaki 876, Astérisque, **276**, 341–368, Soc. Math. France, Paris, 2002.

[26] ______, *Derivatives of Eisenstein series and arithmetic geometry*, in Proc. Intl. Cong. Mathematicians, Vol II (Beijing, 2002), 173–183, Higher Education Press, Beijing, 2002.

[27] ______, *Integrals of Borcherds forms*, Compositio Math., **137** (2003), 293–349.

[28] ______, *Special cycles and derivatives of Eisenstein series*, in Heegner points and Rankin L-Series, Math. Sci. Res. Inst. Publ., **49**, 243–270, Cambridge Univ. Press, Cambridge, 2004.

[29] ______, *Modular forms and arithmetic geometry*, in Current Developments in Mathematics, 2002, 135–179, International Press, Somerville, MA, 2003.

[30] ______, *Notes on the local theta correspondence for* $(\widetilde{\mathrm{SL}}_2, O(3))$, preprint, 2005.

[31] S. Kudla and J. Millson, *The theta correspondence and harmonic forms I*, Math. Annalen, **274** (1986), 353–378.

[32] ______, *The theta correspondence and harmonic forms II*, Math. Annalen, **277** (1987), 267–314.

[33] ______, *Intersection numbers of cycles on locally symmetric spaces and Fourier coefficients of holomorphic modular forms in several complex variables*, Publ. Math. IHES, **71** (1990), 121–172.

[34] S. Kudla and M. Rapoport, *Arithmetic Hirzebruch–Zagier cycles*, J. reine angew. Math., **515** (1999), 155–244.

[35] ______, *Cycles on Siegel 3-folds and derivatives of Eisenstein series*, Ann. Sci. École. Norm. Sup., **33** (2000), 695–756

[36] ______, *Height pairings on Shimura curves and p-adic uniformization*, Invent. math., **142** (2000), 153–223.

[37] S. Kudla, M. Rapoport, and T. Yang, *On the derivative of an Eisenstein series of weight 1*, Int. Math. Res. Notices (1999), 347–385.

[38] ______, *Derivatives of Eisenstein series and Faltings heights*, Compositio Math., **140** (2004), 887–951.

[39] U. Kühn, *Generalized arithmetic intersection numbers*, J. reine angew. Math., **534** (2001), 209–236.

[40] S. Lang, Introduction to Arakelov Theory, Springer-Verlag, New York, 1988.

[41] J.-S. Li, *Nonvanishing theorems for the cohomology of certain arithmetic quotients*, J. reine angew. Math., **428** (1992), 177–217.

[42] V. Maillot and D. Roessler, *Conjectures sur les dérivées logarithmiques des fonctions L d'Artin aux entiers négatifs*, Math. Res. Lett., **9** (2002), no. 5-6, 715–724.

[43] S. Niwa, *Modular forms of half integral weight and the integral of certain theta-functions*, Nagoya Math. J., **56** (1975), 147–161.

[44] T. Oda, *On modular forms associated to quadratic forms of signature* $(2, n-2)$, Math. Annalen., **231** (1977), 97–144.

[45] I. I. Piatetski-Shapiro and S. Rallis, *L-functions for classical groups*, Lecture Notes in Math., **1254**, 1–52, Springer-Verlag, New York, 1987.

[46] S. Rallis, *Injectivity properties of liftings associated to Weil representations*, Compositio Math., **52** (1984), 139–169.

[47] T. Shintani, *On construction of holomorphic cusp forms of half integral weight*, Nagoya Math. J., **58** (1975), 83–126.

[48] C. Soulé, D. Abramovich, J.-F. Burnol, and J. Kramer, Lectures on Arakelov Theory, Cambridge Studies in Advanced Math. **33**, Cambridge Univ. Press, Cambridge, 1992.

[49] Y. Tong and S. P. Wang, *Construction of cohomology of discrete groups*, Trans. AMS, **306** (1988), 735–763.

[50] J.-L. Waldspurger, *Correspondance de Shimura*, J. Math. Pures Appl., **59** (1980), 1–132.

[51] ______, *Sur les coefficients de Fourier des formes modulaires de poids demi-entier*, J. Math. Pures Appl., **60** (1981), 375–484.

[52] ______, *Sur les valeurs de certaines fonctions L automorphes en leur centre de symétrie*, Compositio Math., **54** (1985), 173–242.

[53] ______, *Correspondances de Shimura et quaternions*, Forum Math., **3** (1991), 219–307.

[54] T. H. Yang, *An explicit formula for local densities of quadratic forms*, J. Number Theory, **72** (1998), 309–356.

[55] ______, *Local densities of 2-adic quadratic forms*, J. Number Theory, **108** (2004), 287–345.

[56] ______, *The second term of an Eisenstein series*, Proc. Intl. Cong. Chinese Mathematicians, forthcoming.

[57] ______, *Faltings heights and the derivative of Zagier's Eisenstein series*, in Heegner points and Rankin L-Series, Math. Sci. Res. Inst. Publ., **49**, 271–284, Cambridge Univ. Press, Cambridge, 2004.

[58] D. Zagier, *Nombres de classes et formes modulaires de poids* $3/2$, C. R. Acad. Sci. Paris, **281** (1975), 883–886.

[59] ______, *Modular points, modular curves, modular surfaces and modular forms*, in Lecture Notes in Math. **1111**, 225–248, Springer-Verlag, Berlin, 1985.

[60] Shou-Wu Zhang, *Gross–Zagier formula for* GL_2, Asian J. Math., **5** (2001), 183–290.

Chapter Two

Arithmetic intersection theory on stacks

The aim of the present chapter is to outline the (arithmetic) intersection theory on Deligne-Mumford (DM) stacks that will be relevant to us. The stacks $\mathcal{M}$ we consider will satisfy the following conditions:

- $\mathcal{M}$ is regular of dimension 2 and is proper and flat over $S = \operatorname{Spec} \mathbb{Z}$, and is a relative complete intersection over $\operatorname{Spec} \mathbb{Z}$. Also we assume $\mathcal{M}$ to be connected (and later even geometrically connected).
- Let $M = \mathcal{M}_{\mathbb{C}} = \mathcal{M} \times_{\operatorname{Spec} \mathbb{Z}} \operatorname{Spec} \mathbb{C}$ be the complex fiber of $\mathcal{M}$. Then M is given by an orbifold presentation,

  $$M = [\Gamma \setminus X],$$

  where X is a compact Riemann surface (not necessarily connected) and Γ is a finite group acting on X.

2.1 THE ONE-DIMENSIONAL CASE

As a preparation for later developments we start with the one-dimensional case.

First we consider a DM-stack $\mathcal{Z}$ which is reduced and proper of relative dimension 1 over an algebraically closed field k. Let $\mathcal{L}$ be an invertible sheaf on $\mathcal{Z}$. Before defining the degree of $\mathcal{L}$ we recall [10] that if R is an integral domain of dimension 1, with fraction field K, we put for $f = \frac{a}{b} \in K^{\times}$ with $a, b \in R$,

$$\operatorname{ord}_R(f) = \lg(R/a) - \lg(R/b). \tag{2.1.1}$$

This is extended in the obvious way to define $\operatorname{ord}_L(s)$ for an element $s \in L \otimes_R K$ of a free R-module L of rank one.

Now let s be a rational section of $\mathcal{L}$. If x is a closed geometric point of $\mathcal{Z}$ and $\tilde{\mathcal{O}}_{\mathcal{Z},x}$ is the strictly local henselian ring of $\mathcal{Z}$ in x, we get a direct sum decomposition into integral domains according to the formal branches of $\mathcal{Z}$ through x,

$$\tilde{\mathcal{O}}_{\mathcal{Z},x} = \bigoplus_i \mathcal{O}_i. \tag{2.1.2}$$

We put

$$\deg_x s = \Sigma_i \mathrm{ord}_{\mathcal{O}_i}(s_i), \tag{2.1.3}$$

where s_i is the image of s in $\mathcal{L} \otimes_{\mathcal{O}_{\mathcal{Z},x}} \mathcal{O}_i$.

As in [3], VI, 4.3, we put

$$\deg(\mathcal{L}) = \deg(\mathcal{Z}, \mathcal{L}) = \sum_{x \in \mathcal{Z}(k)} \frac{1}{|\mathrm{Aut}(x)|} \cdot \deg_x(s). \tag{2.1.4}$$

If k is not algebraically closed, one defines the degree after extension of scalars to $\bar{k}$. This definition is independent of the choice of s and coincides with the usual definition when $\mathcal{Z}$ is a scheme. It satisfies

(i) *additivity in* $\mathcal{L}$*:* $\deg(\mathcal{L} \otimes \mathcal{L}') = \deg(\mathcal{L}) + \deg(\mathcal{L}')$

(ii) *coverings:* If $f : \mathcal{Z}' \to \mathcal{Z}$ is a finite flat morphism of constant degree, then

$$\deg(f^*\mathcal{L}) = \deg(f) \cdot \deg(\mathcal{L}). \tag{2.1.5}$$

In particular, let $\pi : \tilde{\mathcal{Z}} \to \mathcal{Z}$ be the normalization of $\mathcal{Z}$. This is the relatively representable morphism such that for any étale presentation $X \to \mathcal{Z}$, the resulting morphism $X \times_{\mathcal{Z}} \tilde{\mathcal{Z}} \to X$ is the normalization of X. Then

$$\deg(\mathcal{Z}, \mathcal{L}) = \deg(\tilde{\mathcal{Z}}, \pi^*(\mathcal{L})). \tag{2.1.6}$$

The calculation of the RHS is somewhat easier since if $\mathcal{Z} = \tilde{\mathcal{Z}}$ is normal, for a rational section s of $\mathcal{L}$ we have

$$\deg_x(s) = \mathrm{ord}_x(s). \tag{2.1.7}$$

(If $\mathcal{Z}$ is normal, then $\tilde{\mathcal{O}}_{\mathcal{Z},x}$ is a discrete valuation ring and $\mathrm{ord}_x(s)$ is the valuation of s.)

In Arakelov theory it is more convenient to use the *Arakelov degree* which is defined as

$$\widehat{\deg}(\mathcal{Z}, \mathcal{L}) = \deg(\mathcal{Z}, \mathcal{L}) \cdot \log p, \tag{2.1.8}$$

when $\mathcal{Z}$ is of finite type over $\mathbb{F}_p$. Here $\mathcal{Z}$ is considered as a stack over $\mathbb{F}_p$. If $\Gamma(\mathcal{Z}, \mathcal{O}) = \mathbb{F}_q$ and $\mathcal{Z}$ is considered as a stack *over* $\mathbb{F}_q$, then the RHS equals $\deg(\mathcal{Z}, \mathcal{L}) \cdot \log q$.

Next we consider the case where $\mathcal{Z}$ is a reduced irreducible DM-stack of dimension 1 which is proper and flat over Spec $\mathbb{Z}$. In this case we want to consider *metrized line bundles*. There are two ways to define the concept of a metrized line bundle on $\mathcal{Z}$. First, one can define a metrized line bundle to be a rule which associates, functorially, to any S-valued point $S \to \mathcal{Z}$ a line bundle $\mathcal{L}_S$ on S equipped with a C^∞-metric on the line bundle $\mathcal{L}_{S,\mathbb{C}}$ on $S \times_{\operatorname{Spec}\mathbb{Z}} \operatorname{Spec}\mathbb{C}$. Second, one can define a metrized line bundle on $\mathcal{Z}$ to be a metrized line bundle on an étale presentation $X \to \mathcal{Z}$, equipped with a descent datum which respects the metric.

We denote by $\widehat{\operatorname{Pic}}(\mathcal{Z})$ the set of isomorphism classes of metrized line bundles on $\mathcal{Z}$. This is an abelian group under the tensor product operation.

Let $\Gamma(\mathcal{Z}, \mathcal{O}_{\mathcal{Z}})$ be the ring of regular functions on $\mathcal{Z}$. This may be identified with the ring of regular functions on the coarse moduli space of $\mathcal{Z}$. Then $\Gamma(\mathcal{Z}, \mathcal{O}_{\mathcal{Z}})$ is an order O in a number field K with $\Gamma(\mathcal{Z} \otimes_{\mathbb{Z}} \mathbb{Q}, \mathcal{O}) = K$. If $\nu : \tilde{\mathcal{Z}} \to \mathcal{Z}$ is the normalization of $\mathcal{Z}$, then $\Gamma(\tilde{\mathcal{Z}}, \mathcal{O}_{\tilde{\mathcal{Z}}})$ is the ring of integers O_K. We now put for a rational section s of the metrized line bundle $\hat{\mathcal{L}}$,

$$(2.1.9)\quad \widehat{\deg}(\mathcal{Z}, \hat{\mathcal{L}}) = \sum_p \Big(\sum_{x \in \mathcal{Z}(\bar{\mathbb{F}}_p)} \frac{\deg_x(s)}{|\operatorname{Aut}(x)|} \Big) \log p - \frac{1}{2} \int_{\mathcal{Z}(\mathbb{C})} \log \|s\|^2 .$$

Here the integral is defined as

$$(2.1.10)\qquad \int_{\mathcal{Z}(\mathbb{C})} \log \|s\|^2 = \sum_{x \in \mathcal{Z}(\mathbb{C})} \frac{1}{|\operatorname{Aut}(x)|} \cdot \log \|s(x)\|^2 .$$

Let us check that (2.1.9) is independent of the choice of s. This comes down to checking for a function $f \in K^\times$ that

(2.1.11)

$$0 = \sum_p \Big(\sum_{x \in \mathcal{Z}(\bar{\mathbb{F}}_p)} \frac{\deg_x(f)}{|\operatorname{Aut}(x)|} \Big) \log p - \frac{1}{2} \sum_{\sigma : K \to \mathbb{C}} \frac{1}{|\operatorname{Aut}(\sigma)|} \log |\sigma(f)|^2 .$$

For $x \in \mathcal{Z}(\bar{\mathbb{F}}_p)$, let $[x]$ be the corresponding geometric point of the coarse moduli scheme $Z = \operatorname{Spec} O$ of $\mathcal{Z}$. Then

$$(2.1.12)\qquad \tilde{\mathcal{O}}_{Z,[x]} = (\tilde{\mathcal{O}}_{\mathcal{Z},x})^{\operatorname{Aut}(x)/\operatorname{Aut}(\bar{\eta})},$$

where $\bar{\eta}$ is any generic geometric point of $\mathcal{Z}$. It follows that

$$(2.1.13)\qquad \deg_x f = |\operatorname{Aut}(x)| / |\operatorname{Aut}(\bar{\eta})| \cdot \deg_{[x]} f .$$

Inserting this into (2.1.11) we obtain for the right-hand side the expression

$$(2.1.14)\quad \frac{1}{|\operatorname{Aut}(\bar{\eta})|} \cdot \Big(\sum_p \sum_{x \in (\operatorname{Spec} O)(\bar{\mathbb{F}}_p)} \deg_x(f) \cdot \log p - \sum_\sigma \log |\sigma(f)|^2 \Big).$$

Using the normalization $\nu : \tilde{\mathcal{Z}} \to \mathcal{Z}$ we may rewrite this as

$$(2.1.15)\quad \frac{1}{|\operatorname{Aut}(\bar{\eta})|} \cdot \Big(\sum_p \sum_{\tilde{x} \in (\operatorname{Spec} O_K)(\bar{\mathbb{F}}_p)} \operatorname{ord}_{\tilde{x}}(f) \cdot \log p - \sum_\sigma \log |\sigma(f)|^2 \Big),$$

which is zero by the product formula for $f \in K^\times$.

The definition of $\widehat{\deg}(\hat{\mathcal{L}})$ is again additive in $\hat{\mathcal{L}}$ and compatible with passing to a finite covering; see (2.1.5).

An important line bundle is the relative dualizing sheaf $\omega_{\mathcal{Z}/S}$. It is characterized by the fact that its pullback to any étale presentation $X \to \mathcal{Z}$ is the relative dualizing sheaf $\omega_{X/\mathrm{Spec}\,\mathbb{Z}}$. Recall that by Grothendieck duality we have (we identify the sheaves with the $\mathbb{Z}$-modules they define),

$$\omega_{X/\mathbb{Z}} = \mathrm{Hom}_{\mathcal{O}_X}(\mathcal{O}_X, \omega_{X/\mathbb{Z}}) = \mathrm{Hom}_{\mathbb{Z}}(\mathcal{O}_X, \mathbb{Z}), \tag{2.1.16}$$

the inverse different of the order $\Gamma(X, \mathcal{O}_X)$ in $\Gamma(X \otimes_{\mathbb{Z}} \mathbb{Q}, \mathcal{O})$. In particular we obtain a natural homomorphism

$$\Gamma(X, \omega_{X/\mathbb{Z}}) \hookrightarrow \Gamma(X \otimes_{\mathbb{Z}} \mathbb{Q}, \mathcal{O}). \tag{2.1.17}$$

It follows that $\omega_{X/\mathbb{Z}}$ is equipped with a natural metric $\|\ \|$. For this metric we have for any complex embedding

$$\sigma : \Gamma(X \otimes_{\mathbb{Z}} \mathbb{Q}, \mathcal{O}) \hookrightarrow \mathbb{C} \tag{2.1.18}$$

that $\|\sigma(1)\| = 1$. By naturality, this metric descends to a metric on $\omega_{\mathcal{Z}/S}$. We define

$$\mathfrak{d}_{\mathcal{Z}} = \widehat{\deg}\,(\omega_{\mathcal{Z}/S}, \|\ \|). \tag{2.1.19}$$

It therefore follows that

$$\mathfrak{d}_{\mathcal{Z}} = \log|\omega_{\mathcal{Z}/S} : O| = \log|\mathcal{D}^{-1} : O|, \tag{2.1.20}$$

where $\mathcal{D}^{-1} = O^*$ is the dual module with respect to the trace form of O (inverse of the absolute different of the order O). Indeed the trace map $\mathrm{tr}_{O/\mathbb{Z}}$ defines an element $t \in O^* = \Gamma(\mathcal{Z}, \omega_{\mathcal{Z}/S})$ which goes to $1 \in \mathbb{C}$ under every σ. Using the global section t to calculate (2.1.9) for $\hat{\omega}_{\mathcal{Z}/S}$, we see that the first summand gives $\log|\mathcal{D}^{-1} : O|$ while the second summand vanishes.

2.2 $\mathrm{Pic}(\mathcal{M})$, $\mathrm{CH}^1_{\mathbb{Z}}(\mathcal{M})$, AND $\mathrm{CH}^2_{\mathbb{Z}}(\mathcal{M})$

In this section we take up the study of our two-dimensional stack $\mathcal{M}$. Let $\mathrm{Pic}(\mathcal{M})$ be the set of isomorphism classes of line bundles on $\mathcal{M}$, an abelian group under the tensor product operation. This is related to the Chow group $\mathrm{CH}^1_{\mathbb{Z}}(\mathcal{M})$ as follows.

By a *prime divisor on* $\mathcal{M}$ we mean a closed irreducible reduced substack $\mathcal{Z}$ of $\mathcal{M}$ which is locally for the étale topology a Cartier divisor. Let $Z^1_{\mathbb{Z}}(\mathcal{M})$ be the free abelian group generated by the prime divisors on $\mathcal{M}$. Let $f \in$

$\mathbb{Q}(\mathcal{M})^\times$ be a rational function. In other words, f is the germ of a morphism $\mathcal{U} \to \mathbb{A}^1$ defined on a dense open substack $\mathcal{U}$ of $\mathcal{M}$. Equivalently, since $\mathcal{M}$ is irreducible, f is an element of the function field of the coarse moduli scheme of $\mathcal{M}$. Then to f there is associated a principal divisor

$$\mathrm{div}(f) = \sum_{\mathcal{Z}} \mathrm{ord}_{\mathcal{Z}}(f) \cdot \mathcal{Z}, \tag{2.2.1}$$

where the sum is over all prime divisors $\mathcal{Z}$ of $\mathcal{M}$ and where we note that, since $\mathcal{M}$ is regular, the strict henselization of the local ring of $\mathcal{Z}$, $\tilde{\mathcal{O}}_{\mathcal{M},\mathcal{Z}}$, is a discrete valuation ring so that $\mathrm{ord}_{\mathcal{Z}}(f)$ has a meaning. The factor group of $Z^1_{\mathbb{Z}}(\mathcal{M})$ by the group of principal divisors is the first Chow group $\mathrm{CH}^1_{\mathbb{Z}}(\mathcal{M})$.

The groups $\mathrm{Pic}(\mathcal{M})$ and $\mathrm{CH}^1_{\mathbb{Z}}(\mathcal{M})$ are isomorphic. Under this isomorphism, an element $\mathcal{L}$ goes to the class of $\Sigma_{\mathcal{Z}} \mathrm{ord}_{\mathcal{Z}}(s) \cdot \mathcal{Z}$, where s is a meromorphic section of $\mathcal{L}$. Conversely, if $\mathcal{Z} \in \mathrm{Z}^1_{\mathbb{Z}}(\mathcal{M})$, then its preimage under this isomorphism is $\mathcal{O}(\mathcal{Z})$.

We denote by $Z^1(\mathcal{M}) = Z^1_{\mathbb{Z}}(\mathcal{M}) \otimes \mathbb{R}$ the space of real divisors (i.e., the formal sums of prime divisors with coefficients in $\mathbb{R}$), and by $\mathrm{CH}^1(\mathcal{M})$ the factor group by the $\mathbb{R}$-subspace generated by the principal divisors.

We will also have use for the *second* Chow group $\mathrm{CH}^2_{\mathbb{Z}}(\mathcal{M})$. By a *0-cycle on* $\mathcal{M}$ we mean a formal sum $\Sigma_{\mathcal{P}} n_{\mathcal{P}} \mathcal{P}$ where $\mathcal{P}$ ranges over the irreducible reduced closed substacks of $\mathcal{M}$ of dimension 0. We denote by $Z^2_{\mathbb{Z}}(\mathcal{M})$ the abelian group of 0-cycles on $\mathcal{M}$.

We have the homomorphism of abelian groups,

$$\begin{array}{ccc} \bigoplus_{\mathcal{Z}} \mathbb{Q}(\mathcal{Z})^\times & \longrightarrow & Z^2_{\mathbb{Z}}(\mathcal{M}) \\ f_{\mathcal{Z}} & \longmapsto & \sum_{\mathcal{P} \subset \mathcal{Z}} \deg_{\mathcal{P}}(f) \cdot \mathcal{P}. \end{array} \tag{2.2.2}$$

(Here, as in (2.1.3), the degree function $\deg_{\mathcal{P}}$ is the sum of the corresponding degree functions over all formal branches of $\mathcal{Z}$ at $\mathcal{P}$.) Then $\mathrm{CH}^2_{\mathbb{Z}}(\mathcal{M})$ is the factor group of $Z^2_{\mathbb{Z}}(\mathcal{M})$ by the image of (2.2.2).

We also denote by $\mathrm{CH}^2(\mathcal{M})$ the $\mathbb{R}$-version of $\mathrm{CH}^2_{\mathbb{Z}}(\mathcal{M})$, i.e. the factor space of $Z^2(\mathcal{M}) = Z^2_{\mathbb{Z}}(\mathcal{M}) \otimes \mathbb{R}$ by the $\mathbb{R}$-subspace generated by the image of (2.2.2).

2.3 GREEN FUNCTIONS

In this section, we review the theory of Green functions needed to define the arithmetic Chow groups in the next section. Due to the assumptions we made at the beginning of this chapter about the stack $\mathcal{M}$, we can restrict ourselves to the following situation. Let X be a compact Riemann surface (not necessarily connected), and let Γ be a finite group which acts on X by holomorphic automorphisms. We do not assume that the action is effective. The quotient stack $M = [\Gamma\backslash X]$ then has a presentation

$$\Gamma \times X \rightrightarrows X, \tag{2.3.1}$$

where one arrow is the projection onto the second factor and the other is the group action. There is a holomorphic projection

$$\mathrm{pr} : X \longrightarrow [\Gamma\backslash X] = M. \tag{2.3.2}$$

A C^∞ (resp. meromorphic, resp. ...) function on M is given by a Γ-invariant C^∞ (resp. meromorphic, resp. ...) function on X. Similarly, a measure (resp. 2-form) μ on M is given by a Γ-invariant measure (resp. 2-form) on X, and we have

$$\int_M f \cdot \mu = \int_{[\Gamma\backslash X]} f \cdot \mu := |\Gamma|^{-1} \int_X f \cdot \mu. \tag{2.3.3}$$

If $z \in X$ is a point, there is a corresponding point, i.e., a closed irreducible substack of dimension 0,

$$P = [\Gamma_z \backslash z] \longrightarrow [\Gamma\backslash X] = M \tag{2.3.4}$$

of M. Of course, two points z and z' in X define the same point of M if and only if they are in the same Γ-orbit. Note that there is an alternative presentation

$$[\Gamma\backslash \mathrm{pr}^{-1}(P)] = [\Gamma_z \backslash z] = P. \tag{2.3.5}$$

We define the *delta distribution* δ_P of a point $P \in \mathcal{M}$. For a function f on M,

$$\langle \delta_P, f \rangle_M := |\Gamma|^{-1} \cdot \langle \delta_{\mathrm{pr}^{-1}(P)}, f \rangle_X = |\Gamma_z|^{-1} \cdot f(z). \tag{2.3.6}$$

A divisor Z on M is an element of the free abelian group on the points of M. Associated to

$$Z = \sum_{P \in X} n_P \cdot P \tag{2.3.7}$$

is a Γ-invariant divisor

$$\tilde{Z} = \sum_{P \in M} \sum_{z \in \mathrm{pr}^{-1}(P)} n_P \cdot z \tag{2.3.8}$$

on X. The degree of Z is given by

$$\deg_M(Z) = \langle \delta_Z, 1 \rangle_M = \sum_{P \in Z} n_P \cdot |\Gamma_P|^{-1} = |\Gamma|^{-1} \cdot \deg_X(\tilde{Z}). \tag{2.3.9}$$

Definition 2.3.1. A Green function for a divisor Z on M is a Γ-invariant Green function g for the divisor $\tilde{Z}$ on X. In particular, g satisfies the Green equation

$$dd^c g + \delta_{\tilde{Z}} = [\omega]$$

of currents on X, where ω is a smooth, Γ-invariant $(1,1)$-form on X.

To see that this is the correct definition, we check that the corresponding Green equation holds for currents on M. By linearity, we may assume that $Z = P$ is a single point on M, so that $\tilde{Z} = \mathrm{pr}^{-1}(P)$. Then

$$\begin{aligned} \langle dd^c g, f \rangle_M &= |\Gamma|^{-1} \cdot \langle dd^c g, f \rangle_X \\ &= |\Gamma|^{-1} \cdot \Big(-\langle \delta_{\mathrm{pr}^{-1}(P)}, f \rangle_X + \int_X f \cdot \omega \Big) \\ &= -\langle \delta_x, f \rangle_M + \int_M f \cdot \omega. \end{aligned} \tag{2.3.10}$$

Thus

$$dd^c g + \delta_Z = [\omega] \tag{2.3.11}$$

as currents on M. Next, if Z_1 and Z_2 are divisors on M with disjoint supports, then the supports of the Γ-invariant divisors $\tilde{Z}_1$ and $\tilde{Z}_2$ on X are also disjoint. If g_1 and g_2 are Green functions for Z_1 and Z_2, one may view them as Γ-invariant Green functions for $\tilde{Z}_1$ and $\tilde{Z}_2$ on X and form their usual star product

$$g_1 * g_2 = g_1\,\delta_2 + g_2\,\omega_1, \tag{2.3.12}$$

where δ_2 is the delta current for the divisor $\tilde{Z}_2$ on X. This is a Γ-invariant distribution on X, and, as above, we can view it as a distribution on M. Again, we may suppose that $Z_2 = P_2$ is a single point and compute

$$\begin{aligned} \langle g_1 * g_2, f \rangle_M :&= |\Gamma|^{-1} \cdot \langle g_1 * g_2, f \rangle_X \\ &= |\Gamma|^{-1} \cdot \Big(\sum_{z \in \mathrm{pr}^{-1}(P_2)} g_1(z)\, f(z) + \int_X f \cdot g_2\,\omega_1 \Big) \\ &= |\Gamma|^{-1}\, |\Gamma/\Gamma_z| \cdot g_1(z)\, f(z) + \int_M f \cdot g_2\,\omega_1 \\ &= \langle g_1\,\delta_{P_2} + g_2\,\omega_1, f \rangle_M \end{aligned} \tag{2.3.13}$$

Thus, we may view the formula (2.3.12) for the star product above as an identity of currents on M, where the δ_2 on the right side is the delta function of the divisor Z_2. We obtain the symmetry

$$\langle g_1 * g_2, 1 \rangle_M = \langle g_2 * g_1, 1 \rangle_M \tag{2.3.14}$$

by appealing to the corresponding symmetry on X, [1]. When M is not connected, the same identity (2.3.14) holds with 1 replaced by the characteristic function of any connected component.

One can check that the notions described so far are intrinsic to M, i.e., do not depend on the particular presentation (2.3.1). For example, suppose that $\Gamma_0 \subset \Gamma$ is a normal subgroup which acts without fixed points on X, let $X_1 = \Gamma_0 \backslash X$ be the quotient Riemann surface and let $\Gamma_1 = \Gamma/\Gamma_0$. Then, the orbifolds $[\Gamma\backslash X]$ and $[\Gamma_1\backslash X_1]$ are isomorphic. A Green function g for a divisor Z on M can be given as a Γ-invariant Green function g on X for the divisor $\mathrm{pr}^{-1}(Z)$. But since such a g is then Γ_0-invariant, it may, in turn, be viewed as a Γ_1-invariant Green function on X_1 for the Γ_1-invariant divisor $\mathrm{pr}_1^{-1}(Z)$ on X_1.

Eventually, we will be in the situation where Γ is a discrete co-compact subgroup of $\mathrm{GL}_2(\mathbb{R})$ acting on $D = \mathbb{P}^1(\mathbb{C}) \setminus \mathbb{P}^1(\mathbb{R})$. Moreover, there will be a normal subgroup $\Gamma_0 \subset \Gamma$ of finite index which acts without fixed points on D. Then, we can work with the orbifold $M = [\Gamma_1\backslash X_1] = [\Gamma\backslash D]$, where $X_1 = \Gamma_0\backslash D$ and $\Gamma_1 = \Gamma/\Gamma_0$. Our Green functions for divisors Z on M will be given as Γ-invariant Green functions on D for the divisor $\mathrm{pr}^{-1}(Z)$. Here, of course, we mean that the corresponding function g on X_1 is a Green function for the Γ_1-invariant divisor $\mathrm{pr}_1^{-1}(Z)$ on X_1 as discussed above. By the previous remark, the construction is independent of the choice of Γ_0.

2.4 $\widehat{\mathrm{Pic}}(\mathcal{M})$, $\widehat{\mathrm{CH}}^1_{\mathbb{Z}}(\mathcal{M})$, AND $\widehat{\mathrm{CH}}^2_{\mathbb{Z}}(\mathcal{M})$

By $\widehat{\mathrm{Pic}}(\mathcal{M})$ we mean, as in the one-dimensional case, the abelian group of isomorphism classes of metrized line bundles on $\mathcal{M}$. Let $\mathcal{Z} \in Z^1_{\mathbb{Z}}(\mathcal{M})$. In the previous section we explained what is meant by a Green function for $\mathcal{Z}$. We denote by $\hat{\mathcal{Z}}^1_{\mathbb{Z}}(\mathcal{M})$ the group of Arakelov divisors, i.e., of pairs $(\mathcal{Z}, g)$ consisting of a divisor $\mathcal{Z}$ and a Green function for $\mathcal{Z}$, with componentwise addition.

If $f \in \mathbb{Q}(\mathcal{M})^\times$, then $f|\mathcal{M}_\mathbb{C}$ corresponds to a Γ-invariant meromorphic function $\tilde{f}_\mathbb{C}$ on X, and we define the *associated principal Arakelov divisor*

$$\widehat{\mathrm{div}}(f) = (\,\mathrm{div}(f), -\log|\tilde{f}_\mathbb{C}|^2\,). \tag{2.4.1}$$

The factor group of $\hat{\mathcal{Z}}^1_{\mathbb{Z}}(\mathcal{M})$ by the group of principal Arakelov divisors is the *arithmetic Chow group* $\widehat{\mathrm{CH}}^1_{\mathbb{Z}}(\mathcal{M})$. The groups $\widehat{\mathrm{CH}}^1_{\mathbb{Z}}(\mathcal{M})$ and $\widehat{\mathrm{Pic}}(\mathcal{M})$

are isomorphic. Under this isomorphism, an element $\hat{\mathcal{L}}$ goes to the class of

$$(2.4.2)\qquad \Big(\sum_{\mathcal{Z}} \mathrm{ord}_{\mathcal{Z}}(s)\,\mathcal{Z},\ -\log\|s\|^2\Big),$$

where s is a meromorphic section of $\mathcal{L}$. Conversely, if $(\mathcal{Z}, g) \in \hat{Z}^1_{\mathbb{Z}}(\mathcal{M})$, then its preimage under this isomorphism is

$$(2.4.3)\qquad (\mathcal{O}(\mathcal{Z}), \|\ \|),$$

where $-\log\|\mathbf{1}\|^2 = g$, with $\mathbf{1}$ the canonical Γ-invariant section of the pullback of $\hat{\mathcal{O}}(\mathcal{Z})$ to X.

We also introduce the $\mathbb{R}$-version $\widehat{\mathrm{CH}}^1(\mathcal{M})$ of $\widehat{\mathrm{CH}}^1_{\mathbb{Z}}(\mathcal{M})$; see [1], 5.5. In its definition one starts with $\hat{Z}^1(\mathcal{M})$, which is the $\mathbb{R}$-vector space of pairs $(\mathcal{Z}, g)$, where $\mathcal{Z} \in Z^1(\mathcal{M})$ is an $\mathbb{R}$-divisor and g is a Green function for $\mathcal{Z}$, and divides out by the $\mathbb{R}$-subspace generated by the Arakelov principal divisors. As Bost points out [1], whereas $\mathrm{CH}^1(\mathcal{M}) = \mathrm{CH}^1_{\mathbb{Z}}(\mathcal{M}) \otimes \mathbb{R}$, the space $\widehat{\mathrm{CH}}^1(\mathcal{M})$ cannot be identified with $\widehat{\mathrm{CH}}^1_{\mathbb{Z}}(\mathcal{M}) \otimes \mathbb{R}$.

We next turn to $\widehat{\mathrm{CH}}^2_{\mathbb{Z}}(\mathcal{M})$. Let

$$(2.4.4)\qquad \hat{Z}^2_{\mathbb{Z}}(\mathcal{M}) = \{(\mathcal{Z}, g);\ \mathcal{Z} \in Z^2_{\mathbb{Z}}(M),\ g \in D^{1,1}(\mathcal{M}_{\mathbb{C}})\}.$$

Here $D^{1,1}(\mathcal{M}_{\mathbb{C}})$ is the $\mathbb{R}$-vector space of $(1,1)$-currents on $\mathcal{M}_{\mathbb{C}}$ (i.e., the space of Γ-invariant currents of type $(1,1)$ on X) which are *real* in the sense that

$$(2.4.5)\qquad F_\infty^*(g) = -g$$

where F_∞ denotes complex conjugation. Note that $\mathcal{Z}$ is 'in the top degree' so that there is no Green equation linking $\mathcal{Z}$ to g. Then $\hat{Z}^2_{\mathbb{Z}}(\mathcal{M})$ is a group under componentwise addition. Let $\hat{R}^2_{\mathbb{Z}}(\mathcal{M})$ be the subgroup of $\hat{Z}^2_{\mathbb{Z}}(\mathcal{M})$ generated by elements of the form

$$(2.4.6)\qquad \Big(\sum_{\mathcal{P}\subset\mathcal{Z}} \deg_{\mathcal{P}}(f)\cdot\mathcal{P},\, i_{\mathcal{Z}*}(-\log|f^2|)\Big),$$

where $f = f_{\mathcal{Z}} \in \mathbb{Q}(\mathcal{Z})^\times$ for some prime divisor $\mathcal{Z}$ on $\mathcal{M}$, and by elements of the form $(0, \partial u + \bar{\partial} v)$ for currents u of type $(0,1)$ and v of type $(1,0)$. Here $i_{\mathcal{Z}*}(-\log|f|^2)$ is the current with

$$(2.4.7)\qquad \langle \phi, i_{\mathcal{Z}*}(-\log|f|^2)\rangle = -\sum_{x\in\mathcal{Z}(\mathbb{C})} \frac{1}{|\mathrm{Aut}(x)|}\cdot\phi(x)\cdot\log|f(x)|^2.$$

Then $\widehat{\mathrm{CH}}^2_{\mathbb{Z}}(\mathcal{M})$ is the factor group $\hat{Z}^2_{\mathbb{Z}}(\mathcal{M})/\hat{R}^2_{\mathbb{Z}}(\mathcal{M})$. Similarly, we let $\hat{Z}^2(\mathcal{M})$ be the $\mathbb{R}$-vector space of pairs $(\mathcal{Z}, g)$ where $\mathcal{Z} \in Z^2(\mathcal{M}) \otimes_{\mathbb{Z}} \mathbb{R}$

and where g is as before. We let $\widehat{\mathrm{CH}}^2(\mathcal{M})$ be the quotient of $\hat{Z}^2(\mathcal{M})$ by the $\mathbb{R}$-subvector space generated by $\hat{R}^2_{\mathbb{Z}}(\mathcal{M})$.

There is the Arakelov degree map

$$\begin{array}{lccc} (2.4.8) & \widehat{\mathrm{CH}}^2_{\mathbb{Z}}(\mathcal{M}) & \longrightarrow & \mathbb{R}, \\ & (\mathcal{Z}, g) & \longmapsto & \widehat{\deg}\, \mathcal{Z} + \int_{\mathcal{M}(\mathbb{C})} g. \end{array}$$

Here for $\mathcal{Z} = \sum_{\mathcal{P}} m_{\mathcal{P}}\, \mathcal{P}$ we have put

$$\widehat{\deg}\, \mathcal{Z} = \sum_p \sum_{\mathcal{P}} m_{\mathcal{P}} \Big(\sum_{x \in \mathcal{P}(\bar{\mathbb{F}}_p)} \frac{1}{|\mathrm{Aut}(x)|} \Big) \log p. \tag{2.4.9}$$

The integral $\int_{\mathcal{M}(\mathbb{C})} g = |\Gamma|^{-1} \cdot \int_X g$ is to be understood as in the previous section.

That this map is well defined follows from Stokes's theorem for modifications of the form $\partial u + \bar{\partial} v$. For elements of the form (2.4.6), for $\mathcal{Z}$ horizontal, irreducible and reduced, it follows from the product formula, cf. section 1.

The degree map obviously factors through the $\mathbb{R}$-arithmetic Chow group,

$$\widehat{\deg} : \ \widehat{\mathrm{CH}}^2(\mathcal{M}) \longrightarrow \mathbb{R}. \tag{2.4.10}$$

If $\mathcal{M}$ is geometrically irreducible, this last map is an isomorphism. Indeed, in this case the $\mathbb{R}$-vector space

$$D^{1,1}(\mathcal{M}_{\mathbb{C}}) / \ (\ \mathrm{Im}\, \partial + \mathrm{Im}\, \bar{\partial}\ )$$

has dimension 1. On the other hand, let $\mathcal{P} \in Z^2_{\mathbb{Z}}(\mathcal{M})$ and choose an irreducible horizontal divisor $\mathcal{Z} \in Z^1_{\mathbb{Z}}(\mathcal{M})$ with $\mathcal{P} \subset \mathcal{Z}$. Now $\mathrm{Pic}(\mathcal{Z})$ is finite, hence there exists $n \in \mathbb{Z}$ and $f \in \Gamma(\mathcal{Z} \otimes \mathbb{Q}, \mathcal{O})^\times$ such that $n\mathcal{P} = \mathrm{div}(f)$ as divisors on $\mathcal{Z}$. But then $(n\mathcal{P}, 0) \equiv (0, i_{\mathcal{Z}*}(\log |f|^2)$ in $\widehat{\mathrm{CH}}^2_{\mathbb{Z}}(\mathcal{M})$.

The same argument shows that when $\mathcal{M}$ is not geometrically connected, then

$$\widehat{\mathrm{CH}}^2(\mathcal{M}) \xrightarrow{\sim} \mathbb{R}^{\pi_0(\mathcal{M}_{\mathbb{C}})}.$$

2.5 THE PAIRING $\widehat{\mathrm{CH}}^1(\mathcal{M}) \times \widehat{\mathrm{CH}}^1(\mathcal{M}) \to \widehat{\mathrm{CH}}^2(\mathcal{M})$

Let

(2.5.1)

$$(\hat{\mathcal{Z}}^1(\mathcal{M}) \times \hat{\mathcal{Z}}^1(\mathcal{M}))_o = \{(\mathcal{Z}_1, g_1), (\mathcal{Z}_2, g_2);\ \mathrm{supp}\mathcal{Z}_{1\mathbb{Q}} \cap \mathrm{supp}\mathcal{Z}_{2\mathbb{Q}} = \emptyset\}$$

be the set of pairs of Arakelov divisors with disjoint support on the generic fiber. On this subset we define the intersection pairing by setting

$$(2.5.2)\qquad (\mathcal{Z}_1, g_1).(\mathcal{Z}_2, g_2) = (\mathcal{Z}_1.\mathcal{Z}_2, g_1 * g_2).$$

Here the first component is the 0-cycle defined by bilinear extension from the case where $\mathcal{Z}_1$ and $\mathcal{Z}_2$ are irreducible and reduced. In this case, if $\mathcal{Z}_1 \neq \mathcal{Z}_2$, the definition of $\mathcal{Z}_1.\mathcal{Z}_2$ is clear (each 'point' in the intersection is weighted with the length of the local ring of $\mathcal{Z}_1 \cap \mathcal{Z}_2$). If $\mathcal{Z}_1 = \mathcal{Z}_2 = \mathcal{Z}$, then $\mathcal{Z}$ is a vertical divisor lying in the special fiber $\mathcal{M}_p = \mathcal{M} \otimes_{\mathbb{Z}} \mathbb{F}_p$ for some p. We write

$$(2.5.3)\qquad \operatorname{div}(p) = a \cdot \mathcal{Z} + R \ \text{ in } Z^1(\mathcal{M}),$$

where R is prime to $\mathcal{Z}$. Then we set

$$(2.5.4)\qquad \mathcal{Z}.\mathcal{Z} = -\frac{1}{a} \cdot (R.\mathcal{Z}).$$

We claim that the induced pairing

$$(2.5.5)\qquad (\hat{\mathrm{Z}}^1(\mathcal{M}) \times \hat{\mathrm{Z}}^1(\mathcal{M}))_o \longrightarrow \widehat{\mathrm{CH}}^2(\mathcal{M})$$

is symmetric. This is obvious as far as the symmetry in the first component is concerned. For the second component it follows from the symmetry (2.3.14) which shows that

$$(2.5.6)\qquad g_1 * g_2 \equiv g_2 * g_1 \mod (\ \operatorname{im} \partial + \operatorname{im} \bar{\partial}\ ).$$

We now want to show that the pairing above descends to a symmetric pairing

$$(2.5.7)\qquad \langle\ ,\ \rangle : \widehat{\mathrm{CH}}^1(\mathcal{M}) \times \widehat{\mathrm{CH}}^1(\mathcal{M}) \longrightarrow \widehat{\mathrm{CH}}^2(\mathcal{M}).$$

Since any pair of two classes in $\widehat{\mathrm{CH}}^1(\mathcal{M})$ can be represented by an element in $(\hat{Z}^1(\mathcal{M}) \times \hat{Z}^1(\mathcal{M}))_o$ the assertion comes down to proving that if

$$((\mathcal{Z}_1, g_1), (\mathcal{Z}_2, g_2)) \quad \text{and} \quad ((\mathcal{Z}_1', g_1'), (\mathcal{Z}_2', g_2'))$$

in $(\hat{Z}^1(\mathcal{M}) \times \hat{Z}^1(\mathcal{M}))_o$ represent the same element in $\widehat{\mathrm{CH}}^1(\mathcal{M}) \times \widehat{\mathrm{CH}}^1(\mathcal{M})$, then

$$(2.5.8)\qquad (\mathcal{Z}_1, g_1).(\mathcal{Z}_2, g_2) = (\mathcal{Z}_1', g_1').(\mathcal{Z}_2', g_2').$$

This in turn is reduced to the following statement. Let $(\mathcal{Z}, g) \in \hat{Z}^1_{\mathbb{Z}}(\mathcal{M})$ with $\mathcal{Z}$ irreducible and reduced. Let $f \in \mathbb{Q}(\mathcal{M})^\times$ such that $\mathcal{Z}_{\mathbb{Q}} \cap \operatorname{div}(f)_{\mathbb{Q}} = \emptyset$. Then

$$(2.5.9)\qquad \widehat{\operatorname{div}}(f).(\mathcal{Z}, g) = 0 \ \text{ in } \widehat{\mathrm{CH}}^2(\mathcal{M}).$$

If $\mathcal{Z}$ is horizontal, then the LHS of (2.5.9) is the image of $f|\mathcal{Z}$ under the map (2.4.6) since for the Green function part

$$-\log|f|^2 * g = -\log|f|^2 \cdot \delta_{\mathcal{Z}} \tag{2.5.10}$$

by the Lelong formula. Therefore, the claim (2.5.9) follows in this case. If $\mathcal{Z}$ is vertical, of the form (2.5.3), let $\mathrm{div}(f) = m\mathcal{Z} + \mathcal{Z}'$, where $\mathcal{Z}'$ is relatively prime to $\mathcal{Z}$. Then

$$\mathrm{div}(f^a \cdot p^{-m}) = a\mathcal{Z}' - mR \tag{2.5.11}$$

is relatively prime to $\mathcal{Z}$. Hence $f^a \cdot p^{-m}|\mathcal{Z}$ is a nonzero rational function $\mathcal{Z}$ and

$$\mathrm{div}(f^a \cdot p^{-m}).\mathcal{Z} = i_{\mathcal{Z}^*}(\mathrm{div}(f^a \cdot p^{-m}|\mathcal{Z})) \equiv 0 \tag{2.5.12}$$

in $\widehat{\mathrm{CH}}^2(\mathcal{M})$. We are therefore reduced to proving $\mathrm{div}(p).\mathcal{Z} \equiv 0$ (recall that $\widehat{\mathrm{CH}}^2(\mathcal{M})$ is the $\mathbb{R}$-version of the arithmetic Chow group). But this is exactly the content of the definition (2.5.4) of $\mathcal{Z}.\mathcal{Z}$.

Composing (2.5.7) with the arithmetic degree map $\widehat{\deg} : \widehat{\mathrm{CH}}^2(\mathcal{M}) \to \mathbb{R}$, we obtain the Arakelov intersection pairing

$$\langle\, ,\, \rangle : \widehat{\mathrm{CH}}^1(\mathcal{M}) \times \widehat{\mathrm{CH}}^1(\mathcal{M}) \longrightarrow \mathbb{R}. \tag{2.5.13}$$

2.6 ARAKELOV HEIGHTS

Let $\hat{\mathcal{L}} = (\mathcal{L}, \|\ \|)$ be a metrized line bundle on $\mathcal{M}$. Let $\mathcal{Z} \in Z^1_{\mathbb{Z}}(\mathcal{M})$ be an irreducible and reduced divisor. We then define the *height of* $\mathcal{Z}$ with respect to $\mathcal{L}$ by

$$h_{\hat{\mathcal{L}}}(\mathcal{Z}) = \begin{cases} \widehat{\deg}(\mathcal{Z}, i^*_{\mathcal{Z}}(\hat{\mathcal{L}})), & \text{if } \mathcal{Z} \text{ is horizontal,} \\ \widehat{\deg}(\mathcal{Z}, i^*_{\mathcal{Z}}(\mathcal{L})) & \text{if } \mathcal{Z} \text{ is vertical.} \end{cases} \tag{2.6.1}$$

We extend $h_{\hat{\mathcal{L}}}$ to all of $Z^1_{\mathbb{Z}}(\mathcal{M})$ by linearity. Using this concept we have the following expression for the Arakelov intersection pairing. Let $(\mathcal{Z}', g') \in \hat{Z}^1_{\mathbb{Z}}(\mathcal{M})$ be a representative of $\hat{\mathcal{L}}$ under the isomorphism $\widehat{\mathrm{Pic}}(\mathcal{M}) = \widehat{\mathrm{CH}}^1_{\mathbb{Z}}(\mathcal{M})$. Then by [1], (5.11),

$$\langle(\mathcal{Z}, g), (\mathcal{Z}', g')\rangle = h_{\hat{\mathcal{L}}}(\mathcal{Z}) + \frac{1}{2} \cdot \int_{[\Gamma\backslash X]} g \cdot c_1(\hat{\mathcal{L}}). \tag{2.6.2}$$

Here $c_1(\hat{\mathcal{L}}) = \omega'$ is the RHS of the Green equation for $(\mathcal{Z}', g')$, i.e.,

$$dd^c g' + \delta_{\tilde{\mathcal{Z}}'} = c_1(\hat{\mathcal{L}}). \tag{2.6.3}$$

This formula is obvious from our definitions when the supports of $\mathcal{Z}$ and $\mathcal{Z}'$ are disjoint on the generic fiber. The general case follows since all terms are unchanged when $(\mathcal{Z}', g')$ is replaced by $(\mathcal{Z}', g') + \widehat{\text{div}} f$ for some $f \in \mathbb{Q}(\mathcal{M})^{\times}$.

2.7 THE ARITHMETIC ADJUNCTION FORMULA

In this section, we review the adjunction formula for arithmetic surfaces that we will need. We follow the treatment in [7], which has the advantage of allowing us sufficient flexibility in the choice of the metrics.

Let X be a compact Riemann surface with a smooth $(1,1)$-form ν. Let Δ be the diagonal divisor on $X \times X$ and let $\mathcal{O}_{X\times X}(\Delta)$ be the associated line bundle with its canonical section s_Δ with divisor $\text{div}(s_\Delta) = \Delta$.

Definition 2.7.1. A weakly ν-biadmissible Green function g is a function $g : X \times X - \Delta \to \mathbb{R}$, satisfying the following conditions:

(i) g is C^∞ and has a logarithmic singularity along Δ, i.e., on an open neighborhood $U \times U$,

$$g(z_1, z_2) = -\log|\zeta_1 - \zeta_2|^2 + \text{smooth}$$

with respect to a local coordinate ζ on U, where $\zeta_1 = \zeta(z_1)$ and $\zeta_2 = \zeta(z_2)$.

(ii) (symmetry)

$$g(z_1, z_2) = g(z_2, z_1).$$

(iii) (Green equation) There is a C^∞ function

$$\phi : X \times X \longrightarrow \mathbb{R}$$

such that $\phi(z_1, z_2) = \phi(z_2, z_1)$,

$$\int_X \phi(z_1, z_2)\, d\nu(z_2) = 1,$$

and, for any fixed $z_1 \in X$,

$$d_2 d_2^c g(z_1, z_2) + \delta_{z_1} = \phi(z_1, z_2)\, \text{pr}_2^*(\nu).$$

Here, in (iii), pr_2 is the projection on the second factor of $X \times X$ and, as usual, $d^c = (\partial - \bar\partial)/4\pi i$, so that $dd^c = -\partial\bar\partial/2\pi i$.

The definition above generalizes the definition of a ν-biadmissible metric in [8] (*bipermise*). In [8], ϕ is required to be the constant 1, and it is explained that in the case of the Arakelov volume form ν, these conditions arise in a very natural way from the relation between X and its Jacobian. Our variant will allow us to make a connection with the peculiar metrics introduced in [6] and is also suitable when X is no longer compact. Note that, for fixed $z_1 \in X$, the function $z_2 \mapsto g(z_1, z_2)$ is a Green function of logarithmic type in the terminology of Gillet and Soulé [5], [10] for the point $z_1 \in X$.

A weakly ν-biadmissible Green function g determines a metric $|| \; ||$ on $\mathcal{O}_{X\times X}(\Delta)$, defined by

$$g(z_1, z_2) = -\log ||s_\Delta(z_1, z_2)||^2. \tag{2.7.1}$$

It therefore also defines a metric on each point bundle

$$\mathcal{O}_X(z) = i_z^* \, \mathcal{O}_{X\times X}(\Delta), \tag{2.7.2}$$

where, for a fixed $z \in X$,

$$i_z : X \to X \times X, \qquad z' \mapsto (z, z'). \tag{2.7.3}$$

The associated Green function is given by $z' \mapsto g(z, z')$. For any divisor $Z = \sum_i z_i$, the bundle $\mathcal{O}_X(Z) = \otimes_i \mathcal{O}_X(z_i)$ gets the tensor product metric. Of course, this metric can depend on the given divisor Z and not just on the isomorphism class of $\mathcal{O}_X(Z)$ in $\mathrm{Pic}(X)$.

A metric on the canonical bundle Ω^1_X is determined by the canonical isomorphism

$$\Omega^1_X \simeq i_\Delta^* \, \mathcal{O}_{X\times X}(-\Delta). \tag{2.7.4}$$

Note that the Green function on $X \times X - \Delta$ associated to the metric on $\mathcal{O}_{X\times X}(-\Delta)$ used here is $-g$. The key point is that, by construction, for these metrics, the residue map

$$\mathrm{res}_z : (\Omega^1_X \otimes \mathcal{O}_X(z))|z \xrightarrow{\sim} \mathbb{C} \tag{2.7.5}$$

is an *isometry* for any $z \in X$, where $\mathbb{C}$ is given the standard metric with $||1|| = 1$.

Now suppose that Γ is a finite group of automorphisms of X and let $M = [\Gamma\backslash X]$, as in section 2.3. Suppose that the weakly ν-biadmissible Green function g and associated function ϕ, as in Definition 2.7.1, satisfy

$$g(\gamma z_1, \gamma z_2) = g(z_1, z_2) \qquad \text{and} \qquad \phi(\gamma z_1, \gamma z_2) = \phi(z_1, z_2) \tag{2.7.6}$$

for all $\gamma \in \Gamma$. For a divisor $Z = \sum_{P \in M} n_P \cdot P$ in M, consider the function

$$(2.7.7) \qquad g_Z(\zeta) := \sum_{z \in \mathrm{pr}^{-1}(P)} n_P \cdot g(z, \zeta) = \sum_{P \in Z} n_P \cdot e_P^{-1} \sum_{\gamma \in \Gamma} g(z, \gamma\zeta),$$

where, in this last expression, $\mathrm{pr}(z) = P$, and where $e_P = |\Gamma_z|$. Then g_Z is a Γ-invariant Green function for the divisor $\tilde{Z} = \mathrm{pr}^{-1}(Z)$ in X, and hence, as explained in section 2.3, g_Z is a Green function for Z on M. Thus, g defines metrics on all of the bundles $\mathcal{O}_M(Z)$ on M. Similarly, the diagonal invariance property (2.7.6) implies that the metric on Ω^1_X determined by the isomorphism (2.7.4) is, in fact, Γ-invariant, and hence defines a metric on Ω^1_M.

We now return to our stack $\mathcal{M}$ with 'uniformization' $M = \mathcal{M}_{\mathbb{C}} = [\Gamma \backslash X]$. We suppose that X is equipped with a weakly ν-biadmissible Green function satisfying the diagonal invariance property. We denote by $\widehat{\omega}_{\mathcal{M}/S}$ the relative dualizing sheaf $\omega_{\mathcal{M}/S}$ with the metric determined by g. Since we are assuming that $\mathcal{M}$ is a relative complete intersection over S, the dualizing sheaf $\omega_{\mathcal{M}/S}$ is a line bundle on $\mathcal{M}$.

Suppose that $\mathcal{Z}$ is an irreducible horizontal divisor on $\mathcal{M}$, and let $\widehat{\mathcal{O}}_{\mathcal{M}}(\mathcal{Z})$ be the bundle $\mathcal{O}_{\mathcal{M}}(\mathcal{Z})$ with the metric determined by g. Writing $j : \mathcal{Z} \to \mathcal{M}$ for the closed immersion of $\mathcal{Z}$ into $\mathcal{M}$, we have a canonical isomorphism

$$(2.7.8) \qquad j^*(\widehat{\omega}_{\mathcal{M}/S} \otimes \widehat{\mathcal{O}}_{\mathcal{M}}(\mathcal{Z})) = \widehat{\omega}_{\mathcal{Z}/S},$$

where $\widehat{\omega}_{\mathcal{Z}/S}$ is the dualizing sheaf of $\mathcal{Z}$ over S, equipped with the metric which makes this an isometry. The first version of the adjunction formula is then

$$(2.7.9) \qquad h_{\widehat{\mathcal{O}}_{\mathcal{M}}(\mathcal{Z})}(\mathcal{Z}) = -h_{\widehat{\omega}_{\mathcal{M}/S}}(\mathcal{Z}) + \widehat{\deg}_{\mathcal{Z}}(\widehat{\omega}_{\mathcal{Z}/S}).$$

As explained in [7], Proposition 5.2, Chapter 4, the second term on the RHS here can be written as

$$(2.7.10) \qquad \widehat{\deg}_{\mathcal{Z}}(\widehat{\omega}_{\mathcal{Z}/S}) = \mathfrak{d}_{\mathcal{Z}} + \frac{1}{2}|\Gamma|^{-1} \sum_{P, P' \in \mathcal{Z}(\mathbb{C})} \sum_{\substack{z \in \mathrm{pr}^{-1}(P) \\ z' \in \mathrm{pr}^{-1}(P') \\ z \neq z'}} g(z, z'),$$

where $\mathfrak{d}_{\mathcal{Z}}$ is the degree of $\omega_{\mathcal{Z}/S}$ with trivial metric, see (2.1.19) above, and where the second term comes from the use of the metric determined by g, via (2.7.8). Thus,

(2.7.11)

$$h_{\widehat{\mathcal{O}}_{\mathcal{M}}(\mathcal{Z})}(\mathcal{Z}) = -h_{\widehat{\omega}_{\mathcal{M}/S}}(\mathcal{Z}) + \eth_{\mathcal{Z}} + \frac{1}{2}|\Gamma|^{-1} \sum_{P,P' \in \mathcal{Z}(\mathbb{C})} \sum_{\substack{z \in \mathrm{pr}^{-1}(P) \\ z' \in \mathrm{pr}^{-1}(P') \\ z \neq z'}} g(z,z').$$

Recalling that g is invariant under the diagonal action of Γ, we can rewrite the last term in (2.7.11) as follows:

$$\frac{1}{2}|\Gamma|^{-1} \sum_{P,P' \in \mathcal{Z}(\mathbb{C})} \sum_{\substack{z \in \mathrm{pr}^{-1}(P) \\ z' \in \mathrm{pr}^{-1}(P') \\ z \neq z'}} g(z,z')$$

$$(2.7.12) \qquad = \frac{1}{2}|\Gamma|^{-1} \sum_{P,P' \in \mathcal{Z}(\mathbb{C})} e_P^{-1} e_{P'}^{-1} \sum_{\substack{\gamma,\gamma' \in \Gamma \\ \gamma z \neq \gamma' z'}} g(\gamma z, \gamma' z')$$

$$= \frac{1}{2} \sum_{P,P' \in \mathcal{Z}(\mathbb{C})} e_P^{-1} e_{P'}^{-1} \sum_{\substack{\gamma \in \Gamma \\ z \neq \gamma z'}} g(z, \gamma z'),$$

where z and $z' \in X$ are points with $\mathrm{pr}(z) = P$ and $\mathrm{pr}(z') = P'$, and $e_P = |\Gamma_z|$, $e_{P'} = |\Gamma_{z'}|$.

If the cycle $\mathcal{Z}$ is equipped with the Green function $g_{\mathcal{Z}}$ and if $(\mathcal{Z}, g_{\mathcal{Z}}) \in \widehat{\mathrm{CH}}^1(\mathcal{M})$ is the corresponding class, then we also have the arithmetic intersection pairing (2.6.2),

$$(2.7.13) \qquad \langle (\mathcal{Z}, g_{\mathcal{Z}}), (\mathcal{Z}, g_{\mathcal{Z}}) \rangle_{\mathcal{M}} = h_{\widehat{\mathcal{O}}_{\mathcal{M}}(\mathcal{Z})}(\mathcal{Z}) + \frac{1}{2}\int_M \phi_{\mathcal{Z}} \cdot g_{\mathcal{Z}} \cdot \nu,$$

where

$$\frac{1}{2}\int_M \phi_{\mathcal{Z}} \cdot g_{\mathcal{Z}} \cdot \nu = \sum_{P,P' \in \mathcal{Z}(\mathbb{C})} e_P^{-1} e_{P'}^{-1} \sum_{\gamma \in \Gamma} \frac{1}{2}\int_X \phi(z,\zeta) \cdot g(\gamma z', \zeta)\, d\nu(\zeta).$$

Combining these expressions, we obtain the version of the formula which will be needed later.

Theorem 2.7.2. [Arithmetic Adjunction Formula] *Let g be a weakly ν-biadmissible Green function on $X \times X - \Delta_X$ for a presentation $\mathcal{M}_{\mathbb{C}} = [\Gamma \backslash X]$ of $\mathcal{M}_{\mathbb{C}}$, satisfying the diagonal invariance condition (2.7.6). Suppose that $\mathcal{Z}$ is an irreducible horizontal divisor on $\mathcal{M}$ with Green function $g_{\mathcal{Z}}$ determined by g. Let $\widehat{\mathcal{O}}_{\mathcal{M}}(\mathcal{Z})$ be the corresponding metrized line bundle*

and let $\hat{\omega}_{\mathcal{M}/S}$ be the relative dualizing sheaf of $\mathcal{M}$ with the metric determined by g. Then
(i)

$$h_{\widehat{\mathcal{O}}_{\mathcal{M}}(\mathcal{Z})}(\mathcal{Z}) = -h_{\hat{\omega}_{\mathcal{M}/S}}(\mathcal{Z}) + \mathfrak{d}_{\mathcal{Z}} + \frac{1}{2} \sum_{P,P' \in \mathcal{Z}(\mathbb{C})} e_P^{-1} e_{P'}^{-1} \sum_{\substack{\gamma \in \Gamma \\ z \neq \gamma z'}} g(z, \gamma z'),$$

where $\mathfrak{d}_{\mathcal{Z}}$ is the discriminant degree of $\mathcal{Z}$; see (2.1.19).
(ii)

$$\begin{aligned} \langle (\mathcal{Z}, g_{\mathcal{Z}}), (\mathcal{Z}, g_{\mathcal{Z}}) \rangle_{\mathcal{M}} &= -h_{\hat{\omega}_{\mathcal{M}/S}}(\mathcal{Z}) + \mathfrak{d}_{\mathcal{Z}} \\ &\quad + \sum_{P,P' \in \mathcal{Z}(\mathbb{C})} e_P^{-1} e_{P'}^{-1} \sum_{\substack{\gamma \in \Gamma \\ z \neq \gamma z'}} \frac{1}{2} \Big(g(z, \gamma z') \\ &\qquad\qquad + \int_M \phi(z, \zeta) \cdot g(\gamma z', \zeta)\, d\nu(\zeta) \Big) \\ &\quad + \sum_{P \in \mathcal{Z}(\mathbb{C})} e_P^{-1} \frac{1}{2} \int_M \phi(z, \zeta) \cdot g(z, \zeta)\, d\nu(\zeta). \end{aligned}$$

Here in the last sums, z and $z' \in X$ are points with $\mathrm{pr}(z) = P$ *and* $\mathrm{pr}(z') = P'$*, and the metric on $\omega_{\mathcal{M}/S}$ depends on the archimedean presentation* $\mathrm{pr} : X \longrightarrow [\Gamma\backslash X] = M$.

When we use this formula in Chapter 7, we will, in fact, pass to a presentation $M = [\Gamma\backslash D]$, where $D = \mathfrak{H}^+ \cup \mathfrak{H}^-$ and Γ is an infinite discrete group. The expressions in the previous theorem will still be valid for such a presentation provided the, now infinite, sums over Γ are absolutely convergent.

Bibliography

[1] J.-B. Bost, *Potential theory and Lefschetz theorems for arithmetic surfaces*, Ann. Sci. École Norm. Sup., **32** (1999), 241–312.

[2] J.-B. Bost, H. Gillet, and C. Soulé, *Heights of projective varieties and positive Green forms*, J. Amer. Math. Soc., **7** (1994), 903–1027.

[3] P. Deligne and M. Rapoport, *Les schémas de modules de courbes elliptiques*, Modular Functions of One Variable II (Proc. Intl. Summer

School, Univ. Antwerp, Antwerp, 1972), Lecture Notes in Math. **349**, 143–316, Springer-Verlag, Berlin, 1973.

[4] G. Faltings, *Calculus on arithmetic surfaces*, Annals of Math., **119** (1984), 387–424.

[5] H. Gillet and C. Soulé, *Arithmetic intersection theory*, Publ. Math. IHES, **72** (1990), 93–174.

[6] S. Kudla, *Central derivatives of Eisenstein series and height pairings*, Annals of Math., **146** (1997), 545–646.

[7] S. Lang, Introduction to Arakelov Theory, Springer-Verlag, New York, 1988.

[8] L. Moret-Bailly, *Métriques permises*, Séminaire sur les Pinceaux Arithmétiques: La Conjecture de Mordell, Astérisque **127**, 29–87, Société Mathématique de France, Paris, 1985.

[9] J. Neukirch, Algebraische Zahlentheorie, Springer-Verlag, Berlin, 1992.

[10] C. Soulé, D. Abramovich, J.-F. Burnol and J. Kramer, Lectures on Arakelov Theory, Cambridge Studies in Advanced Math. **33**, Cambridge Univ. Press, Cambridge, 1992.

Chapter Three

Cycles on Shimura curves

In this chapter we review the objects which will be our main concern. We introduce the moduli problem attached to a quaternion algebra B over $\mathbb{Q}$ solved by the DM-stack $\mathcal{M}$, which has relative dimension 1 over $\operatorname{Spec}\mathbb{Z}$, has semistable reduction everywhere, and is smooth outside the ramification locus of B. This stack has a complex uniformization by $\mathbb{C}\setminus\mathbb{R} = \mathfrak{H}^+ \cup \mathfrak{H}^-$ and, for every prime p in the ramification locus of B, a p-adic uniformization by the Drinfeld upper half space.

We then recall from [12] the construction of classes $\hat{\mathcal{Z}}(t,v)$ in $\widehat{\mathrm{CH}}^1(\mathcal{M})$ for every $t \in \mathbb{Z}$ and every $v \in \mathbb{R}_{>0}$. This is done in three steps. In a first step we define for every $t > 0$ a divisor $\mathcal{Z}(t)$ on $\mathcal{M}$ by imposing a 'special endomorphism' of degree t, and we set $\mathcal{Z}(t) = 0$ for $t < 0$. In a second step we define for every $t \neq 0$ a Green function for the divisor $\mathcal{Z}(t)$ which depends on the positive real parameter v (for $t < 0$, this is just a smooth function on $\mathcal{M}_{\mathbb{C}}$). Finally, to define $\hat{\mathcal{Z}}(0,v)$, we consider the Hodge line bundle ω, equipped with a suitable metric depending on v. We then define $\hat{\mathcal{Z}}(0,v)$ to be the image of the corresponding class of $\widehat{\mathrm{Pic}}(\mathcal{M})$ under the natural map from $\widehat{\mathrm{Pic}}(\mathcal{M})$ to $\widehat{\mathrm{CH}}^1(\mathcal{M})$; see Section 2.4. The classes $\hat{\mathcal{Z}}(t,v)$ will be the coefficients of our generating series $\hat{\phi}_1$ with coefficients in $\widehat{\mathrm{CH}}^1(\mathcal{M})$, which will be considered in the next chapter.

Finally, we define 0-cycles $\mathcal{Z}(T)$ on $\mathcal{M}$ for positive definite 'good' semi-integral matrices T. The classes of these 0-cycles in $\widehat{\mathrm{CH}}^2(\mathcal{M})$ will be part of our generating series $\hat{\phi}_2$ (the remaining coefficients will be defined in Chapter 6). The 0-cycle $\mathcal{Z}(T)$ (if nonempty) is concentrated in the supersingular locus of a fiber $\mathcal{M}_p$ for a unique prime number p depending on T. The lengths of all local rings of $\mathcal{Z}(T)$ are identical and are given by the Gross-Keating formula [6]. More precisely, this is the special case of this formula, in which the first entry of the Gross-Keating invariant (a_1, a_2, a_3) is equal to zero. The proof of this special case is easier than the general case (as is already pointed out in the original paper). Because of the importance of this formula for us, we give an exposition of the proof.

3.1 SHIMURA CURVES

Let B be an indefinite quaternion algebra over $\mathbb{Q}$. Let $D(B)$ be the product of primes p for which $B_p = B \otimes \mathbb{Q}_p$ is a division algebra. For the moment we allow the case $B = M_2(\mathbb{Q})$, where $D(B) = 1$, but later on this case will be excluded. Let O_B be a maximal order in B.

We consider the following moduli problem over $\operatorname{Spec} \mathbb{Z}$. The moduli problem $\mathcal{M}$ associates to a scheme S the category of pairs (A, ι) where A is an abelian scheme over S and where ι is a ring homomorphism

$$\iota : O_B \longrightarrow \operatorname{End} A. \tag{3.1.1}$$

The morphisms in this category are the isomorphisms of such objects. We impose the condition on the characteristic polynomial

$$\operatorname{char}(\iota(b) \mid \operatorname{Lie} A)(T) = (T - b) \cdot (T - b^\iota), \quad b \in O_B. \tag{3.1.2}$$

In other words, (A, ι) is *special* in the sense of Drinfeld. Here $b \mapsto b^\iota$ is the main involution of B. If $S = \mathbb{C}$, such an (A, ι) is simply an abelian surface with an action of O_B, but if S is a scheme of characteristic p, then the condition (3.1.2) is stronger and excludes some pathologies.

Proposition 3.1.1. *(i) The moduli problem $\mathcal{M}$ is representable by a DM-stack which is flat of relative dimension one over* $\operatorname{Spec} \mathbb{Z}$. *The stack $\mathcal{M}$ is regular and in fact is smooth over* $\operatorname{Spec} \mathbb{Z}[D(B)^{-1}]$ *and semistable over all of* $\operatorname{Spec} \mathbb{Z}$. *If B is a division algebra (i.e., $D(B) > 1$), then $\mathcal{M}$ is proper over* $\operatorname{Spec} \mathbb{Z}$. *Furthermore, $\mathcal{M}$ is geometrically connected and so are all its fibers over* $\operatorname{Spec} \mathbb{Z}$.
(ii) Let $D(B) > 1$. The coarse moduli scheme of $\mathcal{M}$ is proper and flat of relative dimension one over $\operatorname{Spec} \mathbb{Z}$ *and is in fact a projective scheme over* $\operatorname{Spec} \mathbb{Z}$, *with all fibers geometrically connected curves.*

We will not give a proof of this proposition but will give instead some references to the literature. For the representability of $\mathcal{M}$, the main point is to establish the existence and uniqueness of a polarization of (A, ι) of a certain type. This is explained in [2], Chapter 3, Section 3. Once this is done, the construction of $\mathcal{M}$ can be performed along familiar lines. One approach is to use the Artin representability theorem for stacks [13], Chapter 10. Another approach is to impose a level structure and use the relative representability over the moduli space of principally polarized abelian varieties of dimension 2 with level structure (cf. [2], Chapter 3). By varying the level structure, one obtains $\mathcal{M}$.

That $\mathcal{M}$ is smooth over $\operatorname{Spec} \mathbb{Z}[D(B)^{-1}]$ can be shown using deformation theory and the Serre-Tate theorem. In this way one sees that the formal deformation space of a point $(A, \iota) \in \mathcal{M}(\bar{\mathbb{F}}_p)$ for $p \not| D(B)$ is isomorphic to the formal deformation space of the corresponding p-divisible group $(A(p), \iota)$. Now using the idempotents in $O_B \otimes \mathbb{Z}_p \simeq \mathrm{M}_2(\mathbb{Z}_p)$ to write $A(p)$ as the square of a p-divisible group G of dimension 1 and height 2, one sees that this formal deformation space is isomorphic to the formal deformation space of G and hence is formally smooth of relative dimension 1. The semistability of $\mathcal{M}$ at primes p dividing $D(B)$ relates to the p-adic uniformization; cf. Corollary 3.2.4 below. If B is a division algebra, the properness of $\mathcal{M}$ is checked through the valuative criterion for properness, using the semistable reduction theorem. That the fiber of $\mathcal{M}$ over $\mathbb{C}$ is connected follows from the complex uniformization; cf. Corollary 3.2.2 below. Then all other fibers are geometrically connected by Zariski's connectedness theorem. The assertions (ii) about the coarse moduli scheme follow from the corresponding properties of $\mathcal{M}$, except the projectivity of the coarse moduli space, for which one has to use the second approach to the representability of $\mathcal{M}$ and the quasi-projectivity of the moduli space of principally polarized abelian varieties.

From now on we will assume that B is a division algebra and will call $\mathcal{M}$ a *Shimura curve*, even though $\mathcal{M}$ is a stack and not a scheme.

3.2 UNIFORMIZATION

We first recall from [12] the *complex uniformization* of $\mathcal{M}$. Let

$$V = \{x \in B \mid \operatorname{tr}(x) = 0\}, \tag{3.2.1}$$

with quadratic form $Q(x) = \mathrm{Nm}(x) = -x^2$. Let $H = B^\times$, viewed as an algebraic group over $\mathbb{Q}$. Then H acts on V by conjugation and this induces an isomorphism

$$H \xrightarrow{\sim} \mathrm{GSpin}(V, Q), \tag{3.2.2}$$

where $\mathrm{GSpin}(V, Q)$ is the spinor similitude group of V. Since B is indefinite, i.e., $B \otimes_{\mathbb{Q}} \mathbb{R} \simeq \mathrm{M}_2(\mathbb{R})$, the quadratic space $V_{\mathbb{R}}$ has signature $(1, 2)$. Let

$$D = \{w \in V(\mathbb{C}) \mid (w, w) = 0,\ (w, \bar{w}) < 0\}/\mathbb{C}^\times. \tag{3.2.3}$$

Here $(x, y) = \operatorname{tr}(x^\iota y)$ is the bilinear form associated to Q. Then D is an open subset of $\mathbb{P}^1(\mathbb{C})$, stable under the action of $H(\mathbb{R})$. If we fix an isomorphism $B_{\mathbb{R}} \simeq \mathrm{M}_2(\mathbb{R})$, we have an identification

$$\mathbb{C} \backslash \mathbb{R} = \mathfrak{H}^+ \cup \mathfrak{H}^- \xrightarrow{\sim} D, \quad z \longmapsto w(z) = \begin{pmatrix} z & -z^2 \\ 1 & -z \end{pmatrix} \mod \mathbb{C}^\times. \tag{3.2.4}$$

This identification is equivariant for the action of $H(\mathbb{R}) \simeq \mathrm{GL}_2(\mathbb{R})$ by fractional linear transformations on the lefthand side, and by conjugation on the righthand side.

Let $\Gamma = O_B^\times$ and $K = \hat{O}_B^\times$ in $H(\mathbb{A}_f)$. By strong approximation we have

$$H(\mathbb{Q}) \setminus H(\mathbb{A})/K = \Gamma \setminus H(\mathbb{R}). \tag{3.2.5}$$

The complex uniformization of $\mathcal{M}$ is now given by the following statement.

Proposition 3.2.1. *We have an isomorphism of stacks*

$$[\Gamma \setminus D] \xrightarrow{\sim} \mathcal{M}_{\mathbb{C}}.$$

Proof. For $z \in \mathbb{C} \setminus \mathbb{R}$ we obtain a lattice L_z in $\mathbb{C}^2$ as the image of O_B under the isomorphism

$$\lambda_z : B_{\mathbb{R}} = M_2(\mathbb{R}) \longrightarrow \mathbb{C}^2, \;\; b \longmapsto b \cdot \binom{z}{1}. \tag{3.2.6}$$

Then $A_z = \mathbb{C}^2/L_z$ is an abelian variety with O_B-action ι_z given by left multiplication. We obtain a map

$$D \longrightarrow \mathcal{M}_{\mathbb{C}}, \;\; z \longmapsto (A_z, \iota_z). \tag{3.2.7}$$

Two points in D give the same lattice if and only if they are in the same orbit under $O_B^\times = \Gamma$. Furthermore, the automorphisms of (A_z, ι_z) are given by elements in Γ_z. Since every (A, ι) over $\mathbb{C}$ is isomorphic to some (A_z, ι_z), the resulting morphism of stacks $[\Gamma \setminus D] \to \mathcal{M}_{\mathbb{C}}$ is an isomorphism. □

Corollary 3.2.2. $\mathcal{M}_{\mathbb{C}}$ *is connected.*

Proof. Γ contains elements b with $\mathrm{Nm}(b) < 0$. □

We next turn to *p-adic uniformization* for $p \mid D(B)$. We fix such a prime and denote by $\hat{\Omega}$ the Drinfeld upper half plane. This is an adic formal scheme over $\mathbb{Z}_p$ with semistable reduction [2], [16], equipped with an action of $\mathrm{PGL}_2(\mathbb{Q}_p)$. The name derives from the remarkable property that for any finite extension K of $\mathbb{Q}_p$ with ring of integers O_K we have $\hat{\Omega}(O_K) = \mathbb{P}^1(K) \setminus \mathbb{P}^1(\mathbb{Q}_p)$. We let $\mathrm{GL}_2(\mathbb{Q}_p)$ act on $\hat{\Omega} \times \mathbb{Z}$ by

$$g : (z, i) \longmapsto (gz, i + \mathrm{ord}_p(\det g)). \tag{3.2.8}$$

Let $B' = B^{(p)}$ be the definite quaternion algebra over $\mathbb{Q}$ whose invariants agree with those of B at all primes $\ell \neq p$, and denote by H' the algebraic group over $\mathbb{Q}$ with $H'(\mathbb{Q}) = B'^\times$. We fix isomorphisms

$$H'(\mathbb{Q}_p) \simeq \mathrm{GL}_2(\mathbb{Q}_p) \;\text{ and }\; H'(\mathbb{A}_f^p) \simeq H(\mathbb{A}_f^p). \tag{3.2.9}$$

We write $K = K^p.K_p$ with $K^p \subset H(\mathbb{A}_f^p)$ and $K_p \subset H(\mathbb{Q}_p)$. Let $K'^p \subset H'(\mathbb{A}_f^p)$ be the image of K^p under this last isomorphism and let $\Gamma' = H'(\mathbb{Q}) \cap (H'(\mathbb{Q}_p).K'^p)$. Then $\Gamma' = O_B\left[p^{-1}\right]^\times$. By strong approximation we have

$$H'(\mathbb{Q}) \setminus H'(\mathbb{A})/K'^p = \Gamma' \setminus H'(\mathbb{Q}_p). \tag{3.2.10}$$

Let $\hat{\mathcal{M}}_W$ be the base change to $W = W(\bar{\mathbb{F}}_p)$ of the formal completion of $\mathcal{M}$ along its fiber at p. We denote by $\hat{\Omega}_W$ the base change of $\hat{\Omega}$ to W. The Cherednik-Drinfeld uniformization theorem may now be formulated as follows.

Theorem 3.2.3. *There is an isomorphism of formal stacks over* $W(\bar{\mathbb{F}}_p)$,

$$\hat{\mathcal{M}}_W \simeq H'(\mathbb{Q})\backslash\left[(\hat{\Omega}_W \times \mathbb{Z}) \times H'(\mathbb{A}_f^p)/K'^p\right] = \Gamma'\backslash(\hat{\Omega}_W \times \mathbb{Z}) = \Gamma'^1\backslash\hat{\Omega}_W,$$

where $\Gamma'^1 = \{g \in \Gamma' \mid \operatorname{ord}_p \det\ g = 0\}$.

For the proof we refer to [2] and [16]. In these references one also finds a comparison of the descent data from W to $\mathbb{Z}_p$ of both sides. An informal discussion of this theorem appears in [10]. The principal reason for this theorem to hold is that all elements $(A, \iota) \in \mathcal{M}(\bar{\mathbb{F}}_p)$ are isogenous to one another.

Corollary 3.2.4. $\mathcal{M} \times_{\mathrm{Spec}\ \mathbb{Z}} \mathrm{Spec}\ \mathbb{Z}_p$ *is semistable over* $\mathrm{Spec}\ \mathbb{Z}_p$.

3.3 THE HODGE BUNDLE

Let $(\mathcal{A}, \iota)$ be the universal abelian scheme over $\mathcal{M}$. The *Hodge line bundle* (or *Hodge bundle* for short) is the following line bundle on $\mathcal{M}$,

$$\omega = \epsilon^*(\Omega^2_{\mathcal{A}/\mathcal{M}}) = \wedge^2\mathrm{Lie}(\mathcal{A}/\mathcal{M})^*. \tag{3.3.1}$$

Here $\epsilon : \mathcal{M} \to \mathcal{A}$ is the zero section. The Hodge bundle is related to the relative dualizing sheaf $\omega_{\mathcal{M}/\mathbb{Z}}$ by the following proposition; see [12], Proposition 3.2.

Proposition 3.3.1. *The Hodge bundle* ω *is isomorphic to the relative dualizing sheaf* $\omega_{\mathcal{M}/\mathbb{Z}}$.

The complex fiber $\omega_{\mathbb{C}}$ is a line bundle on $\mathcal{M}_{\mathbb{C}} = [\Gamma \setminus D]$. It may be given by a line bundle on D, equipped with a descent datum with respect to the action of Γ on D as follows ([12], Section 3): The line bundle on D is associated to the trivial vector bundle $D \times \mathbb{C}$, and the action of Γ on D is lifted to $D \times \mathbb{C}$ by the automorphy factor $(cz+d)^2$, i.e.,

$$\gamma = \begin{pmatrix} a & b \\ c & d \end{pmatrix} : \ (z, \zeta) \longmapsto (\gamma \cdot z, (cz+d)^2 \cdot \zeta). \tag{3.3.2}$$

More precisely, the pullback of $\omega_{\mathbb{C}}$ to D is trivialized by associating to $z \in \mathbb{C} \setminus \mathbb{R}$ the section $\alpha_z = D(B)^{-1} \cdot (2\pi i)^2 \cdot d\omega_1 \wedge d\omega_2$ of $\Omega^2_{A_z/\mathbb{C}}$. Here $A_z = \mathbb{C}^2/L_z$ with $L_z = \lambda_z(O_B)$ and w_1, w_2 are the coordinates on $\mathbb{C}^2 = B_{\mathbb{R}}$ via λ_z; see (3.2.6). It follows that on $\mathcal{M}_{\mathbb{C}}$ the Hodge bundle is isomorphic to the canonical bundle $\Omega^1_{\mathcal{M}_{\mathbb{C}}/\mathbb{C}}$ under the map which sends α_z to dz.

On the Hodge bundle $\omega_{\mathbb{C}}$ there is a natural metric $\| \ \|_{\mathrm{nat}}$. If $s : z \mapsto s_z$ is a section of $\omega_{\mathbb{C}}$, then

$$\|s\|^2_{\mathrm{nat}} = \left| \left(\frac{i}{2\pi}\right)^2 \cdot \int_{A_z} s_z \wedge \bar{s}_z \right|. \tag{3.3.3}$$

We prefer to use the normalized metric defined by

$$\| \ \|^2 = e^{-2C} \cdot \| \ \|^2_{\mathrm{nat}}. \tag{3.3.4}$$

Here

$$2C = \log(4\pi) + \gamma, \tag{3.3.5}$$

where γ is Euler's constant. The reason for this renormalization is explained in the introduction to [12]. Further justification is given in Chapter 7 in connection with the adjunction formula. The metric on $\Omega^1_{\mathcal{M}_{\mathbb{C}}}$ resulting under the above isomorphism with $\omega_{\mathbb{C}}$ is then

(3.3.6)

$$\begin{aligned} \|dz\|^2 = \|\alpha_z\|^2 &= e^{-2C} \cdot \left| \left(\frac{i}{2\pi}\right)^2 \cdot \int_{A_z} \alpha_z \wedge \bar{\alpha}_z \right| \\ &= e^{-2C} \cdot (2\pi)^{-2} \cdot (2\pi)^4 \cdot D(B)^{-2} \cdot \mathrm{vol}(\mathrm{M}_2(\mathbb{R})/O_B) \cdot \mathrm{Im}(z)^2 \\ &= e^{-2C} \cdot (2\pi)^2 \cdot \mathrm{Im}(z)^2. \end{aligned}$$

Here we have used the fact that the pullback under λ_z of the form $dw_1 \wedge dw_2 \wedge d\bar{w}_1 \wedge d\bar{w}_2$ on $\mathbb{C}^2$ is $4\,\mathrm{Im}(z)^2$ times the standard volume form on $\mathrm{M}_2(\mathbb{R})$, and that $\mathrm{vol}(\mathrm{M}_2(\mathbb{R})/O_B) = D(B)^2$.

3.4 SPECIAL ENDOMORPHISMS

Let $(A, \iota) \in \mathcal{M}(S)$. The *space of special endomorphisms of* (A, ι) is the $\mathbb{Z}$-module

$$V(A, \iota) = \{x \in \mathrm{End}_S(A, \iota) \mid \mathrm{tr}(x) = 0\}. \tag{3.4.1}$$

When S is connected, $V(A, \iota)$ is a free $\mathbb{Z}$-module of finite rank equipped with the $\mathbb{Z}$-valued quadratic form Q given by

$$-x^2 = Q(x) \cdot \mathrm{id}_A. \tag{3.4.2}$$

Proposition 3.4.1. *The quadratic form Q is positive-definite, i.e., $Q(x) \geq 0$ and $Q(x) = 0$ only for $x = 0$.*

Proof. We may assume that S is the spectrum of an algebraically closed field. But then it follows from the classification of $\mathrm{End}(A, \iota) \otimes_{\mathbb{Z}} \mathbb{Q}$ that any nonscalar $x \in \mathrm{End}(A, \iota) \otimes \mathbb{Q}$ generates an imaginary quadratic field extension $\boldsymbol{k}$. For $x \in V(A, \iota) \otimes \mathbb{Q}$ we obtain $-x^2 = \mathrm{Nm}_{\boldsymbol{k}/\mathbb{Q}}(\mathrm{x}) \cdot \mathrm{id}_{\mathrm{A}}$, which proves the claim. □

Definition 3.4.2. Let $t \neq 0$ be an integer. Let $\mathcal{Z}(t)$ be the moduli stack of triples (A, ι, x), where (A, ι) is an object of $\mathcal{M}$ and where $x \in V(A, \iota)$ with $Q(x) = t$.

Then $\mathcal{Z}(t)$ is a DM-stack which is relatively representable over $\mathcal{M}$ by an unramified finite morphism (rigidity theorem),

$$\mathcal{Z}(t) \longrightarrow \mathcal{M}. \tag{3.4.3}$$

If $t < 0$, then $\mathcal{Z}(t) = \emptyset$.

The degree of the generic fiber $\mathcal{Z}(t)_{\mathbb{Q}}$ is given by the following formula [12]: Let $4t = n^2 d$, where $-d$ is the fundamental discriminant of $\boldsymbol{k}_t$. Then

$$\deg\ \mathcal{Z}(t)_{\mathbb{Q}} = 2 \cdot \delta(d, D(B)) \cdot H_0(t, D(B)), \tag{3.4.4}$$

where

$$\delta(d, D) = \prod_{p \mid D} (1 - \chi_d(p)) \tag{3.4.5}$$

(zero or a power of 2) and

$$H_0(t, D) = \sum_{c \mid n} \frac{h(c^2 d)}{w(c^2 d)} = \frac{h(d)}{w(d)} \Bigg(\sum_{\substack{c \mid n \\ (c, D) = 1}} c \cdot \prod_{\ell \mid c} (1 - \chi_d(\ell) \ell^{-1}) \Bigg). \tag{3.4.6}$$

Here $h(c^2d)$ is the class number of the order O_{c^2d} of conductor c in $\pmb{k}_t$, $w(c^2d)$ is the number of units in O_{c^2d}, and χ_d is the Dirichlet character for $\pmb{k}_t$. Note that in defining the degree of $\mathcal{Z}(t)_{\mathbb{Q}}$ over $\mathbb{Q}$, each point $\eta = (A, \iota, x)$ in $\mathcal{Z}(t)(\mathbb{C})$ counts with multiplicity $1/|\mathrm{Aut}(A, \iota, x)|$.

The proof of (3.4.6) in [12] uses the *complex uniformization* of the stack $\mathcal{Z}(t)$. If $(A, \iota, x) \in \mathcal{Z}(t)(\mathbb{C})$, then A is an abelian surface with an action of $O_B \otimes_{\mathbb{Z}} \mathbb{Z}[\sqrt{-t}]$. If $A = A_z = \mathbb{C}^2/L_z$ as in (3.2.6), then the action of x on Lie $A = B_{\mathbb{R}} \simeq \mathbb{C}^2$ commutes with the action of O_B and hence is given by right multiplication by an element $j_x \in O_D \cap V$ with $\mathrm{Nm}(j_x) = t$. The map given by right multiplication with j_x,

$$\tilde{x} = r(j_x) : \mathbb{C}^2 \longrightarrow \mathbb{C}^2 \tag{3.4.7}$$

is holomorphic and preserves D. The point z is fixed by $\tilde{x}$. Denote by D_x the fixed locus of $r(j_x)$. Let

$$L(t) = \{x \in O_B \cap V \mid Q(x) = t\}, \tag{3.4.8}$$

and put

$$D_t = \coprod_{x \in L(t)} D_x. \tag{3.4.9}$$

We thus obtain a morphism

$$D_t \longrightarrow \mathcal{Z}(t)_{\mathbb{C}}. \tag{3.4.10}$$

One checks, cf. [12], that this induces an isomorphism of stacks over $\mathbb{C}$,

$$[\Gamma \setminus D_t] \simeq \mathcal{Z}(t)_{\mathbb{C}}, \tag{3.4.11}$$

compatible with the isomorphism $[\Gamma \setminus D] \simeq \mathcal{M}_{\mathbb{C}}$ of Proposition 3.2.1.

By (9.5) and (9.6) of [12], there is a 2–1 map of orbifolds,

$$[\,\Gamma \backslash D_{\mathcal{Z}(t)}\,] \longrightarrow [\,\Gamma \backslash L(t)\,]. \tag{3.4.12}$$

This arises as follows. For any $w \in D$ the negative 2-plane $U(w) = (\mathbb{C}w + \mathbb{C}\bar{w}) \cap V$ has a natural orientation determined by $iw \wedge \bar{w}$. Fix an orientation of V. For $x \in L(t)$, with $t > 0$, let $D_x^0 = \{z_0\}$, where $z_0 \in D_x$ is determined by the condition that $[x, U(w(z_0))]$ is properly oriented. Note that $D_{-x}^0 = \{\,\bar{z}_0\,\}$, and $\mathrm{pr}(D_x) = \mathrm{pr}(D_x^0) \cup \mathrm{pr}(D_{-x}^0)$.

Lemma 3.4.3. *Let $x \in L(t)$ with $D_x^0 = \{z_0\}$. Then*
(i) $-x \notin \Gamma \cdot x$.
(ii) $\bar{z}_0 \notin \Gamma \cdot z_0$.

Proof. To prove (i), suppose that $\gamma \in \Gamma$ is such that $\gamma \cdot x = \gamma x \gamma^{-1} = -x$. It follows that γ and x generate B and hence that γ^2 is central. Thus $\gamma^2 = \pm 1$. The case $\gamma^2 = 1$ is excluded, since B is a division algebra. If $\gamma^2 = -1$, then $B \simeq (-1, -t)$ as a cyclic algebra over $\mathbb{Q}$. But this cannot happen, since $(-1, -t)_\infty = -1$ whereas B is indefinite. For (ii), if $\gamma \cdot z_0 = \bar{z}_0$, then γ reverses the orientation on $U(w(z_0))$ and hence acts by -1 on x, which is excluded by (i). □

By (ii) of the lemma, $\mathrm{pr}(D_x)$ consists of two distinct points $\mathrm{pr}(D_x^0)$ and $\mathrm{pr}(D_{-x}^0)$, while, by (i), the vectors x and $-x$ both contribute the same pair of points to the sum

$$\mathcal{Z}(t)(\mathbb{C}) = \sum_{\substack{x \in L(t) \\ \mathrm{mod}\ \Gamma}} \mathrm{pr}(D_x) = 2 \sum_{\substack{x \in L(t) \\ \mathrm{mod}\ \Gamma}} \mathrm{pr}(D_x^0) \tag{3.4.13}$$

Thus,

$$\deg \mathcal{Z}(t)_\mathbb{Q} = 2 \sum_{\substack{\mathbf{x} \in L(t) \\ \mathrm{mod}\ \Gamma}} e_\mathbf{x}^{-1} \tag{3.4.14}$$

so that the computation of $\deg \mathcal{Z}(t)_\mathbb{Q}$ is reduced to a counting problem; see [12], Proposition 9.1.

Remark 3.4.4. The factor of 2 on the righthand side of (3.4.14) is also due to the fact that the morphism $\mathcal{Z}(t) \to \mathcal{M}$ is *not* a closed immersion. In fact, $\mathcal{Z}(t)_\mathbb{Q} \to \mathcal{M}_\mathbb{Q}$ is of degree 2 over its image ((A, ι, x) and $(A, \iota, -x)$ are the two nonisomorphic preimages of (A, ι)); see [12], Remark 9.3). In the sequel we denote by $\mathcal{Z}(t)$ both the DM-stack defined above and its image divisor in $\mathcal{M}$. At those places where this notational ambiguity can cause a problem, we have pointed out explicitly which meaning is intended.

Proposition 3.4.5. *The* 0*-cycle* $\mathcal{Z}(t)_\mathbb{C}$ *is nonempty if and only if the imaginary quadratic field* $\mathbf{k}_t = \mathbb{Q}(\sqrt{-t})$ *embeds in* B. *In this case the stack* $\mathcal{Z}(t)$ *is flat over* $\mathrm{Spec}\, \mathbb{Z}[D(B)^{-1}]$.

Proof. The first assertion is clear by the above description of $\mathcal{Z}(t)_\mathbb{C}$. Let $p \nmid D(B)$ and let $(A, \iota) \in \mathcal{M}_p(\bar{\mathbb{F}}_p)$. The second assertion follows from the fact that the locus in the formal deformation space of (A, ι) where a nonzero special endomorphism of (A, ι) deforms is a relative divisor over $\mathrm{Spf}\, W(\bar{\mathbb{F}}_p)$. Using the idempotents in $O_B \otimes \mathbb{Z}_p \simeq \mathrm{M}_2(\mathbb{Z}_p)$ to write the p-divisible group of A as the square of a p-divisible group G of dimension 1 and height 2, we are reduced to showing that the locus in the formal deformation space of G, where a nonscalar endomorphism of G deforms, is a relative divisor. This is well known; see [17], Proposition 6.1 and also [19], [6]. □

Just as $\mathcal{M}$ has a complex uniformization as well as a p-adic uniformization, so do the special cycles $\mathcal{Z}(t)$; see [12], Section 11. We fix $p \mid D(B)$ and introduce, as in Section 3.2, the definite quaternion algebra $B' = B^{(p)}$, and isomorphisms as in (3.2.9). We fix $x \in O_{B'}$ with $\mathrm{tr}(x) = 0$ and with $x^2 = -t$. Put

$$I(x) = \{gK'^p \in H'(\mathbb{A}_f^p)/K'^p \mid g^{-1}xg \in \hat{O}_{B'}^p\}. \tag{3.4.15}$$

Also put $\tilde{x} = x$ if $\mathrm{ord}_p(t) = 0$, resp. $\tilde{x} = 1 + x$ if $\mathrm{ord}_p(t) > 0$. Then in all cases $\mathrm{ord}_p\,\mathrm{Nm}(x) = 0$. Let H'_x be the stabilizer of x in H' and denote by

$$\hat{\mathcal{C}}(x) = (\hat{\Omega}_W \times \mathbb{Z})^{\tilde{x}} \tag{3.4.16}$$

the fixed locus of $\tilde{x}$. Since $\mathrm{ord}_p\,\mathrm{Nm}(x) = 0$, this is equal to the product $\hat{\Omega}_W^{\tilde{x}} \times \mathbb{Z}$.

Then the p-adic uniformization is given as an isomorphism of stacks over W,

$$\mathcal{Z}(t)_W = \left[H'_x(\mathbb{Q}) \setminus I(x) \times \hat{\mathcal{C}}(x)\right] . \tag{3.4.17}$$

A closer analysis of the situation for $p \mid D(B)$ gives the following facts, proved in [11] for $p \neq 2$ and in [12] for $p = 2$.

Proposition 3.4.6. *Let $t > 0$ and let $p \mid D(B)$. Then $\mathcal{Z}(t)$ has vertical components of the fiber $\mathcal{M}_p = \mathcal{M} \otimes_{\mathbb{Z}} \mathbb{F}_p$ if $\mathrm{ord}_p(t) \geq 2$ and the field $\Bbbk_t = \mathbb{Q}(\sqrt{-t})$ embeds into $B' = B^{(p)}$.*

For $\Bbbk_t$ there are therefore the following two alternatives. If $\Bbbk_t$ embeds into B, then it also embeds into $B^{(p)}$ and the cycle $\mathcal{Z}(t)$ will, by Proposition 3.4.5, have a nonempty generic fiber and, by Proposition 3.4.6, have vertical components at p, as soon as $\mathrm{ord}_p(t) \geq 2$. In this case p is either inert or ramified in $\Bbbk_t$. If, on the other hand, p splits in $\Bbbk_t$, then $\Bbbk_t$ does not embed into B, and the generic fiber of $\mathcal{Z}(t)$ is empty. In this case, if $\Bbbk_t$ embeds into $B^{(p)}$, then all prime divisors of $D(B)$ other than p are non-split in $\Bbbk_t$ and hence the cycle $\mathcal{Z}(t)$ has support in the fiber $\mathcal{M}_p$.

We therefore see that, while $\mathcal{Z}(t)$ is flat over $\mathrm{Spec}\,\mathbb{Z}[D(B)^{-1}]$, the stack $\mathcal{Z}(t)$ is not flat over $\mathrm{Spec}\,\mathbb{Z}$ (at least when $\mathrm{ord}_p(t) \geq 2$ for some $p \mid D(B)$). This seems to be unavoidable if one insists on a modular definition of $\mathcal{Z}(t)$.

We will define away another unpleasant feature of our cycles. Namely, by [11], the fiber at $p \mid D(B)$ of $\mathcal{Z}(t)$ contains embedded components when $p \mid t$, and it has the 'wrong dimension' 0, when $\mathrm{ord}_p(t) = 0$ and p splits in

$\Bbbk_t$, and embeds into $B' = B^{(p)}$. We therefore redefine $\mathcal{Z}(t)$ by replacing it by its Cohen-Macauleyfication (denoted $\mathcal{Z}(t)^{\text{pure}}$ in [11]). By Proposition 3.4.5 this does not change $\mathcal{Z}(t)$ over $\operatorname{Spec}\mathbb{Z}[D(B)^{-1}]$, and now $\mathcal{Z}(t)$ is pure of dimension 1 (unless it is empty) and without embedded components. We consider $\mathcal{Z}(t)$ as a divisor although, strictly speaking, $\mathcal{Z}(t)$ is not a closed substack of $\mathcal{M}$; cf. Remark 3.4.4.

To form our generating series, we will also need to define $\mathcal{Z}(0)$. We take the image of the class of ω under the composition of the isomorphism $\operatorname{Pic}(\mathcal{M}) \simeq \widehat{\operatorname{CH}}^1_{\mathbb{Z}}(\mathcal{M})$ and the natural map $\widehat{\operatorname{CH}}^1_{\mathbb{Z}}(\mathcal{M}) \longrightarrow \widehat{\operatorname{CH}}^1(\mathcal{M})$.

Remark 3.4.7. In fact, it is possible to slightly refine the definition of the cycle $\mathcal{Z}(t)$ as follows. Recall that there is an element $\delta \in O_B$ such that $\delta^2 = -D(B)$. For a point (A, ι, x) of $\mathcal{Z}(t)$, the special endomorphism x defines an action on A of the order $\mathbb{Z}[\sqrt{-t}]$ of discriminant $-4t$ in $\Bbbk_t = \mathbb{Q}(\sqrt{-t})$. If we write $4t = n^2 d$, as above, and let n_0 be the prime to $D(B)$ part of n, then the action of $\mathbb{Z}[\sqrt{-t}]$ extends to an action of the order $\mathcal{O}_{n_0^2 d}$ of discriminant $-n_0^2 d$. Note that this order is maximal at each p dividing $D(B)$. Since x commutes with the action of O_B, $X := \ker(\iota(\delta)) \subset A$ is a finite group scheme of order $D(B)^2$ equipped with an action of

$$\mathcal{O}_{n_0^2 d}/(D(B)) \xrightarrow{\sim} \prod_{\substack{p \mid D(B) \\ p \text{ inert}}} \mathbb{F}_{p^2} \times \prod_{\substack{p \mid D(B) \\ p \text{ ramified}}} \mathbb{F}_p. \tag{3.4.18}$$

It also carries an action of

$$O_B/(\delta) \xrightarrow{\sim} \prod_{p \mid D(B)} \mathbb{F}_{p^2}. \tag{3.4.19}$$

In our situation, the two actions are related by a commutative diagram

$$\begin{array}{ccc} \prod_{p \text{ inert}} \mathbb{F}_{p^2} & \hookrightarrow & \operatorname{End}(X) \\ (*) \downarrow & & \| \\ \prod_{p \text{ inert}} \mathbb{F}_{p^2} & \hookrightarrow & \operatorname{End}(X). \end{array} \tag{3.4.20}$$

There are $2^\nu = \delta(d, D(B))$ possibilities, which we call types, for the isomorphism (*), where ν is the number of prime factors of $D(B)$ which are inert in $\Bbbk_t$. For each type η, we can define a component $\mathcal{Z}(t, \eta)$ of $\mathcal{Z}(t)$ by requiring that $\ker(\iota(\delta))$ be of type η. Note that this construction explains the occurrence of the factor $\delta(d, D(B))$ in the formula (3.5.8) for $\deg_{\mathbb{C}} \mathcal{Z}(t)$. This decomposition according to types is analogous to the decomposition of Heegner cycles in [4] and [7]. There, prime divisors of the level N must be split or ramified in $\Bbbk_t$, and the role of η is played by the parameter r, which determines the isomorphism class of the kernel of the cyclic N-isogeny. In both cases, the group of Atkin-Lehner involutions permutes the components transitively. We will make no further use of this refinement in the present work so that, in effect, we consider only cycles which are invariant under the group of Atkin-Lehner involutions.

3.5 GREEN FUNCTIONS

To obtain arithmetic cycle classes in $\widehat{\mathrm{CH}}^1_{\mathbb{Z}}(\mathcal{M})$ from the cycles $\mathcal{Z}(t)$ introduced in Section 3.4, we will equip them with Green functions following [9]. For $x \in V(\mathbb{R})$ with $Q(x) \neq 0$ and $z \in D$ set

$$R(x,z) = |(x, w(z))|^2 \, |(w(z), \overline{w(z)})|^{-1} = -(\mathrm{pr}_z(x), \mathrm{pr}_z(x)). \tag{3.5.1}$$

where $\mathrm{pr}_z(x)$ is the orthogonal projection of x to the negative 2-plane obtained by intersecting $(\mathbb{C}\cdot w(z)+\mathbb{C}\cdot\overline{w(z)})$ with $V(\mathbb{R})$. Here $w(z) \in V(\mathbb{C})$ is any vector with image z in $\mathbb{P}(V(\mathbb{C}))$. Then this function for fixed x vanishes precisely when $(x, w(z)) = 0$, i.e., $z \in D_x$. Let

$$\beta_1(r) = \int_1^\infty e^{-ru} u^{-1} du = -\mathrm{Ei}(-r) \tag{3.5.2}$$

be the exponential integral. Then

$$\beta_1(r) = \begin{cases} -\log(r) - \gamma + O(r) & \text{as } r \to 0, \\ O(e^{-r}) & \text{as } r \to \infty. \end{cases} \tag{3.5.3}$$

As in [9], we form the functions

$$\xi(x,z) = \beta_1(2\pi R(x,z)) \tag{3.5.4}$$

and

$$\varphi(x,z) = \left[\, 2(R(x,z) + 2Q(x)) - \frac{1}{2\pi}\,\right] \cdot e^{-2\pi R(x,z)}. \tag{3.5.5}$$

Then $\xi(x,z)$ has a logarithmic singularity along D_x and decays exponentially as z goes to the boundary of D. The following result is Proposition 11.1 of [9] (where the functions ξ and φ were denoted by ξ^0 and φ^0).

Proposition 3.5.1. *Let $\mu = \frac{i}{2} y^{-2}\, dz \wedge d\bar{z}$ be the hyperbolic volume form on $D = \mathbb{C} \setminus \mathbb{R}$. Then*

$$dd^c \xi(x) + \delta_{D_x} = [\varphi(x)\mu]$$

as currents on D. In particular, $\xi(x, \cdot)$ is a Green function for D_x.

Because of the rapid decay of $\xi(x,\cdot)$, we can average over lattice points.

Corollary 3.5.2. *For $v \in \mathbb{R}_{>0}$ let*

$$\Xi(t,v) = \sum_{\substack{x \in O_B \cap V \\ Q(x)=t}} \xi(v^{1/2}x, z).$$

(i) For $t > 0$, $\Xi(t,v)$ defines a Green function for $\mathcal{Z}(t)$.
(ii) For $t < 0$, $\Xi(t,v)$ defines a smooth function on $\mathcal{M}_{\mathbb{C}} = [\Gamma \setminus D]$.

This allows us to define classes in the arithmetic Chow group $\widehat{\mathrm{CH}}^1(\mathcal{M})$:

$$\hat{\mathcal{Z}}(t,v) = \begin{cases} (\mathcal{Z}(t), \Xi(t,v)) & \text{if } t > 0, \\ (0, \Xi(t,v)) & \text{if } t < 0. \end{cases} \tag{3.5.6}$$

Remark 3.5.3. We note that the function $\Xi(t,v)$ vanishes when the quadratic field $k_t = \mathbb{Q}(\sqrt{-t})$ is not embeddable in B, since then the summation in the definition is empty.

Finally, we have to define the class $\hat{\mathcal{Z}}(0,v)$ in $\widehat{\mathrm{CH}}^1(\mathcal{M})$. We set, following [12],

$$\hat{\mathcal{Z}}(0,v) = -\hat{\omega} - (0, \log(v)) + (0, \mathbf{c}). \tag{3.5.7}$$

Here the first summand is the image of $\hat{\omega}$ under the natural map from $\widehat{\mathrm{Pic}}(\mathcal{M})$ into $\widehat{\mathrm{CH}}^1(\mathcal{M})$, and $\mathbf{c}$ is the real constant determined by the identity

(3.5.8)

$$\frac{1}{2}\deg_{\mathbb{Q}}(\hat{\omega})\cdot\mathbf{c} = \langle\hat{\omega},\hat{\omega}\rangle - \zeta_D(-1)\left[2\frac{\zeta'(-1)}{\zeta(-1)} + 1 - 2C - \sum_{p|D(B)} \frac{p\log(p)}{p-1}\right].$$

In Chapter 7, we will prove that, in fact, $\mathbf{c} = -\log D(B)$.

3.6 SPECIAL 0-CYCLES

Let

$$\mathrm{Sym}_2(\mathbb{Z})^\vee = \{T \in \mathrm{Sym}_2(\mathbb{Q}) \mid \mathrm{tr}(Tb) \in \mathbb{Z},\ \forall b \in \mathrm{Sym}_2(\mathbb{Z})\} \tag{3.6.1}$$

be the space of *semi-integral symmetric matrices.* For $T \in \mathrm{Sym}_2(\mathbb{Z})^\vee$ we define

$$\mathcal{Z}(T) = \{(A,\iota;x_1,x_2);\ x_1,x_2 \in V(A,\iota);\ Q(\mathbf{x}) = T\}. \tag{3.6.2}$$

Here (A, ι) is an object of $\mathcal{M}$, and for special endomorphisms $x_1, x_2 \in V(A, \iota)$ we set

$$Q(\mathbf{x}) = Q(x_1, x_2) = \frac{1}{2}\begin{pmatrix}(x_1, x_1) & (x_1, x_2)\\ (x_2, x_1) & (x_2, x_2)\end{pmatrix}, \tag{3.6.3}$$

where $(x, y) = Q(x + y) - Q(x) - Q(y)$ is the associated bilinear form.

Then $\mathcal{Z}(T)$ is a DM-stack equipped with an unramified morphism

$$\mathcal{Z}(T) \longrightarrow \mathcal{M}. \tag{3.6.4}$$

By Proposition 3.4.1, $\mathcal{Z}(T) = \emptyset$ unless T is positive-semidefinite.

We next explain when $\mathcal{Z}(T)$ is nonempty of dimension 0 for T with $\det T \neq 0$. We introduce the *incoherent* collection of quadratic spaces $\mathcal{C} = (\mathcal{C}_p)$, where $\mathcal{C}_p = V_p$ for $p < \infty$ and where $\mathcal{C}_\infty$ is the positive-definite quadratic space of dimension 3. Note that, since the quaternion algebra B is indefinite, the signature type of V_∞ is equal to (1,2), so that indeed the collection $\mathcal{C}$ is incoherent, i.e., the product of the Hasse invariants of the local quadratic spaces $\mathcal{C}_p$ is equal to -1. Let $T \in \mathrm{Sym}_2(\mathbb{Q})$ with $\det(T) \neq 0$. Let V_T be the unique ternary quadratic space over $\mathbb{Q}$ with discriminant $-1 = \mathrm{discr(V)}$ which represents T. Then V_T is isometric to the space of trace zero elements of a well-determined quaternion algebra B_T over $\mathbb{Q}$. We then define following [9], Definition 5.1,

$$\mathrm{Diff}(T, B) = \{\, p \leq \infty \mid \mathrm{inv}_p(B_T) \neq \mathrm{inv}(\mathcal{C}_p)\,\}. \tag{3.6.5}$$

Note that, since $\mathcal{C}$ is an incoherent collection, the cardinality $|\mathrm{Diff}(T, B)|$ is a positive odd number. Furthermore, $\infty \in \mathrm{Diff}(T, B)$ if and only if the signature of T is equal to (1,2) or (0,2), i.e., if T is not positive-definite.

Theorem 3.6.1. *Let $T \in \mathrm{Sym}_2(\mathbb{Z})^\vee$ be nonsingular.*
(i) If $|\mathrm{Diff}(T, B)| > 1$, then $\mathcal{Z}(T) = \emptyset$.
(ii) If $\mathrm{Diff}(T, B) = \{p\}$, with $p < \infty$, and $p \nmid D(B)$, then $\mathcal{Z}(T)$ is a finite stack in characteristic p with support in the supersingular locus. The lengths of the local rings at all geometric points of $\mathcal{Z}(T)$ are identical and given by the Gross-Keating formula

$$\nu_p(T) = \begin{cases} \sum_{j=0}^{\frac{a_2-1}{2}} (a_2 + a_3 - 4j)p^j & \text{if } a_2 \text{ is odd},\\[2ex] \sum_{j=0}^{\frac{a_2}{2}-1} (a_2 + a_3 - 4j)p^j + \frac{1}{2}(a_3 - a_2 + 1)p^{\frac{a_2}{2}} & \text{if } a_2 \text{ is even}. \end{cases}$$

Here $(0, a_2, a_3)$ with $0 \leq a_2 \leq a_3$ are the Gross-Keating invariants[1] *of $\tilde{T} = \mathrm{diag}(1, \mathrm{T})$.*

---

[1]If $p \neq 2$, this means that T is $\mathrm{GL}_2(\mathbb{Z}_p)$-equivalent to $\mathrm{diag}(\varepsilon_1 p^{a_2}, \varepsilon_2 p^{a_3})$ with $\varepsilon_1, \varepsilon_2 \in \mathbb{Z}_p^\times$.

(iii) If $\mathrm{Diff}(T, B) = \{p\}$, *with* $p < \infty$, *and* $p \mid D(B)$, *then* $\mathcal{Z}(T)$ *is concentrated in characteristic* p. *Furthermore,* $\dim \mathcal{Z}(T) = 1$ *if* $p^2 | T$ *in* $\mathrm{Sym}_2(\mathbb{Z})^\vee$.

Remark 3.6.2. If $p \neq 2$, the converse holds in (iii), i.e., if $p^2 \nmid T$, then $\dim\ \mathcal{Z}(T) = 0$, and in this case the lengths of the local rings at all geometric points of $\mathcal{Z}(T)$ are identical and given by

$$\nu_p(T) = \begin{cases} \beta & \text{if } \alpha = 1, \\ \beta & \text{if } \alpha = 0 \text{ and } \chi(\varepsilon_1) = -1, \\ 0 & \text{if } \alpha = 0 \text{ and } \chi(\varepsilon_1) = 1. \end{cases}$$

Here T is $\mathrm{GL}_2(\mathbb{Z}_p)$-equivalent to $\mathrm{diag}(\varepsilon_1 p^\alpha, \varepsilon_2 p^\beta)$ with $\varepsilon_1, \varepsilon_2 \in \mathbb{Z}_p^\times$ and χ is the quadratic residue character modulo p.

If $p = 2$ in (iii), the situation is more complicated and can be read off from the results in the appendix to Chapter 6. Let us call $T \in \mathrm{Sym}_2(\mathbb{Z}_p)^\vee$ *primitive,* if an equation $T = AT'A^t$ with $T' \in \mathrm{Sym}_2(\mathbb{Z}_p)^\vee$ and $A \in \mathrm{M}_2(\mathbb{Z}_p)$ implies $A \in \mathrm{GL}_2(\mathbb{Z}_p)$. Obviously, if T is primitive, then $p^2 \nmid T$. If $p = 2$, then $\dim\ \mathcal{Z}(T) = 0$ if and only if T is primitive. If $\dim\ \mathcal{Z}(T) = 0$, then the lengths of the local rings at all geometric points of $\mathcal{Z}(T)$ are the same and can be read off from Theorem 6A.1.1 in the appendix to Chapter 6.

Proof. Let $(A, \iota; y_1, y_2)$ be a point of $\mathcal{Z}(T)$. Then $B' = \mathrm{End}(A, \iota) \otimes \mathbb{Q}$ contains the three linearly independent elements $1, y_1, y_2$. Hence (A, ι) is a supersingular point in positive characteristic p. Since T is represented by the space V' of trace zero elements of B', it follows that $V' = V_T$. On the other hand, B' is a definite quaternion algebra whose invariants agree with those of B at all finite primes $\ell \neq p$. It follows that $\mathrm{Diff}(T, B) = \{p\}$, which proves the first assertion.

Now consider (i). Let (A, ι) be a supersingular point and let $\mathbf{y} = (y_1, y_2)$ be a pair of special endomorphisms of (A, ι) with $Q(\mathbf{y}) = T$. We need to determine the length $\lg\ \hat{\mathcal{O}}_x / J(\mathbf{y})_x$. Here $\hat{\mathcal{O}}_x = \hat{\mathcal{O}}_{\mathcal{M},x}$ is the formal completion of the local ring of x and $J(\mathbf{y})_x$ denotes the minimal ideal in $\hat{\mathcal{O}}_x$ such that y_1 and y_2 extend to endomorphisms of $(\mathcal{A}, \iota) \bmod J(\mathbf{y})$. Here $(\mathcal{A}, \iota)$ denotes the universal object over $\mathcal{M}$.

We use the Serre-Tate theorem. According to this theorem, the universal deformation space of $(A_x, \iota_x, \mathbf{y})$ coincides with the universal deformation space of the corresponding p-divisible group $(A_x(p), \iota)$ with its special endomorphisms. This p- divisible group is independent of x. More precisely, let G be the p-divisible formal group of dimension 1 and height 2 over $\bar{\mathbb{F}}_p$ (it is unique up to isomorphism). Then there is an isomorphism $\hat{A}_x \simeq G^2$

compatible with an isomorphism $O_B \otimes \mathbb{Z}_p \simeq \mathrm{M}_2(\mathbb{Z}_p)$. The special endomorphisms in $\mathrm{End}(A_x(p), \iota_x)$ induced by y_1 and y_2 can be identified with endomorphisms $x_1, x_2 \in \mathrm{End}(G)$ with $\mathrm{tr}(x_i) = 0$, for $i = 1, 2$. We therefore obtain isomorphisms

$$\hat{\mathcal{O}}_x \simeq W[[t]], \quad \hat{\mathcal{O}}_x/J(\mathbf{y})_x \simeq W[[t]]/J(\mathbf{x}). \tag{3.6.6}$$

Here $W = W(\bar{\mathbb{F}}_p)$ is the ring of Witt vectors, Spf $W[[t]]$ is the base of the universal deformation space of G, and $J(\mathbf{x})$ is the minimal ideal in $W[[t]]$ such that the special endomorphisms x_1 and x_2 of G deform to endomorphisms of the universal deformation Γ modulo $J(\mathbf{x})$,

$$x_1, x_2 : \Gamma \longrightarrow \Gamma \pmod{J(\mathbf{x})} . \tag{3.6.7}$$

We therefore have reduced the statement (i) to an assertion about the p-divisible group G, with its pair of special endomorphisms x_1, x_2, given by Theorem 3.6.3 below.

For (ii) we refer to [11], Proposition 2.10 and Theorem 6.1 in the case $p \neq 2$ and to the appendix to Chapter 6 below in the case $p = 2$. □

Theorem 3.6.3. *Let G be the formal group of dimension 1 and height 2 over $\bar{\mathbb{F}}_p$ and let Γ be its universal deformation over $W[[t]]$. Let x_1, x_2 be a pair of special endomorphisms of G ($\mathrm{tr}(x_i) = 0$ for $i = 1, 2$) which are linearly independent. Let $J(\mathbf{x})$ be the minimal ideal such that the x_i deform to endomorphisms of $\Gamma \pmod{J(\mathbf{x})}$. Then $\lg W[[t]]/J(\mathbf{x})$ only depends on the $\mathrm{GL}_2(\mathbb{Z}_p)$-equivalence class of $T = \frac{1}{2}(\mathbf{x}, \mathbf{x})$ and is equal to*

$$\nu_p(T) = \begin{cases} \sum_{j=0}^{\frac{a_2-1}{2}} (a_2 + a_3 - 4j)p^j & \text{if } a_2 \text{ is odd,} \\ \sum_{j=0}^{\frac{a_2}{2}-1} (a_2 + a_3 - 4j)p^j + \frac{1}{2}(a_3 - a_2 + 1)p^{\frac{a_2}{2}} & \text{if } a_2 \text{ is even.} \end{cases}$$

Here $(0, a_2, a_3)$ with $0 \leq a_2 \leq a_3$ are the Gross-Keating invariants of $\tilde{T} = \mathrm{diag}(1, \mathrm{T})$.

Theorem 3.6.3 is the special case $a_1 = 0$ of [6], Proposition 5.4. However, this special case is easier to prove and follows from the results in [8], as we now proceed to show.

Remark 3.6.4. Theorem 5.1 of [8] is not proved there. Instead, the reader is referred to Keating's unpublished Harvard thesis. A detailed account of Keating's result can be found in [1], which also contains a proof of the general result of Gross-Keating. We also point out that Theorem 3.6.3 is overstated as Proposition 14.6 in [9].

Proof. Let $L = \mathbb{Z}_p x_1 + \mathbb{Z}_p x_2 \subset \operatorname{End}(G)$. It is obvious that

$$J(L) = J(\mathbf{x}), \tag{3.6.8}$$

where $J(L)$ is the minimal ideal in $W[[t]]$ such that

$$L \subset \operatorname{End}(\Gamma(\operatorname{mod} J(L))). \tag{3.6.9}$$

We may therefore choose any basis φ, ψ of L and then

$$J(x_1, x_2) = J(\varphi, \psi) = J(L).$$

We choose φ, ψ, as we may, such that $1, \varphi, \psi$ form an *optimal basis* of $\tilde{L} = \mathbb{Z}_p \cdot 1 + \mathbb{Z}_p x_1 + \mathbb{Z}_p x_2$, with $v(\varphi) = a_2$ and $v(\psi) = a_3$. If $p \neq 2$, this simply means that the bilinear form on L is diagonalized by φ, ψ and that $a_2 = v(\varphi)$ and $a_3 = v(\psi)$. Here, as elsewhere, we denote by v the valuation on the quaternion division algebra $D = \operatorname{End}(G) \otimes_{\mathbb{Z}_p} \mathbb{Q}_p$.

Lemma 3.6.5. *Let $\boldsymbol{k} = \mathbb{Q}_p(\varphi) \subset D$. Then $\boldsymbol{k}/\mathbb{Q}_p$ is unramified if a_2 is even and is ramified if a_2 is odd. The order $\mathbb{Z}_p[\varphi]$ has conductor p^s with $s = [a_2/2]$.*

Proof for $p \neq 2$. We have $\operatorname{tr}(\varphi) = 0$ and $\operatorname{Nm} \varphi = \varepsilon_2 p^{a_2}$ with $\varepsilon_2 \in \mathbb{Z}_p^\times$. If a_2 is odd, then $\boldsymbol{k} = \mathbb{Q}_p(\pi)$ with $\pi = \sqrt{-\varepsilon_2 p}$ and $O_{\boldsymbol{k}} = \mathbb{Z}_p[\pi]$. Since $\varphi = p^s \cdot \pi$, the claim follows in this case. If a_2 is even, then $\boldsymbol{k} = \mathbb{Q}_p(\xi)$ with $\xi = \sqrt{-\varepsilon_2}$. Then $O_{\boldsymbol{k}} = \mathbb{Z}_p[\xi]$. Since $\varphi = p^s \cdot \xi$, the claim follows. □

The proof for $p = 2$ is more complicated and is given at the end of this section.

Lemma 3.6.6.

$$\psi \in \Pi^{a_3} O_D \setminus (O_{\boldsymbol{k}} + \Pi^{a_3+1} O_D).$$

Here Π denotes a uniformizer of O_D.

Proof for $p \neq 2$. In this case ψ anticommutes with $\boldsymbol{k}$, i.e., conjugation with ψ induces the nontrivial automorphism of $\boldsymbol{k}$. Since $\boldsymbol{k}/\mathbb{Q}_p$ is unramified or tamely ramified, we may apply [14], Lemma 2.2. Let $v = v(\psi) = a_3$. Then [14] tells us that, under the ramification hypothesis made, we have $\psi \in \Pi^v O_D \setminus (O_{\boldsymbol{k}} + \Pi^{v+1} O_D)$. □

The proof for $p = 2$ is given at the end of this section.

Let us now continue the proof of Theorem 3.6.3, assuming (even for $p = 2$) the validity of Lemmas 3.6.5 and 3.6.6. By Lemma 3.6.5 we may write the locus Spf $W[[t]]/J(\varphi)$ where φ deforms as a sum of quasi-canonical divisors in Spf $W[[t]]$,

$$\text{Spf } W[[t]]/J(\varphi) = \sum_{r=0}^{s} \mathcal{W}_r(\varphi). \tag{3.6.10}$$

Recall [5], (cf. also Section 7.7), that the quasi-canonical divisor of level r is a reduced irreducible regular divisor in Spf $W[[t]]$ such that the pullback of the universal p-divisible group on Spf $W[[t]]$ to $\mathcal{W}_r(\varphi)$ has the order O_r of conductor p^r in $\boldsymbol{k}$ (embedded via φ into D) as its endomorphism algebra. We have $\mathcal{W}_r(\varphi) \simeq$ Spf $W_r(\boldsymbol{k})$, where $W_r(\boldsymbol{k})$ is the ring of integers in a ramified extension M_r of $M = \text{Frac } W$.

Now the locus inside $\mathcal{W}_r(\varphi)$ to which ψ deforms is defined by an ideal I_ℓ in $W_r(\boldsymbol{k})$ which only depends on the integer ℓ with $\psi \in (O_r + \Pi^\ell O_D) \setminus (O_r + \Pi^{\ell+1} O_D)$. By Lemma 3.6.6, we have $\ell = a_3$. Since $a_3 \geq a_2 \geq 2s \geq 2r - 1$, we are in the 'stable range' of the formula of [15], Theorem 1.1. Hence

$$\lg\ W_r(\boldsymbol{k})/I_{a_3} = \frac{(p^{r-1}-1)(p+1)}{p-1} + p^{r-1} + \left(\frac{a_3+1}{2} - r\right) e_r + 1, \tag{3.6.11}$$

where e_r is the ramification index of M_r over M.

Now from the explicit description of $W_r(\boldsymbol{k})$, as described, e.g., in Section 7.7, we have

$$e_r = \begin{cases} 2p^r & \text{if } \boldsymbol{k}/\mathbb{Q}_p \text{ is ramified,} \\ p^r + p^{r-1} & \text{if } \boldsymbol{k}/\mathbb{Q}_p \text{ is unramified and } r \geq 1, \\ 1 & \text{if } \boldsymbol{k}/\mathbb{Q}_p \text{ is unramified and } r = 0. \end{cases} \tag{3.6.12}$$

Therefore we have obtained a completely explicit expression for

$$\lg\ W[[t]]/J(\varphi, \psi) = \sum_{r=0}^{s} \lg\ W_r(\boldsymbol{k})/I_{a_3}. \tag{3.6.13}$$

Taking into account the first statement of Lemma 3.6.5, which allows us to distinguish the ramified case from the unramified case by the parity of a_2, we get the following values for lg $W[[t]]/J(\varphi, \psi)$:

$$\sum_{r=0}^{s} \left(\frac{(p^{r-1}-1)(p+1)}{p-1} + p^{r-1} + \left(\frac{a_3+1}{2} - r\right) \cdot 2p^r + 1 \right), \tag{3.6.14}$$

if $a_2 = 2s+1$ is odd, and

$$\frac{a_3+1}{2} + \sum_{r=1}^{s} \left(\frac{(p^r-1)(p+1)}{p-1} + p^{r-1} + \left(\frac{a_3+1}{2} - r\right)(p^r + p^{r-1}) + 1 \right), \tag{3.6.15}$$

if $a_2 = 2s$ is even. An elementary (but tedious) calculation now gives the formula in Theorem 3.6.3. □

It remains to prove Lemmas 3.6.5 and 3.6.6 for $p = 2$. For this we use the classification of the possible $\tilde{T}$ and the construction of optimal bases of $\tilde{L}$ in each case; see [18], [3].

$\tilde{T}$'s and Optimal Bases

(A) $\tilde{T} = \mathrm{diag}\left(1, 2^{\beta}\begin{pmatrix}2 & 1\\ 1 & 2\end{pmatrix}\right)$ with $\beta \geq 0$ even.[2] Then $1 = e_1, \varphi = e_2, \psi = e_3$ is an optimal basis, and

$$\mathrm{GK}(\tilde{T}) = (0, \beta + 1, \beta + 1).$$

(B) $\tilde{T} = \mathrm{diag}(1, \varepsilon_2 2^{\beta_2}, \varepsilon_3 2^{\beta_3})$ with $0 \leq \beta_2 \leq \beta_3$. Then $\tilde{T}$ is anisotropic if and only if

$$(-1, \varepsilon_2\varepsilon_3) = (\varepsilon_2, \varepsilon_3) \cdot (2, \varepsilon_2)^{\beta_3} \cdot (2, \varepsilon_3)^{\beta_2} \ .$$

(1) β_2 odd. Then $1 = e_1, \varphi = e_2, \psi = c_1 e_1 + c_2 e_2 + e_3$ for suitable $c_1, c_2 \in \mathbb{Z}_2$, and

$$\mathrm{GK}(\tilde{T}) = (0, \beta_2, \beta_3 + 2).$$

(2) β_2 even and $\beta_3 \leq \beta_2 + 1$.

(a) $\beta_2 = \beta_3$. Then $1 = e_1, \varphi = 2^{\beta_2/2} e_1 + e_2, \psi = 2^{\beta_2/2} e_1 + e_3$, and

$$\mathrm{GK}(\tilde{T}) = (0, \beta_2 + 1, \beta_3 + 1).$$

(b) $\beta_3 = \beta_2 + 1$ and $\varepsilon_1 \equiv 1 (\mathrm{mod}\ 4)$. Then $1 = e_1, \varphi = 2^{\beta_2/2} e_1 + e_2, \psi = 2^{\beta_2/2} e_1 + e_3$, and

$$\mathrm{GK}(\tilde{T}) = (0, \beta_2 + 1, \beta_3 + 1).$$

(c) $\beta_3 = \beta_2 + 1$ and $\varepsilon_2 \equiv -1 (\mathrm{mod}\ 4)$. Then $1 = e_1, \varphi = 2^{\beta_2/2} e_1 + e_2 + e_3, \psi = 2^{\beta_2/2} e_1 + e_2 + 2e_3$, and

$$\mathrm{GK}(\tilde{T}) = (0, \beta_2 + 1, \beta_3 + 1).$$

(3) β_2 even and $\beta_3 \geq \beta_2 + 2$.

---

[2] The parity condition on β comes from the fact that T is anisotropic.

(a) $\varepsilon_2 \equiv -1 \pmod 4$. Then $1 = e_1, \varphi = 2^{\beta_2/2}e_1 + e_2, \psi = e_3$, and

$$\mathrm{GK}(\tilde{T}) = (0, \beta_2 + 2, \beta_3).$$

(b) $\varepsilon_2 \equiv 1 \pmod 4$. Then $1 = e_1$, $\varphi = 2^{\beta_2/2}e_1 + e_2$, $\psi = c_1e_1 + c_2e_2 + e_3$ for suitable $c_1, c_2 \in \mathbb{Z}_2$, and

$$\mathrm{GK}(\tilde{T}) = (0, \beta_2 + 1, \beta_3 + 1).$$

Proof of Lemma 3.6.5. We distinguish cases.

Case A: Here $\mathrm{tr}(\varphi) = (e_1, e_2) = 0$ and $\mathrm{Nm}(\varphi) = 2^{\beta+1}$. Hence $\boldsymbol{k} = \mathbb{Q}_2(\pi)$ where $\pi = \sqrt{-2}$. So $\boldsymbol{k}$ is ramified, $O_{\boldsymbol{k}} = \mathbb{Z}_p[\pi]$ and $\varphi = 2^{\beta/2} \cdot \pi$, which proves the claim.

Case B1: Here $\mathrm{tr}\,\varphi = (e_1, e_2) = 0$ and $\mathrm{Nm}(\varphi) = \varepsilon_2 2^{\beta_2}$. Hence $\boldsymbol{k} = \mathbb{Q}_2(\pi)$ with $\pi = \sqrt{-\varepsilon_2 2}$. So $\boldsymbol{k}$ is ramified, $O_{\boldsymbol{k}} = \mathbb{Z}_p[\pi]$ and $\varphi = 2^{(\beta_2-1)/2} \cdot \pi$, which proves the claim.

Case B2a: Here $\mathrm{tr}(\varphi) = (e_1, 2^{\beta_2/2}e_1 + e_2) = 2^{\beta_2/2+1}$ and $\mathrm{Nm}\,\varphi = 2^{\beta_2}(1 + \varepsilon_2)$. Now by the anisotropy of $\tilde{T}$ we have $\varepsilon_2 \equiv 1 \pmod 4$. Hence $\boldsymbol{k} = \mathbb{Q}_2[X]/(X^2 - 2X + (1 + \varepsilon_2))$ is defined by an Eisenstein polynomial. Denoting by π the residue class of X we have $O_{\boldsymbol{k}} = \mathbb{Z}_2[\pi]$ and $\varphi = 2^{\beta_2/2} \cdot \pi$, which proves the claim.

Case B2b: This is identical to the previous case.

Case B2c: Here $\mathrm{tr}(\varphi) = 2^{\beta_2/2+1}$ and $\mathrm{Nm}(\varphi) = 2^{\beta_2} + \varepsilon_2 2^{\beta_2} + \varepsilon_3 2^{\beta_3} = 2^{\beta_2}(2\varepsilon_3 + 1 + \varepsilon_2)$. Since $\varepsilon_2 \equiv -1 \pmod 4$, we have $\boldsymbol{k} = \mathbb{Q}_2[X]/(X^2 - 2X + (2\varepsilon_3 + 1 + \varepsilon_2))$, which is defined by an Eisenstein equation. Denoting by π the residue class of X, we have $O_{\boldsymbol{k}} = \mathbb{Z}_2[\pi]$ and $\varphi = 2^{\beta_2/2} \cdot \pi$, which proves the claim.

Case B3a: Here $\mathrm{tr}\,\varphi = 2^{\beta_2/2+1}$ and $\mathrm{Nm}(\varphi) = 2^{\beta_2}(1 + \varepsilon_2)$. Now since $\tilde{T}$ is anisotropic, $1 + \varepsilon_2 \equiv 4 \pmod 8$. Hence writing $1 + \varepsilon_2 = 4\eta$ we have $\eta \in \mathbb{Z}_2^\times$ and $\boldsymbol{k} = \mathbb{Q}_2[X]/(X^2 - X + \eta)$. Hence $\boldsymbol{k}/\mathbb{Q}_2$ is unramified as asserted and, denoting by ξ the residue class of X, we have $O_{\boldsymbol{k}} = \mathbb{Z}_2[\xi]$ and $\varphi = 2^{\beta_2/2+1} \cdot \xi$, which proves the claim.

Case B3b: Here $\mathrm{tr}(\varphi)$ and $\mathrm{Nm}(\varphi)$ are as in the previous case but this time $1 + \varepsilon_2 \equiv 2 \pmod 4$. Hence $\boldsymbol{k} = \mathbb{Q}_2[X]/(X^2 - 2X + (1 + \varepsilon_2))$ is defined by an Eisenstein polynomial. Denoting by π the residue class of X, we have $O_{\boldsymbol{k}} = \mathbb{Z}_2[\pi]$ and $\varphi = 2^{\beta_2/2} \cdot \pi$, which proves the claim.

The lemma is now proved in all cases. □

Proof of Lemma 3.6.6. Again we go through all cases. When $p = 2$, the endomorphism ψ does not in general anticommute with φ (and even if it did, Lemma 2.2 of [14] does not apply when $\boldsymbol{k}/\mathbb{Q}_2$ is ramified). Still, the commuting properties of ψ and φ can be used to prove Lemma 3.6.6.

Case A: Here

$$\begin{aligned}\varphi\psi &= (e_1^\iota \circ e_2) \circ (e_1^\iota \circ e_3) = -e_1^\iota \circ e_2 \circ (e_3^\iota \circ e_1) \\ &= -e_1^\iota \circ (-e_3 \circ e_2^\iota + 2^{\beta+1}) \circ e_1 = e_1^\iota \circ e_3 \circ e_2^\iota \circ e_1 - 2^{\beta+1} \\ &= -(e_1^\iota \circ e_3) \circ (e_1^\iota \circ e_2) - 2^{\beta+1} \\ &= -\psi\varphi - 2^{\beta+1}.\end{aligned}$$

Here, as usual, $x \mapsto x^\iota$ denotes the main involution of D. Hence

$$\varphi\psi + \psi\varphi = -2^{\beta+1}. \tag{3.6.16}$$

Now we can write, following [5], Proposition 4.3,

$$D = \boldsymbol{k} \oplus \boldsymbol{k}j, \tag{3.6.17}$$

where j anticommutes with $\boldsymbol{k}$ and with $j^2 \equiv 1 \mod 2^{e-1}$, where e is the valuation of the different of $\boldsymbol{k}/\mathbb{Q}_2$, and then

$$O_D = O_{\boldsymbol{k}} \oplus O_{\boldsymbol{k}}\alpha, \tag{3.6.18}$$

where $\alpha = \pi^{1-e}(1 + j) \in O_D^\times$.

In the case at hand, the different of $\boldsymbol{k}/\mathbb{Q}_p$ has valuation 3, hence $\alpha = \pi^{-2}(1+j)$. Writing $\psi = a + b\alpha$, the assertion of Lemma 3.6.6 comes down to $v(b) = a_3$. But the LHS of (3.6.16) is equal to (recall $\varphi = 2^s\pi$),

$$2^s(a\pi + b\pi^{-1}(1+j) + a\pi + b\pi^{-1}(1-j)) = 2^{s+1} \cdot (a\pi + b\pi^{-1}). \tag{3.6.19}$$

Noting $\beta = 2s$, we get from (3.6.16) the identity

$$b = -2^s\pi - a\pi^2. \tag{3.6.20}$$

Comparing valuations of the summands and recalling $v(\psi) = \beta + 1 = a_3$, we obtain $v(b) = a_3$ as asserted.

Case B1: In this case

$$\varphi\psi - \psi\varphi = 2 \cdot e_3^\iota \circ e_2. \tag{3.6.21}$$

Again $D = \boldsymbol{k} + \boldsymbol{k}j$ and $O_D = O_{\boldsymbol{k}} \oplus O_{\boldsymbol{k}}\alpha$ with $\alpha = \pi^{-2}(1 + j) \in O_D^\times$. Writing $\psi = a + b\alpha$, we get for the LHS of (3.6.21), noting $\varphi = 2^s\pi$,

$$2^s\left((a\pi + b\pi^{-1}(1+j)) - (a\pi + b\pi^{-1}(1-j))\right) = 2^{s+1} \cdot b\pi^{-1}j. \tag{3.6.22}$$

Hence equation (3.6.21) gives

$$2^s b\pi^{-1} j = e_3^\iota \circ e_2. \tag{3.6.23}$$

Comparing valuations on both sides gives

$$2s + v(b) - 1 = \beta_2 + \beta_3, \tag{3.6.24}$$

i.e. $v(b) = \beta_3 + 2 = a_3$, as claimed.

Case B2a: In this case

$$\varphi\psi - \psi\varphi = 2 \cdot e_3^\iota \circ e_2. \tag{3.6.25}$$

Again $D = \boldsymbol{k} + \boldsymbol{k}j$ and $O_D = O_{\boldsymbol{k}} \oplus O_{\boldsymbol{k}}\alpha$, but this time, since the different has valuation 2, we have $\alpha = \pi^{-1}(1+j)$. Writing $\psi = a + b\alpha$, we get for the LHS of (3.6.25) and noting $\varphi = 2^s \cdot \pi$,

$$2^s \left((a\pi + b(1+j)) - (a\pi + b + b\pi^\iota/\pi \cdot j)\right) = 2^s jb \cdot (2 - 2/\pi). \tag{3.6.26}$$

Hence equation (3.6.25) gives

$$2^s jb \cdot (2 - 2/\pi) = 2 \cdot e_3^\iota \circ e_2. \tag{3.6.27}$$

Comparing valuations on both sides gives

$$2s + v(b) + 1 = \beta_2 + \beta_3 + 2, \tag{3.6.28}$$

i.e., $v(b) = \beta_3 + 1 = a_3$, as claimed.

Case B2b: identical with the previous one.

Case B2c: In this case

$$\varphi\psi - \psi\varphi = 2 \cdot e_3^\iota \circ e_2. \tag{3.6.29}$$

Again $D = \boldsymbol{k} + \boldsymbol{k}j$ and $O_D = O_{\boldsymbol{k}} \oplus O_{\boldsymbol{k}}\alpha$ with $\alpha = \pi^{-1}(1+j)$. Writing $\psi = a + b\alpha$ we obtain by an identical reasoning as in Case B2a that $v(b) = \beta_3 + 1 = a_3$, as desired.

Case B3a: In this case

$$\varphi\psi - \psi\varphi = 2 \cdot e_3^\iota \circ e_2. \tag{3.6.30}$$

In this case $\boldsymbol{k}/\mathbb{Q}_2$ is unramified. We may choose a uniformizer Π of O_D which anticommutes with $\boldsymbol{k}$ and such that $\Pi^2 = 2$, and then

$$O_D = O_{\boldsymbol{k}} \oplus O_{\boldsymbol{k}}\Pi. \tag{3.6.31}$$

Now writing $\psi = a + b\Pi$ we need to show that $v(b\Pi) = a_3$. Now $\varphi = 2^s \cdot \xi$ and the LHS of (3.6.30) is equal to

$$(3.6.32)\quad 2^s((a\xi + b\xi\Pi) - (a\xi + b\xi^\iota\Pi)) = 2^s b(\xi - \xi^\iota) = 2^s b(1 - 2\xi) \cdot \Pi.$$

Hence equation (3.6.30) gives

$$(3.6.33)\qquad 2^s b \cdot (1 - 2 \cdot \xi^\iota) \cdot \Pi = 2 \cdot e_3^\iota \circ e_2.$$

Comparing valuations gives

$$(3.6.34)\qquad 2s + v(b\Pi) = 2 + \beta_2 + \beta_3.$$

Since $2s = \beta_2 + 2$, we get $v(b\Pi) = \beta_3 = a_3$, as desired.

Case B3b: Again

$$(3.6.35)\qquad \varphi\psi - \psi\varphi = 2 \cdot e_3^\iota \circ e_2.$$

Here $D = \boldsymbol{k} + \boldsymbol{k}j$ and $O_D = O_{\boldsymbol{k}} \oplus O_{\boldsymbol{k}}\alpha$ with $\alpha = \pi^{-1}(1 + j)$. Writing $\psi = a + b\alpha$, we get as in Case B2a

$$(3.6.36)\qquad 2^{\beta_2/2} \cdot j \cdot b \cdot (2 - 2/\pi) = 2 \cdot e_3^\iota \circ e_2.$$

Hence comparing valuations we get

$$(3.6.37)\qquad \beta_2 + v(b) + 1 = 2 + \beta_2 + \beta_3,$$

i.e., $v(b) = \beta_3 + 1 = a_3$, as desired. □

In Chapter 6, when we form the generating series $\hat{\phi}_2$ for 0-cycles on $\mathcal{M}$, we will need to associate to every $T \in \mathrm{Sym}_2(\mathbb{Z})^\vee$ and every $v \in \mathrm{Sym}_2(\mathbb{R})_{>0}$ an element

$$\hat{\mathcal{Z}}(T, v) \in \widehat{\mathrm{CH}}^2(\mathcal{M}).$$

When T is positive definite and $\mathrm{Diff}(T, B) = \{p\}$ for some $p \nmid D(B)$, we take the class of $(\mathcal{Z}(T), 0) \in \hat{Z}^2_{\mathbb{Z}}(\mathcal{M})$, and if T is positive definite and $|\mathrm{Diff}(T, B)| > 1$, we take the zero class in $\widehat{\mathrm{CH}}^2(\mathcal{M})$. In all other cases, when T is not positive definite, or when T is positive definite and $\mathrm{Diff}(T, B) = \{p\}$ with $p \mid D(B)$, the definition is less obvious. This issue is dealt with in Chapter 6.

Bibliography

[1] ARGOS (Arithmetische Geometrie Oberseminar), Proceedings of the Bonn seminar 2003/04, forthcoming.

[2] J.-F. Boutot and H. Carayol, *Uniformisation p-adique des courbes de Shimura*, in Courbes Modulaires et Courbes de Shimura, Astérisque, **196–197**, 45–158, Soc. Math. France, Paris, 1991.

[3] I. Bouw, *Invariants of Ternary Quadratic Forms*, in [1].

[4] B. Gross, *Heegner points on $X_0(N)$*, in Modular Forms, Ellis Horwood Ser. Math. and Its Appl.: Statist. Oper. Res., 87–105, Horwood, Chichester, 1984.

[5] ______, *On canonical and quasi-canonical liftings*, Invent. math., **84** (1986), 321–326.

[6] B. Gross and K. Keating, *On the intersection of modular correspondences*, Invent. math., **112** (1993), 225–245.

[7] B. H. Gross, and D. Zagier, *Heegner points and the derivatives of L-series*, Invent. math., **84** (1986), 225–320.

[8] K. Keating, *Lifting endomorphisms of formal A-modules*, Compositio Math., **67** (1988), 211–239.

[9] S. Kudla, *Central derivatives of Eisenstein series and height pairings*, Annals of Math., **146** (1997), 545–646.

[10] ______, *Modular forms and arithmetic geometry*, in Current Developments in Mathematics, 2002, 135–179, International Press, Somerville, MA, 2003.

[11] S. Kudla and M. Rapoport, *Height pairings on Shimura curves and p-adic uniformization*, Invent. math., **142** (2000), 153–223.

[12] S. Kudla, M. Rapoport, and T. Yang, *Derivatives of Eisenstein series and Faltings heights*, Compositio Math., **140** (2004), 887–951.

[13] G. Laumon and L. Moret-Bailly, Champs algébriques, Ergebnisse der Mathematik und ihrer Grenzgebiete, **39**, Springer-Verlag, Berlin, 2000.

[14] M. Rapoport, *Deformations of isogenies of formal groups*, in [1].

[15] I. Vollaard, *Endomorphisms of quasi-canonical lifts*, in [1].

[16] M. Rapoport and T. Zink, Period Spaces for p-Divisible Groups, Annals of Mathematics Studies, **141**, Princeton Univ. Press, Princeton, NJ, 1996.

[17] S. Wewers, *Canonical and quasi-canonical liftings*, in [1]

[18] T. H. Yang, *Local densities of 2-adic quadratic forms*, J. Number Theory, **108** (2004), 287–345.

[19] T. Zink, *The display of a formal p-divisible group*, in Cohomologies p-adiques et applications arithmétiques, I. Astérisque **278**, 127–248, Soc. Math. France, Paris, 2002.

Chapter Four

An arithmetic theta function

In this chapter, we consider the generating function $\hat{\phi}_1$ whose coefficients are the special divisors $\hat{\mathcal{Z}}(t,v) \in \widehat{\mathrm{CH}}^1(\mathcal{M})$ defined in the previous chapter. We prove that $\hat{\phi}_1$ is a modular form of weight $\frac{3}{2}$ (Theorem A). In the first section we introduce the decomposition

$$\widehat{\mathrm{CH}}^1(\mathcal{M}) = \widetilde{\mathrm{MW}} \oplus \mathbb{R}\hat{\omega} \oplus \mathrm{Vert} \oplus a(A^0(\mathcal{M}_{\mathbb{R}})_0) \tag{4.0.1}$$

of $\widehat{\mathrm{CH}}^1(\mathcal{M})$ into its 'Mordell-Weil' component, its 'Hodge bundle' component, its 'vertical' component, and its 'C^∞'-component. We then reduce the proof of Theorem A to an assertion about the various components of $\hat{\phi}_1$ with respect to this decomposition. The modularity of the Hodge component follows from [16]. In Section 4.3, we prove the modularity of the vertical component of $\hat{\phi}_1$ by identifying it with the theta function of a positive definite ternary form. In Section 4, we prove the modularity of the C^∞-component of $\hat{\phi}_1$ by identifying it with the theta lift of a Maass form. In Sections 4.5 and 4.6, we show that the modularity of the Mordell-Weil component of $\hat{\phi}_1$ follows from Borcherds' theorem [2]. In the last section, we check an intertwining property of the arithmetic theta function which will be crucial in the definition of the arithmetic theta lift in Chapter 9.

4.1 THE STRUCTURE OF ARITHMETIC CHOW GROUPS

Let $\mathcal{M}$ be an arithmetic surface over Spec $\mathbb{Z}$, as in Chapter 2. We assume that $\mathcal{M}$ is geometrically irreducible. In this section, we review the structure of the arithmetic Chow group of $\mathcal{M}$ and set up a convenient decomposition of this group with respect to the arithmetic intersection pairing or Arakelov-Gillet-Soulé height pairing.

Let $\widehat{\mathrm{CH}}^1(\mathcal{M})$ be the arithmetic Chow group with real coefficients as defined in Chapter 2. Recall that this is the quotient of the real vector space $\widehat{Z}^1(\mathcal{M})_{\mathbb{R}}$ spanned by pairs $(\mathcal{Z}, g)$, where $\mathcal{Z} \in Z^1(\mathcal{M})_{\mathbb{R}}$ is a divisor on $\mathcal{M}$ with real coefficients and g is a Green function for $\mathcal{Z}$, by the subspace of relations spanned over $\mathbb{R}$ by the elements $\widehat{\mathrm{div}}(f) = (\mathrm{div}(f), -\log|f|^2)$ for $f \in \mathbb{Q}(\mathcal{M})^\times$. We will use the notation $g(\hat{\mathcal{Z}})$ (resp. $\omega(\hat{\mathcal{Z}})$) for the Green function (resp. $(1,1)$-form) associated to a class $\hat{\mathcal{Z}}$. These are related by

$$dd^c g(\hat{\mathcal{Z}}) + \delta_{\mathcal{Z}} = \omega(\hat{\mathcal{Z}}).$$

We will only consider Green functions of C^∞-regularity, in the terminology of [1], p. 18, as in [5]. As explained in Chapter 1, the group $\widehat{\mathrm{CH}}^1(\mathcal{M})$ comes equipped with an arithmetic intersection pairing, the Arakelov-Gillet-Soulé height pairing $\langle\ ,\ \rangle$, which we will also refer to as the height pairing. In addition, there is a geometric degree map $\deg_{\mathbb{Q}} : \widehat{\mathrm{CH}}^1(\mathcal{M}) \to \mathbb{R}$ obtained as a composition

$$\deg_{\mathbb{Q}} : \widehat{\mathrm{CH}}^1(\mathcal{M}) \xrightarrow{\mathrm{res}_{\mathbb{Q}}} \mathrm{CH}^1(\mathcal{M}_{\mathbb{Q}}) \otimes \mathbb{R} \xrightarrow{\deg} \mathbb{R}, \tag{4.1.1}$$

where $\mathrm{CH}^1(\mathcal{M}_{\mathbb{Q}})$ is the usual Chow group of the generic fiber $\mathcal{M}_{\mathbb{Q}}$ of $\mathcal{M}$.

We fix a volume form μ_1 on $\mathcal{M}(\mathbb{C})$, with $\mathrm{vol}(\mathcal{M}(\mathbb{C}), \mu_1) = 1$. We also fix a metrized line bundle $\hat{\omega} = (\omega, ||\ ||)$ on $\mathcal{M}$ with first Chern form $c_1(\hat{\omega}) = \deg_{\mathbb{Q}}(\omega) \cdot \mu_1$ with $\deg_{\mathbb{Q}}(\omega) > 0$, and denote by the same symbol

$$\hat{\omega} \in \widehat{\mathrm{CH}}^1(\mathcal{M}) \tag{4.1.2}$$

the associated class under the map $\widehat{\mathrm{Pic}}(\mathcal{M}) \to \widehat{\mathrm{CH}}^1(\mathcal{M})$. Let $A^0(\mathcal{M}_{\mathbb{R}})$ be the space of C^∞-functions on $\mathcal{M}(\mathbb{C})$ invariant under the archimedean Frobenius F_∞, and let

$$a : A^0(\mathcal{M}_{\mathbb{R}}) \longrightarrow \widehat{\mathrm{CH}}^1(\mathcal{M}), \qquad f \mapsto (0, f), \tag{4.1.3}$$

be the inclusion. Note that this map is $\mathbb{R}$-linear. Let

$$1\!\!1 := a(1) \in \widehat{\mathrm{CH}}^1(\mathcal{M}). \tag{4.1.4}$$

Let $A^0(\mathcal{M}_{\mathbb{R}})_0$ be the subspace of $f \in A^0(\mathcal{M}_{\mathbb{R}})$ such that

$$\int_{\mathcal{M}(\mathbb{C})} f \cdot c_1(\hat{\omega}) = 0. \tag{4.1.5}$$

Let

$$\mathrm{Vert} = \mathrm{Vert}(\mathcal{M}) \tag{4.1.6}$$

be the subspace of $\widehat{\mathrm{CH}}^1(\mathcal{M})$ generated by the classes of the form $(Y_p, 0)$, where Y_p is an irreducible component of a fiber $\mathcal{M}_p$. The relation $\widehat{\mathrm{div}}(p) = (\mathcal{M}_p, -\log(p^2)) \equiv 0$, implies that $(\mathcal{M}_p, 0) \equiv 2\log(p)\, 1\!\!1$, so that $1\!\!1 \in \mathrm{Vert}$. Let

$$\mathrm{Vert}_p = \oplus_i \mathbb{R}\, Y_{p,i} \tag{4.1.7}$$

be the real vector space with basis the irreducible components $Y_{p,i}$ of the fiber $\mathcal{M}_p$, and let

$$\overline{\text{Vert}}_p = \text{Vert}_p/(\mathbb{R}\cdot\mathcal{M}_p) \qquad \text{and} \qquad \overline{\text{Vert}} = \text{Vert}/(\mathbb{R}\cdot 1\!\!1). \tag{4.1.8}$$

Then we have a commutative diagram with exact rows and columns

$$\begin{array}{ccccccc} R & \longrightarrow & \oplus_p\mathbb{R} & \longrightarrow & \mathbb{R} & 1 \\ \| & & \downarrow & & \downarrow & \downarrow \\ R & \longrightarrow & \oplus_p\text{Vert}_p & \longrightarrow & \text{Vert} & 1\!\!1 \\ & & \downarrow & & \downarrow & \\ & & \oplus_p\overline{\text{Vert}}_p & \longrightarrow & \overline{\text{Vert}} & \end{array} \tag{4.1.9}$$

where

$$R = \{(\lambda_p)\in\oplus_p\mathbb{R} \mid \sum_p \lambda_p \log(p) = 0\}. \tag{4.1.10}$$

The inclusion map on R in the middle line is given by

$$(\lambda_p)\mapsto \sum_p \lambda_p\,\mathcal{M}_p, \tag{4.1.11}$$

while the second arrow in the top line is given by $(\lambda_p)\mapsto 2\sum_p \lambda_p \log(p)$. Of course, Vert lies in the kernel of $\deg_{\mathbb{Q}}$.

The intersection pairing $\langle\,\cdot,\cdot\,\rangle$ on $\widehat{\text{CH}}^1(\mathcal{M})$ can be written in general as

$$\langle\,\hat{\mathcal{Z}}_1,\hat{\mathcal{Z}}_2\,\rangle = h_{\hat{\mathcal{L}}_1}(\mathcal{Z}_2) + \frac{1}{2}\int_{\mathcal{M}(\mathbb{C})} g(\hat{\mathcal{Z}}_2)\,\omega(\hat{\mathcal{Z}}_1), \tag{4.1.12}$$

where $h_{\hat{\mathcal{L}}_1}(\mathcal{Z}_2)$ is the height of the cycle $\mathcal{Z}_2$ with respect to the metrized line bundle $\hat{\mathcal{L}}_1$ corresponding to $\hat{\mathcal{Z}}_1$; see Section 2.6 or (5.11) of [1]. The alternative expression

$$\begin{aligned} &\langle\,\hat{\mathcal{Z}}_1,\hat{\mathcal{Z}}_2\,\rangle \\ &= \widehat{\deg}(\mathcal{Z}_1\cdot\mathcal{Z}_2) + \frac{1}{2}\int_{\mathcal{M}(\mathbb{C})} g(\hat{\mathcal{Z}}_1) * g(\hat{\mathcal{Z}}_2) \\ &= \widehat{\deg}(\mathcal{Z}_1\cdot\mathcal{Z}_2) + \frac{1}{2}\int_{\mathcal{M}(\mathbb{C})} g(\hat{\mathcal{Z}}_1)\,\delta_{\mathcal{Z}_2} + \frac{1}{2}\int_{\mathcal{M}(\mathbb{C})} g(\hat{\mathcal{Z}}_2)\,\omega(\hat{\mathcal{Z}}_1), \end{aligned} \tag{4.1.13}$$

from [1] (5.8), is valid when the supports $|\mathcal{Z}_1|$ and $|\mathcal{Z}_2|$ are disjoint on the generic fiber; see Section 2.5. In particular, using (4.1.13), we have

$$\langle\,\hat{\mathcal{Z}},1\!\!1\,\rangle = \frac{1}{2}\deg_{\mathbb{Q}}(\hat{\mathcal{Z}}), \tag{4.1.14}$$

so that

$$\langle \hat{\omega}, \mathbb{1} \rangle = \frac{1}{2} \deg_{\mathbb{Q}}(\omega) \qquad \text{and} \qquad \langle \mathbb{1}, \mathbb{1} \rangle = 0. \tag{4.1.15}$$

For $f \in A^0(\mathcal{M}_{\mathbb{R}})_0$, we have

$$\langle \hat{\omega}, a(f) \rangle = \langle \mathbb{1}, a(f) \rangle = 0, \tag{4.1.16}$$

since, via (4.1.13),

$$\langle \hat{\omega}, a(f) \rangle = \frac{1}{2} \int_{\mathcal{M}(\mathbb{C})} f \cdot c_1(\hat{\omega}) = 0. \tag{4.1.17}$$

Also

$$\langle a(f_1), a(f_2) \rangle = \frac{1}{2} \int_{\mathcal{M}(\mathbb{C})} f_2 \cdot dd^c f_1. \tag{4.1.18}$$

Again using (4.1.13), on Vert we have

$$\langle \hat{\omega}, (Y_p, 0) \rangle = \deg(\omega | Y_p) \cdot \log(p), \tag{4.1.19}$$

where $\deg(\omega|Y_p)$ is the degree of the restriction of the line bundle ω to Y_p. For fixed p, and denoting by m_i the multiplicities of the irreducible components Y_i of $\mathcal{M}_p$,

$$\sum_i m_i \cdot \langle \hat{\omega}, (Y_{p,i}, 0) \rangle = \langle \hat{\omega}, (\mathcal{M}_p, 0) \rangle = \deg_{\mathbb{Q}} \omega \cdot \log(p), \tag{4.1.20}$$

and

$$\langle \mathbb{1}, \text{Vert} \rangle = 0. \tag{4.1.21}$$

Let

$$\text{MW} = \text{MW}(\mathcal{M}) := \text{Jac}(M)(\mathbb{Q}) \otimes_{\mathbb{Z}} \mathbb{R} \tag{4.1.22}$$

be the Mordell-Weil space of the generic fiber $\mathcal{M}_{\mathbb{Q}}$. Here $\text{Jac}(M)$ denotes the neutral component of the Picard variety of M.

Definition 4.1.1. Let

$$\widetilde{\text{MW}} := \Big(\mathbb{R}\,\hat{\omega} \oplus \text{Vert} \oplus a(A^0(\mathcal{M}_{\mathbb{R}})_0) \Big)^{\perp} \quad \subset \quad \widehat{\text{CH}}^1(\mathcal{M})$$

be the orthogonal complement of $\mathbb{R}\,\hat{\omega} \oplus \text{Vert} \oplus a(A^0(\mathcal{M}_{\mathbb{R}})_0)$ with respect to the height pairing.

The following result is well known [8], [4], [1].

Proposition 4.1.2. *(i)*

$$\widehat{\mathrm{CH}}^1(\mathcal{M}) = \widetilde{\mathrm{MW}} \oplus (\,\mathbb{R}\hat{\omega} \oplus \mathrm{Vert}\,) \oplus a(A^0(\mathcal{M}_{\mathbb{R}})_0),$$

and the three summands are orthogonal with respect to the height pairing.
(ii) The restriction $\mathrm{res}_{\mathbb{Q}}$ *to the generic fiber induces an isomorphism*

$$\mathrm{res}_{\mathbb{Q}} : \widetilde{\mathrm{MW}} \xrightarrow{\sim} \mathrm{MW}$$

which is an isometry for the Gillet-Soulé height pairing on $\widetilde{\mathrm{MW}}$ *and the negative of the Neron-Tate height pairing on* MW.

Remark 4.1.3. (i) Note that the finite dimensional real vector space

$$\mathrm{CH}^1(\mathcal{M}, \mu_1) = \widetilde{\mathrm{MW}} \oplus \mathbb{R}\hat{\omega} \oplus \mathrm{Vert} \tag{4.1.23}$$

is the Arakelov Chow group with respect to the normalized volume form $\mu_1 = c_1(\hat{\omega})/\deg(\omega)$, except that we have taken real coefficients. Thus we have the decomposition

$$\widehat{\mathrm{CH}}^1(\mathcal{M}) = \mathrm{CH}^1(\mathcal{M}, \mu_1) \oplus a(A^0(\mathcal{M}_{\mathbb{R}})_0). \tag{4.1.24}$$

(ii) If Z is a 0-cycle of degree zero on the generic fiber $\mathcal{M}_{\mathbb{Q}}$, let $\underline{Z}$ be a divisor with rational coefficients on $\mathcal{M}$ with $\underline{Z}_{\mathbb{Q}} = Z$ and with $\underline{Z} \cdot Y_p = 0$ for all irreducible components Y_p of closed fibers $\mathcal{M}_p$. Such an extension of Z is unique up to the addition of a finite linear combination of $\mathcal{M}_p$'s. Let $\mathfrak{g}_Z$ be the μ_1-admissible antiharmonic Green function for Z. Then

$$\hat{Z} = (\underline{Z}, \mathfrak{g}_Z) \in \widehat{\mathrm{CH}}^1(\mathcal{M}) \tag{4.1.25}$$

lies in

$$\Big(\,\mathrm{Vert} \oplus a(A^0(\mathcal{M}_{\mathbb{R}})_0)\,\Big)^{\perp}. \tag{4.1.26}$$

There is a real scalar $\kappa \in \mathbb{R}$ such that

$$\langle\, \hat{\omega}, \hat{Z} + a(\kappa)\,\rangle = 0. \tag{4.1.27}$$

Once the extension $\underline{Z}$ has been chosen, then

$$\kappa = -2\,\langle\, \hat{\omega}, \hat{Z}\,\rangle. \tag{4.1.28}$$

The resulting class

$$\tilde{Z} := \hat{Z} + a(\kappa) \in \widetilde{\mathrm{MW}} \tag{4.1.29}$$

is the preimage in $\widetilde{\mathrm{MW}}$ of the class of Z in MW. It is independent of the choice of $\underline{Z}$.

We can use the height pairing to write the components in the decomposition (i) in Proposition 4.1.2 more explicitly.

For each prime p for which the fiber $\mathcal{M}_p$ is not irreducible, choose $Y_{p,i}$, $i = 1, \dots, r_p$ such that the images $\overline{Y}_{p,i}$ are a basis of $\overline{\mathrm{Vert}}_p$. Recall that $\mathbb{R}\mathcal{M}_p$ is the radical of the intersection form on Vert_p. Let $\overline{Y}^{\vee}_{p,i}$ be a dual basis for $\overline{\mathrm{Vert}}_p$ with respect to the intersection form and let $Y^{\vee}_{p,i}$ be the preimage of $\overline{Y}^{\vee}_{p,i}$ in Vert_p such that

$$\langle \hat{\omega}, (Y^{\vee}_{p,i}, 0) \rangle = 0. \tag{4.1.30}$$

For convenience, we write

$$y_{p,i} = (Y_{p,i}, 0), \qquad y^{\vee}_{p,i} = (Y^{\vee}_{p,i}, 0) \in \widehat{\mathrm{CH}}^1(\mathcal{M}), \tag{4.1.31}$$

and

$$\hat{\omega}_1 = \deg(\omega)^{-1}\, \hat{\omega}. \tag{4.1.32}$$

Then, if $\hat{z}$ is any class in $\mathbb{R}\hat{\omega} \oplus \mathrm{Vert}$, we have the decomposition

$$\hat{z} = \deg_{\mathbb{Q}}(\hat{z}) \cdot \hat{\omega}_1 + \sum_p \sum_{i=1}^{r_p} \langle \hat{z}, y_{p,i} \rangle \cdot y^{\vee}_{p,i} + 2\,\kappa(\hat{z}) \cdot \mathbb{1}, \tag{4.1.33}$$

where $\deg_{\mathbb{Q}}(\hat{z}) = 2 \langle \hat{z}, \mathbb{1} \rangle$ and

$$\kappa(\hat{z}) = \langle \hat{z}, \hat{\omega}_1 \rangle - \deg_{\mathbb{Q}}(\hat{z}) \langle \hat{\omega}_1, \hat{\omega}_1 \rangle. \tag{4.1.34}$$

Next consider the archimedean component. Suppose that there is a 'uniformization'

$$\mathcal{M}(\mathbb{C}) \simeq [\Gamma \backslash \mathfrak{H}], \tag{4.1.35}$$

where $\mathfrak{H}$ is the upper half plane and $\Gamma \subset SL_2(\mathbb{R})$ is a Fuchsian group. If Γ has elements of finite order, we understand the quotient as an orbifold.[1] In addition, we assume that $\mu_1 = \mathrm{vol}(\mathcal{M}(\mathbb{C}), \mu)^{-1}\, \mu$ where $\mu = (2\pi)^{-1}\, y^{-2}\, dx \wedge dy$ is the hyperbolic volume form on $\mathfrak{H}$. Recall that, for $f \in A^0(\mathcal{M}(\mathbb{C}))$,

$$dd^c f = \frac{1}{2}\, \Delta f \cdot \mu \tag{4.1.36}$$

[1] This explains the quotation marks on the term 'uniformization'.

where Δ is the hyperbolic Laplacian. The space $A^0(\mathcal{M}_{\mathbb{R}})$ is spanned by eigenfunctions of the Laplacian f_λ satisfying

$$\Delta f_\lambda + \lambda\, f_\lambda = 0, \tag{4.1.37}$$

with eigenvalues $0 = \lambda_0 < \lambda_1 \le \lambda_2 \le \dots$ and orthonormal with respect to μ. In particular, $f_0 = \mathrm{vol}(\mathcal{M}(\mathbb{C}), \mu)^{-\frac{1}{2}} \cdot \mathbb{1}$. For the height pairing, we have, by (4.1.18),

$$\langle\, a(f_{\lambda_i}), a(f_{\lambda_j})\,\rangle = -\frac{1}{4}\int_{\mathcal{M}(\mathbb{C})} f_{\lambda_i}\cdot \lambda_j f_{\lambda_j}\, \mu = -\frac{1}{4}\,\lambda_j \delta_{ij}. \tag{4.1.38}$$

In summary,

Proposition 4.1.4. *Any $\hat{z} \in \widehat{\mathrm{CH}}^1(\mathcal{M})$ has a unique decomposition:*

$$\hat{z} = \hat{z}_{\mathrm{MW}} + \frac{\deg_{\mathbb{Q}}(\hat{z})}{\deg_{\mathbb{Q}}(\hat{\omega})}\cdot\hat{\omega} + \sum_p \sum_{i=1}^{r_p} \langle\, \hat{z}, y_{p,i}\,\rangle \cdot y_{p,i}^{\vee}$$
$$+ 2\,\kappa(\hat{z})\cdot \mathbb{1} - 4\sum_{\lambda>0} \lambda^{-1}\,\langle\, \hat{z}, a(f_\lambda)\,\rangle \cdot a(f_\lambda).$$

where

$$\kappa(\hat{z}) = \langle\, \hat{z}, \hat{\omega}_1\,\rangle - \deg_{\mathbb{Q}}(\hat{z})\,\langle\, \hat{\omega}_1, \hat{\omega}_1\,\rangle,$$

and $\hat{z}_{\mathrm{MW}} \in \widetilde{\mathrm{MW}}$.

4.2 THE ARITHMETIC THETA FUNCTION

In this section, we consider the case of the arithmetic surface $\mathcal{M}$ attached to an indefinite division quaternion algebra B over $\mathbb{Q}$, and we define a generating function $\hat{\phi}_1(\tau)$ valued in $\widehat{\mathrm{CH}}^1(\mathcal{M})$. We use the setup and notation of Chapter 3. In particular, $D(B) > 1$ is the product of the primes p at which B is ramified and $\mathcal{M}$ has good reduction for $p \nmid D(B)$. We let $\hat{\omega}$ be the Hodge line bundle on $\mathcal{M}$ with metric normalized as in Section 3.3.

For each $t \in \mathbb{Z}$ and positive real number v, there is a class

$$\widehat{\mathcal{Z}}(t, v) \in \widehat{\mathrm{CH}}^1(\mathcal{M}), \tag{4.2.1}$$

defined in Section 3.5; see also [16]. For example, if $t > 0$, then

$$\widehat{\mathcal{Z}}(t, v) = (\mathcal{Z}(t), \Xi(t, v)), \tag{4.2.2}$$

where $\mathcal{Z}(t)$ is defined in terms of special endomorphisms and $\Xi(t, v)$ is the Green function given in Corollary 3.5.2. On the other hand, for $t < 0$,

$$\widehat{\mathcal{Z}}(t, v) = (0, \Xi(t, v)), \tag{4.2.3}$$

for a smooth function $\Xi(t, v)$. Finally,

$$\widehat{\mathcal{Z}}(0, v) = -\hat{\omega} - a(\log(v)) + a(\mathbf{c}), \tag{4.2.4}$$

where $\mathbf{c}$ is the real constant determined by the identity

(4.2.5)

$$\frac{1}{2}\deg_{\mathbb{Q}}(\hat{\omega})\cdot\mathbf{c} = \langle\,\hat{\omega}, \hat{\omega}\,\rangle - \zeta_D(-1)\left[2\frac{\zeta'(-1)}{\zeta(-1)} + 1 - 2C - \sum_{p|D(B)} \frac{p\log(p)}{p-1}\right].$$

In Chapter 7, we will prove that, in fact, $\mathbf{c} = -\log D(B)$.

We form the generating series

$$\widehat{\phi}_1 := \sum_t \widehat{\mathcal{Z}}(t, v)\, q^t \quad \in \widehat{\mathrm{CH}}^1(\mathcal{M})[[q]]. \tag{4.2.6}$$

As explained in [12] and [14], we refer to this series as an arithmetic theta function. The main result of this chapter is the following:

Theorem A. *The series $\widehat{\phi}_1$ is the q-expansion of a nonholomorphic elliptic modular form of weight $\frac{3}{2}$ for the subgroup $\Gamma' = \Gamma_0(4D(B)_o)$ of $\mathrm{SL}_2(\mathbb{Z})$ valued in $\widehat{\mathrm{CH}}^1(\mathcal{M}) \otimes_{\mathbb{R}} \mathbb{C}$.*

This means that there is a smooth function ϕ_{Ar} of $\tau \in \mathfrak{H}$, valued in the finite dimensional complex vector space $\mathrm{CH}^1(\mathcal{M}, \mu_1)_{\mathbb{C}}$, and a smooth function $\phi_{\mathrm{an}}(\tau, z)$ on $\mathfrak{H} \times \mathcal{M}(\mathbb{C})$, with

$$\int_{\mathcal{M}(\mathbb{C})} \phi_{\mathrm{an}}(\tau, z)\, d\mu(z) = 0,$$

such that the sum $\phi(\tau) = \phi_{\mathrm{Ar}}(\tau) + \phi_{\mathrm{an}}(\tau, z)$ satisfies the usual transformation law for a modular form of weight $\frac{3}{2}$ for $\Gamma_0(4D(B)_o)$ and such that the q-expansion of $\phi(\tau)$ is the formal generating series $\widehat{\phi}_1$ defined above. Of course, the coefficients in the q-expansion of $\phi(\tau)$ are functions of v. We will abuse notation and write $\widehat{\phi}_1(\tau)$ for both the function $\phi(\tau)$ and for the q-expansion $\widehat{\phi}_1$.

The proof of Theorem A is based on the application of the decomposition of Proposition 4.1.2 and Proposition 4.1.4 to $\widehat{\phi}_1$. This gives rise to the component functions

$$\phi_{\deg} = \deg_{\mathbb{Q}}(\widehat{\phi}_1) = \sum_t \deg_{\mathbb{Q}}(\mathcal{Z}(t))\, q^t, \tag{4.2.7}$$

(4.2.8)
$$\langle \widehat{\phi}_1, \hat{\omega} \rangle = \sum_t \langle \widehat{\mathcal{Z}}(t,v), \hat{\omega} \rangle \, q^t,$$

(4.2.9)
$$\langle \widehat{\phi}_1, y_{p,i} \rangle = \sum_t \langle \widehat{\mathcal{Z}}(t,v), y_{p,i} \rangle \, q^t,$$

(4.2.10)
$$\langle \widehat{\phi}_1, a(f_\lambda) \rangle = \sum_t \langle \widehat{\mathcal{Z}}(t,v), a(f_\lambda) \rangle \, q^t,$$

and, finally, the component in $\widetilde{\mathrm{MW}}$, or, equivalently, in MW,

(4.2.11)
$$\phi_{\mathrm{MW}} = \mathrm{res}_{\mathbb{Q}}(\widehat{\phi}_1(\tau) - \phi_{\mathrm{deg}}(\tau) \cdot \hat{\omega}_1) \in \mathrm{MW} \otimes \mathbb{C}.$$

It will be shown that each of these component functions is modular of weight $\frac{3}{2}$ for Γ', and hence so is $\widehat{\phi}_1$.

The modularity of (4.2.7) and (4.2.8) is proved in [16]. More precisely, let $\mathcal{E}_1(\tau, s; D(B))$ be the Eisenstein series of weight $\frac{3}{2}$ defined in Section 6 of [16]. Then, Proposition 7.1 of [16] asserts that

(4.2.12)
$$\phi_{\mathrm{deg}}(\tau) = \mathcal{E}_1(\tau, \frac{1}{2}; D(B)),$$

and hence gives the modularity of (4.2.7). Here and elsewhere, the meaning of this equation is that the q-expansion of $\mathcal{E}_1(\tau, \frac{1}{2}; D(B))$ is given by ϕ_{deg}.

The main result, Theorem 7.2 of [16], asserts that

(4.2.13)
$$\langle \widehat{\phi}_1(\tau), \hat{\omega} \rangle = \mathcal{E}_1'(\tau, \frac{1}{2}; D(B)).$$

This gives the modularity of (4.2.8). The vertical, archimedean, and Mordell-Weil components will be handled in the following three sections.

4.3 THE VERTICAL COMPONENT: DEFINITE THETA FUNCTIONS

In this section, for a prime $p \mid D(B)$ and a vertical element $y_{p,i} \in \widehat{\mathrm{CH}}^1(\mathcal{M})$, as in (4.1.31), we prove that the component function $\langle \widehat{\phi}_1(\tau), y_{p,i} \rangle$ is a modular form of weight $\frac{3}{2}$. Our tool will the p-adic uniformization of $\mathcal{M}$ and of the special divisors $\mathcal{Z}(t)$ recalled in Section 3.2.

Recall that $B' = B^{(p)}$ denotes the definite quaternion algebra over $\mathbb{Q}$ whose local invariants coincide with those of B at all places $v \neq p, \infty$, and $V' = \{x \in B' \mid \mathrm{tr}(x) = 0\}$. For $H' = (B')^\times$, we have fixed isomorphisms

(4.3.1)
$$H'(\mathbb{Q}_p) \simeq \mathrm{GL}_2(\mathbb{Q}_p) \text{ and } H'(\mathbb{A}_f^p) \simeq H(\mathbb{A}_f^p).$$

We have $K = \widehat{O}_B^\times = K^p K_p$ with $K^p \subset H(\mathbb{A}_f^p)$ and $K_p \subset H(\mathbb{Q}_p)$, and also $K' = K'^p K'_p$, where $K'^p = K^p$ under the isomorphism (4.3.1), and $K'_p = \mathrm{GL}_2(\mathbb{Z}_p)$. Let $\hat{\mathcal{M}}_W$ be the base change to $W = W(\bar{\mathbb{F}}_p)$ of the formal completion of $\mathcal{M}$ along its special fiber at p. We also denote by $\hat{\Omega}_W$ the base change of the Drinfeld upper half space to W. For convenience of notation, we introduce $\hat{\Omega}_W^\bullet = \hat{\Omega}_W \times \mathbb{Z}$, the disjoint sum of copies of $\hat{\Omega}_W$ parametrized by $\mathbb{Z}$. We then have

$$\hat{\mathcal{M}}_W \simeq H'(\mathbb{Q}) \setminus [\, \hat{\Omega}_W^\bullet \times H'(\mathbb{A}_f^p)/K'^p \,]. \tag{4.3.2}$$

From the uniformization (4.3.2), one obtains the following well-known description of the irreducible components over $\overline{\mathbb{F}}_p$ of the special fiber $\mathcal{M}_p$ of $\mathcal{M}$. Note that these correspond to the irreducible components of the special fiber of $\hat{\mathcal{M}}_W$. Recall that $\gamma' \in H'(\mathbb{Q})$ acts on $\hat{\Omega}_W^\bullet \times H(\mathbb{A}_f^p)/K^p$ by

$$\gamma' : [\,\xi\,]^i \times hK \mapsto [\,\gamma\,\xi\,]^{i+\mathrm{ord}_p(\nu(\gamma))} \times \gamma hK, \tag{4.3.3}$$

where $[\xi]^i$ is the image in $\hat{\Omega}_W^i$ of $\xi \in \hat{\Omega}_W$. The components of the special fiber $\hat{\Omega}_p$ of $\hat{\Omega}_W$ are projective lines $\mathbb{P}_{[\Lambda]}$ indexed by the vertices $[\Lambda]$ of the building $\mathcal{B}$ for $\mathrm{PGL}_2(\mathbb{Q}_p)$, where $[\Lambda]$ is the homothety class of a $\mathbb{Z}_p$-lattice Λ in $\mathbb{Q}_p^2$. Thus, $\mathrm{GL}_2(\mathbb{Q}_p)$ acts transitively on the components of $\hat{\Omega}_W^\bullet$, and, if $\Lambda_0 = \mathbb{Z}_p^2$ is the standard lattice, the stabilizer in $\mathrm{GL}_2(\mathbb{Q}_p)$ of $[\,\mathbb{P}_{[\Lambda_0]}\,]^0$ is $K'_p = \mathrm{GL}_2(\mathbb{Z}_p)$.

Proposition 4.3.1. *There is a bijection*

$$H'(\mathbb{Q})\backslash H'(\mathbb{A}_f)/K' \xrightarrow{\sim} \begin{pmatrix}\text{irreducible components}\\ \text{of } \mathcal{M}_p\end{pmatrix},$$

$$H'(\mathbb{Q})gK' \longmapsto H'(\mathbb{Q})\text{-orbit of} \quad g_p[\,\mathbb{P}_{[\Lambda_0]}\,]^0 \times g^p K^p.$$

In particular, the irreducible components of the special fiber $\mathcal{M}_p$ are indexed by the double cosets in the decomposition

$$H'(\mathbb{A}_f) = \coprod_j H'(\mathbb{Q})g_j K'.$$

We will frequently refer to the pair $[\Lambda]^i \times gK^p$ as the data associated to the component $[\mathbb{P}_{[\Lambda]}]^i \times gK^p$.

We next recall from [16] the p-adic uniformization of the cycles $\mathcal{Z}(t)$; cf. also Section 6.2. Fix $t \in \mathbb{Z}_{>0}$ and let $\mathcal{Z}(t)$ be the corresponding divisor on $\mathcal{M}$; see Section 3.4. Loosely speaking, $\mathcal{Z}(t)$ is the locus where the O_B-abelian variety is equipped with a special endomorphism of square $-t$. Let

$$\mathcal{C}(t) = \mathcal{Z}(t) \times_{\mathrm{Spec}\ \mathbb{Z}} \mathrm{Spec}\ \mathbb{Z}_{(p)}, \tag{4.3.4}$$

and let

$$\hat{\mathcal{C}} = \widehat{\mathcal{C}(t)} \times_{\mathrm{Spf}\,\mathbb{Z}_p} \mathrm{Spf}\, W, \tag{4.3.5}$$

be the cycle in $\hat{\mathcal{M}}_W$ determined by $\mathcal{C}(t)$. Let

$$V_t'(\mathbb{Q}) = \{\, x \in V'(\mathbb{Q}) \mid Q(x) = -x^2 = t \,\}.$$

Then, (8.17) of [15] gives a p-adic uniformization of $\hat{\mathcal{C}}$:

$$\begin{array}{ccc} \hat{\mathcal{C}} & \hookrightarrow & H'(\mathbb{Q})\backslash\Big(V_t'(\mathbb{Q}) \times \hat{\Omega}^\bullet_W \times H(\mathbb{A}_f^p)/K^p\Big) \\ \downarrow & & \downarrow \\ \hat{\mathcal{M}}_W & \xrightarrow{\sim} & H'(\mathbb{Q})\backslash\Big(\hat{\Omega}^\bullet_W \times H(\mathbb{A}_f^p)/K^p\Big), \end{array} \tag{4.3.6}$$

where the image of $\hat{\mathcal{C}}$ in the upper right corner is the set of triples

$$\left\{ (x, (X,\rho), gK^p) \,\middle|\, \begin{array}{l} \text{(i) } g^{-1}xg \in V(\mathbb{A}_f^p) \cap \widehat{O}_B^p, \text{ and} \\ \text{(ii) for } j = j(x),\ (X,\rho) \in Z^\bullet(j) \end{array} \right\}. \tag{4.3.7}$$

Here we have identified $\hat{\Omega}^\bullet_W$ with the formal moduli space of quasi-isogenies $\rho : \mathbb{X} \longrightarrow X$ of special formal O_B-modules, where $\mathbb{X}$ is the base point and where we have fixed an identification of $V'(\mathbb{Q}_p)$ with the traceless elements in $\mathrm{End}_{O_B}(\mathbb{X})$. Then $j(x)$ is the special endomorphism of the p-divisible group $\mathbb{X}$, determined by x, and $Z^\bullet(j)$ is the corresponding cycle in $\hat{\Omega}^\bullet_W$, as defined in [15] (the locus of ρ's where the isogeny $j(x)$ of $\mathbb{X}$ extends to an endomorphism of X).

Using the p-adic uniformization just described, we compute the intersection number $(\mathcal{C}(t), Y_p)$ for an irreducible component Y_p of the special fiber $\mathcal{M}_p$.

Proposition 4.3.2. *Write $t = -\varepsilon p^\alpha$, with $\varepsilon \in \mathbb{Z}_p^\times$ and $\alpha \in \mathbb{Z}$. Then, for an irreducible component Y_p of $\mathcal{M}_p$ determined by data $[\Lambda]^i \times gK^p$, as above,*

$$(\mathcal{C}(t), Y_p) = \sum_{\substack{x \in V'(\mathbb{Q}) \\ Q(x)=t}} \varphi^p(g^{-1}x)\, \mu_{[\Lambda]}(x).$$

Here φ^p denotes the characteristic function of $V(\mathbb{A}_f^p) \cap \widehat{O}_B^p$ and the multiplicity $\mu_{[\Lambda]}(x)$ is given as follows:
Let $j = j(x)$ be the endomorphism of $\mathbb{Q}_p^2$ associated to x. Let $e \geq f$, with $e, f \in \mathbb{Z}$ be the elementary divisors of the lattice pair $(\Lambda, j(\Lambda))$. Note that

$e+f=\alpha=\mathrm{ord}_p(\det(j))$, $f\le[\frac{\alpha}{2}]$, and that $r=\frac{1}{2}(e-f)=d([\Lambda],\mathcal{B}^j)$ is the distance from $[\Lambda]$ to the fixed point set $\mathcal{B}^j$ of j on $\mathcal{B}$. Then, for $\alpha>0$,

$$\mu_{[\Lambda]}(x)=\begin{cases}1-p & \text{if } f>0,\\ 1 & \text{if } f=0,\\ 0 & \text{if } f<0.\end{cases}$$

For $\alpha=0$, and $p\neq 2$,

$$\mu_{[\Lambda]}(x)=\begin{cases}1-\chi_p(\varepsilon) & \text{if } f=0,\\ 0 & \text{if } f<0.\end{cases}$$

For $\alpha=0$, and $p=2$,

$$\mu_{[\Lambda]}(x)=\begin{cases}1-p & \text{if } f=0 \text{ and } (1+j)(\Lambda)\subset 2\Lambda,\\ 1 & \text{if } f=0 \text{ and } (1+j)(\Lambda)\not\subset 2\Lambda,\\ 0 & \text{otherwise.}\end{cases}$$

Here $\chi_p(\varepsilon)=(\varepsilon,p)_p$ where $(\ ,\ )_p$ is the quadratic Hilbert symbol for $\mathbb{Q}_p$.

Proof. Let $\hat{Y}$ be the irreducible component of $\hat{\mathcal{M}}_W$ corresponding to Y_p. Fixing the preimage $[\mathbb{P}_{[\Lambda]}]^i\times gK^p$ of $\hat{Y}$, we must calculate the total intersection number of this curve with the full preimage of $\hat{C}$ in $\hat{\Omega}^\bullet_W\times H(\mathbb{A}^p_f)/K^p$, as described by the incidence set (4.3.7) for the upper right corner in (4.3.6) above. First, for a given $(x,(X,\rho),g_0K^p)$ to contribute, we must have $g_0K^p=gK^p$ and the height of the quasi-isogeny ρ must be the given i. The first incidence condition becomes $\varphi^p(g^{-1}x)\neq 0$. It remains to calculate the intersection number of $\mathbb{P}_{[\Lambda]}$ and $Z(j)$, $j=j(x)$ on $\hat{\Omega}_W$. We write $Z(j)=Z(j)^h+Z(j)^v$ as the sum of horizontal and vertical parts. For $p\neq 2$, Lemma 4.9 of [15] gives $(Z(j)^h,\mathbb{P}_{[\Lambda]})$, and Lemma 6.2 of [15] gives $(Z(j)^v,\mathbb{P}_{[\Lambda]})$, taking into account the relation between the elementary divisors e,f and the distance r, described above; see also Lemma 2.4 of [15]. This yields the result in the case $p\neq 2$.

For $p=2$, we use the results of the appendix to Chapter 6. There the cycle $Z(j)$ is described in terms of 'cycle data' (S,μ,Z^h); see Section 6A.2 of the appendix to Chapter 6. Here S is a subset of the building (parametrizing the 'central vertical components'), the integer μ is the common multiplicity of the central vertical components, and Z^h denotes the horizontal part of the divisor $Z(j)$. Remark (I) right before (6A.4.12) gives

(4.3.8)

$$(Z(j)^v, \mathbb{P}_{[\Lambda]}) = \begin{cases} 1-p & \text{if } 1 \leq d([\Lambda], S) \leq \mu - 1, \\ & \qquad \text{i.e., } [\Lambda] \text{ is regular in } Z(j), \\ 1 & \text{if } d([\Lambda], S) = \mu, \\ 0 & \text{if } d([\Lambda], S) > \mu, \end{cases}$$

provided $\mu > 0$. If $\mu > 0$ and $d([\Lambda], S) = 0$, i.e., $[\Lambda] \in S$, then

$$(Z(j)^v, \mathbb{P}_{[\Lambda]}) = \chi_2(j) - p, \tag{4.3.9}$$

where $\chi_2(j)$ is equal to 1, -1, or 0, depending on whether the extension of $\mathbb{Q}_2$ generated by j is split, unramified, or ramified. Finally, for $\mu = 0$, we have $(Z(j)^v, \mathbb{P}_{[\Lambda]}) = 0$.

As for the intersection number $(Z(j)^h, \mathbb{P}_{[\Lambda]})$, we have, by a case by case inspection of the cases (**1**) to (**4**) in Section A.2 of the appendix,

$$(Z(j)^h, \mathbb{P}_{[\Lambda]}) = \begin{cases} 1-\chi_2(j) & \text{if } [\Lambda] \in S, \\ 0 & \text{otherwise.} \end{cases} \tag{4.3.10}$$

Summing, we obtain

$$(Z(j), \mathbb{P}_{[\Lambda]}) = \begin{cases} 1-p & \text{if } \mu \geq 1 \text{ and } d([\Lambda], S) \leq \mu - 1, \\ 1 & \text{if } \mu \geq 1 \text{ and } d([\Lambda], S) = \mu, \\ 1-\chi_2(j) & \text{if } \mu = d([\Lambda], S) = 0, \\ 0 & \text{in all other cases.} \end{cases} \tag{4.3.11}$$

We now note, by checking case by case in Section A.2 of the appendix, that

$$d([\Lambda], S) \leq \mu \iff d([\Lambda], \mathcal{B}^j) \leq \frac{\alpha}{2} \iff f \geq 0. \tag{4.3.12}$$

Similarly, $d([\Lambda], S) \leq \mu - 1$ is equivalent to $f > 0$. When $\alpha > 0$, then either $\mu > 0$ and the assertion in the proposition follows from formula (4.3.11), or $\alpha = 1$ and $j^2 = \varepsilon \cdot 2$ (i.e., j^2 is of type (**4**) in the terminology of the appendix). In this case, $\mu = 0$, and $d([\Lambda], S) = 0$ is equivalent to $d([\Lambda], \mathcal{B}^j) = \frac{1}{2}$, i.e., $(e, f) = (1, 0)$, and formula (4.3.11) gives $\mu_{[\Lambda]}(x) = 1$. Now, in this case, the two vertices in S correspond to the lattices given by the ring of integers $\mathcal{O}$ in the ramified extension $\mathbb{Q}_2(j)$ and to $\pi\mathcal{O}$, where π denotes a uniformizer in $\mathcal{O}$. For either of them $(1 + j)(\Lambda) = \Lambda$. Hence the value for $\mu_{[\Lambda]}(x)$ obtained above confirms the statement in the proposition.

Suppose that $\alpha = 0$. Then, when ε is of type (**1**) or (**2**), we have $\mu = 1$ and formula (4.3.11) gives the value $1 - p$, 1, or 0, depending on whether

$d([\Lambda], S) = 0$, $d([\Lambda], S) = 1$, or $d([\Lambda], S) > 1$. The first case is characterized by $j(\Lambda) = \Lambda$, $(1+j)\Lambda \subset 2\Lambda$. The second case is characterized by $j(\Lambda) = \Lambda$, $(1+j)\Lambda \not\subset 2\Lambda$. This confirms the claim of the proposition in this case. When $\alpha = 0$ and ε is of type (**3**), we have $\mu = 0$ and formula (4.3.11) gives 1 if $[\Lambda] \in S$ and 0 otherwise. The first alternative is characterized by $j(\Lambda) = \Lambda$. Again the two lattices with $j(\Lambda) = \Lambda$ are given, up to scalar multiples, by the ring of integers $\mathcal{O}$ in the ramified extension $\mathbb{Q}_2(j)$ and by $\pi\mathcal{O}$. Hence $(1+j)\mathcal{O} = \pi\mathcal{O}$ so that $(1+j)(\Lambda) \not\subset 2\Lambda$, which again confirms the assertion of the proposition in this case. □

The multiplicity $\mu_{[\Lambda]}(x)$ can be expressed in terms of a Schwartz function on $V'(\mathbb{Q}_p)$ as follows. Let $\varphi_{[\Lambda]}$ be the characteristic function of the set

$$\{x \in V'(\mathbb{Q}_p) \mid j(\Lambda) \subset \Lambda, \ \text{ for } j = j(x)\}. \tag{4.3.13}$$

Note that, with the notation above,

$$\varphi_{[\Lambda]}(x) \neq 0 \iff f \geq 0, \qquad \varphi_{[\Lambda]}(p^{-1}x) \neq 0 \iff f > 0, \tag{4.3.14}$$

and $\varphi_{[\Lambda]}$ is invariant under the action of $K'_{[\Lambda]} = g_p GL_2(\mathbb{Z}_p) g_p^{-1}$ on $V'(\mathbb{Q}_p)$ by conjugation, where $\Lambda = g_p\Lambda_0$. Thus, if $x \in V'(\mathbb{Q}_p)$ with $Q(x) = t$, and excluding the case $\alpha = f = 0$, we can write

$$\mu_{[\Lambda]}(x) = \varphi_{[\Lambda]}(x) - p\,\varphi_{[\Lambda]}(p^{-1}x). \tag{4.3.15}$$

Also, let $\varphi^+_{[\Lambda]}$, $\varphi^-_{[\Lambda]}$, and $\varphi^0_{[\Lambda]}$ be the characteristic functions of the following sets,

$$\begin{aligned} \mathcal{X}^{\pm}_{\Lambda} &= \{x \in V'(\mathbb{Q}_p) \mid j(\Lambda) = \Lambda, \text{ and } \chi_p(\det(j)) = \pm 1\} \\ \mathcal{X}^{0}_{\Lambda} &= \{x \in V'(\mathbb{Q}_p) \mid j(\Lambda) = \Lambda, \ (1+j)(\Lambda) \subset 2\Lambda\}. \end{aligned} \tag{4.3.16}$$

These are again Schwartz functions on $V'(\mathbb{Q}_p)$, invariant under the action of $K'_{[\Lambda]}$ on $V'(\mathbb{Q}_p)$ by conjugation.

Then, we have,

Lemma 4.3.3. *(i) For all $x \in V'(\mathbb{Q}_p)$, the function $\mu_{[\Lambda]} \in S(V'(\mathbb{Q}_p))$ is given by*

$$\mu_{[\Lambda]}(x) = \begin{cases} \varphi_{[\Lambda]}(x) - p\,\varphi_{[\Lambda]}(p^{-1}x) - \varphi^+_{[\Lambda]}(x) + \varphi^-_{[\Lambda]}(x) & \text{if } p \neq 2, \\ \varphi_{[\Lambda]}(x) - p\,\varphi_{[\Lambda]}(p^{-1}x) - p\,\varphi^0_{[\Lambda]} & \text{if } p = 2. \end{cases}$$

(ii) If $\Lambda = g_p\Lambda_0$, then

$$\mu_{[\Lambda]}(x) = \mu_{[\Lambda_0]}(g_p^{-1}x).$$

Finally, we can consider the component function of (4.2.9):

$$\begin{aligned}\langle \widehat{\phi}_1(\tau), y_p \rangle &= \sum_t \langle \widehat{\mathcal{Z}}(t,v), y_p \rangle\, q^t, \\ &= -(\omega, Y_p)\log(p) + \log(p) \sum_{t>0} (\mathcal{Z}(t), Y_p)\, q^t\end{aligned} \tag{4.3.17}$$

where, for the constant term, we recall (4.2.4) and (4.1.17). Cancelling the $\log(p)$, we define

$$\phi_{\text{Vert}}(\tau; Y_p) := \sum_{t\geq 0} (\mathcal{C}(t), Y_p)\, q^t, \tag{4.3.18}$$

where we have set $\mathcal{C}(0) = -\omega$.

Theorem 4.3.4. *Let Y_p be the component of $\mathcal{M}_p$ corresponding to the double coset $H'(\mathbb{Q})gK'$ under the bijection of Proposition 4.3.1. Let*

$$\varphi'_p = \mu_{[\Lambda_0]} \in S(V'(\mathbb{Q}_p))$$

be the Schwartz function described in Lemma 4.3.3, and let $\varphi' = \varphi'_p\varphi^p \in S(V'(\mathbb{A}_f))$, where φ^p is the characteristic function of $V'(\mathbb{A}_f^p) \cap \widehat{O}_B^p$. Then

$$\phi_{\text{Vert}}(\tau; Y_p) = \theta(\tau, g; \varphi') := \sum_{x\in V'(\mathbb{Q})} \varphi'(g^{-1}x)\, q^{Q(x)}$$

is the theta function of weight $\frac{3}{2}$ for the data φ' and $g \in H'(\mathbb{A}_f)$ for the positive definite quadratic space V'.

Remark 4.3.5. It remains to check that the theta function $\theta(\tau, g; \varphi')$ is a modular form for $\Gamma_0(4D(B)_o)$, as claimed in Theorem A. Equivalently, we must determine the compact open subgroup of the metaplectic extension of $\mathrm{SL}_2(\mathbb{Q}_p)$ which fixes the Schwartz function φ'_p. This will be done in Section 8.5, where the necessary information about the Weil representation is reviewed.

Proof. For $t > 0$, Proposition 4.3.2 and (ii) of Lemma 4.3.3 give

$$(\mathcal{C}(t), Y_p) = \sum_{\substack{x\in V'(\mathbb{Q}) \\ Q(x)=t}} \varphi'(g^{-1}x). \tag{4.3.19}$$

On the other hand, Proposition 11.1 of [16] gives

$$(\omega, Y_p) = -(p-1) = \varphi'(0). \tag{4.3.20}$$

□

Corollary 4.3.6. *The component functions $\langle\, \widehat{\phi}_1(\tau), y_{p,i} \,\rangle$ of (4.2.9) are holomorphic theta functions of weight $\frac{3}{2}$ attached to $V' = V^{(p)}$.*

Remark 4.3.7. (i) The special fiber of $\mathcal{M}$ is reduced. Hence we have

$$\sum_i \langle\, \widehat{\phi}_1(\tau), (Y_{p,i}, 0) \,\rangle = \langle\, \widehat{\phi}_1(\tau), \mathcal{M}_p \,\rangle$$

$$(4.3.21) \qquad = 2\log(p) \cdot \langle\, \widehat{\phi}_1(\tau), \mathbb{1} \,\rangle$$

$$= \log(p) \cdot \phi_{\mathrm{deg}}(\tau)$$

$$= \log(p) \cdot \mathcal{E}(\tau, \frac{1}{2}, D(B)).$$

On the other hand, by the Siegel-Weil formula for the anisotropic ternary space $V^{(p)}$, the sum of the theta functions is the Eisenstein series,

$$(4.3.22) \qquad \sum_i \theta(\tau, g; \varphi') = \mathcal{E}_1(\tau, \frac{1}{2}, D(B)).$$

The coincidence of the two formulas is an example of the 'matching' identity for the Siegel-Weil formula described in [13]. Up to the factor of $\log(p)$, the modular forms $\langle\, \widehat{\phi}_1(\tau), (Y_{p,i}, 0) \,\rangle$ are the theta series of weight $\frac{3}{2}$ for the classes, as i varies, in the K'-genus for the positive definite ternary space $V' = V^{(p)}$.

(ii) Note that the key fact in this proof is that the degree $-(p-1)$ of the restriction of ω to a line $\mathbb{P}_{[\Lambda]}$ is equal to the intersection number $(\mathbb{P}_{[\Lambda]}, \mathcal{C}(t))$, provided that the vertex $[\Lambda]$ is regular in $\mathcal{C}(t)$ in the sense of the appendix to Chapter 6. This equality seems quite mysterious.

(iii) Theorem 4.3.4 carries over to the situation where a level structure away from p is imposed, as in [15]. More precisely, let $K^p \subset H(\mathbb{A}_f^p)$ be *any* open compact subgroup, but keep $K_p = O_{B_p}^\times$, as before. Let $K = K^p.K_p$. Then we have the moduli space $\mathcal{A}_K$ over Spec $\mathbb{Z}_{(p)}$ as in [15], which again admits a p-adic uniformization. Let $\varphi^p \in S(V(\mathbb{A}_f^p))$ be the characteristic function of a K^p-invariant compact open subset of $V(\mathbb{A}_f^p)$. (This subset was denoted by ω in [15], but here we avoid this notation because it is already in use.) For $t \in \mathbb{Z}_{(p),>0}$ there is a cycle $\mathcal{C}(t, \varphi^p)$ on $\mathcal{A}_K$. Note the slight shift in notation from [15], where the quadratic form had the opposite sign and ω was written in place of φ^p. In the case of $\mathcal{M}$, i.e., when $K^p = \widehat{O}_B^{p,\times}$, the function φ^p is the characteristic function of $V(\mathbb{A}_f^p) \cap \widehat{O}_B$, and

$$(4.3.23) \qquad \mathcal{C}(t, \varphi^p) = \mathcal{Z}(t) \times_{\mathrm{Spec}\ \mathbb{Z}} \mathrm{Spec}\ \mathbb{Z}_{(p)}.$$

We can then form the generating series associated to an irreducible component Y_p of $(\mathcal{A}_K)_p$, as in (4.3.18). Let

$$\mathcal{C}(0, \varphi^p) = -\varphi^p(0) \cdot \omega, \tag{4.3.24}$$

and define the generating function

$$\phi_{\text{Vert}}(\tau, \varphi^p; Y_p) := \sum_{t \geq 0} (\mathcal{C}(t, \varphi^p), Y_p)\, q^t. \tag{4.3.25}$$

Then we can again identify this function with a theta function. Let Y_p correspond to the double coset $H'(\mathbb{Q})gK'$, and let as before $\varphi'_p = \mu_{[\Lambda_0]} \in S(V'(\mathbb{Q}_p))$ be the Schwartz function described in Lemma 4.3.3, and let $\varphi' = \varphi'_p\varphi^p \in S(V'(\mathbb{A}_f))$. Then

$$\phi_{\text{Vert}}(\tau, \varphi^p; Y_p) = \theta(\tau, g; \varphi') := \sum_{x \in V'(\mathbb{Q})} \varphi'(g^{-1}x)\, q^{Q(x)}. \tag{4.3.26}$$

(iv) In his Montreal article [7], Gross gave an analogue of the Gross-Zagier formula for the central value of the L-function, in the case of root number $+1$. This should have a natural interpretation here in terms of the geometry of components of the fiber $\mathcal{M}_p$.

4.4 THE ANALYTIC COMPONENT: MAASS FORMS

In this section, we consider the component functions $\langle\, \widehat{\phi}_1(\tau), a(f_\lambda)\,\rangle$ of (4.2.10) associated to eigenfunctions of the Laplacian. In this section we allow an arbitrary level structure $K \subset (B \otimes \mathbb{A}_f)^\times$, although for Theorem A only the case $K = \widehat{O}_B^\times$ is needed.

First we review the definition of the Green function of Section 3.5 and [11], where more details can be found. For $H = B^\times$, as in Chapter 3, and for any compact open subgroup $K \subset H(\mathbb{A}_f)$, the complex points of the associated Shimura curve M_K over $\mathbb{Q}$ are given by

$$M_K(\mathbb{C}) \simeq [\, H(\mathbb{Q})\backslash D \times H(\mathbb{A}_f)/K\,], \tag{4.4.1}$$

where the right side is, as usual, understood in the sense of stacks or orbifolds. Here D is as in (3.2.3). Recall that, for any K, the irreducible components of $M_K(\mathbb{C})$ are indexed by the double cosets

$$H(\mathbb{A}_f) = \coprod_j H(\mathbb{Q})^+ h_j K, \tag{4.4.2}$$

so that

$$M_K(\mathbb{C}) \simeq \coprod_j [\, \Gamma_j\backslash D^+\,], \tag{4.4.3}$$

where $\Gamma_j = H(\mathbb{Q})^+ \cap h_j K h_j^{-1}$. Here $H(\mathbb{Q})^+ = H(\mathbb{Q}) \cap H(\mathbb{R})^+$, where $H(\mathbb{R})^+$ is the identity component of $H(\mathbb{R}) \simeq \mathrm{GL}_2(\mathbb{R})$, and D^+ is a fixed connected component of D. Thus, in general, M_K is not geometrically connected, and the irreducible components are defined over a cyclotomic extension [20], [18].

For $t \in \mathbb{Q}_{>0}$ and for a weight function $\varphi \in S(V(\mathbb{A}_f))^K$, there is a divisor $Z(t,\varphi) = Z(t,\varphi;K)$ on M_K, which is rational over $\mathbb{Q}$ [11]. We suppress the K to lighten the notation. If $t \in \mathbb{Z}_{>0}$, $K = \widehat{O}_B^\times$, and φ is the characteristic function of $V(\mathbb{A}_f) \cap \widehat{O}_B$, then $Z(t,\varphi;K) = \mathcal{Z}(t)_{\mathbb{Q}}$ coincides with the generic fiber of the cycle $\mathcal{Z}(t)$ defined in Chapter 3. A Green function of logarithmic type for $Z(t,\varphi)$ is constructed as follows; see Section 11 of [11] for more details.

Recall that, for $z \in D$ and $x \in V(\mathbb{R})$,

$$\xi(x,z) = -\mathrm{Ei}(-2\pi R(x,z)), \tag{4.4.4}$$

where Ei is the exponential integral. For a fixed $x \neq 0$, this function is smooth on $D \setminus D_x$ and satisfies

$$dd^c\xi(x,z) + \delta_{D_x} = e^{2\pi Q(x)}\,\varphi^*_\infty(x,z)\cdot\mu, \tag{4.4.5}$$

where, by [13], Proposition 4.10, and [11],

$$\varphi^*_\infty(x,z) = \big[\,4\pi(\,R(x,z) + 2Q(x)\,) - 1\,\big]\cdot\varphi_\infty(x,z) \tag{4.4.6}$$

for

$$\varphi_\infty(x,z) := e^{-2\pi R(x,z)}\,e^{-2\pi Q(x)}, \tag{4.4.7}$$

the Gaussian defined by $z \in D$. Here, under an isomorphism $D \simeq \mathbb{P}^1(\mathbb{C}) \setminus \mathbb{P}^1(\mathbb{R})$,

$$\mu = \frac{1}{2\pi}\,\frac{i}{2}\,\frac{dz \wedge d\bar{z}}{y^2}. \tag{4.4.8}$$

For fixed $z \in D$, $\varphi^*_\infty(\cdot,z) \in S(V(\mathbb{R}))$ is a Schwartz function, and, for fixed x, $\xi(x,\cdot)$ is a Green function of logarithmic type for the point $D_x \subset D$, in the sense of Gillet-Soulé.

For $t > 0$, $\varphi \in S(V(\mathbb{A}_f))^K$, $v \in \mathbb{R}_{>0}$, and $[z,h] \in D \times H(\mathbb{A}_f)$, let

$$\Xi(v;t,\varphi)(z,h) = \sum_{\substack{x \in V(\mathbb{Q}) \\ Q(x)=t}} \xi(v^{\frac{1}{2}}x,z)\,\varphi(h^{-1}x). \tag{4.4.9}$$

This function is well defined on $M_K(\mathbb{C})$ with logarithmic singularities on $Z(t,\varphi)$ and satisfies the Green equation

$$dd^c\,\Xi(v;t,\varphi) + \delta_{Z(t,\varphi)} = \Psi(v;t,\varphi)\cdot\mu, \tag{4.4.10}$$

where

$$\Psi(v;t,\varphi)(z,h)\cdot e^{-2\pi tv} = \sum_{\substack{x\in V(\mathbb{Q})\\ Q(x)=t}} \varphi_\infty^*(v^{\frac{1}{2}}x,z)\,\varphi(h^{-1}x). \tag{4.4.11}$$

In particular, $\Xi(v;t,\varphi)$ is a Green function of logarithmic type for the cycle $Z(t,\varphi)$. Note that the function $\Xi(v;t,\varphi)$ can be defined by the same formula when $t<0$ and, in this case, it is a smooth function on $M_K(\mathbb{C})$, again satisfying (4.4.10), but without the delta current on the left side. The function $\Psi(v;t,\varphi)$ is defined for all t, with, for example,

$$\Psi(v;0,\varphi) = -\varphi(0). \tag{4.4.12}$$

Let $G=\mathrm{SL}(2)$ and let $G'_\mathbb{R}$ be the metaplectic cover of $G(\mathbb{R})$. Then $G'_\mathbb{R}$ acts on the space $S(V(\mathbb{R}))$ by the Weil representation ω for the additive character $\psi_\infty(x)=e(x)=e^{2\pi i x}$. For $\tau=u+iv\in H$, let

$$g'_\tau = [\begin{pmatrix} v^{\frac{1}{2}} & uv^{-\frac{1}{2}} \\ & v^{-\frac{1}{2}}\end{pmatrix}, 1] \in G'_\mathbb{R}. \tag{4.4.13}$$

Then

(4.4.14)

$$v^{-\frac{3}{4}}(\omega(g'_\tau)\varphi_\infty^*)(x,z) = \left[\,4\pi v(w,\bar w)^{-1}\det\begin{pmatrix}(x,x) & (x,\bar w)\\ (w,x) & (w,\bar w)\end{pmatrix} - 1\right]\cdot e^{-2\pi v R(x,z)}\,q^{Q(x)},$$

where $q=e(\tau)$ and $w=w(z)$ is as in (3.2.4). Thus, the generating function

$$\sum_t \Psi(v;t,\varphi)(z,h)\cdot q^t = v^{-\frac{3}{4}}\sum_{x\in V(\mathbb{Q})}\omega(g'_\tau)\varphi_\infty^*(x,z)\varphi(h^{-1}x) =: \theta^*(\tau,z,h;\varphi) \tag{4.4.15}$$

is the weight $\frac{3}{2}$ theta function associated to $\varphi_\infty^*(\cdot,z)\otimes\varphi\in S(V(\mathbb{A}))$. Here the $*$ distinguishes this function from the weight $-\frac{1}{2}$ Siegel theta function

$$\theta(\tau,z,h;\varphi) = v^{-\frac{3}{4}}\sum_{x\in V(\mathbb{Q})}\omega(g'_\tau)\varphi_\infty(x,z)\varphi(h^{-1}x) = \sum_{x\in V(\mathbb{Q})} e^{-2\pi v R(x,z)}\,q^{Q(x)}\,\varphi(h^{-1}x) \tag{4.4.16}$$

defined using the Gaussian φ_∞ in place of φ_∞^*.

Remark 4.4.1. In fact, this relation of the right side of the Green equation (4.4.10) to the 'geometric' theta function $\theta^*(\tau, z, h; \varphi)$ was the original motivation for the definition of the Green function $\Xi(v; t, \varphi)$ in [11].

For the moment, we assume that the image of the compact open subgroup K under the reduced norm is $\nu(K) = \hat{\mathbb{Z}}^\times$. This is equivalent to the assumption that M_K is geometrically irreducible, so that

$$X := M_K(\mathbb{C}) = [\Gamma \backslash D^+] \tag{4.4.17}$$

in (4.4.3), where $D^+ \simeq \mathfrak{H}$ is one component of D and $\Gamma = H(\mathbb{Q}) \cap H(\mathbb{R})^+ K$, where $H(\mathbb{R})^+ \simeq \mathrm{GL}_2(\mathbb{R})^+$ is the identity component.

Let $\mu_1 = \mathrm{vol}(X)^{-1}\mu$. Recall that there is a unique μ_1-admissible Green function $\mathfrak{g}(t, \varphi)$ for the cycle $Z(t, \varphi)$ satisfying

$$dd^c \mathfrak{g}(t, \varphi) + \delta_{Z(t,\varphi)} = \deg(Z(t, \varphi))\, \mu_1, \tag{4.4.18}$$

and

$$\int_X \mathfrak{g}(t, \varphi)\, \mu_1 = 0. \tag{4.4.19}$$

The two Green functions, $\mathfrak{g}(t, \varphi)$ and $\Xi(v; t, \varphi)$ differ by a smooth function on X, which is best described in terms of spectral theory. Let $0 = \lambda_0 < \lambda_1 \leq \lambda_2 \leq \ldots$ be the spectrum of the hyperbolic Laplacian on X, with associated smooth eigenfunctions functions f_λ satisfying (4.1.37) and orthonormal with respect to μ. For $\lambda > 0$, these are the weight 0 Maass forms with respect to Γ. For any smooth function f on X, let

$$\theta^*(\tau; \varphi; f) := \int_X \theta^*(\tau, z; \varphi)\, f(z)\, d\mu(z) \tag{4.4.20}$$

be the associated theta integral. We then can write the spectral expansion of the theta function, as a function of z, as

$$\theta^*(\tau, z; \varphi) = \frac{1}{\mathrm{vol}(X)}\, \theta^*(\tau; \varphi; 1) + \sum_{\lambda>0} \theta^*(\tau; \varphi; f_\lambda) \cdot f_\lambda(z), \tag{4.4.21}$$

where the 'spectral' coefficients are the theta lifts of the Maass forms. Taking the Fourier expansion with respect to τ of both sides and recalling (4.4.15), we obtain

$$\sum_t \Psi(v; t, \varphi) \cdot q^t = \sum_t \sum_{\lambda \geq 0} \theta_t^*(\tau; \varphi; f_\lambda)\, f_\lambda \tag{4.4.22}$$

where $\theta_t^*(\tau; \varphi; f_\lambda)$ is the t-th Fourier coefficient of $\theta^*(\tau; \varphi; f_\lambda)$. This yields the spectral expansion of the $\Psi(v; t, \varphi)$'s:

Proposition 4.4.2. *(i) The Fourier expansion of the theta integral of the constant function* 1 *is*

$$\theta^*(\tau;\varphi;1) = -\mathrm{vol}(X)\,\varphi(0) + \sum_{t>0} \deg(Z(t,\varphi))\,q^t.$$

Also, for $\lambda > 0$,

$$\theta_0^*(\tau;\varphi;f_\lambda) = 0,$$

so that the $\theta^*(\tau;\varphi;f_\lambda)$*'s are cuspidal Maass forms of weight* $\frac{3}{2}$.
(ii) For $t > 0$,

$$\Psi(v;t,\varphi)(z)\,q^t = \frac{1}{\mathrm{vol}(X)}\deg(Z(t,\varphi))\,q^t + \sum_{\lambda>0}\theta_t^*(\tau;\varphi;f_\lambda)\cdot f_\lambda(z).$$

(iii) For $t = 0$,

$$\Psi(v;0,\varphi) = -\varphi(0)$$

is a constant.
(iv) For $t < 0$,

$$\Psi(v;t,\varphi)(z)\,q^t = \sum_{\lambda>0}\theta_t^*(\tau;\varphi;f_\lambda)\cdot f_\lambda(z).$$

Here $\theta_t^*(\tau;\varphi;f_\lambda)$ *denotes the t-th Fourier coefficient of the theta integral of the Maass form* f_λ.

Proof. We only need to prove (i). An easy estimate shows that the theta function $\theta^*(\tau,z;\varphi)$ can be integrated termwise on $X = \Gamma\backslash D^+$, and (4.4.15) together with a standard Stoke's Theorem argument [11], pp. 606–608, yields

$$\int_X \Psi(v;t,\varphi)\,\mu = \sum_{P\in Z(t,\varphi)} 1 = \deg(Z(t,\varphi)). \tag{4.4.23}$$

Also note that the constant term $\theta_0^*(\tau,z;\varphi)$ of the whole theta function is the constant function $-\varphi(0)$ on X, so that, for $\lambda > 0$,

$$\theta_0^*(\tau;\varphi;f_\lambda) = <\theta_0^*(\tau,\cdot;\varphi), f_\lambda> = 0. \tag{4.4.24}$$

□

Remark 4.4.3. In fact, it is possible to give an explicit formula for the Fourier coefficients $\theta_t^*(\tau;\varphi;f_\lambda)$. These involve sums of values of f_λ over the CM points in $Z(t,\varphi)$ when $t > 0$ and integrals of f_λ over certain geodesics when $t < 0$; see [9] for the case of modular curves.

The spectral expansion of $\Psi(v;t,\varphi)$ can be 'lifted' to give the spectral expansion of the Green function $\Xi(v;t,\varphi)$. Here recall (4.1.36): $dd^c f = \frac{1}{2}\Delta f\cdot\mu$.

Theorem 4.4.4. *(i) For $t > 0$,*

$$\Xi(v;t,\varphi)\,q^t = \big(\,\mathfrak{g}(t,\varphi) + \kappa(v;t,\varphi)\,\big)\,q^t - 2\sum_{\lambda>0}\frac{1}{\lambda}\,\theta_t^*(\tau;\varphi;f_\lambda)\cdot f_\lambda,$$

where

$$\kappa(v;t,\varphi) = \frac{1}{\mathrm{vol}(X)}\int_X \Xi(v;t,\varphi)\,\mu.$$

(ii) For $t < 0$,

$$\Xi(v;t,\varphi)\,q^t = \kappa(v;t,\varphi)\,q^t - 2\sum_{\lambda>0}\frac{1}{\lambda}\,\theta_t^*(\tau;\varphi;f_\lambda)\cdot f_\lambda.$$

Proof. For $t \neq 0$, the function $\Xi(v;t,\varphi) - \mathfrak{g}(t,\varphi)$ is smooth with

$$dd^c\Big(\,\Xi(v;t,\varphi) - \mathfrak{g}(t,\varphi)\Big)\cdot q^t = \Psi(t,v;\varphi)\,q^t\,\mu - \deg(Z(t,\varphi))\,q^t\,\mu_1$$

$$= \sum_{\lambda>0}\theta_t^*(\tau;f_\lambda;\varphi)\,f_\lambda \tag{4.4.25}$$

by (ii) of Proposition 4.4.2. Thus

$$\Xi(v;t,\varphi) - \mathfrak{g}(t,\varphi) = \text{constant} - 2\sum_{\lambda>0}\lambda^{-1}\theta_t^*(\tau;f_\lambda;\varphi)\,f_\lambda, \tag{4.4.26}$$

and the constant is determined by integrating against μ_1. □

Let $\mathcal{M}_K$ be a regular model of M_K over $\operatorname{Spec}\mathbb{Z}$ and let $\mathcal{Z}(t,\varphi)$ be a (weighted sum of) divisor(s) on $\mathcal{M}_K$ with $\mathcal{Z}(t,\varphi)_{\mathbb{Q}} = Z(t,\varphi)$. Let

$$\widehat{\mathcal{Z}}(t,v,\varphi) = (\mathcal{Z}(t,\varphi), \Xi(t,v;\varphi)) \in \widehat{\mathrm{CH}}^1(\mathcal{M}_K)_{\mathbb{C}}. \tag{4.4.27}$$

Corollary 4.4.5. *For $\lambda > 0$,*

$$\langle\,\widehat{\mathcal{Z}}(t,v), a(f_\lambda)\,\rangle\cdot q^t = \theta_t^*(\tau;\varphi;f_\lambda).$$

Proof. By (4.1.38), we have

$$\langle\,\widehat{\mathcal{Z}}(t,v;\varphi), a(f_\lambda)\,\rangle\cdot q^t = \langle\, a(\Xi(t,v;\varphi) - \mathfrak{g}(t,\varphi)), a(f_\lambda)\,\rangle\cdot q^t$$

$$= -2\lambda^{-1}\theta_t^*(\tau;\varphi;f_\lambda)\langle\, a(f_\lambda), a(f_\lambda)\,\rangle \tag{4.4.28}$$

$$= \theta_t^*(\tau;\varphi;f_\lambda).$$

□

In the case of $\mathcal{M}$ as in Section 4.2, i.e., $K = \widehat{O}_B^\times$, by summing on $t \in \mathbb{Z}$, we obtain

Corollary 4.4.6. *For the generating function $\hat{\phi}_1(\tau)$ of Section 4.2, the component*

$$\langle\, \hat{\phi}_1(\tau), a(f_\lambda)\,\rangle = \theta^*(\tau;\varphi; f_\lambda).$$

is a Maass form of weight $\frac{3}{2}$.

We now return to the situation at the beginning of this section so that K is an arbitrary compact open subgroup of $H(\mathbb{A}_f)$ and the components of $M_K(\mathbb{C})$ are described by (4.4.3). For $\varphi \in S(V(\mathbb{A}_f))^K$, the analogue of (i) of Proposition 4.2 gives

$$\text{(4.4.29)}\quad \int_{M_K(\mathbb{C})} \theta^*(\tau;\varphi)\,\mu = -\text{vol}(M_K(\mathbb{C}))\,\varphi(0) + \sum_{t>0} \deg(Z(t,\varphi))\,q^t$$

where $\theta^*(\tau;\varphi)$ is the theta function defined by (4.4.15), and $\deg(Z(t,\varphi))$ is the degree of the weighted 0-cycle, i.e., the sum of its degrees on each of the components of $M_K(\mathbb{C})$. The right side of (4.4.29) is an Eisenstein series. We would like to have similar information about the degrees of the weighted 0-cycles $Z(t,\varphi)$ on individual components of $M_K(\mathbb{C})$. Note that, by strong approximation,

$$\text{(4.4.30)}\qquad \pi_0(M_K(\mathbb{C})) \;\simeq\; \mathbb{Q}^\times\backslash\mathbb{A}^\times/\mathbb{R}_+^\times\nu(K).$$

For any function η on $\pi_0(M_K(\mathbb{C}))$, we define a locally constant function $\tilde{\eta}$ on $D \times H(\mathbb{A}_f)$ by

$$\text{(4.4.31)}\quad \tilde{\eta}(z,h) = \eta((\text{sgn}(z),\nu(h))), \qquad (\text{sgn}(z),\nu(h)) \in \mathbb{R}^\times \times \mathbb{A}_f^\times,$$

where $\text{sgn}(z) = \pm 1$ if $z \in D^\pm$. The integral of the theta function against $\tilde{\eta}$ then gives the generating function
(4.4.32)

$$\int_{M_K(\mathbb{C})} \theta^*(\tau;\varphi)\cdot\tilde{\eta}\cdot\mu = -\text{vol}_\eta(M_K(\mathbb{C}))\,\varphi(0) + \sum_{t>0} \deg_\eta(Z(t,\varphi))\,q^t.$$

for degrees weighted by η.

Proposition 4.4.7. *Suppose that η is orthogonal to the constant function on $\pi_0(M_K)$. Then the generating function for η-degrees is a distinguished cusp form of weight $\frac{3}{2}$, i.e., lies in the space of cusp forms generated by theta functions for quadratic forms in one variable.*

Proof. Let dh be Tamagawa measure on $SO(V)(\mathbb{A}) \simeq Z(\mathbb{A})\backslash H(\mathbb{A})$, where Z is the center of H. Let dz_f be the measure on $Z(\mathbb{A}_f)$ which gives the maximal compact subgroup $\hat{\mathbb{Z}}^\times$ volume 1, and let $dh' = dz_f\,dh$ be the resulting invariant measure on $Z(\mathbb{R})\backslash H(\mathbb{A})$. Write $dh' = dh_\infty \times dh'_f$, where the measure dh_∞ on $SO(V)(\mathbb{R}) = Z(\mathbb{R})\backslash H(\mathbb{R})$ is determined by the identity

$$\int_{Z(\mathbb{R})\backslash H(\mathbb{R})} \phi(h_\infty \cdot z_0)\, dh_\infty = \int_D \phi(z)\, d\mu(z). \tag{4.4.33}$$

By a very slight modification of the proof of Proposition 4.17 of [13] in our current situation, we obtain

Lemma 4.4.8. *For an integrable function ϕ on $D \times H(\mathbb{A}_f)$ which is left invariant under $H(\mathbb{Q})$ and right invariant under K,*

$$\int_{H(\mathbb{Q})Z(\mathbb{R})\backslash H(\mathbb{A})} \phi(h_\infty z_0, h_f) dh' = e_K^{-1} \mathrm{vol}(K) \sum_j \int_{\Gamma_j\backslash D^+} \phi(z, h_j)\, d\mu(z),$$

where $e_K = |Z(\mathbb{Q}) \cap K|$ and $\mathrm{vol}(K) = \mathrm{vol}(K, dh'_f)$.

Applying this to the integral in (4.4.32), we obtain

$$\int_{M_K(\mathbb{C})} \theta^*(\tau; \varphi) \cdot \tilde{\eta} \cdot \mu = \int_{H(\mathbb{Q})Z(\mathbb{R})\backslash H(\mathbb{A})} \theta(\tau, h; \varphi^*_\infty \otimes \varphi)\, \eta(\nu(h))\, dh'. \tag{4.4.34}$$

We may as well assume that η is a nontrivial character of the component group $\pi_0(M_K(\mathbb{C}))$. The theta function θ is invariant under the center $Z(\mathbb{A})$, since it factors through $SO(V)(\mathbb{A})$. Hence, if $\eta^2 \neq 1$, the integral is identically zero. If $\eta^2 = 1$ with $\eta \neq 1$, let $\boldsymbol{k}_\eta$ be the associated quadratic extension of $\mathbb{Q}$. The integral can be unfolded and only terms with $x \in V(\mathbb{Q})$ with $\mathbb{Q}(x) \simeq \boldsymbol{k}_\eta$ can give a nonzero contribution. Thus, the theta integral is a distinguished cusp form, and hence, by the results of [6] and [21], lies in the space generated theta functions coming from O(1)'s. □

4.5 THE MORDELL-WEIL COMPONENT

In this section, we will prove that the Mordell-Weil component ϕ_{MW} of our generating series given by (4.2.11) is the q-expansion of a modular form valued in MW $\otimes$ $\mathbb{C}$. The idea is that this series is very closely related to the generating series considered by Borcherds [2], so that his result can be applied. Again, we work in a more general case than is finally needed for the moduli space $\mathcal{M}$.

For a compact open subgroup $K \subset H(\mathbb{A}_f)$, a weight function $\varphi \in S(V(\mathbb{A}_f))^K$, and $t \in \mathbb{Q}_{>0}$, let $Z(t, \varphi; K)$ be the corresponding weighted 0-cycle on M_K. Recall that is it rational over $\mathbb{Q}$, so it defines a class in the Chow group $\mathrm{CH}^1(M_K)$ which we also denote by $Z(t, \varphi; K)$. If $K' \subset K$ is an open subgroup, and if $\mathrm{pr} : M_{K'} \longrightarrow M_K$ is the natural projection, then by [10],

$$\mathrm{pr}^*(Z(t,\varphi;K)) = Z(t,\varphi;K'), \tag{4.5.1}$$

so that we obtain a class

$$Z(t,\varphi) \in \mathrm{CH}^1(M) := \varinjlim_{K} \mathrm{CH}^1(M_K)_{\mathbb{C}} \tag{4.5.2}$$

in the direct limit. Let $\mathcal{L}$ be the line bundle on D given by the restriction of the bundle $\mathcal{O}(-1)$ on $\mathbb{P}(V(\mathbb{C}))$, and let $\mathcal{L}_K$ be the associated bundle on M_K. Since the $\mathcal{L}_K$'s are compatible with pullbacks, we have the relation $\mathrm{pr}^*(c_1(\mathcal{L}_K)) = c_1(\mathcal{L}_{K'})$ on Chern classes and can define classes

$$Z(0,\varphi;K) = -\varphi(0)\,c_1(\mathcal{L}_K) \in \mathrm{CH}^1(M_K)_{\mathbb{C}}, \tag{4.5.3}$$

at finite level, and

$$Z(0,\varphi) = -\varphi(0)c_1(\mathcal{L}) \in \mathrm{CH}^1(M)_{\mathbb{C}}, \tag{4.5.4}$$

in the direct limit.

The following result will be proved in Section 4.6.

Theorem 4.5.1. (Borcherds [2], $+\epsilon$) *For any $\varphi \in S(V(\mathbb{A}_f))^K$, the generating series*

$$\phi_{\mathrm{Bor}}(q,\varphi;K) := Z(0,\varphi;K) + \sum_{t>0} Z(t,\varphi;K)\,q^t \in \mathrm{CH}^1(M_K)_{\mathbb{C}}[[\,q\,]]$$

is the q-expansion of a holomorphic modular form $\phi_{\mathrm{Bor}}(\tau,\varphi;K)$ of weight $\frac{3}{2}$ valued in $\mathrm{CH}^1(M_K)_{\mathbb{C}}$.

We can pass to the limit on K and conclude that, for any $\varphi \in S(V(\mathbb{A}_f))$, the generating series

$$\phi_{\mathrm{Bor}}(q,\varphi) := Z(0,\varphi) + \sum_{t>0} Z(t,\varphi)\,q^t \in \mathrm{CH}^1(M)_{\mathbb{C}}[[\,q\,]] \tag{4.5.5}$$

is also the q-expansion of a holomorphic modular form $\phi_{\mathrm{Bor}}(\tau)$ of weight $\frac{3}{2}$. Moreover the map

$$\phi_{\mathrm{Bor}}(\tau) : \; S(V(\mathbb{A}_f)) \longrightarrow \mathrm{CH}^1(M), \qquad \varphi \mapsto \phi_{\mathrm{Bor}}(\tau,\varphi) \tag{4.5.6}$$

is equivariant for the natural action of $H(\mathbb{A}_f)$ on the two sides. In effect, this equivariance describes the action of the Hecke operators for the Shimura variety M on the generating series.

The proof of Theorem 4.5.1 will be given in the next section. Here we use it to prove the modularity of the Mordell-Weil component (4.2.11). In the case $K = \widehat{O}_B^\times$, $M_K = \mathcal{M}_\mathbb{Q}$ is geometrically irreducible. Moreover, by (3.12) of [16] and the remark after (1.15) of [13], we have $\mathcal{L} = \omega_\mathbb{Q}$, for ω the Hodge line bundle, as in Section 3.3. Taking φ to be the characteristic function $\varphi^0 := \mathrm{char}(V(\mathbb{A}_f) \cap \widehat{O}_B)$, we have

$$\deg(\phi_{\mathrm{Bor}}(\tau, \varphi^0; K)) = \phi_{\deg}(\tau) = \mathcal{E}_1(\tau, \frac{1}{2}; D(B)). \tag{4.5.7}$$

Thus, we have proved the following.

Proposition 4.5.2. *The generating series $\phi_{\mathrm{MW}}(q) \in \mathrm{CH}^1(\mathcal{M}_\mathbb{Q})^0[[q]]$ of (4.2.11) is the q-expansion of the holomorphic modular form*

$$\phi^0_{\mathrm{Bor}}(\tau, \varphi^0; K) := \phi_{\mathrm{Bor}}(\tau, \varphi^0; K) - \mathcal{E}_1(\tau, \frac{1}{2}; D(B)) \deg(\mathcal{L}_K)^{-1} c_1(\mathcal{L}_K).$$

4.6 BORCHERDS' GENERATING FUNCTION

In this section, we explain how to formulate Borcherds' result [2] in order to obtain the statement of Theorem 4.5.1. A slightly more detailed discussion of some background is given in Section 1 of [13], to which we refer the reader for more information.

As in [16] and Section 5.5, let $G'_\mathbb{A}$ (resp. $G'_\mathbb{R}$ and $G'_{\mathbb{A}_f}$) be the metaplectic extension of $\mathrm{SL}_2(\mathbb{A})$ (resp. $\mathrm{SL}_2(\mathbb{R})$ and $\mathrm{SL}_2(\mathbb{A}_f)$). Let $G'_\mathbb{Q}$ be the image of $\mathrm{SL}_2(\mathbb{Q})$ under the unique splitting homomorphism $\mathrm{SL}_2(\mathbb{Q}) \to G'_\mathbb{A}$. Let $\Gamma' \subset G'_\mathbb{R}$ be the full inverse image of $\mathrm{SL}_2(\mathbb{Z})$ and let $K' \subset G'_{\mathbb{A}_f}$ be the full inverse image of $\mathrm{SL}_2(\hat{\mathbb{Z}})$. Thus Γ' is a central extension of $\mathrm{SL}_2(\mathbb{Z})$ by $\mathbb{C}^1$. For each $\gamma' \in \Gamma'$, there is a unique element $k' \in K'$ such that the product $\gamma = \gamma' k' \in \mathrm{SL}_2(\mathbb{Z})$, identified with a subgroup of $G'_\mathbb{Q}$. The map $\gamma' \mapsto k'$ defines a homomorphism from Γ' to K'. For $\tau = u + iv \in \mathfrak{H}$, let $g'_\tau \in G'_\mathbb{R}$ be given by (4.4.13), and let $K'_\infty \subset G'_\mathbb{R}$ be the full inverse image of $\mathrm{SO}(2)$. For any $\gamma' \in G'_\mathbb{R}$ with projection $\gamma \in \mathrm{SL}_2(\mathbb{R})$, we have

$$\gamma' g'_\tau = g'_{\gamma(\tau)} k'_\infty(\gamma', \tau), \tag{4.6.1}$$

for a unique element $k'_\infty(\gamma', \tau) \in K'_\infty$. For $r \in \frac{1}{2}\mathbb{Z}$, let χ_r be the character of K'_∞ given by (1.26) of [13], and define an automorphy factor

$$j_r(\gamma', \tau) = \chi_{-r}(k'_\infty(\gamma', \tau))) \, |c\tau + d|^r; \tag{4.6.2}$$

see Section 8.5.

Let ψ_f be the unique additive character of $\mathbb{A}_f$ which is trivial on $\widehat{\mathbb{Z}}$ and such that the additive character $\psi = \psi_\infty \psi_f$, with $\psi_\infty(x) = e(x)$, of $\mathbb{A}$

is trivial on $\mathbb{Q}$. The group $G'_{\mathbb{A}_f}$ acts in the space $S(V(\mathbb{A}_f))$ by the Weil representation ω associated to ψ_f.

Let $L \subset V$ be a lattice on which the quadratic form Q is integral, and let

$$L^\sharp = \{\, x \in V \mid (x, L) \subset \mathbb{Z} \,\}$$

be the dual lattice. Let $S_L \subset S(V(\mathbb{A}_f))$ be the space of functions φ with support in $\widehat{L}^\sharp$ and which are invariant under translation by $\widehat{L}$. This space is isomorphic to $\mathbb{C}[L^\sharp/L]$ under the map $\lambda \mapsto \varphi_\lambda$ where, for $\lambda \in L^\sharp$, φ_λ is the characteristic function of $\lambda + \widehat{L}$. The Weil representation action of K' on $S(V(\mathbb{A}_f))$ preserves S_L and yields a representation ρ_L of Γ' on this space via the homomorphism $\Gamma' \to K'$. Following Borcherds [2], let

$$\text{(4.6.3)} \qquad \mathrm{MF}(\Gamma', \frac{1}{2}, \rho_L)$$

be the space of S_L-valued holomorphic functions f on $\mathfrak{H}$ such that

$$\text{(4.6.4)} \qquad f(\gamma(\tau)) = j_{\frac{1}{2}}(\gamma', \tau)\, \rho_L(\gamma') f(\tau),$$

for all $\gamma \in \mathrm{SL}_2(\mathbb{Z})$, and which have a q-expansion of the form

$$\text{(4.6.5)} \qquad f(\tau) = \sum_\lambda \sum_t c_\lambda(t)\, q^t\, \varphi_\lambda,$$

where only a finite number of $c_\lambda(t)$'s with $t < 0$ are nonzero. Let

$$\text{(4.6.6)} \qquad \mathrm{MF}(\Gamma', \frac{1}{2}, \rho_L)_{\mathbb{Z}}$$

be the $\mathbb{Z}$-submodule of f's for which $c_\lambda(t) \in \mathbb{Z}$ for all $t \leq 0$.

Let $\mathrm{ZHeeg}(L)$ be the $\mathbb{Q}$-vector space with generators $y_\lambda(t)$, for $\lambda \in L^\sharp/L$ and $t \in \mathbb{Q}_{>0}$ with $Q(\lambda) \equiv t \mod \mathbb{Z}$, and an additional element $y_0(0)$. For $f \in \mathrm{MF}(\Gamma', \frac{1}{2}, \rho_L)_{\mathbb{Z}}$, let

$$\text{(4.6.7)} \qquad \mathrm{rel}(f) = c_0(0)\, y_0(0) + \sum_\lambda \sum_{t>0} c_\lambda(-t)\, y_\lambda(t) \in \mathrm{ZHeeg}(L).$$

Finally, let

$$\text{(4.6.8)} \qquad \mathrm{CHeeg}(L) = \frac{\mathrm{ZHeeg}(L)}{\langle\, \mathrm{rel}(f) \mid f \in \mathrm{MF}(\Gamma', \frac{1}{2}, \rho_L)_{\mathbb{Z}} \,\rangle}.$$

This group is a kind of formal Chow group for Heegner divisors. In [2], Borcherds proved that the generating series for the $y_\lambda(t)$'s is a holomorphic modular form, assuming the existence of a basis with rational Fourier coefficients of a certain space of vector valued modular forms. The existence of the required basis was subsequently proved by McGraw [17].

Theorem 4.6.1. (Borcherds)
(i) The space $\mathrm{CHeeg}(L)$ *is finite dimensional over* $\mathbb{Q}$.
(ii) The generating series

$$\phi_{\mathrm{Bor}}(q, L) = \sum_{\lambda} \sum_{t \geq 0} y_{\lambda}(t)\, q^t\, \varphi_{\lambda}^{\vee} \quad \in \mathrm{CHeeg}(L) \otimes S_L^{\vee}[[\,q\,]]$$

is the q-expansion of a holomorphic $\mathrm{CHeeg}(L) \otimes S_L^{\vee}$*-valued modular form* $\phi_{\mathrm{Bor}}(\tau, L)$ *of weight* $\frac{3}{2}$ *and type* $\rho_L^{\vee}$ *for* Γ'. *Here* $S_L^{\vee}$ *is the dual space to* S_L *and* $\{\varphi_{\lambda}^{\vee}\}$ *is the dual basis to* $\{\varphi_{\lambda}\}$.

We can view $\phi_{\mathrm{Bor}}(\tau, L)$ as a map

$$\phi_{\mathrm{Bor}}(\tau, L) : S_L \longrightarrow \mathrm{CHeeg}(L)_{\mathbb{C}}, \tag{4.6.9}$$

where, for $\varphi \in S_L$,

$$\phi_{\mathrm{Bor}}(\tau, \varphi; L) := \phi_{\mathrm{Bor}}(\tau, L)(\varphi) = \sum_{t \geq 0} y_{\varphi}(t)\, q^t, \tag{4.6.10}$$

with

$$y_{\varphi}(t) = \sum_{\lambda} y_{\lambda}(t) \ < \varphi_{\lambda}^{\vee}, \varphi > . \tag{4.6.11}$$

This function has the transformation law

$$\phi_{\mathrm{Bor}}(\gamma(\tau), \varphi; L) = j_{\frac{3}{2}}(\gamma', \tau)\, \phi_{\mathrm{Bor}}(\tau, \rho_L(\gamma')^{-1}\varphi; L), \tag{4.6.12}$$

for all $\gamma \in \mathrm{SL}_2(\mathbb{Z})$, where $\gamma' \in \Gamma'$ is any preimage of γ.

To relate this formal generating series to one valued in a geometric Chow group, we proceed as follows. Let $K_0 = \widehat{O}_B^{\times}$, and let

$$K_L = \{\, k \in K_0 \mid kL = L \text{ and } k \text{ acts trivially in } L^{\sharp}/L \,\}. \tag{4.6.13}$$

The key point in Borcherds' construction is then

Proposition 4.6.2. *The map* $\mathrm{ZHeeg}(L) \to \mathrm{CH}^1(M_{K_L})_{\mathbb{C}}$ *defined by sending* $y_{\lambda}(t)$ *to* $Z(t, \varphi_{\lambda}; K_L)$ *and* $y_0(0)$ *to* $-c_1(\mathcal{L}_{K_L})$ *induces a homomorphism*

$$\mathrm{CHeeg}(L)_{\mathbb{C}} \longrightarrow \mathrm{CH}^1(M_{K_L})_{\mathbb{C}}.$$

Proof. To any $f \in \mathrm{MF}(\Gamma', \frac{1}{2}, \rho_L)_{\mathbb{Z}}$, Borcherds constructs a meromorphic section[2] $\Psi(f)$ of $\mathcal{L}_{K_L}^k$, where $k = c_0(0)$, with explicit divisor

$$\mathrm{div}(\Psi(f)) = \sum_{\lambda} \sum_{t > 0} c_{\lambda}(-t)\, Z(t, \varphi_{\lambda}; K_L). \tag{4.6.14}$$

[2]Technically, the transformation law of $\Psi(f)$ may involve a unitary character. Since this character has finite order [3], and since we are taking Chow groups with $\mathbb{Q}$-coefficients, we can ignore it.

But then

$$\text{rel}(f) \longmapsto -c_0(0)\, c_1(\mathcal{L}) + \text{div}(\Psi(f)) \equiv 0. \tag{4.6.15}$$

□

Thus, we have

$$S_L \overset{\phi_{\text{Bor}}(\tau, L)}{\longrightarrow} \text{CHeeg}(L)_{\mathbb{C}} \longrightarrow \text{CH}^1(M_{K_L})_{\mathbb{C}}. \tag{4.6.16}$$

Finally, we pass to the limit over L. Suppose that $L_2 \subset L_1$ are lattices, as above. Then there is a natural inclusion $S_{L_1} \hookrightarrow S_{L_2}$, compatible with the Γ' actions ρ_{L_1} and ρ_{L_2}, since these are, after all, just coming from the action of K' on

$$S(V(\mathbb{A}_f)) = \varinjlim_{L} S_L. \tag{4.6.17}$$

There is a resulting inclusion

$$\text{MF}(\Gamma', \frac{1}{2}, \rho_{L_1}) \longrightarrow \text{MF}(\Gamma', \frac{1}{2}, \rho_{L_2}) \tag{4.6.18}$$

preserving the 'integral' elements. It is easily checked that the map

$$\text{ZHeeg}(L_1) \longrightarrow \text{ZHeeg}(L_2), \tag{4.6.19}$$

$$y_\lambda(t) \mapsto \sum_{\substack{\mu \in L_2^\sharp/L_2 \\ \mu \equiv \lambda \mod L_1}} y_\mu(t), \qquad y_0(0) \mapsto y_0(0),$$

induces a map

$$\text{pr}^* : \text{CHeeg}(L_1) \longrightarrow \text{CHeeg}(L_2). \tag{4.6.20}$$

We let

$$\text{CHeeg} := \varinjlim_{L} \text{CHeeg}(L). \tag{4.6.21}$$

It is easy to check the following compatibility.

Lemma 4.6.3. *The diagram*

$$\begin{array}{ccccc} S_{L_2} & \overset{\phi_{\text{Bor}}(\tau, L_2)}{\longrightarrow} & \text{CHeeg}(L_2)_{\mathbb{C}} & \longrightarrow & \text{CH}^1(M_{K_{L_2}})_{\mathbb{C}} \\ \uparrow & & \uparrow \text{pr}^* & & \uparrow \text{pr}^* \\ S_{L_1} & \overset{\phi_{\text{Bor}}(\tau, L_1)}{\longrightarrow} & \text{CHeeg}(L_1)_{\mathbb{C}} & \longrightarrow & \text{CH}^1(M_{K_{L_1}})_{\mathbb{C}} \end{array}$$

is commutative.

Since the system of subgroups K_L is cofinal in the system of all compact open subgroups K with $K \cap Z(\mathbb{A}_f) \simeq \hat{\mathbb{Z}}^\times$, we pass to the limit and obtain

$$\phi_{\rm Bor}(\tau): \ S(V(\mathbb{A}_f)) \longrightarrow \mathrm{CHeeg}_{\mathbb{C}} \longrightarrow \mathrm{CH}^1(M)_{\mathbb{C}}, \tag{4.6.22}$$

where we write $\phi_{\rm Bor}(\tau)$ for the composite map.

Proposition 4.6.4. *There is a map*

$$\begin{aligned} \phi_{\rm Bor}(\tau): \quad S(V(\mathbb{A}_f)) &\longrightarrow \mathrm{CH}^1(M)_{\mathbb{C}}, \\ \varphi \quad &\longmapsto \quad \phi_{\rm Bor}(\tau,\varphi) = Z(0,\varphi) + \sum_{t>0} Z(t,\varphi)\, q^t \end{aligned}$$

compatible with the $H(\mathbb{A}_f)$ actions. Moreover, $\phi_{\rm Bor}(\tau,\varphi)$ is a holomorphic modular form of weight $\frac{3}{2}$. More precisely, for all $\gamma \in \mathrm{SL}_2(\mathbb{Z})$,

$$\phi_{\rm Bor}(\gamma(\tau),\varphi) = j_{\frac{3}{2}}(\gamma',\tau)\, \phi_{\rm Bor}(\tau, \rho(\gamma')^{-1}\varphi),$$

where ρ is the representation of Γ' on $S(V(\mathbb{A}_f))$ coming from the action of K' and the homomorphism $\Gamma' \to K'$.

Of course, for any compact open subgroup $K \subset H(\mathbb{A}_f)$, there is a resulting map

(4.6.23)

$$\begin{aligned} \phi_{\rm Bor}(\tau): S(V(\mathbb{A}_f))^K &\longrightarrow \mathrm{CH}^1(M_K)_{\mathbb{C}}, \\ \varphi &\mapsto \phi_{\rm Bor}(\tau,\varphi;K) = Z(0,\varphi;K) + \sum_{t>0} Z(t,\varphi;K)\, q^t \end{aligned}$$

as claimed in Theorem 4.5.1.

4.7 AN INTERTWINING PROPERTY

In this section, we show that the function on the metaplectic group defined by the Borcherds generating function $\phi_{\rm Bor}(\tau,\varphi)$ has the same intertwining property as the usual theta function with respect to the right action of $G'_{\mathbb{A}_f}$.

We lift the generating function $\phi_{\rm Bor}(\tau,\varphi)$ to a function $\widetilde{\phi}_{\rm Bor}(g',\varphi)$ on $G'_{\mathbb{A}}$ by setting

$$\widetilde{\phi}_{\rm Bor}(g',\varphi) := j_{\frac{3}{2}}(g'_\infty, i)^{-1}\, \phi_{\rm Bor}(g'_\infty(i), \omega(k')\varphi), \tag{4.7.1}$$

where $g' = \alpha\, g'_\infty\, k'$, for $\alpha \in G'_{\mathbb{Q}}$, $g'_\infty \in G'_{\mathbb{R}}$, and $k' \in K'$. The transformation law of Proposition 4.6.4 implies that this function is well defined; it is left $G'_{\mathbb{Q}}$-invariant by construction. In addition, since $j_r(k'_\infty, i) = \chi_{-r}(k'_\infty)$, we have

$$\widetilde{\phi}_{\mathrm{Bor}}(g' k'_\infty k'_f, \varphi) = \chi_{\frac{3}{2}}(k'_\infty)\, \widetilde{\phi}_{\mathrm{Bor}}(g', \omega(k'_f)\varphi), \tag{4.7.2}$$

for $k'_f \in K'$. Thus, $\widetilde{\phi}_{\mathrm{Bor}}$ has weight $\frac{3}{2}$.

The main result of this section is the following intertwining property.

Proposition 4.7.1. *For $g'_f \in G'_{\mathbb{A}_f}$,*

$$\widetilde{\phi}_{\mathrm{Bor}}(g' g'_f, \varphi) = \widetilde{\phi}_{\mathrm{Bor}}(g', \omega(g'_f)\varphi).$$

Proof. We define another function on $G'_{\mathbb{A}}$ by

$$\widetilde{\widetilde{\phi}}(g'_\infty\, g'_f, \varphi) = j_{\frac{3}{2}}(g'_\infty, i)^{-1}\, \phi(g'_\infty(i), \omega(g'_f)\varphi). \tag{4.7.3}$$

This function agrees with $\widetilde{\phi}$ on $G'_{\mathbb{R}}K'$ and, by construction, satisfies

$$\widetilde{\widetilde{\phi}}(g' g'_0, \varphi) = \widetilde{\widetilde{\phi}}(g', \omega(g'_0)\varphi), \tag{4.7.4}$$

for $g'_0 \in G'_{\mathbb{A}_f}$. In particular, $\widetilde{\widetilde{\phi}}$ is left invariant under $\mathrm{SL}_2(\mathbb{Z})$, and it suffices to prove that it is, in fact, left invariant under all of $G'_{\mathbb{Q}}$.

For $a \in \mathbb{Q}^\times_{>0}$, let $\delta = \begin{pmatrix} a & \\ & a^{-1} \end{pmatrix} \in G'_{\mathbb{Q}}$. Write $\delta = \delta'_\infty\, \delta'_f$ for $\delta'_\infty \in G'_{\mathbb{R}}$ and $\delta'_f \in G'_{\mathbb{A}_f}$.

Lemma 4.7.2. *(i)*

$$Z(t, \omega(\delta'_f)\varphi) = j_{\frac{3}{2}}(\delta'_\infty, \tau)\, Z(a^2 t, \varphi).$$

(ii)

$$\phi_{\mathrm{Bor}}(a^2\tau, \omega(\delta'_f)\varphi) = j_{\frac{3}{2}}(\delta'_\infty, \tau)\, \phi_{\mathrm{Bor}}(\tau, \varphi).$$

Proof. Note that $\varphi \in S(V(\mathbb{A}_f))^K$ for some compact open subgroup $K \subset H(\mathbb{A}_f)$, so (i) can be viewed as an identity between weighted 0-cycles on the generic fiber M_K. Recall from Lemma 10.1 of [11],

$$Z(t, \varphi; K) = \sum_j \sum_{\substack{x \in V(\mathbb{Q}) \\ Q(x)=t \\ \mathrm{mod}\ \Gamma_j}} \varphi(h_j^{-1}x)\, \mathrm{pr}(D_x), \tag{4.7.5}$$

in the notation of Section 4.4 above, especially (4.4.2) and (4.4.3). On the other hand, since $|a|_{\mathbb{A}_f} = a^{-\frac{3}{2}}$, we have

$$\omega(\delta'_f)\varphi(x) = j_{\frac{3}{2}}(\delta'_\infty, \tau)\,\varphi(ax), \tag{4.7.6}$$

which, together with the previous identity, yields (i). Part (ii) is immediate from (i) and the analogous relation for $Z(0,\varphi)$, which is clear from (4.5.4). □

Thus, we have

$$\begin{aligned}\tilde{\tilde{\phi}}(\delta g',\varphi) &= \tilde{\tilde{\phi}}(\delta'_\infty\, g'_\infty\, \delta'_f\, g'_f, \varphi)\\ &= j_{\frac{3}{2}}(\delta'_\infty,\tau)^{-1}\, j_{\frac{3}{2}}(g'_\infty, i)^{-1}\, \phi_{\mathrm{Bor}}(a^2\tau, \omega(\delta'_f)\omega(g'_f)\varphi)\\ &= j_{\frac{3}{2}}(g'_\infty, i)^{-1}\, \phi_{\mathrm{Bor}}(\tau, \omega(g'_f)\varphi)\\ &= \tilde{\tilde{\phi}}(g',\varphi).\end{aligned} \tag{4.7.7}$$

Since $\mathrm{SL}_2(\mathbb{Q})$ is generated by $\mathrm{SL}_2(\mathbb{Z})$ together with the δ's for $a \in \mathbb{Q}^\times_{>0}$, $\tilde{\tilde{\phi}}$ is left $G'_{\mathbb{Q}}$ invariant, and hence coincides with $\tilde{\phi}_{\mathrm{Bor}}$. This finishes the proof of Proposition 4.7.1. □

Propositions 4.6.4 and 4.7.1 show that $\tilde{\phi}_{\mathrm{Bor}}(g',\varphi)$ behaves just like the classical theta function $\theta(g',h;\varphi)$ as far as $G'_{\mathbb{A}_f} \times H(\mathbb{A}_f)$-equivariance is concerned. More precisely, let $\mathcal{A}(G')_{\frac{3}{2},\mathrm{hol}}$ be the space of automorphic forms on $G'_{\mathbb{A}}$ of weight $\frac{3}{2}$ (for the right action of $K'_{\mathbb{R}}$) which are 'holomorphic' in the sense that the corresponding functions on $\mathfrak{H}$ are holomorphic. Then, we have a map

$$\tilde{\phi}_{\mathrm{Bor}} : S(V(\mathbb{A}_f)) \longrightarrow \mathcal{A}(G')_{\frac{3}{2},\mathrm{hol}} \otimes \mathrm{CH}^1(M), \qquad \varphi \mapsto \tilde{\phi}_{\mathrm{Bor}}(g',\varphi), \tag{4.7.8}$$

which is $G'_{\mathbb{A}_f} \times H(\mathbb{A}_f)$-equivariant. Note that since the action of $G'_{\mathbb{A}_f} \times H(\mathbb{A}_f)$ on $S(V(\mathbb{A}_f))$ is smooth, any φ is fixed by some compact open subgroup $K' \times K$, and thus the function $\tilde{\phi}_{\mathrm{Bor}}(g',\varphi)$ takes values in the finite dimensional space $\mathrm{CH}^1(M_K)$. Moreover, $\tilde{\phi}_{\mathrm{Bor}}(g',\varphi)$ is right K'-invariant, so that the component functions lie in the finite dimensional space $\mathcal{A}(G')^{K'}_{\frac{3}{2},\mathrm{hol}}$.

Bibliography

[1] J.-B. Bost, *Potential theory and Lefschetz theorems for arithmetic surfaces*, Ann. Sci. École Norm. Sup., **32** (1999), 241–312.

[2] R. Borcherds, *The Gross-Kohnen-Zagier theorem in higher dimensions*, Duke Math. J., **97** (1999), 219–233.

[3] ______, *Correction to "The Gross-Kohnen-Zagier theorem in higher dimensions,"* Duke Math. J., **105** (2000), 183–184.

[4] G. Faltings, *Calculus on arithmetic surfaces*, Annals of Math., **119** (1984), 387–424.

[5] H. Gillet and C. Soulé, *Arithmetic intersection theory*, Publ. Math. IHES, **72** (1990), 93–174.

[6] S. Gelbart and I. I. Piatetski-Shapiro, *Distinguished representations and modular forms of half integral weight*, Invent. Math., **59** (1980), 145–188.

[7] B. H. Gross, *Heights and special values of L-series*, in Number Theory (Proc. Montreal Conf., 1985), CMS Conf. Proc. **7**, 115–187, AMS, Providence, 1987.

[8] P. Hriljac, *Heights and arithmetic intersection theory*, Amer. J. Math., **107** (1985), 23–38.

[9] S. Katok and P. Sarnak, *Heegner points, cycles and Maass forms*, Israel J. Math., **84** (1993), 193–227.

[10] S. Kudla, *Algebraic cycles on Shimura varieties of orthogonal type*, Duke Math. J., **86** (1997), 39–78.

[11] ______, *Central derivatives of Eisenstein series and height pairings*, Annals of Math., **146** (1997), 545–646.

[12] ______, *Derivatives of Eisenstein series and generating functions for arithmetic cycles*, Séminaire Bourbaki 876, Astérisque, **276**, 341–368, So. Math. de France, Paris, 2002.

[13] ______, *Integrals of Borcherds forms*, Compositio Math., **137** (2003), 293–349.

[14] ______, *Special cycles and derivatives of Eisenstein series*, in Heegner Points and Rankin L-Series, Math. Sci. Res. Inst. Publ., **49**, 243–270, Cambridge Univ. Press, Cambridge, 2004.

[15] S. Kudla and M. Rapoport, *Height pairings on Shimura curves and p-adic uniformization*, Invent. math., **142** (2000), 153–223.

[16] S. Kudla, M. Rapoport, and T. Yang, *Derivatives of Eisenstein series and Faltings heights*, Compositio Math., **140** (2004), 887–951.

[17] W. J. McGraw, *Rationality of vector valued modular forms associated to the Weil representation*, Math. Annalen, **326** (2003), 105–122.

[18] J. Milne, *Canonical models of (mixed) Shimura varieties and automorphic vector bundles*, in Automorphic Forms, Shimura Varieties and L-functions, Vol I., 283–414, Academic Press, Boston, 1990.

[19] S. Rallis and G. Schiffmann, *Représentations supercuspidales du groupe métaplectique*, J. Math. Kyoto Univ., **17** (1977), 567–603.

[20] G. Shimura, Introduction to the Arithmetic Theory of Automorphic Forms, Princeton Univ. Press, Princeton, N. J., 1971.

[21] K. Snitz, *The theta correspondence for characters of* O(3), Ph. D. Thesis, University of Maryland, August 2003.

Chapter Five

The central derivative of a genus two Eisenstein series

In this chapter, we study an incoherent Eisenstein series of genus two in detail and, in particular, compute its derivative at $s = 0$, the central point for the functional equation. This Eisenstein series was first introduced in [4]. In the first few sections, we consider the Fourier coefficients associated to $T \in \mathrm{Sym}_2(\mathbb{Q})$ with $\det(T) \neq 0$. These coefficients, which are given by a product of local factors, were studied in [4] and [8]. In Sections 5.4 through 5.8, we deal with the Fourier coefficients for T's with $\mathrm{rank}(T) = 1$. These coefficients, which are not given as a product of local factors, are of global nature and are closely related to the Fourier coefficients of a genus one Eisenstein series. The main point here is to make this relation precise and explicit. The result of this effort, Theorems 5.8.1 and 5.8.7, express the central derivative of the rank one Fourier coefficients of the genus two Eisenstein series in terms of the derivative of a genus one Eisenstein series at a critical *noncentral* point $s = \frac{1}{2}$. This genus one Eisenstein series was studied in detail in [9] with this application in mind. In the last section, we prove the analogous relation for the constant terms of the two Eisenstein series. In Chapter 6, we will use these results together with those of Chapter 3 to prove one of the main results in this book—the coincidence of the central derivative of the genus two Eisenstein series computed here with the generating function for 0-cycles on the arithmetic surface attached to a Shimura curve.

5.1 GENUS TWO EISENSTEIN SERIES

In this section, we construct the Eisenstein series of genus two and weight $\frac{3}{2}$ attached to an indefinite quaternion algebra B, by specializing the general construction of incoherent Eisenstein series given in part 1 of [4]. For the moment, we allow the case $B = M_2(\mathbb{Q})$. More details can be found in [4].

We begin by fixing some notation and reviewing the structure of the space of Siegel Eisenstein series. Let Sp_2 be the rank 2 symplectic group over $\mathbb{Q}$

and let $G_{\mathbb{A}}$ be the metaplectic extension

$$(5.1.1) \qquad 1 \longrightarrow \mathbb{C}^1 \longrightarrow G_{\mathbb{A}} \xrightarrow{\text{pr}} \mathrm{Sp}_2(\mathbb{A}) \longrightarrow 1.$$

We write

$$(5.1.2) \qquad \underline{P} = \underline{N}\,\underline{M} = \{\, n(b)m(a) \mid b \in \mathrm{Sym}_2,\ a \in \mathrm{GL}_2 \,\},$$

for the Siegel parabolic subgroup of Sp_2, where

$$(5.1.3) \qquad n(b) = \begin{pmatrix} 1 & b \\ & 1 \end{pmatrix}, \qquad \text{and} \qquad m(a) = \begin{pmatrix} a & \\ & {}^t a^{-1} \end{pmatrix}.$$

Let

$$(5.1.4) \qquad \underline{K} = \underline{K}_\infty \cdot \prod_p \underline{K}_p,$$

where $\underline{K}_p = \mathrm{Sp}_2(\mathbb{Z}_p)$, for $p < \infty$, and $\underline{K}_\infty \simeq U(2)$ is the standard maximal compact subgroup. Let

$$(5.1.5) \qquad \begin{aligned} G_{\mathbb{R}} &= \mathrm{pr}^{-1}(\mathrm{Sp}_2(\mathbb{R})), \\ G_p &= \mathrm{pr}^{-1}(\mathrm{Sp}_2(\mathbb{Q}_p)), \\ G_{\mathbb{A}_f} &= \mathrm{pr}^{-1}(\mathrm{Sp}_2(\mathbb{A}_f)), \\ P_{\mathbb{A}} &= \mathrm{pr}^{-1}(\underline{P}(\mathbb{A})), \\ M_{\mathbb{A}} &= \mathrm{pr}^{-1}(\underline{M}(\mathbb{A})), \\ K &= \mathrm{pr}^{-1}(\underline{K}), \quad K_\infty = \mathrm{pr}^{-1}(\underline{K}_\infty), \end{aligned}$$

and let $G_{\mathbb{Q}}$ (resp. $N_{\mathbb{A}}$) be the image of $\mathrm{Sp}_2(\mathbb{Q})$ (resp. $\underline{N}(\mathbb{A})$) in $G_{\mathbb{A}}$ under the unique splitting homomorphism.

Let ψ be the standard character of $\mathbb{A}/\mathbb{Q}$ which is unramified and such that $\psi_\infty(x) = e(x) = e^{2\pi i x}$. As explained in Section 8.5.5, there is then an isomorphism of groups, via the Leray coordinates,

$$(5.1.6) \qquad \underline{P}(\mathbb{A}) \times \mathbb{C}^1 \xrightarrow{\sim} P_{\mathbb{A}} \qquad (p, z) \mapsto [p, z]_{L,\psi} = [p, z]_L.$$

Note that

$$(5.1.7) \qquad P_{\mathbb{Q}} := P_{\mathbb{A}} \cap G_{\mathbb{Q}} = [\underline{P}(\mathbb{Q}), 1]_L.$$

For a character χ of $\mathbb{A}^\times/\mathbb{Q}^\times$, we also write χ for the character of $P_{\mathbb{A}}$ defined by

$$(5.1.8) \qquad \chi([n(b)m(a), z]_L) = z\,\chi(\det(a)).$$

The character χ on $P_{\mathbb{A}}$ depends on the isomorphism (5.1.6) and hence on the choice of ψ, but we suppress this dependence from the notation. For $s \in \mathbb{C}$, let $I(s,\chi)$ be the global degenerate principal series representation of $G_{\mathbb{A}}$ on smooth functions $\Phi(s)$ on $G_{\mathbb{A}}$ satisfying

$$\Phi([n(b)m(a),z]_L g,s) = z\,\chi(\det(a))\,|\det(a)|^{s+\frac{3}{2}}\,\Phi(g,s). \tag{5.1.9}$$

We also require that $\Phi(s)$ be right K_∞-finite, so that $I(s,\chi)$ is a $(\mathfrak{g},K_\infty)\times G_{\mathbb{A}_f}$-module. A section $\Phi(s)$ is called standard if its restriction to K is independent of s. For such a section $\Phi(s)$, the corresponding Siegel Eisenstein series

$$E(g,s,\Phi) = \sum_{\gamma\in P_{\mathbb{Q}}\backslash G_{\mathbb{Q}}} \Phi(\gamma g,s) \tag{5.1.10}$$

converges for $\mathrm{Re}(s) > \frac{3}{2}$. The analytic continuation of $E(g,s,\Phi)$ to the whole s-plane is holomorphic at the point $s=0$ on the unitary axis and hence there is a $(\mathfrak{g},K_\infty)\times G_{\mathbb{A}_f}$-intertwining map

$$E(0): I(0,\chi) \longrightarrow \mathcal{A}(G_{\mathbb{A}}), \qquad \Phi(0)\mapsto E(\cdot,0,\Phi), \tag{5.1.11}$$

where $\mathcal{A}(G_{\mathbb{A}})$ is the space of genuine automorphic forms on $G_{\mathbb{A}}$.

The representation $I(s,\chi)$ is a restricted product

$$I(s,\chi) = \otimes_{p\le\infty} I_p(s,\chi_p) \tag{5.1.12}$$

of local degenerate principal series representations where $\chi = \otimes_{p\le\infty}\chi_p$. Assume that $\chi^2=1$ and write

$$\chi(x) = (x,2\kappa)_{\mathbb{A}} \tag{5.1.13}$$

for $\kappa\in\mathbb{Q}^\times$, where $(\ ,\ )_{\mathbb{A}}$ is the global Hilbert symbol. For any $p\le\infty$, the representation $I_p(0,\chi_p)$ is unitarizable and is the direct sum

$$I_p(0,\chi_p) = R_p(V_p^+)\oplus R_p(V_p^-) \tag{5.1.14}$$

of irreducible representations defined as follows. Let $B_p^\pm$ be the quaternion algebra over $\mathbb{Q}_p$ with invariant $\mathrm{inv}_p(B_p^\pm)=\pm1$ and let

$$V_p^\pm = \{\, x\in B_p^\pm \mid \mathrm{tr}(x)=0\,\}, \tag{5.1.15}$$

with quadratic form $Q(x) = -\kappa\cdot\nu(x) = \kappa x^2$. Soon we will specialize to the case $\kappa=-1$. For the moment, write $V_p = V_p^\pm$. If $p<\infty$, the group G_p acts in the Schwartz space $S(V_p^2)$ by the Weil representation $\omega_{V_p}=\omega_{V_p,\psi_p}$. If $p=\infty$, we let $S(V_\infty^2)$ be the subspace of the Schwartz space of V_∞^2 which corresponds to the space of polynomial functions in a Fock model compatible with K_∞ and some choice of maximal compact subgroup of $O(V_\infty)$, [2]. Then $S(V_\infty^2)$ is a $(\mathfrak{g}, K_\infty)$-module via the Weil representation $\omega_{V_\infty} = \omega_{V_\infty,\psi_\infty}$. The map

(5.1.16)
$$\lambda = \lambda_V : S((V_p^\pm)^2) \longrightarrow I_p(0, \chi_p), \qquad \varphi \mapsto (g \mapsto \omega_{\psi,V}(g)\varphi(0))$$

is G_p-intertwining (resp. $(\mathfrak{g}, K_\infty)$-intertwining), and the image

$$R_p(V_p) := \lambda_{V_p}(S(V_p^2)) \tag{5.1.17}$$

is an irreducible submodule and is isomorphic to the space $S(V_p^2)_{O(V_p)}$ of $O(V_p)$-coinvariants [12] (resp. $(O(V_\infty), K_{O(V_\infty)})$-coinvariants [6]). In the case $p = \infty$, we will sometimes write

$$R_\infty(V_\infty^-) = \begin{cases} R_\infty(3,0) & \text{if } \kappa < 0, \\ R_\infty(0,3) & \text{if } \kappa > 0, \end{cases} \tag{5.1.18}$$

and

$$R_\infty(V_\infty^+) = \begin{cases} R_\infty(1,2) & \text{if } \kappa < 0, \\ R_\infty(2,1) & \text{if } \kappa > 0, \end{cases} \tag{5.1.19}$$

according to the signatures of the quadratic spaces involved. Note that $R_\infty(p,q)$ is a cyclic $(\mathfrak{g}, K_\infty)$-module generated by the vector $\Phi_\infty^\ell(0)$ with scalar K_∞-type $(\det)^\ell$ where $\ell = \frac{1}{2}(p-q)$, [6]. Let $\Phi_\infty^\ell(s)$ be the standard section of $I_\infty(s, \chi_\infty)$ having this scalar K_∞-type and normalized so that $\Phi_\infty^\ell(e, s) = 1$.

For a global quaternion algebra B with corresponding ternary quadratic space V^B, where the quadratic form is defined by $Q(x) = -\kappa \cdot \nu(x) = \kappa x^2$, let

$$\Pi(V^B) = \otimes_p R_p(V_p^B) \tag{5.1.20}$$

be the associated irreducible summand of $I(0, \chi)$. Similarly, for an incoherent collection $\mathcal{C} = \{\mathcal{C}_p\}$, $\mathcal{C}_p = V_p^{\epsilon_p}$, i.e., a collection of spaces which differs from the collection $\{V_p^B\}$ at an odd number of places, there is an irreducible summand

$$\Pi(\mathcal{C}) = \otimes_p R_p(\mathcal{C}_p), \tag{5.1.21}$$

and

$$I(0, \chi) = \Big(\oplus_B \, \Pi(V^B) \Big) \oplus \Big(\oplus_{\mathcal{C}} \, \Pi(\mathcal{C}) \Big) \tag{5.1.22}$$

is the decomposition of $I(0, \chi)$ into irreducibles. The Siegel-Weil formula describes the image of the summands $\Pi(V^B)$ under the Eisenstein map

$E(0)$ as spaces of theta functions [4], [7]. The summands $\Pi(\mathcal{C})$ lie in the kernel of $E(0)$ and occur only as subquotients of the space of automorphic forms on $G_{\mathbb{A}}$ [10].

We now specialize to the case of the Eisenstein series of weight $\frac{3}{2}$ whose central derivative will be related to the genus two generating function to be defined in Chapter 6.

Fix an indefinite quaternion algebra B, and take $\kappa = -1$, so that $\chi(x) = (x, -2)_{\mathbb{A}}$ and the quadratic form on V^B is given by $Q(x) = \nu(x) = -x^2$. Let O_B be a maximal order in B and let $\varphi^B \in S(V^B(\mathbb{A}_f)^2)$ be the characteristic function of the set $(V(\mathbb{A}_f) \cap (O_B \otimes_{\mathbb{Z}} \hat{\mathbb{Z}}))^2$. Then, let $\Phi_f^B(s)$ be the standard section of $I_f(s, \chi_f)$ with

$$\Phi_f^B(0) = \lambda_V(\varphi^B). \tag{5.1.23}$$

Note that $R_\infty(V_\infty^B) = R_\infty(1, 2)$. Thus, the standard section $\Phi_\infty^{-\frac{1}{2}}(s) \otimes \Phi_f^B(s)$ is coherent with

$$\Phi_\infty^{-\frac{1}{2}}(0) \otimes \Phi_f^B(0) \quad \in \Pi(V^B), \tag{5.1.24}$$

whereas the standard section $\Phi_\infty^{\frac{3}{2}}(s) \otimes \Phi_f^B(s)$ is incoherent with

$$\Phi_\infty^{\frac{3}{2}}(0) \otimes \Phi_f^B(0) \quad \in \Pi(\mathcal{C}^B), \tag{5.1.25}$$

for the collection $\mathcal{C}^B = \{\mathcal{C}_p^B\}$ with $\mathcal{C}_p^B = V_p^B$ for $p < \infty$ and with $\mathcal{C}_\infty^B = R_\infty(3, 0)$. This is almost the section we want, but it turns out that a further modification is needed at the primes $p \mid D(B)$.

For any p, let $B_p^\pm$ be the quaternion algebra over $\mathbb{Q}_p$ with $\mathrm{inv}_p(B_p^\pm) = \pm 1$, and let

$$V_p^\pm = \{\, x \in B_p^\pm \mid \mathrm{tr}(x) = 0 \,\}. \tag{5.1.26}$$

For $p < \infty$, let

$$L_p^0 = M_2(\mathbb{Z}_p) \cap V_p^+, \tag{5.1.27}$$

$$L_p^1 = \{\begin{pmatrix} a & b \\ c & d \end{pmatrix} \in M_2(\mathbb{Z}_p) \mid \mathrm{ord}_p(c) > 0\} \cap V_p^+, \tag{5.1.28}$$

and

$$L_p^{\mathrm{ra}} = O_p^{\mathrm{ra}} \cap V_p^-, \tag{5.1.29}$$

where O_p^{ra} is the maximal order in the division quaternion algebra B_p^- over $\mathbb{Q}_p$. Let φ_p^0, $\varphi_p^1 \in S((V_p^+)^2)$ and $\varphi_p^{\mathrm{ra}} \in S((V_p^-)^2)$ be the characteristic functions of $(L_p^0)^2$, $(L_p^1)^2$, and $(L_p^{\rm ra})^2$ respectively. Then, let $\Phi_p^0(s)$, $\Phi_p^1(s)$, and $\Phi_p^{\rm ra}(s)$ be the standard local sections with

$$\Phi_p^0(0) = \lambda(\varphi_p^0), \tag{5.1.30}$$

$$\Phi_p^1(0) = \lambda(\varphi_p^1), \tag{5.1.31}$$

and

$$\Phi_p^{\rm ra}(0) = \lambda(\varphi_p^{\rm ra}). \tag{5.1.32}$$

The local modified section $\tilde{\Phi}_p(s)$ is defined as follows [8]:

$$\tilde{\Phi}_p(s) = \Phi_p^{\rm ra}(s) + A(s)\,\Phi_p^0(s) + B(s)\,\Phi_p^1(s), \tag{5.1.33}$$

where $A_p(s)$ and $B_p(s)$ are rational functions of p^{-s} with the property that

$$A_p(0) = B_p(0) = 0, \tag{5.1.34}$$

$$A_p'(0) = -\frac{2}{p^2-1}\log(p) \qquad \text{and} \qquad B_p'(0) = \frac{1}{2}\cdot\frac{p+1}{p-1}\log(p). \tag{5.1.35}$$

Note that $\tilde{\Phi}_p(s)$ is *not* a standard section and that $\tilde{\Phi}_p(0) = \Phi_p^{\rm ra}(0)$. This particular combination of standard sections was originally defined in [8] to match certain intersection numbers of special cycles on the Drinfeld upper half space. It turns out to have good properties for the local doubling integral as well; see Chapter 8.

Finally, we define the modified global section

$$\tilde{\Phi}^B(s) = \Phi_\infty^{\frac{3}{2}}(s) \otimes \big(\otimes_{p|D(B)}\tilde{\Phi}_p(s)\big) \otimes \big(\otimes_{p\nmid D(B)}\Phi_p^0(s)\big) \tag{5.1.36}$$

and the corresponding normalized Eisenstein series

$$C^{D(B)}(s)\cdot E(g,s,\tilde{\Phi}^B), \tag{5.1.37}$$

where, for any square free positive integer D,

$$C^D(s) = -\frac{1}{2}\cdot(s+1)\,c(D)\,\Lambda_D(2s+2) \tag{5.1.38}$$

with

$$\Lambda_D(2s) = \left(\frac{D}{\pi}\right)^s \Gamma(s)\,\zeta(2s)\prod_{p|D}(1-p^{-2s}), \tag{5.1.39}$$

and

$$c(D) = (-1)^{\mathrm{ord}(D)+1}\,\frac{D}{2\pi}\prod_{p|D}(p+1)^{-1}. \tag{5.1.40}$$

We often abuse notation and write $C^B(s)$ instead of $C^{D(B)}(s)$. Note that, in contrast to the situation in [9], the central derivative $E'(g, 0, \tilde{\Phi}^B)$ only depends on the value

$$C^D(0) = (-1)^{\mathrm{ord}(D)} \frac{1}{24} \prod_{p|D} (p-1) \tag{5.1.41}$$

of the normalizing factor, since

$$E(g, 0, \tilde{\Phi}^B) = 0. \tag{5.1.42}$$

Since the section $\tilde{\Phi}^B(s)$ is invariant under a suitable compact open subgroup of K of $G_{\mathbb{A}_f}$ (see Section 8.5 for details) and is an eigenfunction for K_∞, $E_2(g, s, B)$ is determined by its restriction to $P_{\mathbb{R}}$. For $\tau = u + iv \in \mathfrak{H}_2$, let

$$g_\tau = [\,n(u)m(v^{\frac{1}{2}}), 1\,]_L \in P_{\mathbb{R}} \tag{5.1.43}$$

and let

$$\mathcal{E}_2(\tau, s, B) := \det(v)^{-\frac{3}{4}} \cdot C^B(s)\, E(g_\tau, s, \tilde{\Phi}^B). \tag{5.1.44}$$

Of course, $\mathcal{E}_2(\tau, 0, B) = 0$.

Our next task is to determine the Fourier expansion

$$\mathcal{E}_2'(\tau, 0, B) = \sum_{T \in \mathrm{Sym}_2(\mathbb{Q})} \mathcal{E}_{2,T}'(\tau, 0, B) \tag{5.1.45}$$

of the central derivative.

5.2 NONSINGULAR FOURIER COEFFICIENTS

In this section, we review the description obtained in [4] and [8] for the Fourier coefficients $\mathcal{E}_{2,T}'(\tau, 0, B)$ when $T \in \mathrm{Sym}_2(\mathbb{Q})$ with $\det(T) \neq 0$. First there is a product formula

$$\begin{aligned} E_T(g_\tau, s, \tilde{\Phi}^B) &= \int_{\mathrm{Sym}_2(\mathbb{Q})\backslash \mathrm{Sym}_2(\mathbb{A})} E(n(b)g_\tau, s, \tilde{\Phi}^B)\, \psi(-\mathrm{tr}(Tb))\, db \\ &= \int_{\mathrm{Sym}_2(\mathbb{A})} \tilde{\Phi}^B(w^{-1}n(b)g_\tau, s)\, \psi(-\mathrm{tr}(Tb))\, db \\ &= W_{T,\infty}(g_\tau, s, \Phi_\infty^{\frac{3}{2}}) \cdot \prod_p W_{T,p}(s, \tilde{\Phi}_p^B), \end{aligned} \tag{5.2.1}$$

where
(5.2.2)

$$W_{T,\infty}(g_\tau, s, \Phi_\infty^{\frac{3}{2}}) = \int_{\mathrm{Sym}_2(\mathbb{R})} \Phi_\infty^{\frac{3}{2}}(w_\infty^{-1} n(b) g_\tau, s)\, \psi_\infty(-\mathrm{tr}(Tb))\, d_\infty b$$

and

$$W_{T,p}(s, \tilde{\Phi}_p^B) = \int_{\mathrm{Sym}_2(\mathbb{Q}_p)} \tilde{\Phi}_p^B(w_p^{-1} n(b), s)\, \psi_p(-\mathrm{tr}(Tb))\, d_p b \tag{5.2.3}$$

are local (degenerate) Whittaker integrals. Here, we are writing $n(b)$ for $[n(b), 1]_L = \prod_p [n(b_p), 1]_{L,p}$, since the splitting homomorphism $\underline{N}(\mathbb{A}) \to G_{\mathbb{A}}$ is unique. The global measure db is the Tamagawa measure; this measure is self-dual with respect to the pairing $[b_1, b_2] = \psi(\mathrm{tr}(b_1 b_2))$ determined by ψ. The local measures $d_p b$ are self-dual with respect to the analogous pairing determined local components ψ_p of ψ. Also

$$w = \begin{pmatrix} & 1_2 \\ -1_2 & \end{pmatrix} \in G_{\mathbb{Q}}, \tag{5.2.4}$$

and we choose elements $w_p = [w_p, 1]_L \in K_p$ projecting to w in $\underline{K}_p$ and with $w = \prod_p w_p$ in $G_{\mathbb{A}}$; see Sections 8.5.1 and 8.5.5. Since the local section $\tilde{\Phi}_p(s)$ is right invariant under $\underline{N}(\mathbb{Z}_p)$, we have immediately

Lemma 5.2.1. *For $T \in \mathrm{Sym}_2(\mathbb{Q})$ with*

$$T \notin \mathrm{Sym}_2(\mathbb{Z})^\vee = \{\, T \in \mathrm{Sym}_2(\mathbb{Q}) \mid \mathrm{tr}(Tb) \in \mathbb{Z},\ \forall b \in \mathrm{Sym}_2(\mathbb{Z}) \,\},$$

$E_T(g_\tau, s, \tilde{\Phi}^B) = 0.$

We next discuss the individual factors in the product on the right side of (5.2.1).

For $T \in \mathrm{Sym}_2(\mathbb{Z})^\vee$ with $\det(T) \neq 0$, let

$$S(T, B) = \{\, p \mid p \mid 8D(B)\det(T) \,\}. \tag{5.2.5}$$

By Proposition 4.1 of [4], for a finite prime $p \notin S(B, T)$, we have

$$W_{T,p}(s, \tilde{\Phi}_p^B) = \zeta_p(2s+2)^{-1}. \tag{5.2.6}$$

For $T \in \mathrm{Sym}_2(\mathbb{Z})^\vee$ with $\det(T) \neq 0$ and $S = S(T, B)$, we have
(5.2.7)

$$E_T(g_\tau, s, \tilde{\Phi}^B) = W_{T,\infty}(g_\tau, s, \Phi_\infty^{\frac{3}{2}}) \cdot \prod_{p \in S} W_{T,p}(s, \tilde{\Phi}_p^B) \cdot \zeta^S(2s+2)^{-1}.$$

Note that, since the local Whittaker functions for nonsingular T's are entire functions of s, the finite product on the right side of (5.2.7) gives the analytic continuation of the nonsingular Fourier coefficient.

Since $E_T(g_\tau, 0, \tilde{\Phi}^B) = 0$, at least one of the factors in the product must vanish at $s = 0$, and, if more than one factor vanishes, then $E'_T(g_\tau, 0, \tilde{\Phi}^B) = 0$ as well. The vanishing of local factors is controlled by the following principle.

Lemma 5.2.2. *(i) For any standard local section $\Phi_p(s)$ for which $\Phi_p(0) \in R_p(V_p)$,*

$$W_{T,p}(0, \Phi_p) \neq 0 \implies T \text{ is represented by } V_p.$$

Similarly, for the nonstandard section $\tilde{\Phi}_p(s)$ defined by (5.1.33),

$$W_{T,p}(0, \tilde{\Phi}_p) \neq 0 \implies T \text{ is represented by } V_p^-.$$

(ii) $T \in \mathrm{Sym}_2(\mathbb{Q}_p)$ with $\det(T) \neq 0$ is represented by V_p^ϵ if and only if

$$\epsilon = \epsilon_p(T)\, \chi_V(\det(T)) = \epsilon_p(T)\, (-\det(T), -1)_p,$$

where $\epsilon_p(T)$ is the Hasse invariant of T.

Note that V_p^ϵ has Hasse invariant $\epsilon_p(V_p^\epsilon) = (-1,-1)_p\, \epsilon$, so that there is a twist here when $p = 2$ or ∞, as compared with Lemma 8.2 of [4].

Definition 5.2.3. For $T \in \mathrm{Sym}_2(\mathbb{Z}_p)^\vee$ with $\det(T) \neq 0$, let

$$\mu_p(T) = \epsilon_p(T)\, (-\det(T), -1)_p.$$

Note that, by (ii) of the previous lemma, $\mu_p(T) = 1$ if and only if T is represented by the space V_p^+ of trace 0 elements in $M_2(\mathbb{Q}_p)$, with quadratic form given by the determinant. If p is odd and T is $\mathrm{GL}_2(\mathbb{Z}_p)$-equivalent to the diagonal form $\mathrm{diag}(\varepsilon_1 p^{a_1}, \varepsilon_2 p^{a_2})$ with $0 \leq a_1 \leq a_2$ and $\varepsilon_1, \varepsilon_2 \in \mathbb{Z}_p^\times$, then

$$\mu_p(T) = (\varepsilon_1, p)_p^{a_2}\, (\varepsilon_2, p)_p^{a_1}\, (-1, p)_p^{a_1 a_2}; \tag{5.2.8}$$

cf. Lemma 8.3 of [4].

Recall that $\mathcal{C} = \mathcal{C}^B$ is the incoherent collection with $\mathcal{C}_\infty = V(3,0) = V_\infty^-$ and $\mathcal{C}_p^B = V_p^B$, for $p < \infty$. Let $\mathrm{inv}_\infty(\mathcal{C}^B) = -1$ and $\mathrm{inv}_p(\mathcal{C}^B) = \mathrm{inv}_p(B)$ for $p < \infty$. Then, the set

$$\mathrm{Diff}(T, B) = \{\, p \leq \infty \mid \mathrm{inv}_p(\mathcal{C}^B) \neq \mu_p(T) \,\} \tag{5.2.9}$$

has odd cardinality, and, by the previous lemma,

$$|\mathrm{Diff}(T, B)| > 1 \quad \implies \quad E'_T(g_\tau, 0, \tilde{\Phi}^B) = 0. \tag{5.2.10}$$

Thus, only T's with $|\mathrm{Diff}(T, B)| = 1$ contribute to the central derivative.

The following results concerning the values and derivatives of local factors $W_{T,p}(s)$ are collected from [4] for $p \nmid 2D(B)$, from [17] and [8] for $p \mid D(B)$, $p \neq 2$, and from [18] for $p = 2$. For $p \leq \infty$, let

$$C_p(V) = \gamma_p(V)^2 \, |D(B)|_p^2 \, |2|_p^{\frac{3}{2}}, \tag{5.2.11}$$

where the quantity $\gamma_p(V)$ is a local Weil index defined in (8.5.21). This quantity will frequently appear as a constant of proportionality. Note that

$$\prod_{p \leq \infty} C_p(V) = 1. \tag{5.2.12}$$

Theorem 5.2.4. (Kitaoka [3] for $p \neq 2$, Yang [18], Theorem 5.7)
Suppose that $p \nmid D(B)$, so that $\tilde{\Phi}_p^B(s) = \Phi_p^0(s)$. Let $X = p^{-s}$.
(i) If $T \notin \mathrm{Sym}_2(\mathbb{Z}_p)^\vee$, then $W_{T,p}(s) = 0$.
(ii) Suppose that $T \in \mathrm{Sym}_2(\mathbb{Z}_p)^\vee$ and let $(0, a_1, a_2)$, with $0 \leq a_1 \leq a_2$, be the Gross-Keating invariants of the matrix $\mathrm{diag}(1, T) \in \mathrm{Sym}_3(\mathbb{Z}_p)^\vee$. Then the quantity

$$\frac{W_{T,p}(s, \tilde{\Phi}_p^B)}{C_p(V) \cdot (1 - p^{-2s-2})}$$

is given by

$$\sum_{j=0}^{\frac{a_1-1}{2}} \left(X^{2j} + \mu_p(T) X^{a_1+a_2-2j} \right) p^j$$

if a_1 is odd, and

$$\sum_{j=0}^{\frac{a_1}{2}-1} \left(X^{2j} + \mu_p(T) X^{a_1+a_2-2j} \right) p^j + p^{\frac{a_1}{2}} X^{a_1} \sum_{j=0}^{a_2-a_1} (\epsilon_0 X)^j$$

if a_1 is even, where $\epsilon_0 = \epsilon_0(T)$ is the Gross-Keating ϵ-constant [1], *p. 236, and the constant $C_p(V)$ is given by (5.2.11).*

Note that, when p is odd and T is $\mathrm{GL}_2(\mathbb{Z}_p)$ equivalent to the matrix $\mathrm{diag}(\varepsilon_1 p^{a_1}, \varepsilon_2 p^{a_2})$, as above, then $(0, a_1, a_2)$ are the Gross-Keating invariants of $\mathrm{diag}(1, T)$, and $\epsilon_0 = \epsilon_0(T) = (-\varepsilon_1, p)_p$.

Corollary 5.2.5. *Suppose that $p \nmid D(B)$ and that $T \in \mathrm{Sym}_2(\mathbb{Z}_p)^\vee$ is as in (ii) of Theorem 5.2.4.*

(i) If $\mu_p(T) = +1$, *then*

$$\frac{W_{T,p}(0, \tilde{\Phi}_p^B)}{C_p(V) \cdot (1 - p^{-2})}$$

$$= \begin{cases} 2\sum_{j=0}^{\frac{a_1-1}{2}} p^j & \text{if } a_1 \text{ is odd}, \\ 2\sum_{j=0}^{\frac{a_1}{2}-1} p^j + p^{\frac{a_1}{2}} & \text{if } a_1 \text{ is even and } \epsilon_0 = -1, \\ 2\sum_{j=0}^{\frac{a_1}{2}-1} p^j + p^{\frac{a_1}{2}}(a_2 - a_1 + 1) & \text{if } a_1 \text{ is even and } \epsilon_0 = 1. \end{cases}$$

(ii) If $\mu_p(T) = -1$, *then* $W_{T,p}(0, \tilde{\Phi}_p^B) = 0$ *and*

$$W'_{T,p}(0, \tilde{\Phi}_p^B) = C_p(V) \cdot (1 - p^{-2}) \log(p) \cdot \nu_p(T),$$

where

$$\nu_p(T)$$

$$= \begin{cases} \sum_{j=0}^{\frac{a_1-1}{2}} (a_1 + a_2 - 4j)\, p^j & \text{if } a_1 \text{ is odd}, \\ \sum_{j=0}^{\frac{a_1}{2}-1} (a_1 + a_2 - 4j)\, p^j + \frac{1}{2}(a_2 - a_1 + 1)\, p^{\frac{a_1}{2}} & \text{if } a_1 \text{ is even}. \end{cases}$$

Next we consider the case $p \mid D(B)$ and we recall that $\tilde{\Phi}_p^B(0) = \Phi_p^{\mathrm{ra}}(0)$ and that, by (5.1.33),

(5.2.13)

$$W'_{T,p}(0, \tilde{\Phi}_p^B) = W'_{T,p}(0, \Phi_p^{\mathrm{ra}}) + A'_p(0) \cdot W_{T,p}(0, \Phi_p^0) + B'_p(0) \cdot W_{T,p}(0, \Phi_p^1),$$

where the coefficients $A'_p(0)$ and $B'_p(0)$ are given by (5.1.35). When $p \neq 2$, the following result is a combination of Proposition 8.7 of [4] and Corollary 7.4 of [8]. These results, in turn, rely on the computation of the local densities [11], [17] and their derivatives [17]. In the case $p = 2$, the densities and their derivatives are computed in [18], and it is shown that the statements of Proposition 8.7 of [4] and Corollary 7.4 of [8] remain valid provided one uses the Gross-Keating invariants of $\mathrm{diag}(1, T)$.

Theorem 5.2.6. *Suppose that* $p \mid D(B)$ *and that* $T \in \mathrm{Sym}_2(\mathbb{Z}_p)^\vee$ *is as in (ii) of Theorem 5.2.4.*

(i) If $\mu_p(T) = -1$, *then*

$$W_{T,p}(0, \tilde{\Phi}_p^B) = W_{T,p}(0, \Phi_p^{\mathrm{ra}}) = C_p(V) \cdot 2\,(p+1).$$

(ii) If $\mu_p(T) = 1$, *then* $W_{T,p}(0, \tilde{\Phi}_p^B) = 0$, *and*

$$W'_{T,p}(0, \tilde{\Phi}_p^B) = C_p(V) \cdot (p+1) \log(p) \cdot \frac{1}{2} \cdot \nu_p(T),$$

where

$$\frac{1}{2}\cdot \nu_p(T) = a_1 + a_2 + 1$$

$$- \begin{cases} p^{\frac{a_1}{2}} + 2\,\frac{p^{\frac{a_1}{2}}-1}{p-1} & \text{if } a_1 \text{ is even and } \epsilon_0 = -1, \\ (a_2 - a_1 + 1)\,p^{\frac{a_1}{2}} + 2\,\frac{p^{\frac{a_1}{2}}-1}{p-1} & \text{if } a_1 \text{ is even and } \epsilon_0 = 1, \\ 2\,\frac{p^{\frac{a_1+1}{2}}-1}{p-1} & \text{if } a_1 \text{ is odd.} \end{cases}$$

Note that this result is the analogue, for $p \mid D(B)$, of Corollary 5.2.5. At this point, we omit the analogue of Theorem 5.2.4, i.e., the expressions for $W_{T,p}(s, \tilde{\Phi}_p^B)$ for general s. These are rather messy and will be given in Section 5.7 below.

Finally, we have the case in which $p = \infty$, where we use the calculations of [4], which are based on Shimura's formulas for the confluent hypergeometric functions [15]. Note, however, that in [4], $[w^{-1}, 1]_{\mathrm{R}} = [w^{-1}, -i]_{\mathrm{L}}$ was used as the preimage in $G_{\mathbb{R}}$ of w^{-1}. Thus the values here include an extra factor of i, since we are taking $[w^{-1}, 1]_{\mathrm{L}}$ for preimage of w^{-1}.

Theorem 5.2.7. *(i) If $T \in \mathrm{Sym}_2(\mathbb{R})_{>0}$ is positive definite, then, for $\tau = u + iv \in \mathfrak{H}_2$,*

$$W_{T,\infty}(g_\tau, 0, \Phi_\infty^{\frac{3}{2}}) = -2\sqrt{2}\,(2\pi)^2\,\det(v)^{\frac{3}{4}} \cdot q^T,$$

where $q^T = e(\mathrm{tr}(T\tau))$.
(ii) If $T \in \mathrm{Sym}_2(\mathbb{R})$ has signature $(1,1)$ or $(0,2)$, then $W_{T,\infty}(g_\tau, 0, \Phi_\infty^{\frac{3}{2}}) = 0$, and

$$W'_{T,\infty}(g_\tau, 0, \Phi_\infty^{\frac{3}{2}}) = -2\sqrt{2}\cdot 2\pi^2 \cdot \det(v)^{\frac{3}{4}} \cdot q^T \cdot \nu_\infty(T, v),$$

where $\nu_\infty(T, v)$ is defined as follows.
If T has signature $(1,1)$, then

$$\nu_\infty(T, v) = -\frac{1}{2}\int_{\mathbb{H}} \mathrm{Ei}(-4\pi\delta_- y^{-2}|z|^2) \times \left(\delta_+ y^{-2}(1+|z|^2)^2 - \frac{1}{2\pi}\right) e^{-\pi\delta_+ (y^{-2}(1+|z|^2)^2 - 4)}\,d\mu(z),$$

where

$$v^{\frac{1}{2}} T v^{\frac{1}{2}} = {}^t k(\theta)\cdot \mathrm{diag}(\delta_+, -\delta_-)\cdot k(\theta),$$

for $k(\theta) \in \mathrm{SO}(2)$ *and* $\delta_\pm \in \mathbb{R}_{>0}$. *Here, for* $z = x + iy \in \mathbb{H}$, $d\mu(z) = y^{-2}\, dx\, dy$, *and* Ei *is the exponential integral, as in (3.5.2).*
If T *has signature* $(0, 2)$, *then*

$$\nu_\infty(T, v) = -\frac{1}{2}\int_{\mathbb{H}} \mathrm{Ei}(-4\pi\delta_2\, y^{-2}|z|^2) \times \Big(\, \delta_1\, y^{-2}(1-|z|^2)^2 - \frac{1}{2\pi}\,\Big)\, e^{-\pi\delta_1\,(\,y^{-2}(1-|z|^2)^2+4\,)}\, d\mu(z),$$

where

$$v^{\frac{1}{2}} T v^{\frac{1}{2}} = -{}^t k(\theta)\cdot \mathrm{diag}(\delta_1, \delta_2)\cdot k(\theta),$$

for $k(\theta) \in \mathrm{SO}(2)$ *and* $\delta_i \in \mathbb{R}_{>0}$.

Proof. Part (i) and the signature $(1, 1)$ part of (ii) are Proposition 9.5 and Corollary 9.8 in [4], respectively. For the sake of completeness, we give the calculation in the case of signature $(0, 2)$, which is a variant of that given in [4] for signature $(1, 1)$.

The manipulations on pp. 585–588 of [4] yield the expression

$$W_{T,\infty}(g'_\tau, s, \Phi^{\frac{3}{2}}) = i\sqrt{2}\,\det(v)^{\frac{1}{2}(s+\frac{3}{2})}\,\frac{e(\frac{\beta-\alpha}{2})\,(2\pi)^3}{2\,\Gamma_2(\alpha)\,\Gamma_2(\beta)}\,\eta(2\,v, T; \alpha, \beta)\cdot e(\mathrm{tr}(Tu)),$$

where $\alpha = \frac{1}{2}(s+3)$ and $\beta = \frac{1}{2}s$, and

$$\eta(2\,v, T; \alpha, \beta) = \int_{\substack{U>-\pi T\\ U>\pi T}} e^{-2\mathrm{tr}(Uv)}\,\det(U+\pi T)^{\alpha-\rho}\,\det(U-\pi T)^{\beta-\rho}\,dU,$$

where $\rho = \frac{3}{2}$. Writing $\pi\, T = -{}^t c\, c$ for $c \in \mathrm{GL}_2(\mathbb{R})^+$, and using Lemma 9.6 of [4], we get

$$\begin{aligned}
\eta(2\,v, T; \alpha, \beta)\cdot(\pi^2\,\det(T))^{-(\alpha+\beta-\rho)}
&= \int_{\substack{U-1>0\\ U+1>0}} e^{-2\,\mathrm{tr}(Ucv^tc)}\,\det(U-1)^{\alpha-\rho}\,\det(U+1)^{\beta-\rho}\,dU\\
&= e^{2\pi\,\mathrm{tr}(Tv)}\int_{U>0} e^{-2\pi\,\mathrm{tr}(Uv')}\,\det(U)^{\alpha-\rho}\,\det(U+2)^{\beta-\rho}\,dU\\
&= 2^{2(\alpha+\beta-\rho)}\,e^{2\pi\,\mathrm{tr}(Tv)}\int_{U>0} e^{-4\pi\,\mathrm{tr}(Uv')}\,\det(U)^{\alpha-\rho}\,\det(U+1)^{\beta-\rho}\,dU,
\end{aligned}$$

where $v' = \pi^{-1} cv^t c$. Writing

$$v' = {}^t k(\theta)\, \Delta\, k(\theta),$$

with $\Delta = \mathrm{diag}(\delta_1, \delta_2)$ and $k(\theta) \in \mathrm{SO}(2)$, we obtain

$$\begin{aligned}
&\eta(2\,v, T; \alpha, \beta) \cdot (4\pi^2 \det(T))^{-(\alpha+\beta-\rho)} \\
&\qquad = e^{2\pi\,\mathrm{tr}(Tv)} \int\limits_{U>0} e^{-4\pi\,\mathrm{tr}(U\Delta)} \det(U)^{\alpha-\rho} \det(U+1)^{\beta-\rho}\, dU.
\end{aligned}$$

Note that this integral is finite when $s = 0$. Since

$$\Gamma_2(\alpha)\,\Gamma_2(\beta) = \pi\,\Gamma\left(\frac{s+3}{2}\right)\Gamma\left(\frac{s+2}{2}\right)\Gamma\left(\frac{s}{2}\right)\Gamma\left(\frac{s-1}{2}\right)$$

has a simple pole at $s = 0$ with residue $-2\pi^2$, we obtain

$$\begin{aligned}
&W'_{T,\infty}(g'_\tau, 0, \Phi^{\frac{3}{2}}) \\
&= -i\sqrt{2}\,\det(v)^{\frac{3}{4}} \cdot 2\pi i \cdot q^T \cdot e^{4\pi\,\mathrm{tr}(Tv)} \int\limits_{U>0} e^{-4\pi\,\mathrm{tr}(U\Delta)} \det(U+1)^{-\frac{3}{2}}\, dU.
\end{aligned}$$

To put the integral here in a better form, we write

$$U = \begin{pmatrix} x & z \\ z & y \end{pmatrix}$$

and make the substitution

$$\begin{aligned}
x &= u, \\
z &= u^{\frac{1}{2}}(1+v)^{\frac{1}{2}}\,w, \\
y &= v + (1+v)w^2 = w^2 + (1+w^2)v.
\end{aligned}$$

Then

$$\begin{aligned}
\det(U) &= uv, \\
\det(U+1) &= (1+v)(1+u+w^2), \\
\left|\frac{\partial(x,y,z)}{\partial(u,v,w)}\right| &= u^{\frac{1}{2}}(1+v)^{\frac{1}{2}},
\end{aligned}$$

and we have

$$\int_{U>0} e^{-4\pi\,\mathrm{tr}(U\Delta)}\,\det(U+1)^{-\frac{3}{2}}\,dU$$

$$= \int_{u>0}\int_{v>0}\int_{w} e^{-4\pi\,(\,\delta_1\,u+\delta_2\,(w^2+(1+w^2)v)\,)}\,(1+v)^{-\frac{3}{2}}$$

$$\times\,(1+u+w^2)^{-\frac{3}{2}}\,u^{\frac{1}{2}}(1+v)^{\frac{1}{2}}\,du\,dv\,dw.$$

Now putting $(1+w^2)u$ for u, we obtain

$$\int_{u>0}\int_{v>0}\int_{w} e^{-4\pi\,(\,\delta_1\,(1+w^2)u+\delta_2\,(w^2+(1+w^2)v)\,)}$$

$$\times\,(1+v)^{-1}(1+u)^{-\frac{3}{2}}\,u^{\frac{1}{2}}\,du\,dv\,dw.$$

The integral with respect to v here is

$$\int_{v>0} e^{-4\pi\,\delta_2\,(1+w^2)v}\,(1+v)^{-1}\,dv = -e^{4\pi\,\delta_2\,(1+w^2)}\,\mathrm{Ei}(-4\pi\,\delta_2\,(1+w^2)).$$

Applying integration by parts to the integral with respect to u, we have

$$\int_{u>0} e^{-4\pi\,\delta_1\,(1+w^2)u}\,(1+u)^{-\frac{3}{2}}\,u^{\frac{1}{2}}\,du$$

$$= -2\pi\int_{u>0} e^{-4\pi\,\delta_1\,(1+w^2)u}\,(\,4\,\delta_1\,(1+w^2)\,u - \frac{1}{2\pi}\,)\,u^{-\frac{1}{2}}\,(1+u)^{-\frac{1}{2}}\,du.$$

Combining these facts, we obtain the expression

$$2\pi\,e^{4\pi\,\delta_2}\int_{w}\int_{u>0}\mathrm{Ei}(-4\pi\,\delta_2\,(1+w^2))$$

$$\times\,e^{-4\pi\,\delta_1\,(1+w^2)u}\,(\,4\,\delta_1\,(1+w^2)\,u - \frac{1}{2\pi}\,)\,u^{-\frac{1}{2}}\,(1+u)^{-\frac{1}{2}}\,du\,dw.$$

which is the analogue of (9.57) of [4] in the present case.

Finally, for $z = x+iy \in \mathfrak{H}$, we make the substitution given on p. 593 of [4]:

$$w = \frac{x}{y}, \qquad u = \left(\frac{|z|-|z|^{-1}}{2}\right)^2,$$

so that

$$u^{-\frac{1}{2}}\,(1+u)^{-\frac{1}{2}}\,du\,dw = 2\,y^{-2}\,dx\,dy,$$

$$(1+w^2)\,u = \frac{1}{4}\,y^{-2}(1-|z|^2)^2,$$

$$(1+w^2) = y^{-2}\,|z|^2,$$

and the previous expression becomes

$$2\pi\, e^{4\pi\,\delta_2} \int_{\mathbb{H}} \mathrm{Ei}(-4\pi\,\delta_2\, y^{-2}\,|z|^2)$$
$$\times\; e^{-\pi\,\delta_1\, y^{-2}(1-|z|^2)^2}\,\big(\,\delta_1\, y^{-2}(1-|z|^2)^2 - \frac{1}{2\pi}\,\big)\, y^{-2}\, dx\, dy.$$

Returning to the function, this gives

$$\begin{aligned} &W'_{T,\infty}(g'_\tau, 0, \Phi^{\frac{3}{2}}) \\ &= 2\sqrt{2}\cdot 2\pi^2\cdot \det(v)^{\frac{3}{4}}\cdot q^T\cdot e^{4\pi\,\mathrm{tr}(Tv)}\, e^{4\pi\,\delta_2}\int_{\mathbb{H}} \mathrm{Ei}(-4\pi\,\delta_2\, y^{-2}\,|z|^2) \\ &\qquad\qquad \times\; e^{-\pi\,\delta_1\, y^{-2}(1-|z|^2)^2}\,\big(\,\delta_1\, y^{-2}(1-|z|^2)^2 - \frac{1}{2\pi}\,\big)\, y^{-2}\, dx\, dy \\ &= -2\sqrt{2}\cdot 2\pi^2\cdot \det(v)^{\frac{3}{4}}\cdot q^T\cdot \nu_\infty(T, v), \end{aligned}$$

where

$$\nu_\infty(T, v) = -\frac{1}{2}\cdot e^{-4\pi\,\delta_1}\int_D \mathrm{Ei}(-4\pi\,\delta_2\, y^{-2}\,|z|^2)$$
$$\times\; e^{-\pi\,\delta_1\, y^{-2}(1-|z|^2)^2}\,\big(\,\delta_1\, y^{-2}(1-|z|^2)^2 - \frac{1}{2\pi}\,\big)\, y^{-2}\, dx\, dy.$$

Here we note that $\mathrm{tr}(Tv) = -(\delta_1+\delta_2)$. □

In summary, we obtain the following formulas.

Proposition 5.2.8. *Suppose that* $T \in \mathrm{Sym}_2(\mathbb{Z})^\vee$ *with* $\det(T) \neq 0$.
(i) If $\mathrm{Diff}(T, B) = \{p\}$ *with* $p < \infty$, *then*

$$\mathcal{E}'_{2,T}(\tau, 0, B) = -C^B(0)\cdot 2\sqrt{2}\cdot 4\pi^2\cdot q^T\cdot W'_{T,p}(0, \tilde{\Phi}^B_p)\cdot \prod_{\ell\neq p} W_{T,\ell}(0, \tilde{\Phi}^B_\ell),$$

where

$$W'_{T,p}(0, \tilde{\Phi}^B_p) = C_p(V)\cdot \nu_p(T)\,\log(p)\cdot \begin{cases} (1-p^{-2}) & \text{if } p\nmid D(B), \\ \frac{1}{2}\,(p+1) & \text{if } p\mid D(B). \end{cases}$$

(ii) If $\mathrm{Diff}(T, B) = \{\infty\}$ *and* $\mathrm{sig}(T) = (1, 1)$ *or* $(0, 2)$,

$$\mathcal{E}'_{2,T}(\tau, 0, B) = -C^B(0)\cdot 2\sqrt{2}\cdot 2\pi^2\cdot q^T\cdot \nu_\infty(T, v)\cdot \prod_{\ell} W_{T,\ell}(0, \tilde{\Phi}^B_\ell).$$

(iii) For all other $T \in \mathrm{Sym}_2(\mathbb{Z})^\vee$, *i.e., if* $|\mathrm{Diff}(T, B)| > 1$,

$$\mathcal{E}'_{2,T}(\tau, 0, B) = 0.$$

5.3 THE SIEGEL-WEIL FORMULA

In this section, we suppose that $T \in \mathrm{Sym}_2(\mathbb{Z})^\vee$ with $\det(T) \neq 0$ and with $\mathrm{Diff}(T, B) = \{p\}$ for $p \leq \infty$, and we evaluate the products,

$$\prod_{\ell \neq p} W_{T,\ell}(0, \tilde{\Phi}_\ell^B), \tag{5.3.1}$$

when $p < \infty$, and

$$\prod_{\ell} W_{T,\ell}(0, \tilde{\Phi}_\ell^B), \tag{5.3.2}$$

when $p = \infty$. Note that, both expressions are unchanged if, for each $\ell \mid D(B)$, we replace the section $\tilde{\Phi}_\ell^B$ by Φ_ℓ^{ra}.

5.3.1 Whittaker functions and orbital integrals

We begin with some general remarks. Suppose that V is any three dimensional anisotropic quadratic space over $\mathbb{Q}$. For $\varphi \in S(V(\mathbb{A})^2)$, let $\Phi_\varphi(s) \in I(s, \chi_V)$ be the standard section such that $\Phi_\varphi(0) = \lambda_V(\varphi)$, where $\lambda_V : S(V(\mathbb{A})^2) \to I(0, \chi)$ is the map given by (5.1.16). By the Siegel-Weil formula [5], we know that

$$E(g, 0, \Phi_\varphi) = 2\, I(g, \varphi), \tag{5.3.3}$$

where

$$I(g, \varphi) = \int_{\mathrm{O}(V)(\mathbb{Q})\backslash \mathrm{O}(V)(\mathbb{A})} \theta(g, h; \varphi)\, dh \tag{5.3.4}$$

is the theta integral, with $\mathrm{vol}(\mathrm{O}(V)(\mathbb{Q})\backslash \mathrm{O}(V)(\mathbb{A}), dh) = 1$. In particular, there is an identity of Fourier coefficients:

$$E_T(g, 0, \Phi_\varphi) = 2\, I_T(g, \varphi) \tag{5.3.5}$$

for all $T \in \mathrm{Sym}_2(\mathbb{Q})$. Since the identity (5.3.5) for all nonsingular T's was actually one of the key ingredients in the *proof* of the Siegel-Weil formula in [5] and since we now need a little more information anyway, we return to the proof of such identities given in [13], rather than viewing them as consequences (5.3.3).

If $\det(T) \neq 0$, $g \in G_{\mathbb{R}}$, and φ is factorizable, then

$$E_T(g, s, \Phi_\varphi) = W_{T,\infty}(g, s, \Phi_{\varphi,\infty}) \cdot \prod_p W_{T,p}(s, \Phi_{\varphi,p}), \tag{5.3.6}$$

as in (5.2.1) above. On the other hand, the corresponding Fourier coefficient of the theta integral is given by

$$I_T(g,\varphi) = \int_{O(V)(\mathbb{Q})\backslash O(V)(\mathbb{A})} \sum_{\substack{x\in V(\mathbb{Q})^2 \\ Q(x)=T}} \omega(g)\varphi(h^{-1}x)\,dh$$

$$(5.3.7) \qquad = \int_{O(V)(\mathbb{Q})_{x_0}\backslash O(V)(\mathbb{A})} \omega(g)\varphi(h^{-1}x_0)\,dh$$

$$= \frac{1}{2}\int_{O(V)(\mathbb{A})} \omega(g)\varphi(h^{-1}x_0)\,dh$$

$$= \frac{1}{2}\cdot O_{T,\infty}(\omega(g)\varphi_\infty)\cdot \prod_p O_{T,p}(\varphi_p),$$

where $x_0 \in V^2(\mathbb{Q})$ with $Q(x_0) = T$, and where, in the last step, we have assumed that φ is factorizable and written

$$(5.3.8) \qquad O_{T,p}(\varphi_p) = \int_{O(V)(\mathbb{Q}_p)} \varphi_p(h^{-1}x_0)\,d_{T,p}h,$$

$p \leq \infty$, for the local orbital integral. The measures are normalized as follows ([13], p. 95): Let

$$(5.3.9) \qquad Q : V^2 \longrightarrow \mathrm{Sym}_2, \qquad x \mapsto Q(x) = \frac{1}{2}\,((x_i, x_j)),$$

be the moment map. Let α and β be basis vectors for the 1-dimensional spaces $\bigwedge^6(V^2)^*$ and $\bigwedge^3(\mathrm{Sym}_2)^*$, respectively. We also write α and β for the corresponding translation invariant differential forms on V^2 and Sym_2. Let

$$(5.3.10) \qquad V^2_{reg} = \{\, x \in V^2 \mid \det Q(x) \neq 0 \,\},$$

and note that V^2_{reg} is a subset of the submersive set of the moment map, i.e., the set of $x = [x_1, x_2] \in V^2$ such that x_1 and x_2 span a 2-plane in V.

Lemma 5.3.1. *There is a 3-form ν on V^2_{reg} with the following properties.*
(i)

$$\alpha = Q^*(\beta) \wedge \nu.$$

(ii) For $(h, g) \in \mathrm{SO}(V) \times \mathrm{GL}_2$, acting on V^2 by $x \mapsto hxg^{-1}$,

$$(h, g)^*\nu = \nu.$$

(iii) For all points $x \in V^2_{reg}$, the restriction of ν to $\ker dQ_x$ is nonzero.

Proof. For $x \in V^2_{reg}$ and $s \in \mathrm{Sym}_2$, we make the usual identifications $T_x(V^2_{reg}) = V^2$ and $T_s(\mathrm{Sym}_2) = \mathrm{Sym}_2$. Then, the differential of Q is given by

$$dQ_x(v) = \frac{1}{2}\,(x,v) + \frac{1}{2}\,(v,x). \tag{5.3.11}$$

Here, $v \in V^2$, $Q(x) = \frac{1}{2}\,((x_i,x_j))$, $(x,v) = ((x_i,v_j))$, etc. For $x \in V^2_{reg}$, we define the map

$$j_x : \mathrm{Sym}_2 \longrightarrow V^2 = T_x(V^2_{reg}), \qquad j_x(u) = \frac{1}{2}\,x \cdot Q(x)^{-1} \cdot u. \tag{5.3.12}$$

Then, $(x, j_x(u)) = u$, and

$$dQ_x \circ j_x(u) = u. \tag{5.3.13}$$

Let $S_x = \mathrm{image}(j_x) \subset T_x(V^2_{reg})$, and note that

$$T_x(V^2_{reg}) = S_x \oplus \ker dQ_x. \tag{5.3.14}$$

We define a 3-form ν on V^2_{reg} as follows. Choose a triple of tangent vectors $\underline{u} = [u_1, u_2, u_3]$ with $u_i \in \mathrm{Sym}_2$ with $\beta(\underline{u}) \neq 0$. For a triple $\underline{t} = [t_1, t_2, t_3]$ of tangent vectors in $T_x(V^2)$, let

$$\nu(\underline{t}) = \alpha(j_x(\underline{u}), \underline{t}) \cdot \beta(\underline{u})^{-1}. \tag{5.3.15}$$

This quantity is independent of the choice of $\underline{u}$, and since the components of $j_x(\underline{u})$ span S_x, $\nu(\underline{t})$ depends only on the projection of $\underline{t}$ onto $\ker dQ_x$ with respect to the decomposition (5.3.14). Property (iii) is clear from the definition.

To check (i), it suffices to evaluate both sides on any 6-tuple of tangent vectors whose components span V^2, e.g., on $\underline{v} = [j_x(\underline{u}), \underline{t}]$, where the components of $\underline{t}$ span $\ker dQ_x$. Then,

$$\begin{aligned} Q^*(\beta) \wedge \nu(\underline{v}) &= \beta(dQ_x \circ j_x(\underline{u})) \cdot \nu(\underline{t}) \\ &= \beta(\underline{u}) \cdot \alpha(j_x(\underline{u}), \underline{t}) \cdot \beta(\underline{u})^{-1} \\ &= \alpha(j_x(\underline{u}), \underline{t}). \end{aligned} \tag{5.3.16}$$

To check (ii), note that, if $y = (h,g) \cdot x$, for $(h,g) \in \mathrm{SO}(V) \times \mathrm{GL}_2$, then

$$(h,g) \cdot j_x(u) = j_y({}^t g^{-1} u g^{-1}). \tag{5.3.17}$$

Then

$$\nu((h,g)_*(\underline{t})) = \alpha(j_y(\underline{u}),(h,g)_*(\underline{t}))\cdot\beta(\underline{u})^{-1}$$

$$(5.3.18)\qquad = \alpha((h,g)_*(j_x({}^tg\underline{u}g)),(h,g)_*(\underline{t}))\cdot\beta(\underline{u})^{-1}$$

$$= \det(g)^{-3}\cdot\alpha(j_x({}^tg\underline{u}g),\underline{t})\cdot\beta({}^tg\underline{u}g)^{-1}\det(g)^3$$

$$= \nu(\underline{t}).$$

□

If $x \in V^2_{reg}$ with $Q(x) = T$, then there is an isomorphism

$$(5.3.19)\qquad i_x : \mathrm{SO}(V) \longrightarrow Q^{-1}(T), \qquad h \mapsto h^{-1}x,$$

and by (ii) and (iii) of the previous lemma, $\omega = \omega_T = i_x^*(\nu)$ is a gauge form on $\mathrm{SO}(V)$.

Lemma 5.3.2. *The gauge form ω is independent of T.*

Proof. Since $i_{xg^{-1}} = (1,g)\circ i_x$, the invariance property (ii) yields

$$(5.3.20)\qquad \omega_{T'} = (i_{xg^{-1}})^*(\nu) = i_x^*\circ(1,g)^*(\nu) = i_x^*(\nu) = \omega_T,$$

where $T' = {}^tg^{-1}Tg^{-1}$. Over an algebraically closed field, the action of the group GL_2 on $(\mathrm{Sym}_2)_{reg}$, the open subset of nonsingular elements of Sym_2, is transitive. □

The form ω determines the Tamagawa measure dh_1 on $\mathrm{SO}(V)(\mathbb{A})$. On $\mathrm{O}(V)(\mathbb{A}) = \mathrm{SO}(V)(\mathbb{A})\times\mu_2(\mathbb{A})$, the measure fixed above is $dh = dh_1\times dc$, where $\mathrm{vol}(\mu_2(\mathbb{A}),dc) = 1$. On the other hand, ω determines measures d_ph_1 on the groups $\mathrm{SO}(V)(\mathbb{Q}_p)$, so that $d_ph = d_ph_1\times dc_p$, with $\mathrm{vol}(\mu_2(\mathbb{Q}_p),dc_p) = 1$, and a factorization

$$(5.3.21)\qquad dh = \prod_{p\le\infty} d_ph$$

of the Tamagawa measure. Moreover, by construction, the measure d_ph_1 coincides, under the isomorphism i_{x_0}, with the measure on $Q^{-1}(T)(\mathbb{Q}_p)$ determined by ν.

Our fixed additive character $\psi = \otimes_p\psi_p$, determines factorizations of the Tamagawa measures on $V^2(\mathbb{A})$ and on $\mathrm{Sym}_2(\mathbb{A})$ as follows. On $V^2(\mathbb{Q}_p)$, we have fixed the measure d_px which is self-dual for the pairing $[x,y] = \psi_p(\mathrm{tr}(x,y))$, where $(\ ,\ )$ is the bilinear form associated to Q. On the other hand, the gauge form α on V^2 determines a measure

$$(5.3.22)\qquad d_{\alpha,p}x = c_p(\alpha,\psi)\,d_px,$$

for a positive real constant $c_p(\alpha, \psi)$. Then, since $d_\alpha x = dx$ on $V^2(\mathbb{A})$,

$$\prod_{p \leq \infty} c_p(\alpha, \psi) = 1. \tag{5.3.23}$$

Similarly, we have fixed the measure $d_p b$ on $\mathrm{Sym}_2(\mathbb{Q}_p)$ which is self-dual with respect to the pairing $[b, c] = \psi_p(\mathrm{tr}(bc))$. The gauge form β determines a measure

$$d_{\beta,p} b = c_p(\beta, \psi)\, d_p b, \tag{5.3.24}$$

on $\mathrm{Sym}_2(\mathbb{Q}_p)$, for a positive real constant $c_p(\beta, \psi)$. Again, on $\mathrm{Sym}_2(\mathbb{A})$, $d_\beta b = db$ and so

$$\prod_{p \leq \infty} c_p(\beta, \psi) = 1. \tag{5.3.25}$$

Finally, for the Weil representation $\omega_{V,\psi} = \otimes_p \omega_{V_p,\psi_p}$, we have, for all $\varphi \in S(V^2(\mathbb{A}))$ and $\varphi_p \in S(V^2(\mathbb{Q}_p))$,

$$\omega_{V,\psi}(w^{-1})\varphi(x) = \int_{V^2(\mathbb{A})} \varphi(y)\, \psi(-\mathrm{tr}(x, y))\, dy, \tag{5.3.26}$$

and

$$\omega_{V_p,\psi_p}(w_p^{-1})\varphi_p(x) = \gamma_p(V)^2 \cdot \int_{V^2(\mathbb{Q}_p)} \varphi_p(y)\, \psi_p(-\mathrm{tr}(x, y))\, d_p y; \tag{5.3.27}$$

see (5.6.3) below and Section 8.5. Here, since we are taking $w_p^{-1} = [w^{-1}, 1]_{\mathrm{L},p}$, the constant $\gamma_p(V)^2$ is given by (8.5.21). Again, by (5.3.26),

$$\prod_{p \leq \infty} \gamma_p(V) = 1. \tag{5.3.28}$$

The following result is a very special case of the results of Chapter 4 of [13].

Proposition 5.3.3. *Let*

$$C_p(V, \alpha, \beta, \psi) = \frac{\gamma_p(V)^2\, c_p(\beta, \psi)}{c_p(\alpha, \psi)}.$$

(i) For each $p < \infty$,

$$W_{T,p}(0, \Phi_{\varphi,p}) = C_p(V, \alpha, \beta, \psi) \cdot O_{T,p}(\varphi_p).$$

(ii) For $p = \infty$, *and* $g \in G_{\mathbb{R}}$,

$$W_{T,\infty}(g, 0, \Phi_{\varphi,\infty}) = C_\infty(V, \alpha, \beta, \psi) \cdot O_{T,\infty}(\omega(g)\varphi_\infty).$$

Remark 5.3.4. Note that the constant of proportionality

$$\gamma_p(V)^2\, c_p(\beta,\psi)\, c_p(\alpha,\psi)^{-1}$$

does not depend on T. Moreover, the Fourier coefficient identity (5.3.5) is an immediate consequence of the combination of this result with (5.3.7) and the fact that

$$\prod_{p\le\infty} C_p(V,\alpha,\beta,\psi) = 1. \tag{5.3.29}$$

Corollary 5.3.5. *For any* $\varphi_f = \prod_{p<\infty}\varphi_p \in S(V(\mathbb{A}_f)^2)$,

$$\prod_{p<\infty} W_{T,p}(0,\Phi_{\varphi,p}) = C_\infty(V,\alpha,\beta,\psi)^{-1}\cdot \prod_{p<\infty} O_{T,p}(\varphi_p).$$

Proof of Proposition 5.3.3. For any $p\le\infty$, the Whittaker integral

$$W_{T,p}(s,\Phi_\varphi) = \int_{\mathrm{Sym}_2(\mathbb{Q}_p)} \Phi_\varphi(w_p^{-1}n(b),s)\,\psi_p(-\mathrm{tr}(Tb))\,d_pb, \tag{5.3.30}$$

defined for $\mathrm{Re}(s)>3/2$, has an entire analytic continuation, and the linear functional

$$\lambda_{T,p}:\varphi\mapsto W_{T,p}(0,\Phi_\varphi) \tag{5.3.31}$$

on $S(V^2(\mathbb{Q}_p))$ defines an $\mathrm{O}(V)(\mathbb{Q}_p)$-invariant distribution such that

$$\lambda_p(\omega(n(b)\varphi) = \psi_p(\mathrm{tr}(Tb))\,\lambda_{T,p}(\varphi). \tag{5.3.32}$$

In addition, if $p=\infty$, $\lambda_{T,p}$ satisfies the derivative conditions of Lemma 4.2 of [13]. By that lemma, the space of such distributions has dimension at most 1 and is spanned by the orbital integral $\varphi\mapsto O_{T,p}(\varphi)$. Thus, there is a constant c_p such that

$$W_{T,p}(0,\varphi) = c_p\cdot O_{T,p}(\varphi). \tag{5.3.33}$$

To determine c_p, we evaluate on a function $\varphi\in S(V^2_{reg}(\mathbb{Q}_p))$, which we can assume is even, i.e., invariant under the subgroup $\mu_2(\mathbb{Q}_p)$ of $\mathrm{O}(V_p)$. Recall that $U := Q(V^2_{reg})$ is a Zariski open subset of Sym_2. Then,

(5.3.34)

$$\begin{aligned}
&W_{T,p}(s,\Phi_\varphi) \\
&= \gamma_p(V)^2 \cdot \int_{\mathrm{Sym}_2(\mathbb{Q}_p)} \int_{V^2(\mathbb{Q}_p)} \psi_p(\mathrm{tr}(b\,Q(y)))\,\varphi(y)\,d_py \\
&\qquad\qquad \times |a(w_p^{-1}n(b))|^s\,\psi_p(-\mathrm{tr}(Tb))\,d_pb \\
&= \frac{\gamma_p(V)^2\,c_p(\beta,\psi)}{c_p(\alpha,\psi)} \cdot \int_{\mathrm{Sym}_2(\mathbb{Q}_p)} \int_{\mathrm{Sym}_2(\mathbb{Q}_p)} \psi_p(\mathrm{tr}(bu))\,M_\varphi(u)\,d_pu \\
&\qquad\qquad \times |a(w_p^{-1}n(b))|^s\,\psi_p(-\mathrm{tr}(Tb))\,d_pb \\
&= \frac{\gamma_p(V)^2\,c_p(\beta,\psi)}{c_p(\alpha,\psi)} \cdot \int_{\mathrm{Sym}_2(\mathbb{Q}_p)} \hat{M}_\varphi(b)|a(w_p^{-1}n(b))|^s\,\psi_p(-\mathrm{tr}(Tb))\,d_pb.
\end{aligned}$$

Here

$$M : S(V^2_{reg}(\mathbb{Q}_p)) \longrightarrow S(U(\mathbb{Q}_p)), \qquad \varphi \mapsto M_\varphi, \tag{5.3.35}$$

is the map defined by integration over the fibers with respect to the measure determined by the restriction of the gauge form ν. Since the function $\hat{M}_\varphi$ lies in the Schwartz space $S(\mathrm{Sym}_2(\mathbb{Q}_p))$, we can evaluate the last expression in (5.3.34) at $s = 0$ to obtain

$$\begin{aligned}
W_{T,p}(0,\Phi_\varphi) &= \frac{\gamma_p(V)^2\,c_p(\beta,\psi)}{c_p(\alpha,\psi)} \cdot \int_{\mathrm{Sym}_2(\mathbb{Q}_p)} \hat{M}_\varphi(b)\cdot\psi_p(-\mathrm{tr}(Tb))\,d_pb \\
&= \frac{\gamma_p(V)^2\,c_p(\beta,\psi)}{c_p(\alpha,\psi)} \cdot M_\varphi(T) \qquad (5.3.36) \\
&= C_p(V,\alpha,\beta,\psi)\cdot O_{T,p}(\varphi).
\end{aligned}$$

Here, in the last step, we have used the fact that the measure $d_{T,p}h_1$ on $SO(V)(\mathbb{Q}_p)$ coincides with the measure determined by the restriction of ν to $Q^{-1}(T)(\mathbb{Q}_p)$, as well as the fact that φ is $\mu_2(\mathbb{Q}_p)$ invariant, so that the orbital integral over $\mathrm{O}(V)(\mathbb{Q}_p)$ coincides with the orbital integral over $\mathrm{SO}(V)(\mathbb{Q}_p)$. In the archimedean case, we now replace φ by $\omega(g)\varphi$ to obtain (ii). □

We next derive a global expression for the product of local orbital integrals in the preceding corollary.

Write

$$\text{SO}(V)(\mathbb{A}) = \coprod_j \text{SO}(V)(\mathbb{Q})\,\text{SO}(V)(\mathbb{R})\,h_j\,K_H, \tag{5.3.37}$$

where $K_H \subset \text{SO}(V)(\mathbb{A}_f)$ is a compact open subgroup and we take the representatives $h_j \in \text{SO}(V)(\mathbb{A}_f)$. Let

$$\Gamma_j = \text{SO}(V)(\mathbb{Q}) \cap \big(\text{SO}(V)(\mathbb{R}) \cdot h_j K_H h_j^{-1}\big). \tag{5.3.38}$$

Proposition 5.3.6. *For any factorizable function $\varphi_f \in S(V(\mathbb{A}_f)^2)$ which is K_H-invariant,*

$$\prod_{p<\infty} O_{T,p}(\varphi_p) = \text{vol}(K_H) \cdot \Bigg(\sum_j \sum_{\substack{x \in V(\mathbb{Q})^2 \\ Q(x)=T \\ \text{mod } \Gamma_j}} \varphi_f^{\text{ev}}(h_j^{-1}x) \Bigg),$$

where

$$\varphi_f^{\text{ev}}(x) = \int_{\mu_2(\mathbb{A}_f)} \varphi_f(cx)\,dc$$

is the locally even component of φ_f. Here the measures used to define the local orbital integrals are determined by the gauge form ω and $\text{vol}(K_H) = \text{vol}(K_H, dh_f)$*, where dh_f is the product of these measures.*

Proof. Noting that $\text{vol}(\mu_2(\mathbb{Q})\backslash\mu_2(\mathbb{A}), dc) = 1/2$, we compute the integral $I_T(g_\tau, \varphi)$ in another way:

(5.3.39)

$$\begin{aligned}
&I_T(g_\tau, \varphi) \\
&= \frac{1}{2} \cdot \int_{\text{SO}(V)(\mathbb{Q})\backslash \text{SO}(V)(\mathbb{A})} \theta_T(g_\tau, h; \varphi^{\text{ev}})\,dh \\
&= \frac{1}{2} \cdot \text{vol}(K_H) \cdot \sum_j \int_{\Gamma_j\backslash \text{SO}(V)(\mathbb{R})} \sum_{\substack{x \in V(\mathbb{Q})^2 \\ Q(x)=T}} \omega(g_\tau)\varphi_\infty^{\text{ev}}(h_\infty^{-1}x)\,\varphi_f^{\text{ev}}(h_j^{-1}x)\,dh_\infty \\
&= \frac{1}{2} \cdot \text{vol}(K_H) \cdot \sum_j \sum_{\substack{x \in V(\mathbb{Q})^2 \\ Q(x)=T \\ \text{mod } \Gamma_j}} \varphi_f^{\text{ev}}(h_j^{-1}x) \int_{\text{SO}(V)(\mathbb{R})} \omega(g_\tau)\varphi_\infty^{\text{ev}}(h_\infty^{-1}x)\,dh_\infty
\end{aligned}$$

$$= \frac{1}{2} \cdot \mathrm{vol}(K_H) \cdot \Big(\sum_j \sum_{\substack{x \in V(\mathbb{Q})^2 \\ Q(x)=T \\ \mathrm{mod}\ \Gamma_j}} \varphi_f^{\mathrm{ev}}(h_j^{-1}x) \Big) \cdot O_{T,\infty}(\omega(g_\tau)\varphi_\infty).$$

Here we have used the factorization $dh = dh_\infty \, dh_f$, determined by the gauge form ω, and $\mathrm{vol}(K_H) = \mathrm{vol}(K_H, dh_f)$. Note that $d_\infty h$ is the measure used to define the local orbital integral, as above. Comparing this expression with the last expression in (5.3.7), we find the claimed result. $\square$

We will also need a variant of the previous proposition. Fix a prime p such that V_p is isotropic. By strong approximation, we may then write

$$\mathrm{SO}(V)(\mathbb{A}) = \mathrm{SO}(V)(\mathbb{Q}) \cdot \mathrm{SO}(V)(\mathbb{R}) \, \mathrm{SO}(V)(\mathbb{Q}_p) \, K_H^p \tag{5.3.40}$$

for any compact open subgroup $K_H^p \subset \mathrm{SO}(V)(\mathbb{A}_f^p)$.

Proposition 5.3.7. *Suppose that $\varphi \in S(V(\mathbb{A}_f^p)^2)$ is factorizable and invariant under K_H^p. Let*

$$\Gamma = \mathrm{SO}(V)(\mathbb{Q}) \cap \big(\mathrm{SO}(V)(\mathbb{R}) \, \mathrm{SO}(V)(\mathbb{Q}_p) \, K_H^p \big).$$

Then

$$\prod_{\ell \neq p} O_{T,\ell}(\varphi_\ell) = \mathrm{vol}(K_H^p) \cdot \sum_{\substack{x \in V(\mathbb{Q})^2 \\ Q(x)=T \\ \mathrm{mod}\ \Gamma}} (\varphi_f^p)^{\mathrm{ev}}(x).$$

Proof. For convenience of notation, we temporarily write $H = \mathrm{SO}(V)$. Then, for $g \in G_{\mathbb{R}}$ and T nonsingular, we have

(5.3.41)

$$\begin{aligned}
&I_T(g,\varphi) \\
&= \frac{1}{2} \int_{H(\mathbb{Q}) \backslash H(\mathbb{Q}) \cdot H(\mathbb{R}) \, H(\mathbb{Q}_p) \, K^p} \sum_{\substack{x \in V(\mathbb{Q})^2 \\ Q(x)=T}} \omega(g)\varphi^{\mathrm{ev}}(h^{-1}x) \, dh \\
&= \frac{1}{2} \mathrm{vol}(K_H^p) \int_{\Gamma \backslash H(\mathbb{R}) \, H(\mathbb{Q}_p)} \sum_{\substack{x \in V(\mathbb{Q})^2 \\ Q(x)=T}} \omega(g)\varphi^{\mathrm{ev}}(h^{-1}x) \, dh \\
&= \frac{1}{2} \mathrm{vol}(K_H^p) \Big(\sum_{\substack{x \in V(\mathbb{Q})^2 \\ Q(x)=T \\ \mathrm{mod}\ \Gamma}} (\varphi_f^p)^{\mathrm{ev}}(x) \Big) \cdot O_{T,\infty}(\omega(g)\varphi_\infty) \cdot O_{T,p}(\varphi_p).
\end{aligned}$$

Here note that the stabilizer in $\mathrm{SO}(V)$ is an element $x \in V^2$ with $Q(x) = T$ is trivial. Again comparing this last expression with (5.3.7), we obtain the claimed result. □

A useful formula for the constant $C_p(V, \alpha, \beta, \psi)$ can be obtained as follows. Fix an isomorphism $V(\mathbb{Q}) \simeq \mathbb{Q}^3$ and write

$$Q(x) = {}^t x S x \tag{5.3.42}$$

with $S \in \mathrm{Sym}_3(\mathbb{Q})$. Then $V^2 \simeq M_{3,2}(\mathbb{Q})$, and we can take $\alpha = dx_{1,1} \wedge \cdots \wedge dx_{3,2}$. For our standard additive character ψ, we then find that

$$c_p(\alpha, \psi) = |\det(2S)|_p^{-1}. \tag{5.3.43}$$

If we take $\beta = db_1 \wedge db_2 \wedge db_3$ where $b = \begin{pmatrix} b_1 & b_2 \\ b_2 & b_3 \end{pmatrix} \in \mathrm{Sym}_2$, then it is easy to check that

$$c_p(\beta, \psi) = |2|_p^{-\frac{1}{2}}. \tag{5.3.44}$$

In the particular case in which V is the space of trace zero elements in a quaternion algebra ramified at the primes dividing the square free integer $D(B)$, we can take an isomorphism $V(\mathbb{Q}) \simeq \mathbb{Q}^3$ which identifies $O_B \cap V(\mathbb{Q})$ with $\mathbb{Z}^3$, where O_B is a given maximal order in B. In this case,

$$|\det(2S)|_p = |D(B)|_p^2 \, |2|_p, \tag{5.3.45}$$

and hence, for these choices of α, β and ψ, we have

$$\begin{aligned} C_p(V, \alpha, \beta, \psi) &= \gamma_p(V)^2 \, |D(B)|_p^2 \, |2|_p^{\frac{1}{2}} \\ &= C_p(V) \cdot |2|_p^{-1}, \end{aligned} \tag{5.3.46}$$

where $C_p(V)$ is the constant in (5.2.11).

5.3.2 The nonsingular coefficients of $\mathcal{E}'_{2,T}(\tau, 0, B)$

We now apply the above analysis to refine the expressions for the Fourier coefficients $\mathcal{E}'_{2,T}(\tau, 0, B)$ given in Proposition 5.2.8.

The relevant three-dimensional quadratic spaces, Schwartz functions, etc., are defined as follows. For a finite prime p, let $B^{(p)}$ be the definite quaternion algebra over $\mathbb{Q}$ whose invariants differ from those of B at precisely p and ∞. Let $V^{(p)}$ be the space of elements of $B^{(p)}$ of trace zero with quadratic form given by $Q(x) = -x^2$. Recall that we have fixed an isomorphism

$$B^{(p)}(\mathbb{A}_f^p) \simeq B(\mathbb{A}_f^p) \tag{5.3.47}$$

and hence an identification

$$S(V^{(p)}(\mathbb{A}_f^p)^2) \simeq S(V(\mathbb{A}_f^p)^2). \tag{5.3.48}$$

If $p \nmid D(B)$, the section

$$\Phi^{(p)}(s) = \Phi_\infty^{\frac{3}{2}}(s) \otimes \Phi_p^{\mathrm{ra}}(s) \otimes \bigotimes_{\ell \neq p} \tilde{\Phi}_\ell^B(s), \tag{5.3.49}$$

obtained by replacing $\tilde{\Phi}_p^B(s)$ by $\Phi_p^{\mathrm{ra}}(s)$, coincides at $s = 0$ with a Siegel-Weil section for $V^{(p)}$ given by

$$\Phi^{(p)}(0) = \lambda_{V^{(p)}}(\varphi^{(p)}), \tag{5.3.50}$$

where

$$\varphi^{(p)} = \varphi_\infty \otimes \varphi_p^{\mathrm{ra}} \otimes \varphi_f^p. \tag{5.3.51}$$

Here $\varphi_\infty \in S(V^{(p)}(\mathbb{R})^2)$ is the Gaussian, φ_p^{ra} the characteristic function of $(L_p^{\mathrm{ra}})^2$, and

$$\varphi_f^p = \otimes_{\ell \neq p} \varphi_\ell^B. \tag{5.3.52}$$

Let $O_{B^{(p)}}$ be the maximal order in $B^{(p)}$ determined by the maximal order O_B in B via the isomorphism (5.3.47). Let $K_H^{(p)}$ be the image of $(O_{B^{(p)}} \otimes_{\mathbb{Z}} \hat{\mathbb{Z}})^\times$ in $\mathrm{SO}(V^{(p)})(\mathbb{A}_f)$, and write

$$\mathrm{SO}(V^{(p)})(\mathbb{A}) = \coprod_j \mathrm{SO}(V^{(p)})(\mathbb{Q})\, \mathrm{SO}(V^{(p)})(\mathbb{R})\, h_j\, K_H^{(p)}. \tag{5.3.53}$$

Also let

$$\Gamma_j = \mathrm{SO}(V^{(p)})(\mathbb{Q}) \cap (\, \mathrm{SO}(V^{(p)})(\mathbb{R})\, h_j K_H^{(p)}\, h_j^{-1}\,). \tag{5.3.54}$$

If $p \mid D(B)$, the space $V_p^{(p)}$ is isotropic, and we write

$$\mathrm{SO}(V^{(p)})(\mathbb{A}) = \mathrm{SO}(V^{(p)})(\mathbb{Q})\, \mathrm{SO}(V^{(p)})(\mathbb{R})\, \mathrm{SO}(V^{(p)})(\mathbb{Q}_p)\, K_H^p, \tag{5.3.55}$$

and let

$$\Gamma' = \mathrm{SO}(V^{(p)})(\mathbb{Q}) \cap (\, \mathrm{SO}(V^{(p)})(\mathbb{R})\, \mathrm{SO}(V^{(p)})(\mathbb{Q}_p)\, K_H^p\,). \tag{5.3.56}$$

Here $K_H^p \subset \mathrm{SO}(V^{(p)})(\mathbb{A}_f^p)$ is obtained by removing the factor at p from $K_H^{(p)}$.

Theorem 5.3.8. *Suppose that* $T \in \mathrm{Sym}_2(\mathbb{Z})^\vee$ *with* $\det(T) \neq 0$ *and with* $\mathrm{Diff}(T, B) = \{p\}$.
(i) If $p \nmid D(B)$,

$$\mathcal{E}'_{2,T}(\tau, 0, B) = q^T \cdot \nu_p(T) \log(p) \cdot \frac{1}{2} \cdot \Big(\sum_j \sum_{\substack{x \in V^{(p)}(\mathbb{Q})^2 \\ Q(x)=T \\ \mathrm{mod}\, \Gamma_j}} \varphi_f^{(p)}(h_j^{-1}x) \Big),$$

where h_j *and* Γ_j *are as in (5.3.53) and (5.3.54), respectively, and the function* $\varphi^{(p)} \in S(V^{(p)}(\mathbb{A}_f)^2)$ *is defined by (5.3.51).*
(ii) If $p \mid D(B)$, *then*

$$\mathcal{E}'_{2,T}(\tau, 0, B) = q^T \cdot \nu_p(T) \log(p) \cdot \frac{1}{2} \cdot \Big(\sum_{\substack{x \in V^{(p)}(\mathbb{Q})^2 \\ Q(x)=T \\ \mathrm{mod}\, \Gamma'}} \varphi_f^p(x) \Big),$$

where Γ' *is as in (5.3.56) and the function* $\varphi_f^p \in S(V(\mathbb{A}_f^p)^2)$ *is given by (5.3.52).*
(iii) If $p = \infty$ *and* $\mathrm{sig}(T) = (1,1)$ *or* $(0,2)$, *then*

$$\mathcal{E}'_{2,T}(\tau, 0, B) = q^T \cdot \nu_\infty(T, v) \cdot \frac{1}{2} \cdot \Big(\sum_{\substack{x \in V(\mathbb{Q})^2 \\ Q(x)=T \\ \mathrm{mod}\, \Gamma}} \varphi_f^B(x) \Big),$$

where $\Gamma = \mathrm{SO}(V)(\mathbb{Q}) \cap (\mathrm{SO}(V)(\mathbb{R}) \cdot K_H)$ *and* $\nu_\infty(T, v)$ *is given in Theorem 5.2.7.*

Proof. First suppose that $\mathrm{Diff}(T, B) = \{p\}$ for a finite prime p. Recall that T is then positive-definite.
Case $p \nmid D(B)$. By Corollary 5.3.5, Proposition 5.3.6, and (5.3.46), we have

(5.3.57)

$$\begin{aligned} &W_{T,p}(0, \Phi_p^{\mathrm{ra}}) \cdot \prod_{\ell \neq p} W_{T,p}(0, \tilde{\Phi}_\ell^B) \\ &= C_\infty(V^{(p)})^{-1} \, |2|_\infty \cdot \mathrm{vol}(K_H^{(p)}) \cdot \Big(\sum_j \sum_{\substack{x \in V^{(p)}(\mathbb{Q})^2 \\ Q(x)=T \\ \mathrm{mod}\, \Gamma_j}} \varphi_f^{(p)}(h_j^{-1}x) \Big). \end{aligned}$$

Thus, using the value for $W_{T,p}(0, \Phi_p^{\mathrm{ra}})$ given in (i) of Theorem 5.2.6, we obtain

$$(5.3.58)\quad \prod_{\ell \neq p} W_{T,p}(0, \tilde{\Phi}_\ell^B) = C_p(V^{(p)})^{-1}\,\frac{1}{2}\,(p+1)^{-1}\cdot C_\infty(V^{(p)})^{-1}\,2\cdot \mathrm{vol}(K_H^{(p)}) \times \Bigg(\sum_j \sum_{\substack{x\in V^{(p)}(\mathbb{Q})^2\\ Q(x)=T\\ \bmod \Gamma_j}} \varphi_f^{(p)}(h_j^{-1}x) \Bigg).$$

Inserting this into the expression of Proposition 5.2.8, we have

$$(5.3.59)\quad \mathcal{E}'_{2,T}(\tau, 0, B) = \mathbf{C}\cdot q^T \cdot \nu_p(T)\,\log(p)\cdot \Bigg(\sum_j \sum_{\substack{x\in V^{(p)}(\mathbb{Q})^2\\ Q(x)=T\\ \bmod \Gamma_j}} \varphi_f^{(p)}(h_j^{-1}x) \Bigg),$$

where the constant $\mathbf{C}$ is given by

$$(5.3.60)\quad \begin{aligned}\mathbf{C} &= -C^B(0)\cdot 2\sqrt{2}\cdot 4\pi^2 \cdot C_p(V)\cdot(1-p^{-2})\cdot C_p(V^{(p)})^{-1}\,C_\infty(V^{(p)})^{-1} \\ &\qquad \times (p+1)^{-1}\,\mathrm{vol}(K_H^{(p)}) \\ &= -C^B(0)\cdot 2\sqrt{2}\cdot 4\pi^2\cdot C_\infty(V)^{-1}\cdot(1-p^{-2})\,(p+1)^{-1}\,\mathrm{vol}(K_H^{(p)}),\end{aligned}$$

since

$$(5.3.61)\qquad C_\infty(V^{(p)})\,C_p(V^{(p)}) = C_\infty(V)\,C_p(V).$$

Case $p \mid D(B)$. In this case, we use Proposition 5.3.7 and write

$$(5.3.62)\quad \begin{aligned}&\prod_{\ell\neq p} W_{T,\ell}(0, \tilde{\Phi}_\ell^B) \\ &= C_\infty(V)^{-1}|2|_\infty\, C_p(V)^{-1}|2|_p\cdot \prod_{\ell\neq p} O_{T,\ell}(\varphi_\ell^B) \\ &= C_\infty(V)^{-1}|2|_\infty\, C_p(V)^{-1}|2|_p\cdot \mathrm{vol}(K_H^p)\cdot \Bigg(\sum_{\substack{x\in V^{(p)}(\mathbb{Q})^2\\ Q(x)=T\\ \bmod \Gamma'}} \varphi_f^p(x) \Bigg).\end{aligned}$$

This gives

$$\mathcal{E}'_{2,T}(\tau, 0, B) = \mathbf{C} \cdot q^T \cdot \nu_p(T) \log(p) \cdot \Bigg(\sum_{\substack{x \in V^{(p)}(\mathbb{Q})^2 \\ Q(x)=T \\ \text{mod } \Gamma'}} \varphi_f^p(x) \Bigg), \tag{5.3.63}$$

where $\mathbf{C}$ is given by

(5.3.64)
$$-C^B(0) \cdot 2\sqrt{2} \cdot 4\pi^2 \cdot C_p(V)\,(p+1)\,C_\infty(V)^{-1}\,C_p(V)^{-1}|2|_p\,\mathrm{vol}(K_H^p)$$
$$= -C^B(0) \cdot 2\sqrt{2} \cdot 4\pi^2 \cdot C_\infty(V)^{-1}\,|2|_p\,(p+1)\,\mathrm{vol}(K_H^p).$$

Finally, in the case $p = \infty$, we use Corollary 5.3.5 and Proposition 5.3.6 for the space V to obtain

$$\mathcal{E}'_{2,T}(\tau, 0, B) = \mathbf{C} \cdot q^T \cdot \nu_\infty(T, v) \cdot \Bigg(\sum_{\substack{x \in V(\mathbb{Q})^2 \\ Q(x)=T \\ \text{mod } \Gamma}} \varphi_f^B(x) \Bigg), \tag{5.3.65}$$

where

$$\mathbf{C} = -C^B(0) \cdot 2\sqrt{2} \cdot 4\pi^2 \cdot C_\infty(V)^{-1} \cdot \mathrm{vol}(K_H). \tag{5.3.66}$$

In all of these statements, the measures are chosen as follows. The gauge forms α on V^2 and β on Sym_2 are chosen as in the end of last subsection. We then use the Haar measures on $\mathrm{SO}(V)(\mathbb{A}_f)$ and $\mathrm{SO}(V)(\mathbb{Q}_p)$ given by the gauge form ν determined by α and β. For these measures, we have the following information about volumes:

Lemma 5.3.9. *(i) If $p \nmid D(B)$,*

$$\mathrm{vol}(K_{H,p}) = (1 - p^{-2})|2|_p.$$

(ii) If $p \mid D(B)$, then

$$\mathrm{vol}(K_{H,p}) = (p+1)|2|_p.$$

(iii) Globally,

$$\mathrm{vol}(K_H) = \frac{1}{2}\zeta^{D(B)}(2)^{-1} \cdot \prod_{p|D(B)} (p+1) = \frac{3}{\pi^2}\,D(B)^2 \prod_{p|D(B)} (p-1)^{-1}.$$

Using these values, we see that $\mathbf{C}$ is given by (5.3.66) in all cases. More precisely, when $p \mid D(B)$, we have

$$(1 - p^{-2})\,(p+1)^{-1}\,\mathrm{vol}(K_H^{(p)}) = \mathrm{vol}(K_H), \tag{5.3.67}$$

whereas, when $p \nmid D(B)$,

$$|2|_p\,(p+1)\,\mathrm{vol}(K_H^p) = \mathrm{vol}(K_H). \tag{5.3.68}$$

Then, since

$$C^B(0) \cdot 4\pi^2 \cdot \mathrm{vol}(K_H) = \frac{1}{2}\,D(B)^2, \tag{5.3.69}$$

and

$$C_\infty(V) = \gamma_\infty(V)^2 \cdot D(B)^2\, 2\sqrt{2}, \tag{5.3.70}$$

with

$$\gamma_\infty(V)^2 = \gamma(V_\infty^+)^2 = (-i)^2 = -1, \tag{5.3.71}$$

by (5.7.23), we obtain $\mathbf{C} = \frac{1}{2}$. □

Proof of Lemma 5.3.9. For completeness, we give the computation of these volumes. In Proposition 5.3.3, set $\phi = \mathrm{char}(L_p^2)$ with $L_p = O_p \cap V_p$, the standard maximal lattice of V_p, we obtain

$$W_{T,p}(0, \tilde{\Phi}_p^B) = C(V, \alpha, \beta, \psi) \cdot O_{T,p}^1(\varphi), \tag{5.3.72}$$

where $O_{T,p}^1(\cdot)$ denotes the $\mathrm{SO}(V)$-orbital integral. Since $K_{H,p}$ preserves L_p, one has by definition

$$O_{T,p}^1(\varphi) = \mathrm{vol}(K_{H,p})[K_{H,x_0} : K_{H,p}]. \tag{5.3.73}$$

Here $x_0 = (x_{01}, x_{02}) \in L_p^2$ with $Q(x_0) = T$, and

(5.3.74)

$$K_{H,x_0} = \{\, h \in \mathrm{SO}(V_p) \mid hx_0 \in L_p^2 \,\} = \{\, h \in \mathrm{SO}(V_p) \mid hx_{01}, hx_{02} \in L_p \,\}.$$

Case 1. We first assume that $p \mid D(B)$, i.e., B_p is a division algebra over $\mathbb{Q}_p$. In this case, an element $x \in V_p$ is in L_p if and only if its reduced norm is in $\mathbb{Z}_p$. So $K_{H,x_0} = \mathrm{SO}(V_p) = B_p^*/\mathbb{Q}_p^*$. But, by adjusting by a central element, we can scale any element of $B_p^\times$ to lie in O_p and have reduced norm of valuation 0 or 1. Thus, the set

$$O_p^\times \cup O_p^\times \cdot \Pi$$

maps surjectively onto $\mathrm{SO}(V_p)$. Here Π is a uniformizer of B_p. Thus, $\mathrm{SO}(V_p)$ consists of precisely 2 cosets for $K_{H,p}$. Thus we have

$$W_{T,p}(0,\tilde{\Phi}_p^B) = C_p(V,\alpha,\beta,\psi)\cdot 2\,\mathrm{vol}(K_{H,p}). \tag{5.3.75}$$

On the other hand, Theorem 5.2.6 says

$$W_{T,p}(0,\tilde{\Phi}_p^B) = C_p(V)\cdot 2\,(p+1). \tag{5.3.76}$$

Now (5.3.46) gives

$$\mathrm{vol}(K_{H,p}) = (p+1)|2|_p \tag{5.3.77}$$

in this case.

Case 2. Now assume $p \nmid D(B)$, i.e., $B_p = M_2(\mathbb{Q}_p)$. Let

$$x_{01} = \begin{pmatrix} 1 & 0 \\ 0 & -1 \end{pmatrix}, \qquad x_{02} = \begin{pmatrix} 0 & 1 \\ 1 & 0 \end{pmatrix}, \tag{5.3.78}$$

so that $x_0 = [x_{01}, x_{02}]$ has $Q(x_0) = T$ with $T = -I_2$. In this case, $\mathrm{SO}(V_p) = \mathrm{PGL}_2(\mathbb{Q}_p)$ and $K_{H,p}$ is the image of $\mathrm{GL}_2(\mathbb{Z}_p)$ in $\mathrm{PGL}_2(\mathbb{Q}_p)$. Notice that

$$\mathrm{PGL}_2(\mathbb{Q}_p) = \{\begin{pmatrix} p^a & u \\ 0 & 1 \end{pmatrix} : a \in \mathbb{Z}, u \in Q_p\}\cdot K_{H,p}. \tag{5.3.79}$$

A matrix $g = \begin{pmatrix} p^a & u \\ 0 & 1 \end{pmatrix} \in K_{H,x_0}$ if and only if

$$g x_{01} g^{-1} \in M_2(\mathbb{Z}_p) \qquad \text{and} \qquad g x_{02} g^{-1} \in M_2(\mathbb{Z}_p). \tag{5.3.80}$$

A direct calculation shows that $g \in K_{H,x_0}$ if and only if

$$2u \in \mathbb{Z}_p, \quad a \le 0 \qquad \text{and} \qquad u p^{-a} + p^a \in \mathbb{Z}_p. \tag{5.3.81}$$

Thus,

$$K_{H,x_0}/K_{H,p} = \begin{cases} \{I_2\} & \text{if } p \ne 2, \\ \{I_2, \begin{pmatrix} 1 & 1 \\ 0 & 2 \end{pmatrix}\} & \text{if } p = 2. \end{cases} \tag{5.3.82}$$

This proves that

$$\begin{aligned} &W_{-I_2,p}(0,\tilde{\Phi}_p^B) \\ &\quad = C_p(V,\alpha,\beta,\psi)\,\mathrm{vol}(K_{H,p})\,|2|_p^{-1} = C_p(V)\,|2|_p^{-2}\,\mathrm{vol}(K_{H,p}). \end{aligned} \tag{5.3.83}$$

When $p \neq 2$, the Gross-Keating invariants of the matrix $\mathrm{diag}(1, -I_2)$ are just $(0, 0, 0)$, and Corollary 5.2.5 gives

$$W_{T,p}(0, \Phi_{\varphi,p}) = C_p(V) \cdot (1 - p^{-2}). \tag{5.3.84}$$

This proves that

$$\mathrm{vol}(K_{H,p}) = (1 - p^{-2})|2|_p = 1 - p^{-2} \tag{5.3.85}$$

in this case. When $p = 2$, however, the Gross-Keating invariants of the matrix $\mathrm{diag}(1, -I_2)$ are $(0, 1, 1)$ by [18], Proposition B.5. So Corollary 5.2.5 gives

$$W_{T,p}(0, \Phi_{\varphi,p}) = 2\, C_p(V) \cdot (1 - p^{-2}), \tag{5.3.86}$$

and again

$$\mathrm{vol}(K_{H,p}) = (1 - p^{-2})|2|_p. \tag{5.3.87}$$

This proves claims (i) and (ii). Claim (iii) follows from (i) and (ii). □

5.4 SINGULAR COEFFICIENTS

We now begin the computation of the coefficients $\mathcal{E}_2'(\tau, 0, B)$ for singular T. Let

$$T = \begin{pmatrix} t_1 & m \\ m & t_2 \end{pmatrix} \in \mathrm{Sym}_2(\mathbb{Z})^\vee \tag{5.4.1}$$

with $\det(T) = 0$. When T is of rank 1, there is a unique integer t and a pair of relatively prime integers (n_1, n_2) such that

$$t_1 = n_1^2\, t, \quad t_2 = n_2^2\, t, \quad m = n_1 n_2\, t. \tag{5.4.2}$$

The pair (n_1, n_2) is unique up to sign; for convenience, we require that $n_2 > 0$ or $n_2 = 0$ and $n_1 \geq 0$. We write $4\,t = n^2\, d$ where $-d$ is the fundamental discriminant of a quadratic field or $-d = 1$. Let

$$\gamma_0 = \begin{cases} \begin{pmatrix} 1 & 0 \\ n_1 & n_2 \end{pmatrix} & \text{if } n_2 \neq 0, \\ \begin{pmatrix} 0 & 1 \\ 1 & 0 \end{pmatrix} & \text{if } n_2 = 0. \end{cases} \tag{5.4.3}$$

Then

$$T = {}^t\gamma_0 \begin{pmatrix} 0 & 0 \\ 0 & t \end{pmatrix} \gamma_0. \tag{5.4.4}$$

When $T = 0$, we set $t = 0$, $d = 0$, and $n = 0$.

Let

$$w_1 = \begin{pmatrix} 1 & & 0 & \\ & 0 & & 1 \\ 0 & & 1 & \\ & -1 & & 0 \end{pmatrix} \in G_{\mathbb{Q}}, \tag{5.4.5}$$

and choose elements $w_{1,p} = [w_1, 1]_L \in K_p$ with image w_1 in $\underline{K}_p$ and with $w_1 = \prod_p w_{1,p}$.

Proposition 5.4.1. *Let $\Phi(s) = \otimes \Phi_p(s) \in I(s, \chi)$ be any factorizable section.*
(i) When $\mathrm{rank}(T) = 1$,

$$\begin{aligned} E_T(g, s, \Phi) &= B_T(m(\gamma_0)g, s, \Phi) + W_T(g, s, \Phi) \\ &= \prod_p B_{T,p}(m(\gamma_0)g_p, s, \Phi_p) + \prod_p W_{T,p}(g_p, s, \Phi_p), \end{aligned}$$

where

$$B_{T,p}(g_p, s, \Phi_p) = \int_{\mathbb{Q}_p} \Phi_p(w_{1,p}^{-1} n(\begin{pmatrix} 0 & 0 \\ 0 & b \end{pmatrix}) g_p, s)\, \psi_p(-tb)\, d_p b,$$

and

$$W_{T,p}(g_p, s, \Phi_p) = \int_{\mathrm{Sym}_2(\mathbb{Q}_p)} \Phi_p(w_p^{-1} n(b) g_p, s)\, \psi_p(-\mathrm{tr}(Tb))\, d_p b.$$

Here $d_p b$ is the standard Haar measure on $\mathbb{Q}_p$ (resp. $\mathrm{Sym}_2(\mathbb{Q}_p)$) which is self-dual with respect to ψ_p (resp. $\psi_p \circ \mathrm{tr}$).
(ii) The constant term of $E(g, s, \Phi)$ is

$$E_0(g, s, \Phi) = \Phi(g, s) + \sum_{\gamma \in \Gamma_\infty \backslash \mathrm{SL}_2(\mathbb{Z})} B_0(m(\gamma)g, s, \Phi) + W_0(g, s, \Phi),$$

where W_0 and B_0 are defined as in (i) with $T = 0$.

In fact, $B_T(g, s, \Phi)$ and $W_T(g, s, \Phi)$ are both related to the Fourier coefficients of an Eisenstein series of genus one, which we recall in the next section. The next few sections are devoted to making such relations precise in the case $\Phi(s) = \Phi^D(s) \in I(s, \chi)$, a standard section defined as follows. For every square free positive integer D, not necessarily the discriminant of an indefinite quaternion algebra, we let

$$\Phi^D(s) = \Phi_\infty^{\frac{3}{2}}(s) \otimes \big(\bigotimes_{\ell | D} \Phi_\ell^{\mathrm{ra}}(s) \big) \otimes \big(\bigotimes_{\ell \nmid D} \Phi_\ell^0(s) \big). \tag{5.4.6}$$

Here $\Phi_\infty^{\frac{3}{2}}(s) \in I_\infty(s, \chi_\infty)$ is the 'weight $\frac{3}{2}$' section. For $p \mid D$, we also define a standard section

$$\Phi^{\frac{D}{p},1}(s) = \Phi_\infty^{\frac{3}{2}}(s) \otimes \Phi_p^1(s) \otimes \big(\bigotimes_{\ell \mid D/p} \Phi_\ell^{\mathrm{ra}}(s) \big) \otimes \big(\bigotimes_{\ell \nmid D} \Phi_\ell^0(s) \big). \tag{5.4.7}$$

Finally, if $\Phi(s) = \Phi_\infty(s) \otimes \Phi_f(s)$ where $\Phi_\infty(s) = \Phi_\infty^l(s)$ is of weight l, we recall that the classical Eisenstein series is given by $E(\tau, s, \Phi) = (\det v)^{-\frac{l}{2}} E(g_\tau, s, \Phi)$. In our present case $l = \frac{3}{2}$.

5.5 EISENSTEIN SERIES OF GENUS ONE

In this section, we recall the Eisenstein series of genus one studied in [9]. Let G'_p and $G'_{\mathbb{A}}$ be the metaplectic extension of $\mathrm{SL}_2(\mathbb{Q}_p)$ and $\mathrm{SL}_2(\mathbb{A})$, respectively. We define the induced representation $I(s, \chi) = I_1(s, \chi)$ and Eisenstein series $E(g, s, \Phi)$ for $G'_{\mathbb{A}}$ as in Section 5.1.

Fix a quadratic character $\chi = \otimes \chi_p$ of $\mathbb{Q}^\times \backslash \mathbb{Q}_{\mathbb{A}}^\times$. For a quadratic space V_p over $\mathbb{Q}_p$ of dimension 3 such that $\chi_{V_p}(\cdot) = (-\det V_p, \cdot)_p = \chi_p(\cdot)$, there is a Weil representation $\omega_{V_p} = \omega_{V_p, \psi_p}$ of G'_p on $S(V_p)$ and a G'_p-equivariant map

$$\lambda_{V_p} : S(V_p) \longrightarrow I_p(\frac{1}{2}, \chi_p), \quad \lambda_{V_p}(\varphi)(g) = \omega_{V_p}(g)\varphi(0). \tag{5.5.1}$$

For a lattice L_p of V_p, there is a unique standard section $\Phi_p(s) \in I(s, \chi_p)$ such that

$$\Phi_p(g, \frac{1}{2}) = \lambda_{V_p}(\mathrm{char} L_p). \tag{5.5.2}$$

We say that $\Phi_p(s)$ is associated to L_p. In particular, we have the standard sections Φ_p^0, Φ_p^1, and Φ_p^{ra} in $I_p(s, \chi_p)$ associated to L_p^0, L_p^1, and L_p^{ra}, respectively.

Notice that we have written $I(s, \chi)$ for both genus one and genus two induced representations. We will abuse notation in the same way for sections and Eisenstein series, and we trust that the reader will be able to distinguish by context which genus is involved. When necessary, we use a subscript to indicate the genus. For example, $\Phi_1^D(s) \in I(s, \chi) = I_1(s, \chi)$ is a section in an induced representation for genus one and $E(g, s, \Phi_1^D)$ is the corresponding genus one Eisenstein series.

The following *normalized* Eisenstein series of genus one

$$\mathbb{E}(\tau, s, \Phi_1^D) = -2\, C^D(s - \frac{1}{2})\, E(\tau, s, \Phi_1^D) \tag{5.5.3}$$

was studied in detail in [9]. Here $C^D(s)$ is given by (5.1.38), so, in particular,

$$-2\,C^D(s-\frac{1}{2}) = (s+\frac{1}{2})\,c(D)\,\Lambda_D(2s+1);$$

see (6.23) of [9]. The normalized Eisenstein series of genus one satisfies a functional equation [9], Section 15,

$$\mathbb{E}(\tau, s, \Phi_1^D) = \mathbb{E}(\tau, -s, \Phi_1^D). \tag{5.5.4}$$

As in [9], we actually need a modified Eisenstein series of genus one

$$\mathcal{E}_1(\tau, s, B) = \mathbb{E}(\tau, s, \Phi_1^D) + \sum_{p|D} C_p(s)\,\mathbb{E}(\tau, s, \Phi_1^{\frac{D}{p}}), \tag{5.5.5}$$

for an indefinite quaternion algebra B. Here $D = D(B)$, and $C_p(s)$ is any rational function of p^{-s} with

$$C_p(\frac{1}{2}) = 0, \quad \text{and} \quad C_p'(\frac{1}{2}) = -\frac{p-1}{p+1}\log p. \tag{5.5.6}$$

The main result of [9] is a precise relation between $\mathcal{E}_1'(\tau, \frac{1}{2}, B)$ and an arithmetic theta function. In addition, explicit formulas for the Fourier coefficients of $\mathcal{E}_1'(\tau, \frac{1}{2}, B)$ were obtained in [9]. We will not need these formulas here.

5.6 B_T

In this section, we relate the quantity $B_T(g, s, \Phi)$ to the t-th Fourier coefficient of a certain Eisenstein series of genus one. Fix a prime p. The key point is to relate the local integrals $B_{T,p}(g, s, \Phi_p)$ to local genus one Whittaker integrals.

We consider the more general case where V_p is an arbitrary quadratic space over $\mathbb{Q}_p$ such that $\chi_{V_p} = \chi_p$, with $m = \dim V_p$, and L_p is a lattice in V_p. For clarity, we use a subscript i, $i = 1, 2$, to stand for sections in $I_i(s, \chi_V)$ and frequently drop the subscript p. For $\Phi_2(s) \in I_2(s, \chi_V)$ and $t \in \mathbb{Q}_p$, we let

$$B_{t,p}(g, s, \Phi_2) = \int_{\mathbb{Q}_p} \Phi_2(w_1^{-1} n(\begin{pmatrix} 0 & 0 \\ 0 & b \end{pmatrix}) g, s)\,\psi(-tb)\,db. \tag{5.6.1}$$

Then, for T of rank 1 and with the conventions of (5.4.1), (5.4.2), and (5.4.3),

$$B_{T,p}(g, s, \Phi_2) = B_{t,p}(m(\gamma_0)g, s, \Phi_2). \tag{5.6.2}$$

Lemma 5.6.1. *Let $\varphi_1', \varphi_1'' \in S(V_p)$, and let $\gamma = \begin{pmatrix} a & b \\ c & d \end{pmatrix} \in \mathrm{GL}_2(\mathbb{Q}_p)$. For $i = 1, 2$, let $\varphi_i \in S(V_p^i)$ be given by*

$$\varphi_1(x) = \varphi_1'(cx)\varphi_1''(dx), \qquad \varphi_2 = \varphi_1' \otimes \varphi_1'' \in S(V_p^2).$$

Let $\Phi_i(s) \in I_i(s, \chi_V)$ be the standard section associated to φ_i. Then

$$B_{t,p}(m(\gamma), s, \Phi_2) = \chi_V(\det \gamma)|\det \gamma|_p^{s+\frac{3}{2}} W_{t,p}(1, s + \frac{1}{2}, \Phi_1).$$

Before giving the proof, we recall some basic formulas of the Weil representation; see Lemma 8.5.6. For $\varphi \in S(V^2)$, $x = [x_1, x_2] \in V^2$, $a \in \mathrm{GL}_2(\mathbb{Q}_p)$, and $b \in \mathrm{Sym}_2(\mathbb{Q}_p)$,

$$\omega_V(w_1^{-1})\varphi(x) = \chi_V(-1)\,\gamma_p(V) \int_V \varphi(x_1, y)\,\psi(-(x_2, y))\,dy,$$

$$\omega_V(w^{-1})\varphi(x) = \gamma_p(V)^2 \int_{V^2} \varphi(y)\,\psi(-\mathrm{tr}(x, y))\,dy, \tag{5.6.3}$$

$$\omega_V(m(a))\varphi(x) = \chi_V(\det a)|a|^{\frac{\dim V}{2}} \varphi(xa),$$

$$\omega_V(n(b))\varphi(x) = \psi(\frac{1}{2}\mathrm{tr}(b(x, x)))\,\varphi(x).$$

Here and throughout this chapter, we identify $g \in \mathrm{Sp}_2(\mathbb{Q}_p)$ with $[g, 1]_L$ in G_p, and $g \in \mathrm{SL}_2(\mathbb{Q}_p)$ with $[g, 1]_L \in G_p'$, Section 8.5 for the notation. Similar formulae hold for the Weil representation of G_p' on $S(V)$. As before, the local constant

$$\begin{aligned} \gamma_p(V) &= \chi_V(-1)\,\gamma_p(\eta)\,\gamma_p(\eta \circ V)^{-1} \\ &= \chi_V(-1)\,\gamma_p(\eta)\,\big(\,\epsilon_p(V)\,\gamma_p(\eta)^3\,\gamma_p(\det V, \eta)\,\big)^{-1}, \end{aligned} \tag{5.6.4}$$

is defined in (8.5.21), $\epsilon_p(V)$ is the Hasse invariant of V, $\gamma_p(\eta)$ is the local Weil index [14], and $\eta(x) = \psi(\frac{1}{2}x)$.

Proof of Lemma 5.6.1. Set $s_1 = \frac{m-2}{2}$, and $s_2 = \frac{m-3}{2} = s_1 - \frac{1}{2}$, where $m = \dim V$. For a positive integer r, let $V^{(r)} = V \oplus V_{r,r}$ where $V_{r,r}$ be the direct sum of r copies of the standard hyperbolic plane. Let $L_{r,r}$ be the direct sum of r copies of the standard hyperbolic lattice, and let $\varphi_i^{(r)} = \varphi_i \otimes \mathrm{char}(L_{r,r}^i)$, Then, unfolding the Weil representation and using Lemma A.3 of [4], we have

$$\begin{aligned} &B_{t,p}(m(\gamma), r + s_2, \Phi_2) \\ &= \int_{\mathbb{Q}_p} \omega_\psi(w_1^{-1} n(\begin{pmatrix} 0 & 0 \\ 0 & b \end{pmatrix}) m(\gamma))\varphi_2^{(r)}(0)\,\psi(-tb)\,db. \end{aligned} \tag{5.6.5}$$

This integral is in turn equal to

$$\int_{\mathbb{Q}_p} \chi_V(-1)\,\gamma_p(V)$$

$$\times \int_{V^{(r)}} \psi(\frac{1}{2}\,b(x,x))\,\chi(\det\gamma)\,|\det\gamma|_p^{r+\frac{m}{2}}\,\varphi_2((0,x)\gamma)\,dx\cdot\psi(-tb)\,db$$

$$= \chi(\det\gamma)|\det\gamma|_p^{r+s_2+\frac{3}{2}} \int_{\mathbb{Q}_p} \chi_V(-1)\,\gamma_p(V)$$

$$\times \int_{V^{(r)}} \psi(\frac{1}{2}\,b(x,x))\,\varphi_1^{(r)}(x)\,dx\cdot\psi(-tb)\,db$$

$$= \chi_V(\det\gamma)|\det\gamma|_p^{r+s_2+\frac{3}{2}}\,W_{t,p}(1, r+s_1, \Phi_1).$$

This proves the lemma. □

The following simple corollary will be useful later.

Corollary 5.6.2. *For a lattice L_p in V_p, and for $i = 1$ or 2, let $\Phi_i \in I_i(s, \chi_{V_p})$ be the standard section associated to* $\mathrm{char}(L_p^i)$. *Then, for* $\gamma = \begin{pmatrix} a & b \\ c & d \end{pmatrix} \in \mathrm{GL}_2(\mathbb{Z}_p)$,

$$B_{t,p}(m(\gamma), s, \Phi_2) = \chi_V(\det\gamma)|\det\gamma|_p^{s+\frac{3}{2}}\,W_{t,p}(1, s+\frac{1}{2}, \Phi_1).$$

Proof. In Lemma 5.6.1, set $\varphi' = \varphi'' = \mathrm{char}(L_p)$. Then the fact that $\min(\mathrm{ord}_p c, \mathrm{ord}_p d) = 0$ implies that $\varphi_i = \mathrm{char}(L_p^i)$. □

Next we look at the case $p = \infty$. For $i = 1, 2$, let $\Phi_{i,\infty}^l(s) \in I_i(s,\chi)$ be the unique eigenfunction of K_∞' with eigencharacter $\det^l$ normalized by the condition $\Phi_{i,\infty}^l(1,s) = 1$.

Lemma 5.6.3. *For* $\tau = \begin{pmatrix} \tau_1 & \tau_{12} \\ \tau_{12} & \tau_2 \end{pmatrix} \in \mathfrak{H}_2$,

$$B_{t,\infty}(g_\tau, s, \Phi_{2,\infty}^l) = \left(\frac{\det v}{v_2}\right)^{\frac{1}{2}(s+\frac{3}{2})} W_{t,\infty}(g_{\tau_2}', s+\frac{1}{2}, \Phi_{1,\infty}^l).$$

Proof. Write

$$w_1\,n(\begin{pmatrix} 0 & 0 \\ 0 & b \end{pmatrix})\,g_\tau = n(x)\,m(y)\begin{pmatrix} c & d \\ -d & c \end{pmatrix} \tag{5.6.6}$$

with $c + id \in \mathrm{U}(2)$, $x \in \mathrm{Sym}_2(\mathbb{R})$, and $y \in \mathrm{Sym}_2(\mathbb{R})_{>0}$. Then multiplying both sides on the right by ${}^t(i, 1)$, we obtain

$$\det y = \frac{(\det v)^{\frac{1}{2}}}{|b+\tau_2|}, \qquad \det(c+id) = \frac{|b+\tau_2|}{b+\tau_2}. \tag{5.6.7}$$

and thus

$$\begin{aligned} \Phi^l_{2,\infty}(w_1 n((\begin{smallmatrix} 0 & 0 \\ 0 & b \end{smallmatrix}))\, g_\tau, s) &= (\det v)^{\frac{1}{2}(s+\frac{3}{2})} |b+\tau_2|^{l-s-\frac{3}{2}} (b+\tau_2)^{-l} \\ &= \left(\frac{\det v}{v_2}\right)^{\frac{1}{2}(s+\frac{3}{2})} \Phi^l_{1,\infty}(w_1 n(b)\, g_{\tau_2}, s+\frac{1}{2}). \end{aligned} \tag{5.6.8}$$

Then,

$$\begin{aligned} &B_{t,\infty}(g_\tau, s, \Phi^l_{2,\infty}) \\ &\quad = \left(\frac{\det v}{v_2}\right)^{\frac{1}{2}(s+\frac{3}{2})} \int_{\mathbb{R}} \Phi^l_{1,\infty}(wn(b) g_{\tau_2}, s+\frac{1}{2})\, \psi(-tb)\, db \\ &\quad = \left(\frac{\det v}{v_2}\right)^{\frac{1}{2}(s+\frac{3}{2})} W_{t,\infty}(g_{\tau_2}, s+\frac{1}{2}, \Phi^l_{1,\infty}). \end{aligned}$$

□

Finally, we return to the global situation.

Theorem 5.6.4. *Let $\{L_p\}_{p<\infty}$ be a collection of quadratic lattices*[1] *over $\mathbb{Z}_p$ of rank 3 with $\chi_{L_p} = \chi_p$. Let $\Phi_{i,p}$ be the standard section in $I_i(s, \chi_p)$ associated to* $\mathrm{char}(L_p^i)$, *and $\Phi_{i,f} = \otimes_{p<\infty} \Phi_{i,p}$. Then*

$$\begin{aligned} &B_T(\tau, s, \Phi^l_{2,\infty} \otimes \Phi_{2,f}) \\ &\quad = \left(\frac{\det v}{t^{-1}\mathrm{tr}(Tv)}\right)^{\frac{1}{2}(s+\frac{3}{2}-l)} E_t(t^{-1}\mathrm{tr}(T\tau), s+\frac{1}{2}, \Phi^l_{1,\infty} \otimes \Phi_{1,f}). \end{aligned}$$

Proof. We assume that $t_2 \neq 0$. Corollary 5.6.2 implies that for $p < \infty$

$$\begin{aligned} B_{T,p}(1, s, \Phi_{2,p}) &= B_{t,p}(m(\gamma_0), s, \Phi_{2,p}) \\ &= \chi_{V_p}(n_2) |n_2|_p^{s+\frac{3}{2}} W_{t,p}(1, s+\frac{1}{2}, \Phi_{1,p}). \end{aligned}$$

On the other hand, one has $m(\gamma_0) g_\tau = g_{\tilde{\tau}}$ with $\tilde{\tau} = \gamma_0 \tau {}^t\gamma_0$. In particular,

$$\det \tilde{v} = \det v \cdot \det \gamma_0^2 = (\det v) \cdot n_2^2 \qquad \text{and} \qquad \tilde{\tau}_2 = t^{-1}\mathrm{tr}(T\tau).$$

---

[1] which almost everywhere agree with the completions of a global quadratic lattice L

Thus, Lemma 5.6.3 and Proposition 5.4.1 give

$$B_{T,\infty}(g_\tau, s, \Phi^l_{2,\infty})$$
$$= B_{t,\infty}(g_{\tilde{\tau}}, s, \Phi^l_{2,\infty})$$
$$= n_2^{s+\frac{3}{2}} \left(\frac{\det v}{t^{-1}\mathrm{tr}(Tv)}\right)^{\frac{1}{2}(s+\frac{3}{2})} W_{t,\infty}(g_{t^{-1}\mathrm{tr}(T\tau)}, s+\frac{1}{2}, \Phi_{1,\infty}).$$

Since $n_2 > 0$, we have $\prod_{p<\infty} \chi_{V_p}(n_2) = 1$, and thus

$$B_T(g_\tau, s, \Phi^l_{2,\infty} \otimes \Phi_{2,f})$$
$$= \left(\frac{\det v}{t^{-1}\mathrm{tr}(Tv)}\right)^{\frac{1}{2}(s+\frac{3}{2})} E_t(g_{t^{-1}\mathrm{tr}(T\tau)}, s, \Phi^l_{1,\infty} \otimes \Phi_{1,f}).$$

Multiplying both sides by $(\det v)^{-\frac{l}{2}}$, we prove the proposition. □

5.7 W_T

In this section, we will prove a relation between $W_T(\tau, s, \Phi_2^D)$ and the coefficient $E_t(t^{-1}\mathrm{tr}(T\tau), s-\frac{1}{2}, \Phi_1^D)$ of the genus one Eisenstein series. Recall that, for a standard factorizable section $\Phi(s) = \otimes_p \Phi_p(s) \in I(s, \chi)$,

$$W_T(\tau, s, \Phi) = W_{T,\infty}(\tau, s, \Phi_\infty) \cdot \prod_{p<\infty} W_{T,p}(1, s, \Phi_p). \tag{5.7.1}$$

When $\Phi_p(0) = \lambda_{V_p}(\varphi)$ for some $\varphi \in S(V_p)^2$, (8.5.25) and formula (5.6.3) yield

$$\begin{aligned} W_{T,p}(r, \Phi_p) &= \int_{\mathrm{Sym}_2(\mathbb{Q}_p)} \Phi_p(w^{-1}n(b), r) \cdot \psi_p(-\mathrm{tr}(Tb))\, d_pb \\ &= \int_{\mathrm{Sym}_2(\mathbb{Q}_p)} \omega_{V_p}(w^{-1}n(b))\varphi^{(r)}(0) \cdot \psi_p(-\mathrm{tr}(Tb))\, d_pb \qquad (5.7.2)\\ &= \gamma_p(V)^2 \int_{\mathrm{Sym}_2(\mathbb{Q}_p)} \int_{(V_p^{(r)})^2} \psi_p(\mathrm{tr}(b\,Q(x)))\, \varphi^{(r)}(x)\, d_px \\ &\qquad \times \psi(-\mathrm{tr}(Tb))\, d_pb, \end{aligned}$$

for a positive integer r. Here, as in Section 5.3, the Haar measures d_px on $(V_p^{(r)})^2$ and d_pb on $\mathrm{Sym}_2(\mathbb{Q}_p)$ are taken to be self-dual with respect to the pairings $[x, y] \mapsto \psi_p(\mathrm{tr}(x, y))$ and $[b, c] \mapsto \psi_p(\mathrm{tr}(bc))$, respectively. We fix a lattice L_p in V_p and let

$$L_p^\# = \{x \in V_p \mid \psi_p((x, y)) = 1 \text{ for all } y \in L_p\} \tag{5.7.3}$$

be the dual lattice L_p. We recall from [17] and [18] that there are rational functions $F_p(X,T,L_p)$ and $F_p(X,t,L_p) \in \mathbb{Q}(X)$ such that, for any integer $r \geq 0$,

(5.7.4)

$$F_p(X,T,L_p)|_{X=p^{-r}} = \int\limits_{\mathrm{Sym}_2(\mathbb{Q}_p)} \int\limits_{(L_p^{(r)})^2} \psi_p(\mathrm{tr}(b\,Q(x)))\,\psi_p(-\mathrm{tr}(Tb))\,dx\,db$$

and

$$F_p(X,t,L_p)|_{X=p^{-r}} = \int_{\mathbb{Q}_p} \int_{L_p^{(r)}} \psi(b\,Q(x))\,\psi_p(-tb)\,dx\,db, \tag{5.7.5}$$

where the measure dx on $(V_p^{(r)})^2$ (resp. db on $\mathrm{Sym}_2(\mathbb{Q}_p)$) is normalized so that $\mathrm{vol}((L_p^{(r)})^2, dx) = 1$ (resp. $\mathrm{vol}(\mathrm{Sym}_2(\mathbb{Z}_p), db) = 1$). We refer to the functions $F_p(X,T,L_p)$ and $F_p(X,t,L_p)$ as the *local density functions.*

Lemma 5.7.1. *For a lattice L_p in V_p with $\varphi = \mathrm{char}(L_p)$ and $\Phi_i(s) \in I_i(s,\chi_V)$, the associated standard section, for $i = 1, 2$,*

$$W_{T,p}(s,\Phi_2) = \gamma_p(V_p)^2[L_p^\# : L_p]^{-1}|2|_p^{\frac{1}{2}}\,F_p(X,T,L_p),$$

and

$$W_{t,p}(s+\frac{1}{2},\Phi_1) = \chi_V(-1)\,\gamma_p(V_p)[L_p^\# : L_p]^{-\frac{1}{2}}\,F_p(X,t,L_p),$$

where $X = p^{-s}$.

Notice that $\gamma_p(V_p)^2[L_p^\# : L_p]^{-1}|2|_p^{\frac{1}{2}}$ is exactly the constant $C_p(V)$ in (5.2.11) and Theorem 5.2.4 for the quadratic space considered there.

The local density functions $F_p(X,T,L_p)$ and $F_p(X,t,L_p)$ are computed explicitly in [17] and [18]. We recall the needed formulae here for convenience. We refer to Appendix B of [18] for the definition of the Gross-Keating invariants and the Gross-Keating ϵ-constant.

Theorem 5.7.2. ([18], Theorem 5.7) *Let $(0,a_2,a_3)$ be the Gross-Keating invariants of* $\mathrm{diag}(1,T)$. *Let*

$$\mu_p(T) = \epsilon_p(T)(-\det T, -1)_p,$$

as in Definition 5.2.3, and let

$$\delta_{p,i} = \delta_{p,i}(T) = \begin{cases} 1 & \text{if } i = 0, [\frac{a_2}{2}]+1, \\ 1-p & \text{if } 1 \leq i \leq [\frac{a_2}{2}]. \end{cases}$$

(i) When a_2 is odd,

$$\frac{F_p(X,T,L_p^0)}{1-p^{-2}X^2} = \sum_{0\le i\le \frac{a_2-1}{2}} p^i(X^{2i} + \mu_p(T)X^{a_2+a_3-2i}),$$

and

$$F_p(X,T,L_p^{\mathrm{ra}}) = \sum_{0\le i\le \frac{a_2+1}{2}} \delta_{p,i}\, p^i(X^{2i} - \mu_p(T)X^{a_2+a_3-2i}).$$

(ii) When a_2 is even,

$$\frac{F_p(X,T,L_p^0)}{1-p^{-2}X^2} = \sum_{0\le i\le \frac{a_2}{2}-1} p^i(X^{2i} + \mu_p(T)X^{a_2+a_3-2i}) + p^{\frac{a_2}{2}} \sum_{0\le k\le a_3-a_2} (\epsilon X)^{a_2+k},$$

and

$$F_p(X,T,L_p^{\mathrm{ra}}) = \sum_{0\le i\le \frac{a_2}{2}} \delta_{p,i}\, p^i(X^{2i} - \mu_p(T)X^{a_2+a_3+2-2i}) - \epsilon p^{\frac{a_2}{2}+1}(X^{a_2+1} - \mu_p(T)X^{a_3+1}).$$

Here $\epsilon = \epsilon(T)$ is the Gross-Keating ϵ-constant.

Theorem 5.7.3. ([18], Theorem 5.8) *Let the notation be as in Theorem 5.7.2.*
(i) When a_2 is odd,

$$F_p(X,T,L_p^1) = 1 - \mu_p(T)X^{a_2+a_3+2} + p^{\frac{a_2-1}{2}}\left((p-2)X^{a_2+1} + \mu_p(T)pX^{a_3+1}\right) + (p-1)\sum_{1\le i\le \frac{a_2-1}{2}} p^{i-1}\left((p+2)X^{2i} + \mu_p(T)pX^{a_2+a_3+2-2i}\right).$$

(ii) When a_2 is even,

$$F_p(X,T,L_p^1) = 1 - \mu_p(T)X^{a_2+a_3+2} + p^{\frac{a_2}{2}} \sum_{a_2\le i\le a_3} C_k(\epsilon X)^{k+1} + (p-1)\sum_{1\le i\le \frac{a_2}{2}} p^{i-1}\left((p+2)X^{2i} + \mu_p(T)pX^{a_2+a_3+2-2i}\right),$$

with

$$C_k = \begin{cases} p & \text{if } k = a_2, \\ 2(p-1) & \text{if } a_2 < k < a_3, \\ p-2 & \text{if } k = a_3, \end{cases}$$

when $a_2 < a_3$, *and*

$$C_{a_2} = C_{a_3} = p - 2,$$

when $a_2 = a_3$.

When $T \in \mathrm{Sym}_2(\mathbb{Q})$ is singular of rank 1, let t be given as in (5.4.2), and write $4t = dn^2$ so that $-d$ is a fundamental discriminant. Then the Gross-Keating invariants of $\mathrm{diag}(1, T)$ are $(0, a_2, \infty)$, where $(0, a_2)$ are the Gross-Keating invariants of $\mathrm{diag}(1, t)$. Concretely, for $k = \mathrm{ord}_p(n)$,

$$a_2 = \begin{cases} 2k+1 & \text{if } p \mid d, \\ 2k & \text{if } p \nmid d. \end{cases}$$

Notice that the local Gross-Keating ϵ-constant is $\epsilon(T) = \chi_d(p)$ when defined, i.e., when $p \nmid d$. When $T = 0$, one has $a_2 = \infty$ as well. In the singular case, Theorems 5.7.2 and 5.7.3 imply the following:

Corollary 5.7.4. *Let* T, t, d, *and* n *be as above and let* $k = \mathrm{ord}_p n$. *Then*

$$\frac{F_p(X, T, L_p^0)}{1 - p^{-2}X^2} = \begin{cases} \sum_{0 \le i \le k} (pX^2)^i & \text{if } p \mid d, \\ \sum_{0 \le i \le k-1} (pX^2)^i + \dfrac{(pX^2)^k}{1 - \chi_d(p)X} & \text{if } p \nmid d, \end{cases}$$

and

$$F_p(X, T, L_p^{\mathrm{ra}}) = \begin{cases} \sum_{0 \le i \le k+1} \delta_{p,i} (pX^2)^i & \text{if } p \mid d, \\ \sum_{0 \le i \le k} \delta_{p,i} (pX^2)^i - \chi_d(p) p^{k+1} X^{2k+1} & \text{if } p \nmid d. \end{cases}$$

Finally,

$$F_p(X, T, L_p^1) = 1 + (p-2)p^k X^{2k+2} + (p-1)(p+2) \sum_{1 \le i \le k} p^{i-1} X^{2i}$$

when $p \mid d$, *and*

$$F_p(X, T, L_p^1) = 1 + \chi_d(p) p^{k+1} X^{2k+1} + (p-1)(p+2) \sum_{1 \le i \le k} p^{i-1} X^{2i} + \frac{2(p-1)p^k X^{2k+2}}{1 - \chi_d(p)X},$$

when $p \nmid d$.

Comparing this corollary with Proposition C.2 of [18], one sees immediately

Proposition 5.7.5. *For $p < \infty$, and for $X = p^{-s}$*

$$F_p(X, T, L_p^0) = F_p(pX, t, L_p^0) \cdot \frac{\zeta_p(2s)}{\zeta_p(2s+2)},$$

and

$$F_p(X, T, L_p^{\mathrm{ra}}) = F_p(pX, t, L_p^{\mathrm{ra}}).$$

The following corollary will be used in the next section and follows immediately from Corollary 5.7.4.

Corollary 5.7.6. *Let the notation be as above with $k = \mathrm{ord}_p n$.*
(i)

$$W_{T,p}(s, \Phi_p^0) = \gamma_p(V_p^+)^2 \, |2|_p^{\frac{3}{2}} \, F_p(X, T, L_p^0)$$

and

$$W_{T,p}(s, \Phi_p^1) = \gamma_p(V_p^+)^2 \, |2|_p^{\frac{3}{2}} \, p^{-2} \, F_p(X, T, L_p^1).$$

(ii) When $\chi_d(p) = 0$, i.e, $p \mid d$,

$$F_p(1, T, L^0) = p^{-2}(p+1)(p^{k+1} - 1),$$
$$F_p(1, T, L^1) = 2\, p^{k+1} - p - 1.$$

(iii) When $\chi_d(p) = -1$,

$$F_p(1, T, L^0) = p^{-2}(p+1)\, \frac{p^{k+1} + p^k - 2}{2},$$
$$F_p(1, T, L^1) = (p+1)(p^k - 1).$$

(iv) When $\chi_d(p) = 1$, both $F_p(X, T, L^0)$ and $F_p(X, T. L^1)$ have a simple pole at $X = 1$ with residues

$$Res_{X=1} \, F_p(X, T, L^0) = -p^{k-2}(p^2 - 1),$$
$$Res_{X=1} \, F_p(X, T, L^1) = -2(p-1)\, p^k.$$

We next turn to the archimedean factors.

Proposition 5.7.7. *For the standard sections $\Phi^{\frac{3}{2}}_{2,\infty}(s) \in I_2(s,\chi_\infty)$ and $\Phi^{\frac{3}{2}}_{1,\infty}(s) \in I_1(s,\chi_\infty)$ of weight $\frac{3}{2}$, and for T and t as in (5.4.1) and (5.4.2),*

$$W_{T,\infty}(\tau,s,\Phi^{\frac{3}{2}}_{2,\infty}) = W_{t,\infty}(t^{-1}\mathrm{tr}(T\tau), s-\frac{1}{2}, \Phi^{\frac{3}{2}}_{1,\infty}) \cdot \left(\frac{\det v}{t^{-1}\mathrm{tr}(Tv)}\right)^{-\frac{s}{2}}$$
$$\times\, e(-\frac{1}{4})\,\sqrt{2}\cdot\frac{s}{s+1}\cdot\frac{\zeta_\infty(2s)}{\zeta_\infty(2s+2)},$$

where $\zeta_\infty(2s) = \pi^{-s}\Gamma(s)$. The formula also holds for $t=0$ and $T=0$ with $t^{-1}\mathrm{tr}(T\tau) = i$ and $t^{-1}\mathrm{tr}(Tv) = 1$.

To prove this proposition, we need some preparation.

Following Shimura [15], let

$$\Gamma_n(s) = \pi^{\frac{n(n-1)}{4}} \prod_{k=0}^{n-1} \Gamma(s-\frac{k}{2}) = \int_{x>0} e^{-\mathrm{tr}(x)} \det(x)^{s-\frac{n+1}{2}}\, dx \tag{5.7.6}$$

be the gamma function of degree n. For $g \in \mathrm{Sym}_n(\mathbb{R})_{>0}$, $h \in \mathrm{Sym}_n(\mathbb{R})$, and α, $\beta \in \mathbb{C}$ with real parts are sufficiently large, let

(5.7.7)

$$\eta(g,h,\alpha,\beta) = \int_{x\pm h>0} e^{-\mathrm{tr}(gx)} \det(x+h)^{\alpha-\frac{n+1}{2}} \det(x-h)^{\beta-\frac{n+1}{2}}\, dx.$$

We will sometimes write $\eta(g,h,\alpha,\beta) = \eta_n(g,h,\alpha,\beta)$ to emphasize the degree n. The normalized function

$$\eta^*(g,h,\alpha,\beta) = (\det g)^{\alpha+\beta-\frac{n+1}{2}}\eta(g,h,\alpha,\beta) \tag{5.7.8}$$

satisfies

$$\eta^*(g,-h,\alpha,\beta) = \eta^*(g,h,\beta,\alpha), \tag{5.7.9}$$

and

$$\eta^*(g,{}^t aha,\alpha,\beta) = \eta^*(ag^t a,h,\alpha,\beta), \tag{5.7.10}$$

for $a \in \mathrm{GL}_n(\mathbb{R})$.

When h is singular, the following reduction formula is a special case of Proposition 4.1 of [15].

Lemma 5.7.8. *Assume $g = \mathrm{diag}(g_1,g_2) > 0$ and $h = \mathrm{diag}(h_1,0)$, where $g_1, h_1 \in \mathrm{Sym}_{n_1}(\mathbb{R})$ and $g_2 \in \mathrm{Sym}_{n_2}(\mathbb{R})$ with $n_1+n_2 = n$. Then*

$$\eta^*_n(g,h,\alpha,\beta) = \pi^{\frac{n_1 n_2}{2}}\,\Gamma_{n_1}(\alpha+\beta-\frac{n+1}{2})\cdot\eta^*_{n_1}(g_1,h_1,\alpha-\frac{n_2}{2},\beta-\frac{n_2}{2}).$$

Proof. To avoid more notation, we only recall the proof in the case $n_1 = n_2 = 1$, which is needed in this paper. Write

$$x = \begin{pmatrix} u & v \\ v & w \end{pmatrix},$$

substitute $y = u - v^{-1}w^2$ and leave v and w unchanged. Then the domain $X \pm h > 0$ becomes $y \pm h_1 > 0$ and $w > 0$, and we have

$$\begin{aligned}
&\eta_n(g, h, \alpha, \beta) \\
&= \int\limits_{\substack{y \pm h_1 > 0 \\ w > 0}} e^{-g_1 y - g_1 v^2 w^{-1} - g_2 w}\, w^{\alpha+\beta-3}\, (y + h_1)^{\alpha - \frac{3}{2}}\, (y - h_1)^{\beta - \frac{3}{2}}\, dy\, dv\, dw \\
&= \eta_{n_1}(g_1, h_1, \alpha - \frac{1}{2}, \beta - \frac{1}{2}) \cdot \int_{w>0} e^{-g_2 w} w^{\alpha+\beta-3} \int_{-\infty}^{\infty} e^{-g_1 w^{-1} v^2}\, dv\, dw \\
&= \pi^{\frac{1}{2}}\, g_1^{-\frac{1}{2}}\, \eta_{n_1}(g_1, h_1, \alpha - \frac{1}{2}, \beta - \frac{1}{2}) \cdot \int_{w>0} e^{-g_2 w}\, w^{\alpha+\beta-\frac{5}{2}}\, dw \\
&= \pi^{\frac{1}{2}}\, g_1^{-\frac{1}{2}}\, g_2^{-(\alpha+\beta-\frac{3}{2})}\, \Gamma(\alpha + \beta - \frac{3}{2}) \cdot \eta_{n_1}(g_1, h_1, \alpha - \frac{1}{2}, \beta - \frac{1}{2}).
\end{aligned}$$

Thus,

$$\eta_n^*(g, h, \alpha, \beta) = \pi^{\frac{1}{2}}\, \Gamma(\alpha + \beta - \frac{3}{2}) \cdot \eta_{n_1}^*(g_1, h_1, \alpha - \frac{1}{2}, \beta - \frac{1}{2}),$$

as claimed. □

In general, for a quadratic character χ, let $I_n(s, \chi)$ be the degenerate principal series representation of $\mathrm{Sp}_n(\mathbb{R})$ or of its metaplectic cover. Let $\Phi^l_{n,\infty}(s) \in I_n(s, \chi)$ be the unique eigenfunction of $K = \mathrm{U}(n)$, or the preimage of $\mathrm{U}(n)$ in the metaplectic cover, with character $(\det)^l$, such that $\Phi^l_{n,\infty}(1, s) = 1$ for all s. For $\tau = u + iv \in \mathfrak{H}_n$, the corresponding Whittaker function is

$$\begin{aligned}
(5.7.11)\quad & W_{T,\infty}(\tau, s, \Phi^l_{n,\infty}) \\
&= (\det v)^{-\frac{l}{2}} \int\limits_{\mathrm{Sym}_n(\mathbb{R})} \Phi^l_{n,\infty}(w^{-1} n(b) g_\tau, s)\, e(-\mathrm{tr}(Tb))\, db.
\end{aligned}$$

First, the same calculation as Lemma 9.2 of [4] gives

$$\text{(5.7.12)}\quad \frac{W_{T,\infty}(\tau, s, \Phi^l_{n,\infty})}{e(\mathrm{tr}(Tu))\, e(\frac{n}{8})}$$
$$= (\det v)^{\beta_n} \int_{\mathrm{Sym}_n(\mathbb{R})} e(-\mathrm{tr}(Tb)) \det(b+iv)^{-\alpha_n} \det(b-iv)^{-\beta_n}\, db,$$

where

$$\text{(5.7.13)}\qquad \alpha_n = \frac{1}{2}\,(s + \frac{n+1}{2} + l), \qquad \beta_n = \frac{1}{2}\,(s + \frac{n+1}{2} - l).$$

Note that the factor $e(\frac{n}{8})$ is due to the fact that we are using $w^{-1} = [w^{-1}, 1]_{\mathrm{L}}$ here, whereas $[w^{-1}, 1]_{\mathrm{R}} = [w^{-1}, \beta(w^{-1})]_{\mathrm{L}}$, with $\beta(w^{-1}) = e(-\frac{n}{8})$, was used in [4]; see Section 8.5.3. Next, the same calculation as on pp. 585–586 of [4] (a special case of formula (1.29) of [15]) gives

$$\text{(5.7.14)}\quad \frac{W_{T,\infty}(\tau, s, \Phi^l_{n,\infty})}{e(\mathrm{tr}(Tu))\, e(-\frac{n}{4}(l-\frac{1}{2}))}$$
$$= \frac{2^{-ns}(2\pi)^{\frac{n(n+1)}{2}}}{2^{\frac{n(n-1)}{2}}\Gamma_n(\alpha_n)\,\Gamma_n(\beta_n)} \cdot (\det v)^{\beta_n - s} \cdot \eta^*_n(2v, \pi\, T, \alpha_n, \beta_n).$$

We will only need this for $n = 1, 2$. We remark that this formula, together with Lemma 5.7.8, implies that the singular Fourier coefficients of an Eisenstein series of genus n are finite sums of Fourier coefficients of Eisenstein series of genus less than n.

Proof of Proposition 5.7.7. For $v \in \mathrm{Sym}_2(\mathbb{R})_{>0}$, let $v^{\frac{1}{2}} \in \mathrm{Sym}_2(\mathbb{R})_{>0}$ be the unique positive definite symmetric matrix whose square is v. For $T \in \mathrm{Sym}_2(\mathbb{R})$ with $\det T = 0$, choose $k \in \mathrm{SO}(2)$ such that

$$v^{\frac{1}{2}}\, T v^{\frac{1}{2}} = {}^t k \cdot \mathrm{diag}(\mathrm{tr}(Tv), 0) \cdot k.$$

Then (5.7.14) gives

(5.7.15)

$$\eta^*_2(2v, \pi\, T, \alpha_2, \beta_2) = \eta^*_2(2, \mathrm{diag}(\pi\, \mathrm{tr}(Tv), 0), \alpha_2, \beta_2)$$
$$= \pi^{\frac{1}{2}}\, \Gamma(s) \cdot \eta^*_1(2, \pi\, \mathrm{tr}(Tv), \alpha_2 - \frac{1}{2}, \beta_2 - \frac{1}{2}),$$

so that we have

$$W_{T,\infty}(\tau, s, \Phi^l_{2,\infty}) = e(\mathrm{tr}(Tu))\, e(-\tfrac{1}{2}(l-\tfrac{1}{2}))\frac{2^{-2s+2}\pi^{\frac{7}{2}}\Gamma(s)}{\Gamma_2(\alpha_2)\Gamma_2(\beta_2)}\,(\det v)^{\beta_2 - s}$$
$$\times\, \eta^*(2, \pi\,\mathrm{tr}(Tv), \frac{1}{2}(s+\frac{1}{2}+l), \frac{1}{2}(s+\frac{1}{2}-l)).$$

On the other hand, (5.7.14) gives

(5.7.16)

$$W_{t,\infty}(t^{-1}\mathrm{tr}(T\tau), s, \Phi^l_{1,\infty}) = e(\mathrm{tr}(Tu))\, e(-\tfrac{1}{4}(l-\tfrac{1}{2}))\frac{2^{-s+1}\pi(t^{-1}\mathrm{tr}(Tv))^{\beta_1 - s}}{\Gamma(\alpha_1)\Gamma(\beta_1)}$$
$$\times\, \eta^*(2, \pi\,\mathrm{tr}(Tv), \frac{1}{2}(s+1+l), \frac{1}{2}(s+1-l)).$$

Thus,

$$W_{T,\infty}(\tau, s, \Phi^l_{2,\infty}) = C_\infty(s)\cdot W_{t,\infty}(t^{-1}\mathrm{tr}(T\tau), s-\frac{1}{2}, \Phi^l_{1,\infty}), \tag{5.7.17}$$

with

$$C_\infty(s) = e(-\tfrac{1}{4}(l-\tfrac{1}{2}))\cdot\frac{\pi^{\frac{5}{2}}\,\Gamma(s)\,\Gamma(\frac{1}{2}(s+1/2+l))\,\Gamma(\frac{1}{2}(s+1/2-l))}{2^{s-\frac{1}{2}}\,\Gamma_2(\alpha_2)\,\Gamma_2(\beta_2)}$$
$$\times\left(\frac{\det v}{t^{-1}\mathrm{tr}(Tv)}\right)^{\beta_2 - s}$$
$$= e(-\tfrac{1}{4}(l-\tfrac{1}{2}))\cdot\frac{2^{-s+\frac{1}{2}}\,\pi^{\frac{3}{2}}\Gamma(s)}{\Gamma(\alpha_2)\,\Gamma(\beta_2)}\cdot\left(\frac{\det v}{t^{-1}\mathrm{tr}(Tv)}\right)^{\beta_2 - s}.$$

Here we have used the fact that $\beta_2(s) - s = \beta_1(s-\frac{1}{2}) - (s-\frac{1}{2})$. When $l = \frac{3}{2}$, note that

$$\begin{aligned}\Gamma(\alpha_2)\,\Gamma(\beta_2) &= \Gamma(\frac{s+3}{2})\,\Gamma(\frac{s}{2})\\ &= \sqrt{\pi}\,2^{-s}\,(s+1)\,\Gamma(s)\\ &= \sqrt{\pi}\,2^{-s}\,\frac{s+1}{s}\,\Gamma(s+1)\,\sqrt{\pi},\end{aligned} \tag{5.7.18}$$

so that

$$\begin{aligned}C_\infty(s) &= e(-\frac{1}{4})\cdot\frac{\sqrt{2}\pi}{s+1}\cdot\left(\frac{\det v}{t^{-1}\mathrm{tr}(Tv)}\right)^{-\frac{s}{2}}\\ &= e(-\frac{1}{4})\cdot\frac{\sqrt{2}s\,\zeta_\infty(2s)}{(s+1)\,\zeta_\infty(2s+2)}\cdot\left(\frac{\det v}{t^{-1}\mathrm{tr}(Tv)}\right)^{-\frac{s}{2}}.\end{aligned} \tag{5.7.19}$$

This proves Proposition 5.7.7. □

We remark that, for a general l, one has

$$C_\infty(s) = i^r\,\sqrt{2}\,\pi\cdot\frac{(r+s)(r-2+s)\cdots(r-2[\frac{r}{2}]+s)}{(r-1-s)(r-3-s)\cdots(r+1-2[\frac{r-1}{2}]-s)}$$

$$\times\left(\frac{\det v}{t^{-1}\mathrm{tr}(Tv)}\right)^{-\frac{1}{2}(s-\frac{3}{2}+l)}. \tag{5.7.20}$$

Here $r = l - \frac{1}{2}$.

Theorem 5.7.9. *With the notation above,*

$$W_T(\tau, s, \Phi_2^D) = (-1)^{\mathrm{ord}(D)+1}\left(\frac{\det v}{t^{-1}\mathrm{tr}(Tv)}\right)^{-\frac{s}{2}}$$

$$\times\frac{s\,\Lambda_D(2s)}{(s+1)\,\Lambda_D(2s+2)}\cdot E_t(t^{-1}\mathrm{tr}(T\tau), s-\frac{1}{2}, \Phi_1^D).$$

Here $\mathrm{ord}(D)$ *is the number of prime factors of* D *and*

$$\Lambda_D(2s) = \pi^{-s}\,\Gamma(s)\,\zeta(2s)\cdot D^s\prod_{p|D}(1-p^{-2s}).$$

Proof. By Lemma 5.7.1 and Proposition 5.7.5, we have

$$W_{T,p}(1, s, \Phi_p^0) = \chi_{V_p}(-1)\,\gamma_p(V_p^+)\,|2|_p^{\frac{1}{2}}\cdot W_{t,p}(1, s-\frac{1}{2}, \Phi_p^0)\,\frac{\zeta_p(2s)}{\zeta_p(2s+2)}$$

and

$$W_{T,p}(1, s, \Phi_p^{\mathrm{ra}}) = \chi_{V_p}(-1)\,\gamma_p(V_p^-)\,|2|_p^{\frac{1}{2}}\,p^{-1}\cdot W_{t,p}(1, s-\frac{1}{2}, \Phi_p^{\mathrm{ra}}).$$

Combining these with Proposition 5.7.7, we obtain

$$C^{-1}\,\frac{s+1}{s}\cdot W_T(\tau, s, \Phi^D)$$

$$= \frac{\prod_{p\nmid D}\zeta_p(2s)}{D\prod_{p\nmid D}\zeta_p(2s+2)}\cdot\left(\frac{\det v}{t^{-1}\mathrm{tr}(Tv)}\right)^{-\frac{s}{2}}\cdot E_t(t^{-1}\mathrm{tr}(T\tau), s-\frac{1}{2}, \Phi^D)$$

$$= \frac{\Lambda_D(2s)}{\Lambda_D(2s+2)}\cdot\left(\frac{\det v}{t^{-1}\mathrm{tr}(Tv)}\right)^{-\frac{s}{2}}\cdot E_t(t^{-1}\mathrm{tr}(T\tau), s-\frac{1}{2}, \Phi^D),$$

where the constant C is given by

$$C = e(-\frac{1}{4})\prod_{p|D}\gamma_p(V_p^-)\prod_{p\nmid D\infty}\gamma_p(V_p^+)\prod_{p<\infty}\chi_{V_p}(-1) = (-1)^{o(D)+1}, \tag{5.7.21}$$

since

$$\gamma_p(V_p^-) = -\gamma_p(V_p^+) \tag{5.7.22}$$

and

$$\gamma_\infty(V_\infty^+)\,\chi_{V_\infty}(-1) = i. \tag{5.7.23}$$

□

Remark 5.7.10. There is a direct way to compare W_T with W_t in theory. It can be proved that for any section $\Phi \in I_2(s, \chi)$,

$$W_T(g, s, \Phi) = W_{t_2}(\frac{1}{2} - s) \circ M(s)\Phi(m(\gamma_0)g).$$

Here $M(s)\Phi = W_0(g, s, \Phi)$ is the intertwining operator. So it suffices to compute $M(s)\Phi$.

5.8 THE CENTRAL DERIVATIVE—THE RANK ONE CASE

In this section, we obtain formulas for the singular Fourier coefficients of the normalized and the modified genus two Eisenstein series. The normalized genus two Eisenstein series is defined by

$$\mathbb{E}(\tau, s, \Phi_2^D) = -\frac{1}{2} \cdot c(D)(s+1)\Lambda_D(2s+2) \cdot E(\tau, s, \Phi_2^D), \tag{5.8.1}$$

where the normalizing factor is similar to the one used in [9] in the genus one case. Notice that, when $D = D(B)$ is the discriminant of an indefinite quaternion algebra B, then

$$\begin{aligned} \mathbb{E}'_T(\tau, 0, \Phi_2^D) &= -\frac{1}{2} \cdot c(D)\Lambda_D(2) \cdot E'_T(\tau, 0, \Phi_2^D) \\ &= \frac{1}{2} \cdot \mathrm{vol}(\mathcal{M}(\mathbb{C})) \cdot E'_T(\tau, 0, \Phi_2^D). \end{aligned} \tag{5.8.2}$$

The following main theorem on the singular coefficients follows immediately from Theorems 5.6.4 and 5.7.9 and the functional equation (5.5.4).

Theorem 5.8.1. *Let D be a square free positive integer and let $\Phi_2^D(s) \in I_2(s, \chi)$ be the standard section defined in (5.4.6). Let $\Phi_1^D(s) \in I_1(s, \chi)$ be the analogous standard section, as explained in Section 5.5. Let T be of rank 1, and let t be given as in (5.4.2). Then*

$$\begin{aligned} \mathbb{E}_T(\tau, s, \Phi_2^D) = &-\frac{1}{2} \cdot \left(\frac{\det v}{t^{-1}\mathrm{tr}(Tv)}\right)^{\frac{s}{2}} \cdot \mathbb{E}_t(t^{-1}\mathrm{tr}(T\tau), s + \frac{1}{2}, \Phi_1^D) \\ &+ \frac{1}{2} \cdot (-1)^{\mathrm{ord}(D)} \left(\frac{\det v}{t^{-1}\mathrm{tr}(Tv)}\right)^{-\frac{s}{2}} \cdot \mathbb{E}_t(t^{-1}\mathrm{tr}(T\tau), -s + \frac{1}{2}, \Phi_1^D). \end{aligned}$$

Remark 5.8.2. This formula shows *why* such singular Fourier coefficients vanish at $s = 0$ in the case $D = D(B)$, when $\mathrm{ord}(D)$ is even. Note that this vanishing arises for a global reason, i.e., comes from the functional equation of the genus one Eisenstein series. In contrast, as explained in Section 5.2 above, the vanishing of the nonsingular Fourier coefficients depends on the vanishing of local Whittaker functions and hence is local in nature. In the next chapter, we will see that the geometric quantities corresponding to such coefficients have a similar nature.

We next consider the modified Eisenstein series $\mathcal{E}_2(\tau, s, B)$ defined in (5.1.44). Let $\tilde{\Phi}^B(s)$ be the section defined in (5.1.36). By writing $\tilde{\Phi}^B(s)$ as a sum of products of standard sections with coefficients which are functions of s, we have

$$E(\tau, s, \tilde{\Phi}^B) = E(\tau, s, \Phi^D) + \sum_{p|D} A_p(s)\, E(\tau, s, \Phi^{\frac{D}{p}}) + \sum_{p|D} B_p(s)\, E(\tau, s, \Phi^{\frac{D}{p},1}) + O(s^2), \tag{5.8.3}$$

where $D = D(B)$ and where the $O(s^2)$ term is a combination of standard Eisenstein series with coefficients having a zero of order at least 2 at $s = 0$. The three standard sections $\Phi^{\frac{D}{p},1}(s)$, $\Phi^{\frac{D}{p}}(s)$, and $\Phi^D(s)$ are the same except for their local components at $p \mid D$, and these local components correspond to the lattices L_p^1, L_p^0, and L_p^{ra}, respectively; see (5.1.30)–(5.1.32). The T-th Fourier coefficient of the normalized series

$$\mathcal{E}_2(\tau, s, B) = -\frac{1}{2} \cdot c(D)(s+1)\Lambda_D(2s+2) \cdot E(\tau, s, \tilde{\Phi}^B) \tag{5.8.4}$$

can then be written as

(5.8.5)

$$\begin{aligned} &\mathcal{E}_{2,T}(\tau, s, B) \\ &= \mathbb{E}_T(\tau, s, \Phi^D) + \sum_{p|D} A_p(s)\, \frac{c(D)}{c(D/p)}\, (p^{s+1} - p^{-s-1})\, \mathbb{E}_T(\tau, s, \Phi^{\frac{D}{p}}) \\ &\quad + \sum_{p|D} B_p(s)\, \frac{c(D)}{c(D/p)}\, (p^{s+1} - p^{-s-1})\, \mathbb{E}_T(\tau, s, \Phi^{\frac{D}{p},1}) + O(s^2). \end{aligned}$$

Here $E(\tau, s, \Phi^{\frac{D}{p},1})$ and $E(\tau, s, \Phi^{\frac{D}{p}})$ are normalized by the factor $C^{D/p}(s)$, given by (5.1.38). Since

$$\frac{c(D)}{c(D/p)} = -\frac{p}{p+1}, \tag{5.8.6}$$

for T of rank 1, we have, by Theorem 5.8.1,

$$(5.8.7)\quad \mathcal{E}'_{2,T}(\tau,0,B) = -\,\mathbb{E}'_t(t^{-1}\mathrm{tr}(T\tau),\frac{1}{2},\Phi_1^D)$$
$$-\frac{1}{2}\cdot\log\left(\frac{\det v}{t^{-1}\mathrm{tr}(Tv)}\right)\cdot\mathbb{E}_t(t^{-1}\mathrm{tr}(T\tau),\frac{1}{2},\Phi_1^D)$$
$$+\sum_{p|D}(p-1)\,A'_p(0)\cdot\mathbb{E}_t(t^{-1}\mathrm{tr}(T\tau),\frac{1}{2},\Phi_1^{\frac{D}{p}})$$
$$-\sum_{p|D}(p-1)\,B'_p(0)\cdot\mathbb{E}_T(\tau,0,\Phi^{\frac{D}{p},1}).$$

It remains to identify the value $\mathbb{E}_T(\tau,0,\Phi^{\frac{D}{p},1})$ occurring in the last line. Recall that $\mathbb{E}_T = \mathbb{B}_T + \mathbb{W}_T$, where the bold letters indicate that B_T and W_T have been multiplied by the relevant normalizing factor. Theorem 5.6.4 implies that

(5.8.8)
$$\mathbb{B}_T(\tau,s,\Phi_2^{\frac{D}{p},1}) = -\frac{1}{2}\cdot\left(\frac{\det v}{t^{-1}\mathrm{tr}(Tv)}\right)^{\frac{s}{2}}\cdot\mathbb{E}_t(t^{-1}\mathrm{tr}(T\tau),s+\frac{1}{2},\Phi_1^{\frac{D}{p},1}).$$

At the point $s=0$, a similar relation holds for the $\mathbb{W}_T$ term.

Proposition 5.8.3.

$$\mathbb{W}_T(\tau,0,\Phi_2^{\frac{D}{p},1}) = -\frac{1}{2}\cdot\mathbb{E}_t(t^{-1}\mathrm{tr}(T\tau),\frac{1}{2},\Phi_1^{\frac{D}{p},1}).$$

From this proposition and identities (5.8.7) and (5.8.8), we immediately obtain the following formula.

Proposition 5.8.4. *Let $D = D(B)$ for an indefinite quaternion algebra B.*

$$\mathcal{E}'_{2,T}(\tau,0,B) = -\mathbb{E}'_t(t^{-1}\mathrm{tr}(T\tau),\frac{1}{2},\Phi_1^D)$$
$$-\frac{1}{2}\log\left(\frac{\det v}{t^{-1}\mathrm{tr}(Tv)}\right)\cdot\mathbb{E}_t(t^{-1}\mathrm{tr}(T\tau),\frac{1}{2},\Phi_1^D)$$
$$+\sum_{p|D}(p-1)\,A'_p(0)\cdot\mathbb{E}_t(t^{-1}\mathrm{tr}(T\tau),\frac{1}{2},\Phi_1^{\frac{D}{p}})$$
$$+\sum_{p|D}(p-1)\,B'_p(0)\cdot\mathbb{E}_t(t^{-1}\mathrm{tr}(T\tau),\frac{1}{2},\Phi_1^{\frac{D}{p},1}).$$

To prove Proposition 5.8.3, we need the following lemma which follows immediately from Proposition 8.1 of [9] or Proposition C.2 of [18].

Lemma 5.8.5. *Let* $k = \mathrm{ord}_p(n)$. *Then*

(i)

$$W_{t,p}(s + \frac{1}{2}, \Phi_p^0) = \chi_{V_p}(-1)\,\gamma_p(V_p^+)\,|2|_p^{\frac{1}{2}}\,F_p(X, t, L_p^0),$$
$$W_{t,p}(s + \frac{1}{2}, \Phi_p^1) = \chi_{V_p}(-1)\,\gamma_p(V_p^+)\,|2|_p^{\frac{1}{2}}\,p^{-1}\,F_p(X, t, L_p^1),$$
$$W_{t,p}(s + \frac{1}{2}, \Phi_p^-) = \chi_{V_p}(-1)\,\gamma_p(V_p^-)\,|2|_p^{\frac{1}{2}}\,p^{-1}\,F_p(X, t, L_p^{\mathrm{ra}}).$$

(ii)

$$F_p(1, t, L^0) = \begin{cases} (p+1)\,p^{-k-2}\,(p^{k+1} - 1) & \text{if } \chi_d(p) = 0, \\ p^{-k-1}\,(p^{k+1} + p^k - 2) & \text{if } \chi_d(p) = -1, \\ p^{-1}\,(p+1) & \text{if } \chi_d(p) = 1. \end{cases}$$

(iii)

$$F_p(1, t, L^1) = \begin{cases} p^{-k-1}\,(2\,p^{k+1} - p - 1) & \text{if } \chi_d(p) = 0, \\ 2\,p^{-k}\,(p^k - 1) & \text{if } \chi_d(p) = -1, \\ 2 & \text{if } \chi_d(p) = 1. \end{cases}$$

(iv)

$$F_p(1, t, L^{\mathrm{ra}}) = \begin{cases} p^{-k-1}\,(p+1) & \text{if } \chi_d(p) = 0, \\ 2\,p^{-k} & \text{if } \chi_d(p) = -1, \\ 0 & \text{if } \chi_d(p) = 1. \end{cases}$$

(v)

$$F_p(X, t, L^{\mathrm{ra}}) + F_p(X, t, L^1) = 2.$$

Proof of Proposition 5.8.3. Theorem 5.7.9 and the functional equation (5.5.4) imply that

(5.8.9)

$$\begin{aligned} &\mathbb{W}_T(\tau, s, \Phi_2^{\frac{D}{p},1}) \cdot \left(\frac{\det v}{t^{-1}\mathrm{tr}(Tv)}\right)^{\frac{s}{2}} \\ &= \mathbb{W}_T(\tau, s, \Phi_2^{\frac{D}{p}}) \cdot \frac{W_{T,p}(s, \Phi_{2,p}^1)}{W_{T,p}(s, \Phi_{2,p}^0)} \cdot \left(\frac{\det v}{t^{-1}\mathrm{tr}(Tv)}\right)^{\frac{s}{2}} \\ &= -\frac{1}{2} \cdot \mathbb{E}_t(t^{-1}\mathrm{tr}(T\tau), s - \frac{1}{2}, \Phi_1^{\frac{D}{p}}) \cdot \frac{W_{T,p}(s, \Phi_{2,p}^1)}{W_{T,p}s, \Phi_{2,p}^0)} \end{aligned}$$

$$= -\frac{1}{2} \cdot \mathbb{E}_t(t^{-1}\mathrm{tr}(T\tau), -s + \frac{1}{2}, \Phi_1^{\frac{D}{p}}) \cdot \frac{W_{T,p}(s, \Phi^1_{2,p})}{W_{T,p}(s, \Phi^0_{2,p})}$$

$$= -\frac{1}{2} \cdot \mathbb{E}_t^{(p)}(t^{-1}\mathrm{tr}(T\tau), -s + \frac{1}{2}) \cdot \frac{W_{t,p}(\frac{1}{2} - s, \Phi^0_{1,p}) \, W_{T,p}(s, \Phi^1_{2,p})}{W_{T,p}(s, \Phi^0_{2,p})}.$$

Here $\mathbb{E}_t^{(p)}$ is the non-p-part of the t-th Fourier coefficient of the normalized Eisenstein series. This quantity is the same for $\Phi^{\frac{D}{p}}(s)$ and $\Phi^{\frac{D}{p},1}(s)$. Evaluating at $s = 0$, we have

$$\text{(5.8.10)} \quad \mathbb{W}_T(\tau, 0, \Phi_2^{\frac{D}{p},1})$$
$$= -\frac{1}{2} \cdot \mathbb{E}_t^{(p)}(t^{-1}\mathrm{tr}(T\tau), \frac{1}{2}) \cdot \lim_{s \to 0} \frac{W_{t,p}(\frac{1}{2} - s, \Phi^0_{1,p}) \, W_{T,p}(s, \Phi^1_{2,p})}{W_{T,p}(s, \Phi^0_{2,p})},$$

and so it suffices to verify that

$$\text{(5.8.11)} \qquad \lim_{s \to 0} \frac{W_{t,p}(\frac{1}{2} - s, \Phi^0_{1,p}) \, W_{T,p}(s, \Phi^1_{2,p})}{W_{T,p}(s, \Phi^0_{2,p})} = W_{t,p}(s, \Phi^1_{1,p}). \qquad \square$$

In terms of local density functions, (5.8.11) is the same as the following identity.

Lemma 5.8.6.

$$\lim_{X \to 1} \frac{p^{-1} F_p(X, t, L^0) \, F_p(X, T, L^1)}{F_p(X, T, L^0)} = F_p(1, t, L^1).$$

Proof. This follows from Corollary 5.7.6, Lemma 5.8.3, and a routine calculation. Indeed, when $\chi_d(p) = 0$, one has from these results

$$F_p(1, T, L^0) = p^k \, F_p(1, t, L^0) = p^{-2}(p+1)(p^{k+1} - 1),$$

and

$$F_p(1, T, L^1) = p^{k+1} \, F_p(1, t, L^1) = 2p^{k+1} - p - 1.$$

The claim is thus clear in this case. When $\chi_d(p) = -1$, one has, similarly,

$$(p+1)^{-1} \, F_p(1, T, L^0) = \frac{1}{2} p^{k-1} \, F_p(1, t, L^0) = \frac{1}{2} p^{-2}(p^{k+1} + p^k - 2),$$

and

$$(p+1)^{-1} \, F_p(1, T, L^1) = \frac{1}{2} p^k \, F_p(1, t, L^1) = (p^k - 1).$$

The claim is again clear in this case. Finally, when $\chi_d(p) = 1$, both $F_p(X, T, L^0)$ and $F_p(X, T, L^1)$ have a simple pole at $X = 1$, which is the reason for the limit sign. Then

$$\lim_{X\to 1} \frac{p^{-1} F_p(X, t, L^0) F_p(X, T, L^1)}{F_p(X, T, L^0)} = p^{-1} \frac{-2(p-1)p^k \cdot p^{-1}(p+1)}{-p^{k-2}(p^2-1)}$$
$$= 2 = F_p(1, t, L^1). \qquad \square$$

Our main result expresses the singular coefficients of $\mathcal{E}_2'(\tau, 0, B)$ in terms of the modified genus one Eisenstein series.

Theorem 5.8.7. *When $D = D(B) > 1$ is the discriminant of an indefinite division quaternion algebra B,*

$$\mathcal{E}_{2,T}'(\tau, 0, B) = -\mathcal{E}_{1,t}'(t^{-1}\mathrm{tr}(T\tau), \frac{1}{2}, B) - \frac{1}{2} \cdot \mathcal{E}_{1,t}(t^{-1}\mathrm{tr}(T\tau), \frac{1}{2}, B) \cdot \left(\log\left(\frac{\det v}{t^{-1}\mathrm{tr}(Tv)}\right) + \log(D)\right).$$

Proof. We first write

$$\begin{aligned}\mathbb{E}_t(\tau, s+\frac{1}{2}, \Phi_1^D) &= (-1)^{\mathrm{ord}(D)+1} \frac{D}{2\pi \prod_{p|D}(p+1)} \Lambda_D(2s+2)\, E_t(\tau, s+\frac{1}{2}, \Phi_1^D) \\ &= \frac{1}{2\pi} \mathbb{W}_{t,\infty}(\tau, s+\frac{1}{2}, \Phi_{1,\infty}^{\frac{3}{2}}) \cdot \prod_{p<\infty} \mathbb{W}_{t,p}(s, \Phi_{1,p}),\end{aligned}$$

with

$$\mathbb{W}_{t,p}(s+\frac{1}{2}, \Phi_{1,p}) = \begin{cases} -p^{s+2}(p+1)^{-1} W_{t,p}(s+\frac{1}{2}, \Phi_{1,p}^{\mathrm{ra}}) & \text{if } p \mid D, \\ \zeta_p(2s+2)\, W_{t,p}(s+\frac{1}{2}, \Phi_{1,p}^0) & \text{if } p \nmid D\infty, \end{cases}$$

and

$$\mathbb{W}_{t,\infty}(\tau, s+\frac{1}{2}, \Phi_{1,\infty}^{\frac{3}{2}}) = -\zeta_\infty(2s+2)\, W_{t,\infty}(\tau, s+\frac{1}{2}, \Phi_{1,\infty}^{\frac{3}{2}}).$$

Let $\mathbb{E}_t^{(p)}$ be the non-p-part of $\mathbb{E}_t$, which is the same for Φ_1^D, $\Phi_1^{\frac{D}{p}}$, and $\Phi_1^{\frac{D}{p},1}$. By Proposition 5.8.4, one has

$$-\mathcal{E}_{2,T}'(\tau, 0, B) = \mathcal{E}_{1,t}'(t^{-1}\mathrm{tr}(T\tau), \frac{1}{2}, B) + \frac{1}{2} \mathbb{E}_t(t^{-1}\mathrm{tr}(T\tau), \frac{1}{2}, \Phi_1^D) \cdot \log\left(\frac{\det v}{t^{-1}\mathrm{tr}(Tv)}\right) + \sum_{p|D} \mathrm{Diff}_p \cdot \log(p),$$

with

$$
\begin{aligned}
&\mathrm{Diff}_p \cdot \log(p) \\
&= -\big(\,(p-1)\,A_p'(0) + C_p'(\tfrac{1}{2})\,\big) \cdot \mathbb{E}_t(t^{-1}\mathrm{tr}(T\tau), \tfrac{1}{2}, \Phi_1^{\frac{D}{p}}) \\
&\qquad - (p-1)B_p'(0) \cdot \mathbb{E}_t(t^{-1}\mathrm{tr}(T\tau), \tfrac{1}{2}, \Phi_1^{\frac{D}{p},1}) \\
&= \mathbb{E}_t(t^{-1}\mathrm{tr}(T\tau), \tfrac{1}{2}, \Phi_1^{\frac{D}{p}})\, \log(p) \\
&\qquad - \tfrac{1}{2}(p+1)\,\mathbb{E}_t(t^{-1}\mathrm{tr}(T\tau), \tfrac{1}{2}, \Phi_1^{\frac{D}{p},1})\, \log(p) \\
&= \mathbb{E}_t^{(p)}(t^{-1}\mathrm{tr}(T\tau), \tfrac{1}{2}) \left(\mathbb{W}_{t,p}(\tfrac{1}{2}, \Phi_{1,p}^0) - \tfrac{1}{2}(p+1)\,\mathbb{W}_{t,p}(\tfrac{1}{2}, \Phi_{1,p}^1) \right) \log(p).
\end{aligned}
$$

Now Lemma 5.8.5 implies that

$$
\begin{aligned}
&\mathbb{W}_{t,p}(\tfrac{1}{2}, \Phi_{1,p}^0) - \tfrac{1}{2}(p+1)\,\mathbb{W}_{t,p}(\tfrac{1}{2}, \Phi_{1,p}^1) \\
&\qquad = \chi_{V_p}(-1)\,\gamma_p(V_p^+)|2|_p^{\frac{1}{2}}\,\zeta_p(2) \cdot \left(F_p(1,t,L_p^0) - \frac{p+1}{2p}\, F_p(1,t,L_p^1) \right) \\
&\qquad = \chi_{V_p}(-1)\,\gamma_p(V_p^+)|2|_p^{\frac{1}{2}} \cdot \begin{cases} \frac{1}{2}\, p^{-k} & \text{if } \chi_d(p) = 0, \\ p^{-k+1}(p+1) & \text{if } \chi_d(p) = -1, \\ 0 & \text{if } \chi_d(p) = 1. \end{cases}
\end{aligned}
$$

Recall that $\gamma_p(V_p^-) = -\gamma_p(V_p^+)$. Comparing this with

$$
\mathbb{W}_{t,p}(\tfrac{1}{2}, \Phi_{1,p}^{\mathrm{ra}}) = \chi_{V_p}(-1)\,\gamma_p(V_p^-)|2|_p^{\frac{1}{2}}\,\frac{-p^2}{p+1}\,p^{-1}\,F_p(1,t,L_p^{\mathrm{ra}})
$$

and (iv) of Lemma 5.8.5, we see that

$$
\mathbb{W}_{t,p}(\tfrac{1}{2}, \Phi_{1,p}^0) - \tfrac{1}{2}(p+1)\,\mathbb{W}_{t,p}(\tfrac{1}{2}, \Phi_{1,p}^1) = \tfrac{1}{2}\,\mathbb{W}_{t,p}(\tfrac{1}{2}, \Phi_{1,p}^{\mathrm{ra}}).
$$

Thus,

$$
\mathrm{Diff}_p = \frac{1}{2}\,\mathbb{E}_t(t^{-1}\mathrm{tr}(T\tau), \frac{1}{2}, \Phi_1^D) = \frac{1}{2}\,\mathcal{E}_{1,t}(t^{-1}\mathrm{tr}(T\tau), \frac{1}{2}, B),
$$

and

$$
\sum_{p|D} \mathrm{Diff}_p \cdot \log p = \frac{1}{2}\,\mathcal{E}_{1,t}(t^{-1}\mathrm{tr}(T\tau), \frac{1}{2}, B) \cdot \log(D). \quad \square
$$

5.9 THE CONSTANT TERM

First, we record part of Proposition 5.4.1 on the constant term here.

Proposition 5.9.1. *For any section $\Phi(s) \in I_2(s,\chi)$,*

$$E_0(g,s,\Phi) = \Phi(g,s) + \sum_{\gamma\in\Gamma_\infty\backslash \mathrm{SL}_2(\mathbb{Z})} B_0(m(\gamma)g,s,\Phi) + W_0(g,s,\Phi),$$

where Γ_∞ is the upper triangle subgroup of $\mathrm{SL}_2(\mathbb{Z})$.

Proposition 5.9.2. *Let $Q_v(x,y) = (x,y)v^t(x,y)$ be the quadratic form associated to the matrix $v = \mathrm{Im}(\tau) \in \mathrm{Sym}_2(\mathbb{R})_{>0}$, and let*

$$E(s,v) = \sum_{\left(\begin{smallmatrix} a & b \\ c & d\end{smallmatrix}\right)\in\Gamma_\infty\backslash \mathrm{SL}_2(\mathbb{Z})} \frac{1}{Q_v(c,d)^s}$$

be the associated Eisenstein series. Then

$$\sum_{\gamma\in\Gamma_\infty\backslash \mathrm{SL}_2(\mathbb{Z})} B_0(m(\gamma)\tau,s,\Phi^D)$$
$$= (-1)^{\mathrm{ord}(D)+1}\frac{s\Lambda(2s+1)}{(s+1)\Lambda(2s+2)}\cdot\prod_{p|D}\frac{1-p^{-2s}}{p\,(1-p^{-2s-2})}\cdot E(s+\frac{1}{2},v)$$

Proof. Corollary 5.6.2 asserts that

$$B_{0,p}(m(\gamma),s,\Phi_{2,p}) = W_{0,p}(1,s+\frac{1}{2},\Phi_{1,p}). \tag{5.9.1}$$

Thus, by Proposition 8.1 of [9],

$$B_{0,p}(m(\gamma),s,\Phi_{2,p}) \tag{5.9.2}$$
$$= \chi_{V_p}(-1)\,|2|_p^{\frac{1}{2}}\begin{cases}\gamma_p(V_p^+)\,\dfrac{\zeta_p(2s+1)}{\zeta_p(2s+2)} & \text{if } \Phi_{2,p}=\Phi_{2,p}^0,\\ \gamma_p(V_p^-)p^{-1}\,\dfrac{\zeta_p(2s+1)}{\zeta_p(2s)} & \text{if } \Phi_{2,p}=\Phi_{2,p}^{\mathrm{ra}},\\ \gamma_p(V_p^+)\,p^{-1}\,\dfrac{1+p^{-2s}-2p^{-1-2s}}{1-p^{-2s-1}} & \text{if } \Phi_{2,p}=\Phi_{2,p}^1.\end{cases}$$

On the other hand, if we write $m(\gamma)g_\tau = g_{\tilde\tau}$ with $\tilde\tau = \gamma\tau{}^t\gamma$, then $(\tilde\tau)_{22} =$

$Q_v(c,d)$. Now Lemma 5.6.3 and Proposition 14.1 of [9] imply that

$$\begin{aligned}
&B_{0,\infty}(m(\gamma)\tau, s, \Phi_{2,\infty}^{\frac{3}{2}}) \\
&= \left(\frac{\det v}{Q_v(c,d)}\right)^{\frac{1}{2}s} W_{0,\infty}(i\,Q_v(c,d), s+\frac{1}{2}, \Phi_{1,\infty}^{\frac{3}{2}}) \\
&= \left(\frac{\det v}{Q_v(c,d)}\right)^{\frac{1}{2}s} e(-1/4)\, Q_v(c,d)^{\frac{1}{2}(-1-s)}\, \frac{2\pi\, 2^{-s-\frac{1}{2}}\,\Gamma(s+\frac{1}{2})}{\Gamma(\frac{s+3}{2})\Gamma(\frac{s}{2})} \\
&= e(-1/4)\,(\det v)^{\frac{s}{2}}\, Q_v(c,d)^{-\frac{1}{2}-s}\, \frac{\sqrt{2}\,\Gamma(s+\frac{1}{2})}{(s+1)\sqrt{\pi}\,\Gamma(s)} \\
&= \sqrt{2}\, e(-1/4)\,(\det v)^{\frac{s}{2}}\, Q_v(c,d)^{-\frac{1}{2}-s}\, \frac{s\,\zeta_\infty(2s+1)}{(s+1)\,\zeta_\infty(2s+2)}.
\end{aligned}$$

Recalling the identity (5.7.21)

$$e(-1/4) \prod_{p<\infty} \chi_{V_p}(-1)\,\gamma_p(V_p) = (-1)^{\mathrm{ord}(D)+1},$$

we then have

$$\begin{aligned}
&B_0(m(\gamma)\tau, s, \Phi_2^D) \\
&= (-1)^{\mathrm{ord}(D)+1}\, \frac{\Lambda(2s+1)}{\prod_{p\nmid D}\zeta_p(2s+2)\prod_{p\mid D} p\,\zeta_p(2s)}\, (\det v)^{\frac{s}{2}}\, Q_v(c,d)^{-\frac{1}{2}-s} \\
&= (-1)^{\mathrm{ord}(D)+1}\, \frac{\Lambda(2s+1)}{\Lambda(2s+2)} \prod_{p\mid D} \frac{1-p^{-2s}}{p\,(1-p^{-2-2s})}\, (\det v)^{\frac{s}{2}}\, Q_v(c,d)^{-\frac{1}{2}-s}.
\end{aligned}$$

This proves the proposition. □

From the proof just given, we see that the series

$$\sum_{\gamma\in\Gamma_\infty\backslash \mathrm{SL}_2(\mathbb{Z})} B_0(m(\gamma)g, s, \Phi^{\frac{D}{p},1})$$

has a very similar formula. It is well known that $E(s,v)$ is holomorphic at $s=\frac{1}{2}$, [16], and so the middle sum in the expression in Proposition 5.9.1 vanishes to order at least $\mathrm{ord}(D)$ at $s=0$ when $\Phi=\Phi^D$, and to order at least $\mathrm{ord}(D)-1$ when $\Phi=\Phi^{D/p}$ or $\Phi^{D/p,1}$. Thus, this middle sum does not contribute to our calculation when D has at least two prime factors.

Proposition 5.9.3. *Let* $\Phi(s)=\Phi_2^D(s)$, *and*

$$G_D(s) = (\det v)^{\frac{s}{2}}\,(1+s)\,\Lambda_D(2s+2).$$

Then

$$\Phi(\tau, s) + W_0(\tau, s, \Phi) = \frac{(\det v)^{\frac{s}{2}}}{G_D(s)} \left(G_D(s) - (-1)^{\mathrm{ord}(D)} G_D(-s) \right).$$

Proof. By Propositions 5.7.5 and 5.7.7, we have

(5.9.3) $W_{0,p}(\tau, s, \Phi_p)$

$$= |2|_p \begin{cases} \dfrac{\zeta_p(2s-1)}{\zeta_p(2s+2)} \cdot \gamma_p(V_p^+)^2 & \text{if } p \nmid D\infty, \\ \dfrac{\zeta_p(2s-1)}{p^2\,\zeta_p(2s-2)} \cdot \gamma_p(V_p^-)^2 & \text{if } p \mid D, \\ \dfrac{(s-1)\,\zeta_\infty(2s-1)}{(s+1)\,\zeta_\infty(2s+2)}\,(\det v)^{-\frac{s}{2}} \cdot e(-1/4)^2 & \text{if } p = \infty. \end{cases}$$

Thus

$$W_0(\tau, s, \Phi) = (\det v)^{-\frac{s}{2}}\, \frac{s-1}{s+1} \cdot \frac{\Lambda(2s-1)}{\prod_{p \nmid D} \zeta_p(2s+2) \prod_{p \mid D} p^2\,\zeta_p(2s-2)}$$

$$(5.9.4) \qquad = -(\det v)^{-\frac{s}{2}}\, \frac{(1-s)\Lambda(2-2s)}{(1+s)\Lambda(2+2s)} \prod_{p \mid D} \frac{1-p^{-2s+2}}{p^2\,(1-p^{-2-2s})}$$

$$= (-1)^{\mathrm{ord}(D)+1} (\det v)^{-\frac{s}{2}}\, \frac{(1-s)\Lambda_D(2-2s)}{(1+s)\Lambda_D(2+2s)}$$

$$= (-1)^{\mathrm{ord}(D)+1} (\det v)^{\frac{s}{2}}\, \frac{G_D(-s)}{G_D(s)}.$$

Since $\Phi(\tau, s) = (\det v)^{\frac{s}{2}}$, this proves the proposition. □

The same argument, using the last two formulas in Corollary 5.7.4 also gives

(5.9.5)

$$W_0(\tau, s, \Phi_2^{\frac{D}{p},1}) = (-1)^{\mathrm{ord}(D)}\, (\det v)^{\frac{s}{2}}\, \frac{G_{D/p}(-s)}{G_{D/p}(s)} \cdot \frac{1-2p^{-2s}+p^{2-2s}}{p^2(1-p^{-2-2s})}.$$

Corollary 5.9.4. *Assume that* $\mathrm{ord}(D) \geq 2$ *is even.*

(i)

$$E_0(\tau, 0, \Phi_2^D) = 0,$$

$$E_0(\tau, 0, \Phi_2^{\frac{D}{p}}) = 2\, E_0(i \det v, \frac{1}{2}, \Phi_1^{\frac{D}{p}}) = 2,$$

$$E_0(\tau, 0, \Phi_2^{\frac{D}{p},1}) = 2.$$

(ii)

$$E_0'(\tau, 0, \Phi_2^D) = 2 + 4\,\frac{\Lambda_D'(2)}{\Lambda_D(2)} + \log(\det v).$$

Theorem 5.9.5. *Let $D = D(B) > 1$ for an indefinite quaternion algebra B. Then*

$$\mathbb{E}_0'(\tau, 0, \Phi_2^D) = -\mathbb{E}_0'(i\det(v), \frac{1}{2}, \Phi_1^D),$$

and

$$\begin{aligned}\mathcal{E}_{2,0}'(\tau, 0, B) &= -\mathcal{E}_{1,0}'(i\det v, \frac{1}{2}, B) - \frac{1}{2}\mathcal{E}_{1,0}(i\det(v), \frac{1}{2}, B)\cdot\log(D)\\ &= C^D(0)\cdot\left[\,2 + 4\,\frac{\Lambda_D'(2)}{\Lambda_D(2)} + \log(D\ \det v) + 2\sum_{p|D}\frac{\log p}{p+1}\,\right].\end{aligned}$$

Proof. The first identity follows from (ii) of Corollary 5.9.4 and the identity

$$\mathbb{E}_0'(\tau, \frac{1}{2}, \Phi_1^D) = \frac{1}{2}\cdot c(D)\,\Lambda_D(2)\cdot\left[\,2 + 4\,\frac{\Lambda_D'(2)}{\Lambda_D(2)} + \log(v)\,\right],$$

which is (8.30) of [9]. Since

$$\begin{aligned}E_0(\tau, s, \tilde{\Phi}_2^D) = E_0(\tau, s, \Phi_2^D) &+ \sum_{p|D} A_p(s)E_0(\tau, s, \Phi_2^{\frac{D}{p}})\\ &+ \sum_{p|D} B_p(s)E_0(\tau, s, \Phi_2^{\frac{D}{p},1}) + O(s^2),\end{aligned}$$

we have, by (i) of Corollary 5.9.4,

$$E_0'(\tau, 0, \tilde{\Phi}_2^D) = E_0'(\tau, 0, \Phi_2^D) + 2\sum_{p|D}(A_p'(0) + B_p'(0)).$$

Multiplying both sides by

$$C^D(0) = -\frac{1}{2}\cdot c(D)\Lambda_D(2),$$

we have

$$\begin{aligned}\mathcal{E}_{2,0}'(\tau, 0, B) &= C^D(0)\cdot E_0'(\tau, 0, \tilde{\Phi}_2^D)\\ &= \mathbb{E}_0'(\tau, 0, \Phi_2^D) + 2\,C^D(0)\sum_{p|D}(A_p'(0) + B_p'(0)).\end{aligned}$$

On the other hand, recalling that

$$\mathbb{E}_0(\tau, \frac{1}{2}, \Phi_1^D) = c(D)\,\Lambda_D(2) = -2\,C^D(0)$$

and that

$$c(D/p)\,\Lambda_{D/p}(2) = -\frac{1}{p-1}\,c(D)\,\Lambda_D(2) = \frac{2\,C^D(0)}{p-1},$$

we have

$$\begin{aligned}\mathcal{E}'_{1,0}(i\det v, \frac{1}{2}, \Phi_1^D) &= \mathbb{E}'_0(i\det v, \frac{1}{2}, \Phi_1^D) + \sum_{p|D} C'_p(\frac{1}{2})\,\mathbb{E}_0(i\det v, \frac{1}{2}, \Phi_1^{\frac{D}{p}})\\ &= \mathbb{E}'_0(i\det v, \frac{1}{2}, \Phi_1^D) + 2\,C^D(0)\sum_{p|D}\frac{1}{p-1}\,C'_p(\frac{1}{2}).\end{aligned}$$

Thus, recalling (5.1.35) and (5.5.6), we have

$$\begin{aligned}&\mathcal{E}'_{2,0}(\tau, 0, B) + \mathcal{E}'_{1,0}(i\det v, \frac{1}{2}, B)\\ &\qquad = 2\,C^D(0)\sum_{p|D}(A'_p(0) + B'_p(0) + \frac{1}{p-1}\,C'_p(\frac{1}{2}))\\ &\qquad = C^D(0)\sum_{p|D}\log p\\ &\qquad = C^D(0)\,\log D,\end{aligned}$$

as claimed. □

Bibliography

[1] B. Gross and K. Keating, *On the intersection of modular correspondences*, Invent. math., **112** (1993), 225–245.

[2] R. Howe, *Transcending classical invariant theory*, J. Amer. Math. Soc., **2** (1989), 535–552.

[3] Y. Kitaoka, Arithmetic of Quadratic Forms, Cambridge Tracts in Mathematics, **106**, Cambridge Univ. Press, 1993.

[4] S. Kudla, *Central derivatives of Eisenstein series and height pairings*, Annals of Math., **146** (1997), 545–646.

[5] S. Kudla and S. Rallis, *On the Weil-Siegel formula*, J. reine angew. Math., **387** (1988), 1–68.

[6] ______, *Degenerate principal series and invariant distributions*, Israel J. Math., **69** (1990), 25–45.

[7] ______, *A regularized Siegel-Weil formula: the first term identity*, Annals of Math., **140** (1994), 1–80.

[8] S. Kudla and M. Rapoport, *Height pairings on Shimura curves and p-adic uniformization*, Invent. math., **142** (2000), 153–223.

[9] S. Kudla, M. Rapoport, and T. Yang, *Derivatives of Eisenstein series and Faltings heights*, Compositio Math., **140** (2004), 887–951.

[10] R. P. Langlands, *On the notion of an automorphic representation*, in Automorphic Forms, Representations and L-Functions, Part 1, Proc. Sympos. Pure Math., **33**, 203–207, AMS, Providence, RI, 1979.

[11] B. Myers, *Local representation densities of non-unimodular quadratic forms*, Ph.D. Thesis, University of Maryland, 1994.

[12] S. Rallis, *On the Howe duality conjecture*, Compositio Math., **51** (1984), 333–399.

[13] ______, L-functions and the oscillator representation, Lecture Notes in Mathematics, **1245**, Springer-Verlag, Berlin, 1987.

[14] R. Ranga Rao, *On some explicit formulas in the theory of Weil representation*, Pacific J. Math., **157** (1993), 335–370.

[15] G. Shimura, *Confluent hypergeometric functions on tube domains*, Math. Annalen, **260** (1982), 269–302.

[16] C. L. Siegel, Advanced Analytic Number Theory, 2nd ed., Tata Institute of Fundamental Research Studies in Mathematics, **9**, Tata Institute of Fundamental Research, Bombay, 1980.

[17] T. H. Yang, *An explicit formula for local densities of quadratic forms*, J. Number Theory, **72** (1998), 309–356.

[18] ______, *Local densities of 2-adic quadratic forms*, J. Number Theory, **108** (2004), 287–345.

Chapter Six

The generating function for 0-cycles

In this chapter, we give the definition of a generating function for 0-cycles on the arithmetic surface $\mathcal{M}$. More precisely, we consider a generating series of the form

(6.0.1) $$\hat{\phi}_2(\tau) = \sum_{T \in \mathrm{Sym}_2(\mathbb{Z})^\vee} \widehat{\mathcal{Z}}(T, v)\, q^T,$$

where $\tau = u + iv \in \mathfrak{H}_2$, the Siegel space of genus two, $q^T = e(\mathrm{tr}(T\tau))$, and $\widehat{\mathcal{Z}}(T, v) \in \widehat{\mathrm{CH}}^2(\mathcal{M})$. Since we are working with arithmetic Chow groups with real coefficients, as explained in Chapter 2, there is an isomorphism

(6.0.2) $$\widehat{\deg} : \widehat{\mathrm{CH}}^2(\mathcal{M}) \xrightarrow{\sim} \mathbb{R}.$$

For example, if $\mathcal{Z}$ is a 0-cycle on $\mathcal{M}$ with $\mathcal{Z} \simeq \mathrm{Spec}\,(R)$ for an Artin ring R, then

(6.0.3) $$\widehat{\deg} : (\mathcal{Z}, 0) \longmapsto \log |R|$$

is the usual arithmetic degree of $\mathcal{Z}$. The first step is to define the terms for 'good' positive definite T's.

For a positive definite $T \in \mathrm{Sym}_2(\mathbb{Z})^\vee$, we can consider the moduli stack $\mathcal{Z}(T)$ for triples $(A, \iota, \mathbf{x})$, where (A, ι) is, as usual, an abelian scheme with a special O_B-action ι, and where $\mathbf{x} = [x_1, x_2] \in V(A, \iota)$ is a pair of special endomorphisms of A such that

(6.0.4) $$Q(\mathbf{x}) = \frac{1}{2}\,((x_i, x_j)) = T.$$

As explained in Chapter 3, any nonsingular matrix T determines a set of primes $\mathrm{Diff}(T, B)$ of odd cardinality and, if $T > 0$, then $\infty \notin \mathrm{Diff}(T, B)$. Moreover, $|\mathrm{Diff}(T, B)| > 1$ implies that $\mathcal{Z}(T) = \phi$, so that the only positive definite T's of interest are those for which $\mathrm{Diff}(T, B) = \{p\}$ for a prime p. A matrix $T \in \mathrm{Sym}_2(\mathbb{Z})^\vee_{>0}$ will be called *good* if $\mathrm{Diff}(T, B) = \{p\}$ with $p \nmid D(B)$. For such a T, $\mathcal{Z}(T)$ is a 0-cycle supported in the fiber $\mathcal{M}_p$, and there is an associated class

(6.0.5) $$\widehat{\mathcal{Z}}(T, v) = (\,\mathcal{Z}(T), 0\,) \in \widehat{\mathrm{CH}}^2(\mathcal{M}).$$

The partial generating series

$$\sum_{\substack{T \in \mathrm{Sym}_2(\mathbb{Z})^\vee_{>0} \\ T \text{ good}}} \widehat{\mathcal{Z}}(T, v)\, q^T \tag{6.0.6}$$

thus has a natural geometric definition, and our first goal in this chapter is to complete it by defining terms $\widehat{\mathcal{Z}}(T, v)$ for the remaining $T \in \mathrm{Sym}_2(\mathbb{Z})^\vee$. We then compute the quantities $\widehat{\deg}(\widehat{\mathcal{Z}}(T, v))$ in all cases and compare them with the Fourier coefficients of the central derivative of the Siegel Eisenstein series $\mathcal{E}_2'(\tau, 0; B)$ computed in Chapter 5. In this way, we prove the following result:

Theorem B. *The generating function $\hat{\phi}_2(\tau)$ is a Siegel modular form of genus two and weight $\frac{3}{2}$ for $\Gamma_0(4D(B)_o) \subset \mathrm{Sp}_2(\mathbb{Z})$, where $D(B)_o$ is the odd part of $D(B)$. More precisely*

$$\hat{\phi}_2(\tau) = \mathcal{E}_2'(\tau, 0; B),$$

so that

$$\hat{\phi}_2(\gamma(\tau)) = \mathrm{sgn}(\det d) \cdot j_{\frac{3}{2}}(\gamma, \tau) \cdot \hat{\phi}_2(\tau),$$

for all $\gamma = \left(\begin{smallmatrix} a & b \\ c & d \end{smallmatrix}\right) \in \Gamma_0(4D(B)_o)$.

The transformation law of $\mathcal{E}_2(\tau, s; B)$ used here is determined in Section 8.5.6. Note that $\mathrm{sgn}(\det d) = (\det d, -1)_2$ and that the automorphy factor $j_{\frac{3}{2}}(\gamma, \tau)$, which is described explicitly in that appendix, satisfies

$$j_{\frac{3}{2}}(\gamma, \tau)^2 = \det(c\tau + d)^3.$$

To complete the definition of $\hat{\phi}_2(\tau)$, we must consider T's of the following types:

(i) $T > 0$ with $\mathrm{Diff}(T, B) = \{p\}$ for $p \mid D(B)$, i.e., T is not good.

(ii) T with signature $(1, 1)$ or $(0, 2)$ with $\mathrm{Diff}(T, B) = \{\infty\}$.

(iii) $T \geq 0$ with $\det(T) = 0$, but $T \neq 0$.

(iv) $T \leq 0$ with $\det(T) = 0$, but $T \neq 0$.

(v) $T = 0$.

In some cases, we define $\widehat{\mathcal{Z}}(T,v)$ by giving the real number $\widehat{\deg}(\widehat{\mathcal{Z}}(T,v))$ directly. On the one hand, it is clear that we want these quantities to coincide with the corresponding Fourier coefficients of $\mathcal{E}_2'(\tau,0;B)$, since we would like a generating function which is modular. On the other hand, we would like to give a definition which is as natural as possible from the point of view of arithmetic geometry. The main result of Chapter 7, which identifies the inner product $\langle\,\hat{\phi}_1(\tau_1),\hat{\phi}_1(\tau_2)\,\rangle$ of two genus one generating functions with the restriction to the diagonal $\hat{\phi}_2(\mathrm{diag}(\tau_1,\tau_2))$ gives a further geometric justification for our definitions.

The fact that the generating series $\hat{\phi}_2(\tau)$ is the q-expansion of a Siegel modular form implies that the numbers $\widehat{\mathcal{Z}}(T,v)$ must satisfy some highly nontrivial identities. For example, for $\alpha\in\mathrm{GL}_2(\mathbb{Z})$, the transformation law

$$\hat{\phi}_2(\tau) \;=\; \hat{\phi}_2(\alpha\tau{}^t\alpha) \;=\; \sum_T \widehat{\deg}(\widehat{\mathcal{Z}}(T,\alpha v{}^t\alpha))\,q^{{}^t\alpha T\alpha} \tag{6.0.7}$$

implies that

$$\widehat{\deg}(\widehat{\mathcal{Z}}({}^t\alpha T\alpha,v))=\widehat{\deg}(\widehat{\mathcal{Z}}(T,\alpha v{}^t\alpha)), \tag{6.0.8}$$

for any T and v. For $T\in\mathrm{Sym}_2(\mathbb{Z})^\vee_{>0,\mathrm{good}}$, this amounts to

$$\widehat{\deg}(\mathcal{Z}({}^t\alpha T\alpha))=\widehat{\deg}(\mathcal{Z}(T)). \tag{6.0.9}$$

This identity is immediate from the definition of $\mathcal{Z}(T)$, since the length of the local deformation ring where a pair $\mathbf{x}$ of special endomorphisms of (A,ι) deforms is the same as the one where $\alpha\cdot\mathbf{x}$ deforms, since this length only depends on the $\mathbb{Z}_p$-span of the components x_1 and x_2 of $\mathbf{x}$. On the other hand, if $T\in\mathrm{Sym}_2(\mathbb{Z})^\vee_{>0}$ is not good, then the equality above is true for any $\alpha\in\mathrm{GL}_2(\mathbb{Z}_p)$, but this is one of the main results of [5].

6.1 THE CASE $T>0$ WITH $\mathrm{Diff}(T,B)=\{p\}$ FOR $p\nmid D(B)$

For a positive definite T with $\mathrm{Diff}(T,B)=\{p\}$ for $p\nmid D(B)$, the class $\widehat{\mathcal{Z}}(T,v)=(\mathcal{Z}(T),0)\in\widehat{\mathrm{CH}}^2(\mathcal{M})$ is defined above, and it remains to compute the real number $\widehat{\deg}(\widehat{\mathcal{Z}}(T,v))=\widehat{\deg}(\mathcal{Z}(T))$ and to compare it with the corresponding Fourier coefficient $\mathcal{E}'_{2,T}(\tau,0;B)$ of $\mathcal{E}'_2(\tau,0;B)$.

To compute $\widehat{\deg}(\mathcal{Z}(T))$ we use the Gross-Keating theorem in the form of Theorem 3.6.1. According to this theorem, the stack $\mathcal{Z}(T)$ has support in the supersingular locus of $\mathcal{M}_p$. Let $B^{(p)}$ be the definite quaternion algebra over $\mathbb{Q}$ whose invariants differ from those of B at precisely p and ∞. Let $H^{(p)}=B^{(p),\times}$ and

$$V^{(p)}=\{x\in B^{(p)}\mid \mathrm{tr}(x)=0\}. \tag{6.1.1}$$

We choose a maximal order $O^{(p)}$ in $B^{(p)}$ and an isomorphism $B^{(p)}(\mathbb{A}_f^p) \simeq B(\mathbb{A}_f^p)$ which carries $O^{(p)} \otimes \hat{\mathbb{Z}}^p$ to $O_B \otimes \hat{\mathbb{Z}}^p$. We write

$$H^{(p)}(\mathbb{A}_f) = \coprod_j H^{(p)}(\mathbb{Q}) h_j K^{(p)}, \tag{6.1.2}$$

where

$$K^{(p)} = K_p^{(p)} \cdot K^p = (O^{(p)} \otimes \hat{\mathbb{Z}})^\times. \tag{6.1.3}$$

Also let $V^{(p)}(\hat{\mathbb{Z}}) = V^{(p)}(\hat{\mathbb{Z}}^p) \times V^{(p)}(\mathbb{Z}_p)$, with

$$V^{(p)}(\hat{\mathbb{Z}}^p) = V^{(p)}(\mathbb{A}_f^p) \cap (O^{(p)} \otimes \hat{\mathbb{Z}}^p)$$

and $V^{(p)}(\mathbb{Z}_p) = V^{(p)}(\mathbb{Q}_p) \cap (O^{(p)} \otimes \mathbb{Z}_p)$. Let $\varphi^{(p)} \in S((V^{(p)}(\mathbb{A}_f))^2)$ be the characteristic function of the set $(V^{(p)}(\hat{\mathbb{Z}}))^2$. Finally, for $\mathbf{y} \in V^{(p)}(\mathbb{Q})^2$, we denote by $e_{\mathbf{y},j}$ the stabilizer of $\mathbf{y}$ in the group $\Gamma_j^{(p)} = H^{(p)}(\mathbb{Q}) \cap h_j K^{(p)} h_j^{-1}$.

The following proposition gives the number of points in the stack sense of the finite stack $\mathcal{Z}(T)$ (recall that we are assuming that $T > 0$ with $\mathrm{Diff}(B, T) = p$ is good).

Proposition 6.1.1.

$$\sum_{x \in \mathcal{M}(\bar{\mathbb{F}}_p)^{\mathrm{ss}}} \sum_{\substack{\mathbf{y} \in V(A,\iota)^2 \\ \frac{1}{2}(\mathbf{y},\mathbf{y}) = T}} \frac{1}{|\mathrm{Aut}(A_x, \iota_x, \mathbf{y})|} = \sum_j \sum_{\substack{\mathbf{y} \in V^{(p)}(\mathbb{Q})^2 \\ \frac{1}{2}(\mathbf{y},\mathbf{y}) = T}} e_{\mathbf{y},j}^{-1} \cdot \varphi^{(p)}(h_j^{-1} \cdot \mathbf{y})$$

Proof. The proof is based on the description of $\mathcal{M}(\bar{\mathbb{F}}_p)^{\mathrm{ss}}$ as a double coset space,

$$\mathcal{M}(\bar{\mathbb{F}}_p)^{\mathrm{ss}} = H^{(p)}(\mathbb{Q}) \backslash H^{(p)}(\mathbb{A}_f) / K^{(p)}. \tag{6.1.4}$$

We recall that this is obtained by parametrizing the elements of $\mathcal{M}(\bar{\mathbb{F}}_p)^{\mathrm{ss}}$ by quasi-isogenies with source a fixed base point (A_0, ι_0), chosen so that the stabilizer of its Dieudonne module $D(A_0)$ in $H^{(p)}(\mathbb{Q}_p)$ is equal to $K_p^{(p)}$ and the stabilizer of its Tate module $\hat{T}^p(A_0)$ in $H^{(p)}(\mathbb{A}_f^p)$ is equal to K^p. We also identify $V(A_0, \iota_0) \otimes \mathbb{Q}$ with $V^{(p)}(\mathbb{Q})$. Now if (A, ι) corresponds to the double coset $H^{(p)}(\mathbb{Q}) h K^{(p)}$, then $V(A, \iota)$ can be identified with

$$V(A, \iota) = \{y \in V^{(p)}(\mathbb{Q}) \mid h^{-1} \cdot y \in V^{(p)}(\hat{\mathbb{Z}})\}. \tag{6.1.5}$$

Here $h \cdot y = hyh^{-1}$. This is an immediate consequence of the fact that

(6.1.6)

$$\begin{aligned} &\mathrm{End}(A, \iota) \\ &= \{y \in B^{(p)}(\mathbb{Q}) \mid yh\, D(A_0) \subset h\, D(A_0), yh\, \hat{T}^p(A_0) \subset h\, \hat{T}^p(A_0)\} \\ &= B^{(p)}(\mathbb{Q}) \cap h(O^{(p)} \otimes \hat{\mathbb{Z}})h^{-1}. \end{aligned}$$

It follows that the left-hand side of the identity in the proposition is equal to

$$\sum_j \sum_{\substack{\mathbf{y} \in V^{(p)}(\mathbb{Q})^2 \\ \frac{1}{2}(\mathbf{y},\mathbf{y})=T \\ h_j^{-1}\mathbf{y} \in V^{(p)}(\hat{\mathbb{Z}})^2 \\ \mathrm{mod}\ \Gamma_j^{(p)}}} 1, \tag{6.1.7}$$

which then leads to the expression of the right-hand side of the proposition.
□

We note that the lengths of the local rings of $\mathcal{Z}(T)$ are all identical and are denoted by $\nu_p(T)$. They are given by the Gross-Keating formula of Theorem 3.6.1. We therefore have the following expression for $\widehat{\deg}\, \mathcal{Z}(T)$:

Corollary 6.1.2. *Let T be positive definite with* $\mathrm{Diff}(T, B) = \{p\}$ *for* $p \nmid D(B)$. *Then*

$$\widehat{\deg}\, \mathcal{Z}(T) = \nu_p(T) \log p \cdot \Bigg(\sum_j \sum_{\substack{\mathbf{y} \in V^{(p)}(\mathbb{Q})^2 \\ \frac{1}{2}(\mathbf{y},\mathbf{y})=T}} e_{\mathbf{y},j}^{-1} \cdot \varphi^{(p)}(h_j^{-1} \cdot \mathbf{y}) \Bigg),$$

where $\nu_p(T)$ is given by Theorem 3.6.1 in terms of the Gross-Keating invariants $(0, a_1, a_2)$ of $\tilde{T} = \mathrm{diag}(1, T) \in \mathrm{Sym}_3(\mathbb{Z}_p)^\vee$:

$$\nu_p(T) = \begin{cases} \sum_{j=0}^{\frac{a_1-1}{2}} (a_1 + a_2 - 4j)\, p^j & \text{if } a_1 \text{ is odd,} \\ \sum_{j=0}^{\frac{a_1}{2}-1} (a_1 + a_2 - 4j)\, p^j + \frac{1}{2}(a_2 - a_1 + 1)\, p^{\frac{a_1}{2}} & \text{if } a_1 \text{ is even.} \end{cases}$$

Comparing this expression with the corresponding Fourier coefficient of the central derivative of the Eisenstein series computed in Theorem 5.3.8, we have.

Corollary 6.1.3. *For* $T \in \mathrm{Sym}_2(\mathbb{Z})^\vee$ *with* $\det(T) \neq 0$ *and* $\mathrm{Diff}(T, B) = \{p\}$ *with* $p \nmid D(B)$,

$$\mathcal{E}'_{2,T}(\tau, 0, B) = \widehat{\deg}(\widehat{\mathcal{Z}}(T, v)) \cdot q^T.$$

Proof. Indeed, the stabilizer of $\mathbf{y}$ in $\Gamma_j^{(p)}$ has twice the size of the stabilizer of $\mathbf{y}$ in the group that was denoted Γ_j in Theorem 5.3.8 and (5.3.54). $\square$

Remark 6.1.4. When $p \neq 2$, this computation and comparison is implicitly done in section 14 of [3]. There, the arithmetic intersection number of a pair of classes $\widehat{\mathcal{Z}}(t_1, v_1)$, $\widehat{\mathcal{Z}}(t_2, v_2) \in \widehat{\mathrm{CH}}^1(\mathcal{M})$ was computed in the case when $t_1 t_2$ is not a square, so that the cycles $\mathcal{Z}(t_1)$ and $\mathcal{Z}(t_2)$ do not meet in the generic fiber. The intersection of such cycles in a fiber $\mathcal{M}_p$, for $p \nmid D(B)$, is a finite union

$$\bigcup_{\substack{T \in \mathrm{Sym}_2(\mathbb{Z})^\vee \\ \mathrm{diag}(T) = (t_1, t_2) \\ \mathrm{Diff}(T, B) = \{p\}}} \mathcal{Z}(T)$$

of 0-cycles, and the corresponding contribution to the arithmetic intersection number is the sum of the $\widehat{\deg}(\mathcal{Z}(T))$'s, which are given by the Gross-Keating formula.

6.2 THE CASE $T > 0$ WITH $\mathrm{Diff}(T, B) = \{p\}$ FOR $p \mid D(B)$

In this case, the naive cycle $\mathcal{Z}(T)$ defined by imposing a pair of special endomorphisms with fundamental matrix T is supported in the fiber $\mathcal{M}_p$, but when $p^2 \mid T$, this cycle has components of dimension 1, as described in Chapter 3. To overcome this difficulty, we define the class $\widehat{\mathcal{Z}}(T, v) \in \widehat{\mathrm{CH}}^2(\mathcal{M})$ by giving directly the number $\widehat{\deg}(\widehat{\mathcal{Z}}(T, v))$.

Write

$$T = \begin{pmatrix} t_1 & m \\ m & t_2 \end{pmatrix} \in \mathrm{Sym}_2(\mathbb{Z})_{>0} \tag{6.2.1}$$

and recall that $\mathcal{Z}(T)$ is the union of those connected components of the intersection $\mathcal{Z}(t_1) \times_{\mathcal{M}} \mathcal{Z}(t_2)$ where the fundamental matrix, as defined in [5], is equal to T. After base change to $\mathbb{Z}_p$, we let, by [4],

(6.2.2)

$$\widehat{\deg}(\widehat{\mathcal{Z}}(T, v)) := \chi(\mathcal{Z}(T), \mathcal{O}_{\mathcal{Z}(t_1)} \otimes^{\mathbb{L}} \mathcal{O}_{\mathcal{Z}(t_2)}) \cdot \log p \quad \in \mathbb{R} \simeq \widehat{\mathrm{CH}}^2(\mathcal{M}),$$

where χ is the Euler-Poincaré characteristic of the derived tensor product of the structure sheaves $\mathcal{O}_{\mathcal{Z}(t_1)}$ and $\mathcal{O}_{\mathcal{Z}(t_2)}$; see [5], Section 4. When $p \neq 2$, the number $\widehat{\deg}(\widehat{\mathcal{Z}}(T, v))$ was computed in [5], Theorem 8.6. There the result was expressed as the product of a local multiplicity (which is actually global on the Drinfeld space) and an orbital integral. The calculation of the multiplicity comes down to a combinatorial problem. The same method works when $p = 2$, although, as described in the appendix to this chapter, the combinatorics become considerably more elaborate. Also, here we replace the orbital integral by an expression which is more in analogy with the formula obtained in the previous section in the case when $p \nmid D(B)$. To state the result, we introduce, as in the previous section, the twisted quaternion algebra $B^{(p)}$ over $\mathbb{Q}$ whose invariants differ from those of B at precisely p and ∞. We again introduce $H^{(p)} = B^{(p),\times}$ and $V^{(p)}$. By strong approximation, we have

$$H^{(p)}(\mathbb{A}_f) = H^{(p)}(\mathbb{Q})H^{(p)}(\mathbb{Q}_p)K^p, \tag{6.2.3}$$

where $K^p = (O^{(p)} \otimes \hat{\mathbb{Z}}^p)^\times$. Let $\varphi^{(p)} \in S((V^{(p)}(\mathbb{A}_f^p))^2)$ be the characteristic function of the set $(V^{(p)}(\hat{\mathbb{Z}}^p))^2$. We also set

$$\Gamma' = H^{(p)}(\mathbb{Q}) \cap K^p. \tag{6.2.4}$$

Using this notation, we have

Theorem 6.2.1. *Let* $T \in \mathrm{Sym}_2(\mathbb{Z})^\vee$ *with* $\det(T) \neq 0$ *and* $\mathrm{Diff}(T, B) = \{p\}$ *for* $p \mid D(B)$. *Then*

$$\widehat{\deg}(\widehat{\mathcal{Z}}(T, v)) = \nu_p(T) \cdot \log p \cdot \frac{1}{2} \sum_{\substack{\mathbf{y} \in V^{(p)}(\mathbb{Q})^2 \\ \frac{1}{2}(\mathbf{y},\mathbf{y})=T \\ \mathrm{mod}\ \Gamma'}} \varphi^{(p)}(\mathbf{y}).$$

Furthermore, the multiplicity $\nu_p(T)$, *defined in (6.2.11), is given as follows in terms of the Gross-Keating invariants* $(0, a_1, a_2)$ *and the Gross-Keating* ϵ-*constant* $\epsilon_0 = \epsilon_0(\tilde{T})$, [2], *p. 236, of* $\tilde{T} = \mathrm{diag}(1, T) \in \mathrm{Sym}_3(\mathbb{Z}_p)^\vee$:

$$\frac{1}{2}\nu_p(T) = a_1 + a_2 + 1 - \begin{cases} p^{\frac{a_1}{2}} + 2\,\dfrac{p^{\frac{a_1}{2}} - 1}{p-1} & \textit{if } a_1 \textit{ is even and } \epsilon_0 = -1, \\[2ex] (a_2 - a_1 + 1)\,p^{\frac{a_1}{2}} + 2\,\dfrac{p^{\frac{a_1}{2}} - 1}{p-1} & \textit{if } a_1 \textit{ is even and } \epsilon_0 = 1, \\[2ex] 2\,\dfrac{p^{\frac{a_1+1}{2}} - 1}{p-1} & \textit{if } a_1 \textit{ is odd.} \end{cases}$$

Proof. (Cf. also Section 7.6.) We use the p-adic uniformization of $\mathcal{M} \otimes \mathbb{Z}_p$ and of $\mathcal{Z}(T) \otimes \mathbb{Z}_p$. After base changing to $W(\bar{\mathbb{F}}_p)$, we have the Drinfeld-Cherednik uniformization of $\mathcal{M}$,

$$\mathcal{M} \otimes W(\bar{\mathbb{F}}_p) = H^{(p)}(\mathbb{Q})\backslash(\hat{\Omega}^\bullet \times H^{(p)}(\mathbb{A}_f^p)/K^p), \tag{6.2.5}$$

where $\hat{\Omega}^\bullet = \hat{\Omega} \times \mathbb{Z}$ is the disjoint union of copies of the Drinfeld space parametrized by $\mathbb{Z}$. The isomorphism depends on the choice of a base point $(A_0, \iota_0) \in \mathcal{M}(\bar{\mathbb{F}}_p)$. We also fix an identification of $O^{(p)}$ with $\mathrm{End}(A_0, \iota_0)$. Now the analysis of [5], pp. 214–215, gives a uniformization of the formal completion of $\mathcal{C}_T = \mathcal{Z}(T) \otimes W(\bar{\mathbb{F}}_p)$ as an injection

$$\hat{\mathcal{C}}_T \hookrightarrow H^{(p)}(\mathbb{Q}) \backslash (V^{(p)}(\mathbb{Q})^2_T \times \hat{\Omega}^\bullet \times H^{(p)}(\mathbb{A}^p_f)/K^p), \tag{6.2.6}$$

where

$$V^{(p)}(\mathbb{Q})^2_T = \{\mathbf{y} \in V^{(p)}(\mathbb{Q})^2 \mid \frac{1}{2}(\mathbf{y}, \mathbf{y}) = T\}. \tag{6.2.7}$$

By [5], (8.30), the conditions describing $\hat{\mathcal{C}}_T$ inside the right-hand side of (6.2.6) are the following: $(\mathbf{y}, (X, \rho), gK^p) \in \hat{\mathcal{C}}_T$ if and only if

(i) $g^{-1}\mathbf{y}g \in (V^{(p)}(\hat{\mathbb{Z}}^p))^2$, and

(ii) $(X, \rho) \in \mathcal{Z}^\bullet(\mathbf{j})$.

Here $\mathbf{j}$ is the $\mathbb{Z}_p$-span of the images j_1 and j_2 of y_1 and y_2 under the injection

$$\mathrm{End}(A_0, \iota_0) \to \mathrm{End}(\mathbb{X})$$

into the endomorphism algebra of the p-divisible group $A_0(p) = \mathbb{X}^2$; see [5].

Using strong approximation, we can write the right-hand side of (6.2.6) as

$$\Gamma' \backslash (V^{(p)}(\mathbb{Q})^2_T \times \hat{\Omega}^\bullet). \tag{6.2.8}$$

Taking into account the description of $\hat{\mathcal{C}}_T$ above, we obtain

$$\hat{\mathcal{C}}_T = \coprod_{\substack{\mathbf{y} \in V^{(p)}(\mathbb{Q})^2_T \cap V^{(p)}(\hat{\mathbb{Z}}^p)^2 \\ \mathrm{mod}\ \Gamma'}} [\,\Gamma'_{\mathbf{y}} \backslash \mathcal{Z}^\bullet(\mathbf{j})\,]. \tag{6.2.9}$$

The stabilizer $\Gamma'_{\mathbf{y}}$ is the intersection of Γ' with the center of $B^{(p),\times}$, hence $\Gamma'_{\mathbf{y}} = \mathbb{Z}[p^{-1}]^\times \simeq \{\pm 1\} \times \mathbb{Z}$. The generator of the infinite factor acts by translating the 'sheets' of $\mathcal{Z}^\bullet(\mathbf{j})$ by 2, and the contribution of each sheet to the Euler-Poincaré characteristic (6.2.2) is the same, see [5]. Denoting by $\mathcal{Z}(\mathbf{j})$ the 0-th sheet, we therefore see that the contribution of $\mathbf{y}$ to $\widehat{\deg}(\hat{\mathcal{C}}_T)$ is equal to

$$e_{\mathbf{y}}^{-1} \cdot 2 \cdot \chi(\mathcal{Z}(\mathbf{j}), \mathcal{O}_{\mathcal{Z}(j_1)} \otimes^{\mathbb{L}} \mathcal{O}_{\mathcal{Z}(j_2)}) \cdot \log p, \tag{6.2.10}$$

where $e_{\mathbf{y}} = 2$ is the order of the stabilizer of $\mathbf{y}$ in the first factor of $\Gamma'_{\mathbf{y}}$. Now the quantity

$$\nu_p(T) := 2 \cdot \chi(\mathcal{Z}(\mathbf{j}), \mathcal{O}_{\mathcal{Z}(j_1)} \otimes^{\mathbb{L}} \mathcal{O}_{\mathcal{Z}(j_2)}) = 2 \cdot (\mathcal{Z}(j_1), \mathcal{Z}(j_2)) \tag{6.2.11}$$

only depends on the $\mathrm{GL}_2(\mathbb{Z}_p)$-equivalence class of T; see [5], Section 5. Then, as is proved for $p \neq 2$ in Section 6 of [5], and for $p = 2$ in the appendix to this chapter, $\nu_p(T)$ is given by the expression appearing in the statement of the theorem. In any case we obtain

$$\begin{aligned} \widehat{\deg}(\mathcal{Z}(T)) &= \log p \cdot \sum_{\substack{\mathbf{y} \in V^{(p)}(\mathbb{Q})^2_T \cap (V^{(p)}(\hat{\mathbb{Z}}^p))^2 \\ \mathrm{mod}\ \Gamma'}} e_{\mathbf{y}}^{-1} \cdot \nu_p(T) \\ &= \nu_p(T) \cdot \log p \cdot \frac{1}{2} \sum_{\substack{\mathbf{y} \in V^{(p)}(\mathbb{Q})^2 \\ \frac{1}{2}(\mathbf{y},\mathbf{y}) = T \\ \mathrm{mod}\ \Gamma'}} \varphi^{(p)}(\mathbf{y}) , \end{aligned} \tag{6.2.12}$$

as was to be shown. □

A comparison with Theorem 5.3.8 yields

Corollary 6.2.2. *For $T \in \mathrm{Sym}_2(\mathbb{Z})^\vee$ with $\det(T) \neq 0$ and $\mathrm{Diff}(T, B) = \{p\}$ for $p \mid D(B)$,*

$$\mathcal{E}'_{2,T}(\tau, 0, B) = \widehat{\deg}(\widehat{\mathcal{Z}}(T, v)) \cdot q^T.$$

6.3 THE CASE OF NONSINGULAR T WITH $\mathrm{sig}(T) = (1, 1)$ OR $(0, 2)$

For $T \in \mathrm{Sym}_2(\mathbb{Z})$ nonsingular of signature $(1, 1)$ or $(0, 2)$, the cycle $\mathcal{Z}(T)$ is empty, since the quadratic form on any $V(A, \iota)$ is positive definite. Thus, the term $\widehat{\mathcal{Z}}(T, v)$ arises as a purely archimedean contribution. For a pair of vectors $\mathbf{x} = [x_1, x_2] \in V(\mathbb{Q})^2$ with nonsingular matrix of inner products $Q(\mathbf{x}) = \frac{1}{2}((x_i, x_j))$, the quantity

$$\Lambda(\mathbf{x}) := \frac{1}{2} \int_D \xi(x_1) * \xi(x_2), \tag{6.3.1}$$

where $\xi(x_1) * \xi(x_2)$ is the $*$-product of the Green functions $\xi(x_1)$ and $\xi(x_2)$, [1], is well defined and depends only on $Q(\mathbf{x})$. In addition, $\Lambda(\mathbf{x})$, which was denoted by $Ht(\mathbf{x})_\infty$ in [3], section 11, has the following invariance property.

Theorem 6.3.1. ([3], Theorem 11.6) *For $k \in \mathrm{O}(2)$, $\Lambda(\mathbf{x} \cdot k) = \Lambda(x)$.*

For $T \in \mathrm{Sym}_2(\mathbb{Z})$ of signature $(1,1)$ or $(0,2)$ and for $v \in \mathrm{Sym}_2(\mathbb{R})_{>0}$, choose[1] $v^{\frac{1}{2}} \in \mathrm{GL}_2(\mathbb{R})$ such that $v = v^{\frac{1}{2}} \cdot {}^t v^{\frac{1}{2}}$, and define

$$\widehat{\deg}(\widehat{\mathcal{Z}}(T,v)) := \sum_{\substack{\mathbf{x} \in L^2 \\ Q(\mathbf{x})=T \\ \mathrm{mod}\ \Gamma}} e_{\mathbf{x}}^{-1} \cdot \Lambda(\mathbf{x}\, v^{\frac{1}{2}}) \quad \in \mathbb{R} \simeq \widehat{\mathrm{CH}}^2(\mathcal{M}). \tag{6.3.2}$$

Here $L = O_B \cap V$, $\Gamma = O_B^\times$, and $e_{\mathbf{x}}$ is the order of the stabilizer $\Gamma_{\mathbf{x}}$ of $\mathbf{x}$ in Γ. In fact, $e_{\mathbf{x}} = 2$ for any $\mathbf{x}$ with $T = Q(\mathbf{x})$ nonsingular. Note that the invariance property of Theorem 6.3.1 is required to make the right side independent of the choice of $v^{\frac{1}{2}}$.

Remark 6.3.2. Notice that $\widehat{\deg}(\widehat{\mathcal{Z}}(T,v)) = 0$ if T is not represented by V, since then the summation in the definition is empty; see also Remark 3.5.3. Recall that T is represented by V if and only if $\mathrm{Diff}(T,B) = \{\infty\}$. For example, if $m \neq 0$, the matrix

$$T = \begin{pmatrix} t_1 & m \\ m & 0 \end{pmatrix}$$

has signature $(1,1)$ but is not represented by any anisotropic V_p. Thus the primes dividing $D(B)$ all lie in $\mathrm{Diff}(T,B)$ and the corresponding $\widehat{\mathcal{Z}}(T,v)$ is zero.

Proposition 6.3.3. *(i) Suppose that T has signature $(1,1)$ and write*

$$v^{\frac{1}{2}} T v^{\frac{1}{2}} = {}^t k(\theta) \cdot \mathrm{diag}(\delta_+, -\delta_-) \cdot k(\theta)$$

for $k(\theta) \in \mathrm{SO}(2)$ and $\delta_\pm \in \mathbb{R}_{>0}$. Then

$$\Lambda(\mathbf{x}\, v^{\frac{1}{2}}) = -\frac{1}{2} \int_D \mathrm{Ei}(-4\pi\delta_- y^{-2}|z|^2) \cdot \left(\delta_+ y^{-2}(1+|z|^2)^2 - \frac{1}{2\pi} \right) \times e^{-\pi\delta_+ [y^{-2}(1+|z|^2)^2 - 4]} \cdot d\mu(z),$$

where, for $z = x + iy \in D$, $d\mu(z) = y^{-2}\, dx\, dy$.
(ii) Suppose that T has signature $(0,2)$ and write

$$v^{\frac{1}{2}} T v^{\frac{1}{2}} = -{}^t k(\theta) \cdot \mathrm{diag}(\delta_1, \delta_2) \cdot k(\theta)$$

for $k(\theta) \in \mathrm{SO}(2)$ and δ_1, $\delta_2 \in \mathbb{R}_{>0}$. Then

$$\Lambda(\mathbf{x}\, v^{\frac{1}{2}}) = -\frac{1}{2} \int_D \mathrm{Ei}(-4\pi\delta_1 y^{-2}|z|^2) \cdot \left(\delta_2\, y^{-2}(1-|z|^2)^2 - \frac{1}{2\pi} \right) \times e^{-\pi\delta_2 [y^{-2}(1-|z|^2)^2 + 4]} \cdot d\mu(z).$$

[1] Note that we may choose $v^{\frac{1}{2}} \in \mathrm{Sym}_2(\mathbb{R})$ with $\det(v^{\frac{1}{2}}) > 0$.

Proof. Part (i) is Theorem 6.3.1 together with Lemma 11.7 of [3]. To prove (ii), we write

$$y_1 = \sqrt{\delta_1}\begin{pmatrix}1 & \\ & -1\end{pmatrix} \qquad \text{and} \qquad y_2 = \sqrt{\delta_2}\begin{pmatrix} & 1\\ 1 & \end{pmatrix}.$$

Then, writing $R_i = \frac{1}{2}(y_i, x(z))^2 + 2\delta_i$, we have

$$2\,R_1 = 4\,\delta_1\,y^{-2}|z|^2 \qquad \text{and} \qquad 2\,R_2 = \delta_2\,(y^{-2}(1-|z|^2)^2 + 4).$$

Now we use the expression of Lemma 11.5 of [3], taking into account the fact that there is no point evaluation term when both y_1 and y_2 have negative length. □

Corollary 6.3.4. *For T of signature $(1,1)$ or $(0,2)$, and for any $\mathbf{x} \in V(\mathbb{R})^2$ with $Q(\mathbf{x}) = T$,*

$$\Lambda(\mathbf{x}\,v^{\frac{1}{2}}) = \nu_\infty(T,v),$$

where $\nu_\infty(T,v)$ is as in (ii) of Theorem 5.2.7, and

$$\widehat{\mathcal{Z}}(T,v) = \nu_\infty(T,v)\cdot\frac{1}{2}\sum_{\substack{\mathbf{x}\in V(\mathbb{Q})^2\\ Q(\mathbf{x})=T\\ \text{mod } \Gamma}} \varphi_f^B(\mathbf{x}).$$

Comparing this expression with Theorem 5.3.8, we obtain

Corollary 6.3.5. *For T of signature $(1,1)$ or $(0,2)$,*

$$\mathcal{E}'_{2,T}(\tau,0,B) = \widehat{\deg}(\widehat{\mathcal{Z}}(T,v))\cdot q^T.$$

6.4 SINGULAR TERMS, T OF RANK 1

We now turn to the case

$$T = \begin{pmatrix} t_1 & m\\ m & t_2\end{pmatrix} \in \mathrm{Sym}_2(\mathbb{Z})^\vee \tag{6.4.1}$$

with $\det(T) = 0$. If $T \neq 0$, we may write $t_1 = n_1^2 t$, $t_2 = n_2^2 t$, and $m = n_1 n_2 t$ for relatively prime integers n_1 and n_2 and $t \in \mathbb{Z}_{\neq 0}$. The pair n_1, n_2 is unique up to simultaneous change in sign. Also note that if $t_1 = 0$, then $n_1 = 0$, $n_2 = 1$, and $t = t_2$. Similarly if $t_2 = 0$, then $n_1 = 1$, $n_2 = 0$, and $t = t_1$. For the cycle associated to such a singular T, we have simply

Lemma 6.4.1. $\mathcal{Z}(T) = \mathcal{Z}(t)$.

Proof. We fix a choice of the pair n_1 and n_2. First suppose that $n_1 n_2 \neq 0$. Then if an object $(A, \iota, \mathbf{x})$ is given, with $\mathbf{x} = [x_1, x_2] \in V(A, \iota)^2$, we have $Q(n_2 x_1 - n_1 x_2) = 0$, so that $n_2 x_1 = n_1 x_2 \in V(A, \iota)$. Since n_1 and n_2 are relatively prime, it follows that $y := n_1^{-1} x_1 = n_2^{-1} x_2 \in V(A, \iota)$. Note that $Q(y) = t$, so that (A, ι, y) is an object for $\mathcal{Z}(t)$. Conversely, for a given (A, ι, y), we have $(A, \iota, [n_1 y, n_2 y])$ for $\mathcal{Z}(T)$. If, say, $t_1 = n_1 = 0$ so that $n_2 = 1$ and $t = t_2$, and $(A, \iota, \mathbf{x})$ is given, we have $Q(x_1) = 0$ so that $x_1 = 0$ and (A, ι, x_2) defines an element of $\mathcal{Z}(t)$. Conversely, given (A, ι, y) we can take $(A, \iota, [0, y])$. Of course, if $t < 0$, then both sides are empty. □

We define $\widehat{\mathcal{Z}}(T, v) \in \mathbb{R} \simeq \widehat{\mathrm{CH}}^2(\mathcal{M})$ by setting

$$\widehat{\deg}(\widehat{\mathcal{Z}}(T, v)) := -\langle\, \widehat{\mathcal{Z}}(t, t^{-1}\mathrm{tr}(Tv))\, ,\, \hat{\omega}\, \rangle - \frac{1}{2}\, \deg_{\mathbb{Q}}(\mathcal{Z}(t)) \Big(\, \log\big(\frac{\det(v)}{t^{-1}\mathrm{tr}(Tv)}\big) + \log(D)\, \Big). \tag{6.4.2}$$

As motivation for this definition, note that we are, in some sense, shifting the 'naive' class

$$\widehat{\mathcal{Z}}(t, t^{-1}\mathrm{tr}(Tv)) \in \widehat{\mathrm{CH}}^1(\mathcal{M}), \tag{6.4.3}$$

which occurs in the wrong degree, by taking its pairing with $-\hat{\omega}$. The additional terms involving v are added in order to obtain agreement with the Fourier coefficient of $\mathcal{E}_2'(\tau, 0; B)$. Note that, for all $\alpha \in \mathrm{GL}_2(\mathbb{Z})$,

$$\widehat{\deg}(\widehat{\mathcal{Z}}(T, \alpha v {}^t\alpha)) = \widehat{\deg}(\widehat{\mathcal{Z}}({}^t\alpha T \alpha, v)), \tag{6.4.4}$$

so that our definition has the invariance which must hold for a Fourier coefficient of a Siegel modular form.

Now the results of [6] give explicit expressions for all such singular terms. More precisely, let $\mathcal{E}_1(\tau, s; B)$ be the modified Eisenstein series of genus one associated to B. We then quote from [6] the following statement:

Theorem 6.4.2. *For $t \neq 0$,*

$$\mathcal{E}_{1,t}(\tau, \frac{1}{2}; B) = \deg_{\mathbb{Q}} \mathcal{Z}(t) \cdot q^t,$$

and

$$\mathcal{E}_{1,t}'(\tau, \frac{1}{2}; B) = \langle\, \widehat{\mathcal{Z}}(t, v), \hat{\omega}\, \rangle \cdot q^t.$$

Corollary 6.4.3. *For* $T \in \mathrm{Sym}_2(\mathbb{Z})^\vee$ *of rank* 1,

$$\widehat{\deg}(\widehat{\mathcal{Z}}(T,v)) \cdot q^T = -\mathcal{E}'_{1,t}(t^{-1}\mathrm{tr}(T\tau), \frac{1}{2}; B) - \frac{1}{2}\,\mathcal{E}_{1,t}(t^{-1}\mathrm{tr}(T\tau), \frac{1}{2}; B) \cdot \Big(\ \log\big(\frac{\det(v)}{t^{-1}\mathrm{tr}(Tv)}\big) + \log(D)\ \Big).$$

Comparing this with Theorem 5.8.7, we have

Corollary 6.4.4. *For* $T \in \mathrm{Sym}_2(\mathbb{Z})^\vee$ *of rank* 1,

$$\mathcal{E}'_{2,T}(\tau, 0; B) = \widehat{\deg}(\widehat{\mathcal{Z}}(T,v)) \cdot q^T.$$

6.5 THE CONSTANT TERM, $T = 0$

We complete the definition of $\hat{\phi}_2(\tau)$ by setting

$$\widehat{\deg}(\widehat{\mathcal{Z}}(0,v)) := \langle\, \hat{\omega}, \hat{\omega}\,\rangle + \frac{1}{2}\,\deg_{\mathbb{Q}}(\hat{\omega}) \cdot \Big(\ \log\det(v) - \mathbf{c} + \log D\ \Big), \tag{6.5.1}$$

as a class in $\widehat{\mathrm{CH}}^2(\mathcal{M}) \simeq \mathbb{R}$, where the constant $\mathbf{c}$ is introduced in [6]. We will sometimes write $\widehat{\mathcal{Z}}(0,v) = \widehat{\mathcal{Z}}_2(0,v)$ to distinguish this constant term of the genus two generating function from the constant term $\widehat{\mathcal{Z}}_1(0,v)$ of the genus one generating function. Recall from [6] that

$$\widehat{\mathcal{Z}}_1(0,v) = -\hat{\omega} - (0, \log(v)) + (0, \mathbf{c}) \in \widehat{\mathrm{CH}}^1(\mathcal{M}) \tag{6.5.2}$$

satisfies the identity

$$\begin{aligned}\mathcal{E}'_{1,0}(\tau, \frac{1}{2}; B) &= \langle\, \widehat{\mathcal{Z}}_1(0,v), \hat{\omega}\,\rangle \\ &= -\langle\, \hat{\omega}, \hat{\omega}\,\rangle - \frac{1}{2}\deg_{\mathbb{Q}}\hat{\omega} \cdot (\ \log(v) - \mathbf{c}\ )\end{aligned} \tag{6.5.3}$$

and that this identity determines the constant $\mathbf{c}$. Also,

$$\begin{aligned}\mathcal{E}_{1,0}(\tau, \frac{1}{2}; B) &= \deg_{\mathbb{Q}} \widehat{\mathcal{Z}}_1(0,v) \\ &= -\deg_{\mathbb{Q}}\hat{\omega}.\end{aligned} \tag{6.5.4}$$

Thus we have

$$\widehat{\deg}(\widehat{\mathcal{Z}}_2(0,v)) = -\mathcal{E}'_{1,0}(i\det(v), \frac{1}{2}; B) - \frac{1}{2} \cdot \mathcal{E}_{1,0}(i\det(v), \frac{1}{2}; B) \cdot \log(D). \tag{6.5.5}$$

Comparing this with the result of Theorem 5.9.5, we have the following identity:

Theorem 6.5.1.

$$\mathcal{E}'_{2,0}(\tau, 0; B) = \widehat{\deg}(\widehat{\mathcal{Z}}_2(0, v)).$$

This concludes the proof of Theorem B above.

Bibliography

[1] H. Gillet and C. Soulé, *Arithmetic intersection theory*, Publ. Math. IHES, **72** (1990), 93–174.

[2] B. Gross and K. Keating, *On the intersection of modular correspondences*, Invent. math., **112** (1993), 225–245.

[3] S. Kudla, *Central derivatives of Eisenstein series and height pairings*, Annals of Math., **146** (1997), 545–646.

[4] ______, *Derivatives of Eisenstein series and arithmetic geometry*, in Proc. Intl. Cong. Mathematicians, Vol II (Beijing, 2002), 173–183, Higher Education Press, Beijing, 2002.

[5] S. Kudla and M. Rapoport, *Height pairings on Shimura curves and p-adic uniformization*, Invent. math., **142** (2000), 153–223.

[6] S. Kudla, M. Rapoport, and T. Yang, *Derivatives of Eisenstein series and Faltings heights*, Compositio Math., **140** (2004), 887–951.

Chapter Six: Appendix

The case $p = 2, p \mid D(B)$

In this appendix, we extend the results of [3] concerning intersections of the special cycles on the Drinfeld space to the case $p = 2$. We will denote by $B' = \mathrm{M}_2(\mathbb{Q}_p)$ the matrix algebra over $\mathbb{Q}_p$ and by V' its subspace of traceless elements.To any $j \in V'$ there is associated a special cycle on the Drinfeld space. In the appendix to section 11 of [4] the geometry of an individual special cycle $Z(j)$ is described in the case $p = 2$. What we have to deal with, then, is the relative position of two such special cycles.

We remark that elsewhere in this book we used the quadratic form on the space of traceless elements of a quaternion algebra given by the reduced norm, $Q(x) = \mathrm{Nm}(x) = x^\iota x$. On the other hand, in the local context, we used in the appendix to section 11 of [4] the quadratic form given by squaring, $q(x) = x^2$, which differs from Q by a sign change. This sign change also occurs in [3] when we make the transition from the local results to the global results. It turns out that the main result of this appendix is best expressed in terms of the quadratic form Q; at the same time, to make the relation to the local calculations in our previous papers [3] and [4] easier, in the explicit calculations we use the quadratic form q. The miracle is that, just as in the Gross-Keating formula (Theorem 3.6.1), the same formula for the local intersection index as in the case $p \neq 2$ continues to hold in the case $p = 2$ when the result is expressed in terms of the Gross-Keating invariants of the matrix $\tilde{T} = \mathrm{diag}(1, T) \in \mathrm{Sym}_3(\mathbb{Z})^\vee$.

6A.1 STATEMENT OF THE RESULT

We consider a pair of special endomorphisms j and $j' \in V'$ and assume that they span a 2-dimensional subspace on which the quadratic form is nondegenerate, i.e., such that $\det(T) \neq 0$, where

$$T = \begin{pmatrix} Q(j) & \frac{1}{2}(j, j') \\ \frac{1}{2}(j', j) & Q(j') \end{pmatrix} \in \mathrm{Sym}_2(\mathbb{Q}_2). \tag{6A.1.1}$$

Theorem 5.1 of [3], whose proof involves no restriction on the residue characteristic, implies that for intersection numbers

(6A.1.2) $$(Z(j), Z(j')) = (Z(j)^{\text{pure}}, Z(j')^{\text{pure}}).$$

Moreover, this intersection number depends only on the $\mathbb{Z}_2$-submodule $\mathbf{j}$ of V spanned by j and j'. Our main result supplements Theorem 6.1 of [3].

To formulate our result we consider the 3-dimensional $\mathbb{Z}_p$-lattice given by

(6A.1.3) $$\tilde{T} = \operatorname{diag}(1, T).$$

Then Gross and Keating associate to $\tilde{T}$ its GK-*invariant*

(6A.1.4) $$\mathrm{GK}(\tilde{T}) = (0, a_2, a_3) \in \mathbb{Z}^3$$

where $0 \le a_2 \le a_3$, and furthermore its GK-*constant*

(6A.1.5) $$\epsilon(\tilde{T}) \in \{\pm 1\},$$

provided that $a_2 \equiv 0 \mod 2$ and $a_2 < a_3$, [5], appendix B. These terms only depend on the $\mathrm{GL}_2(\mathbb{Z}_p)$-equivalence class of T. Note that if $p \neq 2$ and T is $\mathrm{GL}_2(\mathbb{Z}_p)$-equivalent to $\operatorname{diag}(\varepsilon_1 p^\alpha, \varepsilon_2 p^\beta)$ with $0 \le \alpha \le \beta$ with $\varepsilon_i \in \mathbb{Z}_p^\times$ for $i = 1, 2$, then

(6A.1.6) $$\mathrm{GK}(\tilde{T}) = (0, \alpha, \beta) \text{ and } \epsilon(\tilde{T}) = \chi(-\varepsilon_1),$$

where χ is the quadratic residue character modulo p.

Theorem 6A.1.1. *The intersection number*

$$(Z(j), Z(j')) = \frac{1}{2}\nu_2(T)$$

is equal to

$$a_2 + a_3 + 1 - \begin{cases} 2\,\dfrac{p^{(a_2+1)/2} - 1}{p - 1} & \text{if } a_2 \text{ is odd,} \\ (a_3 - a_2 + 1)\, p^{a_2/2} + 2\,\dfrac{p^{a_2/2} - 1}{p - 1} & \text{if } a_2 \text{ is even and } \epsilon(\tilde{T}) = 1, \\ p^{a_2/2} + 2\,\dfrac{p^{a_2/2} - 1}{p - 1} & \text{if } a_2 \text{ is even and } \epsilon(\tilde{T}) = -1. \end{cases}$$

Here $\mathrm{GK}(\tilde{T}) = (0, a_2, a_3)$.[1]

[1]The meaning of the formula in the case where $a_2 = a_3$ is even (in which case $\epsilon(\tilde{T})$ is not defined) is that one takes either of the last two (identical) alternatives.

If $p \neq 2$, this formula is exactly the one of Theorem 6.1 of [3] by the remarks preceding the theorem. As a first step of the proof of the above theorem for $p = 2$, we calculate the Gross-Keating invariants and constants of $\tilde{T}$. Before giving the result, we recall from the appendix to section 11 of [4] the case distinction for $q = \varepsilon\, 2^\alpha \in \mathbb{Z}_2$, with $\varepsilon \in \mathbb{Z}_2^\times$. We call q of type **(1)-(4)** according to

(1) α even, $\varepsilon \equiv 1 \mod 8$
(2) α even, $\varepsilon \equiv 5 \mod 8$
(3) α even, $\varepsilon \equiv 3 \mod 4$
(4) α odd.

The values are now given in Table 2, which is organized as follows: if T is $\mathrm{GL}_2(\mathbb{Z}_2)$-equivalent to a diagonal matrix, we write $T \sim \mathrm{diag}(\varepsilon_1 2^\alpha, \varepsilon_2 2^\beta)$ with $0 \leq \alpha \leq \beta$ and call T of type **(i):(j)** if $-\varepsilon_1 2^\alpha$ is of type **(i)** and $-\varepsilon_2 2^\beta$ is of type **(j)**.

Table 2. Values for Gross-Keating Invariants

Diagonal Cases	$\mathrm{GK}(\tilde{T})$	$\epsilon(\tilde{T})$
(1):(*), $\beta \leq \alpha + 1$	$(0, \alpha+1, \beta+1)$	
(1):(*), $\beta \geq \alpha + 2$	$(0, \alpha+2, \beta)$	$+1$
(2):(*), $\beta \leq \alpha + 1$	$(0, \alpha+1, \beta+1)$	
(2):(*), $\beta \geq \alpha + 2$	$(0, \alpha+2, \beta)$	-1
(3):(*)	$(0, \alpha+1, \beta+1)$	
(4):(*)	$(0, \alpha, \beta+2)$	
Other Cases		
$-2^\alpha \begin{pmatrix} b & \frac{1}{2} \\ \frac{1}{2} & b \end{pmatrix}$, $b = 0, 1$	$(0, \alpha, \alpha)$	

In Table 2, **(*)** means that the type of $-\varepsilon_2 2^\beta$ is arbitrary.

Proof. If $T \sim \mathrm{diag}(\varepsilon_1 2^\alpha, \varepsilon_2 2^\beta)$, with $\alpha \leq \beta$, then

$$\tilde{T} \sim \mathrm{diag}(1, \varepsilon_1 2^\alpha, \varepsilon_2 2^\beta). \tag{6A.1.7}$$

Using the notation of [5], Proposition B.5, we have

$$\beta_1 = 0,\ \beta_2 = \alpha,\ \beta_3 = \beta. \tag{6A.1.8}$$

In Cases **(1):(*)**, or **(2):(*)**, one has $\varepsilon_1 \equiv -1 \mod 4$, and $\beta_2 - \beta_1 = \alpha \equiv 0 \mod 2$. So Proposition B.5 says that the GK-invariant of $\tilde{T} = \mathrm{diag}(1, T)$ is

$$\mathrm{GK}(\tilde{T}) = \begin{cases} (0, \alpha+1, \beta+1) & \text{if } \beta \leq \alpha+1, \\ (0, \alpha+2, \beta) & \text{if } \beta \geq \alpha+2. \end{cases} \tag{6A.1.9}$$

Moreover, when $\beta \geq \alpha+2$, then by [2], the Gross-Keating constant is equal to $(-\varepsilon_1, 2)_2$ as claimed.

In Case **(3):(*)**, one has $\varepsilon_1 \equiv 1 \mod 4$, hence [5], Proposition B.5 gives

$$\mathrm{GK}(\tilde{T}) = (0, \alpha+1, \beta+1). \tag{6A.1.10}$$

The Case **(4):(*)** is simply [5], Proposition B.5, (1).

Finally in the nondiagonal cases, the assertion is [5], Proposition B.4. □

Taking into account Table 2, it is now an elementary matter to calculate the expression occurring in Theorem 6A.1.1 in all cases. In the reformulation we have passed to the quadratic form q on V' which absorbs the minus sign occurring in our definition of the type of a diagonal matrix T in Table 2. Therefore the following statement is equivalent to Theorem 6A.1.1. We will prove Theorem 6A.1.1 by going through all cases in the following theorem.

Theorem 6A.1.2. *The intersection number*

$$(Z(j), Z(j')) = \frac{1}{2}\nu_2(T)$$

depends only on the $\mathrm{GL}_2(\mathbb{Z}_2)$*-equivalence class of* T*. The values of* $\frac{1}{2}\nu_2(T)$ *are the following:*

Diagonal cases

Suppose that T *is* $\mathrm{GL}_2(\mathbb{Z}_2)$*-equivalent to the matrix* $\mathrm{diag}(\varepsilon_1 2^\alpha, \varepsilon_2 2^\beta)$*. In the following list, the initial pair of numbers indicates the types of* $q(j)$ *and* $q(j')$*, e.g.,* **(1):(3)** *indicates that* $q(j)$ *has type* **(1)** *and* $q(j')$ *has type* **(3)** *in the notation introduced above. The comment 'classical' indicates that the corresponding configuration, as defined in Section A.3 below, already occurs for* $p \neq 2$*.*

Case (1):(1) *with* $\alpha \leq \beta$*.*

$$\frac{1}{2}\nu_2(T) = \alpha + \beta + 3 - \begin{cases} (\beta - \alpha - 1)2^{\frac{\alpha}{2}+1} - 2(2^{\frac{\alpha}{2}+1} - 1) & \text{if } \alpha < \beta, \\ 2\,(2^{\frac{\alpha}{2}+1} - 1) & \text{if } \alpha = \beta. \end{cases}$$

Case (1):(2).

$$\frac{1}{2}\nu_2(T) = \alpha + \beta + 3 - \begin{cases} (\beta - \alpha - 1)2^{\frac{\alpha}{2}+1} + 2\,(2^{\frac{\alpha}{2}+1} - 1) & \text{if } \alpha < \beta, \\ 2\,(2^{\frac{\alpha}{2}+1} - 1) & \text{if } \alpha = \beta, \\ 2^{\frac{\beta}{2}+1} + 2\,(2^{\frac{\beta}{2}+1} - 1) & \text{if } \beta < \alpha. \end{cases}$$

Case (1):(3).

$$\frac{1}{2}\nu_2(T) = \alpha + \beta + 3 - \begin{cases} (\beta - \alpha - 1)2^{\frac{\alpha}{2}+1} + 2(2^{\frac{\alpha}{2}+1} - 1) & \text{if } \alpha < \beta \\ 2(2^{\frac{\beta}{2}+1} - 1) & \text{if } \beta \leq \alpha. \end{cases}$$

Case (1):(4) *(classical).*

$$\frac{1}{2}\nu_2(T) = \alpha + \beta + 3 - \begin{cases} (\beta - \alpha - 1)2^{\frac{\alpha}{2}+1} + 2(2^{\frac{\alpha}{2}+1} - 1), & \text{if } \alpha < \beta, \\ 2(2^{\frac{\beta+1}{2}} - 1), & \text{if } \alpha > \beta. \end{cases}$$

Case (2):(2) *with* $\alpha \leq \beta$.

$$\frac{1}{2}\nu_2(T) = \alpha + \beta + 3 - \begin{cases} 2^{\frac{\alpha}{2}+1} + 2(2^{\frac{\alpha}{2}+1} - 1) & \text{if } \alpha < \beta, \\ 2(2^{\frac{\alpha}{2}+1} - 1) & \text{if } \alpha = \beta. \end{cases}$$

Case (2):(3).

$$\frac{1}{2}\nu_2(T) = \alpha + \beta + 3 - \begin{cases} 2^{\frac{\alpha}{2}+1} + 2(2^{\frac{\alpha}{2}+1} - 1) & \text{if } \alpha < \beta, \\ 2(2^{\frac{\beta}{2}+1} - 1) & \text{if } \beta \leq \alpha. \end{cases}$$

Case (2):(4) *excluded.*
Case (3):(3) *excluded.*
Case (3):(4) *(classical).*

$$\frac{1}{2}\nu_2(T) = \alpha + \beta + 3 - \begin{cases} 2(2^{\frac{\alpha}{2}+1} - 1) & \text{if } \alpha < \beta, \\ 2(2^{\frac{\beta+1}{2}} - 1) & \text{if } \beta < \alpha. \end{cases}$$

Case (4):(4) *(classical),* $\alpha \leq \beta$.

$$\frac{1}{2}\nu_2(T) = \alpha + \beta + 3 - 2(2^{\frac{\alpha+1}{2}} - 1).$$

Additional cases

$$T \simeq -2^{\alpha}\begin{pmatrix} 0 & \frac{1}{2} \\ \frac{1}{2} & 0 \end{pmatrix}$$

$$\frac{1}{2}\nu_2(T) = 2\alpha + 1 - \begin{cases} 2(2^{\frac{\alpha+1}{2}} - 1) & \text{if } \alpha \text{ is odd}, \\ 2^{\frac{\alpha}{2}} + 2(2^{\frac{\alpha}{2}} - 1) & \text{if } \alpha \text{ is even}. \end{cases}$$

$T \simeq -2^{\alpha}\begin{pmatrix} 1 & \frac{1}{2} \\ \frac{1}{2} & 1 \end{pmatrix}$, α *even*

$$\frac{1}{2}\nu_2(T) = 2\alpha + 1 - 2^{\frac{\alpha}{2}} - 2(2^{\frac{\alpha}{2}} - 1).$$

The case when α is odd is excluded.

6A.2 REVIEW OF THE SPECIAL CYCLES $Z(j)$, FOR $q(j) \in \mathbb{Z}_p \setminus \{0\}$

When computing intersection numbers, the divisors $Z(j)$ can be replaced by the corresponding cycles $Z(j)^{\rm pure}$ and these have a simple combinatorial description.

Definition 6A.2.1. A set of *cycle data* is a triple (S, μ, Z^h), where $S \subset \mathcal{B}$ is a subset of the building, $\mu \in \mathbb{Z} \geq 0$, and Z^h is a horizontal divisor on $\mathcal{M}$.

Note that, for the computation of the intersection multiplicities of Z^h with vertical divisors, it is enough to give a small amount of incidence data about Z^h.

A set of cycle data (S, μ, Z^h) determines a divisor on $\mathcal{M}$,

$$Z(S, \mu, Z^h) = \sum_{[\Lambda]} m_{[\Lambda]}(S, \mu)\, \mathbb{P}_{[\Lambda]} + Z^h, \tag{6A.2.1}$$

where

$$m_{[\Lambda]}(S, \mu) = \max\{\mu - d([\Lambda], S), 0\}. \tag{6A.2.2}$$

Then Z^h is the horizontal part of the divisor $Z(S, \mu, Z^h)$, and the first sum is the vertical part of $Z(S, \mu, Z^h)$. Conversely, the divisor $Z(S, \mu, Z^h)$ determines the set of cycle data (S, μ, Z^h) (except when $\mu = 0$).

In the 'classical' case, $p \neq 2$, for $q(j) = \varepsilon p^{\alpha}$, $Z(j)^{\rm pure}$ has the following description:

(i) α even and $\chi(\varepsilon) = 1$. Then

$$S = \mathcal{A} = \mathcal{B}^j, \qquad \mu = \frac{\alpha}{2}, \qquad Z^h = 0. \tag{6A.2.3}$$

Here $\mathcal{A}$ denotes an apartment in $\mathcal{B}$.

(ii) α even and $\chi(\varepsilon) = -1$. Then

$$S = [\Lambda_0] = \mathcal{B}^j, \qquad \mu = \frac{\alpha}{2}, \qquad Z^h = 2\,{\rm Spf(W)}. \tag{6A.2.4}$$

Here the horizontal component Z^h meets the vertical part only in the central component $\mathbb{P}_{[\Lambda_0]}$, in two ordinary special points.
(iii) α odd. Then

(6A.2.5) $$S = [[\Lambda_0, \Lambda_1]], \qquad \mu = \frac{\alpha - 1}{2}, \qquad Z^h = \mathrm{Spf}(\mathrm{W}').$$

Then $\mathcal{B}^j$ is the midpoint of the edge $[\Lambda_0, \Lambda_1]$, and Z^h meets the vertical part at the superspecial point corresponding to the edge $[\Lambda_0, \Lambda_1]$. Here we use the notation $[[\Lambda_0, \Lambda_1]]$ for the union of the edge $[\Lambda_0, \Lambda_1]$ and its endpoints.

In the case where $p = 2$, again for $q(j) = \varepsilon p^\alpha$, $Z(j)^{\mathrm{pure}}$ is described in the Appendix to section 11 of [4]:[2]
(1) α even and $\varepsilon \equiv 1 \mod (8)$. Then

(6A.2.6) $$S = \mathcal{A} = \mathcal{B}^{\mathcal{O}^\times}, \qquad \mu = \frac{\alpha}{2} + 1, \qquad Z^h = \emptyset,$$

where again $\mathcal{A}$ is an apartment in $\mathcal{B}$. In this case

(6A.2.7) $$\mathcal{B}^j = \{\, x \in \mathcal{B} \mid d(x, \mathcal{A}) \leq 1 \,\}.$$

(2) α even and $\varepsilon \equiv 5 \mod (8)$. Then

(6A.2.8) $$S = [\Lambda_0] = \mathcal{B}^{\mathcal{O}^\times}, \qquad \mu = \frac{\alpha}{2} + 1, \qquad Z^h = 2\,\mathrm{Spf}(\mathrm{W}),$$

where Z^h meets the vertical part only in the central component $\mathbb{P}_{[\Lambda_0]}$ in two ordinary special points. Here

(6A.2.9) $$\mathcal{B}^j = \{x \in \mathcal{B} \mid d(x, [\Lambda_0]) \leq 1\}.$$

(3) α even and $\varepsilon \equiv 3 \mod (4)$. Then
(6A.2.10)
$$S = [[\Lambda_0, \Lambda_1]] = \mathcal{B}^{\mathcal{O}^\times} = \mathcal{B}^j, \qquad \mu = \frac{\alpha}{2}, \qquad Z^h = \mathrm{Spf}(\mathrm{W}'),$$

as in the classical case (iii).
(4) α odd. Then

(6A.2.11) $$S = [[\Lambda_0, \Lambda_1]] = \mathcal{B}^{\mathcal{O}^\times}, \qquad \mu = \frac{\alpha - 1}{2}, \qquad Z^h = \mathrm{Spf}(\mathrm{W}'),$$

---

[2]We take the occasion to correct an error concerning Case **(1)** in the Appendix to Section 11 of [4]. It is erroneously asserted on pp. 934–935 that $\mathbb{Z}_2[j]^\times \subsetneq \mathcal{O}^\times$ and that for the associated apartment $\mathcal{A}$ one has $\mathcal{A} = \mathcal{B}^{\mathcal{O}^\times}$. In fact, in this case $\mathcal{B}^{\mathcal{O}^\times} = \mathcal{B}^j$. In the table on p. 934 the header of the last column should be named $S(j)$ and defined to be $\mathcal{A}$ in Case **(1)** and $\mathcal{B}^{\mathcal{O}^\times}$ in Cases **(2)**–**(4)**. In the rest of the appendix, $\mathcal{B}^{\mathcal{O}^\times}$ should be replaced everywhere by $S(j)$. Then the statements become correct in all cases.

again as in the classical case (iii). Here

(6A.2.12) $$\mathcal{B}^j = \text{midpoint of } [\Lambda_0, \Lambda_1].$$

Note that each of these cycles has the same type of cycle data as one of the classical cases. Also note that S and Z^h are unchanged if j is replaced by a scalar multiple.

6A.3 CONFIGURATIONS

In this section, we suppose that a pair j and j' is given with $q(j)$ and $q(j')$ both nonzero and with matrix of inner products T, as in the previous section, and we describe the possible configurations (S, S'), i.e., relative positions, of the sets S and S' occuring in the cycle data of $Z(j)$ and $Z(j')$. In the diagonal cases, these configurations depend only on the types of j and j', rather than on the full matrix T.

First suppose that $p \neq 2$. Then the matrix T is diagonalizable. When it has diagonal form, then $jj' = -j'j$, and hence j (resp. j') preserves the set $\mathcal{B}^{j'}$ (resp. $\mathcal{B}^j$). From this fact, it is easy to check that the possible configurations are the following (cf. [3]):

(i):(i) S and S' are apartments which meet at a unique vertex $[\Lambda_0]$.
(i):(ii) S is an apartment containing the unique vertex S' fixed by j'.
(i):(iii) S is an apartment containing the edge S' whose midpoint is the unique fixed point of j'.
(ii):(ii) $S = S' = [\Lambda_0]$ is the common fixed vertex of j and j'.
(ii):(iii) excluded.
(iii):(iii) $S = S' = [[\Lambda_0, \Lambda_1]]$ with midpoint the common fixed point of j and j'.

We will refer to these possibilities as the *classical configurations.*

Proposition 6A.3.1. *Suppose that $p = 2$. Then the possible configurations are the following:*
When the matrix T is diagonal,

(1):(1) *S and S' are apartments with $d(S, S') = 1$.*
(1):(2) *S is an apartment and S' is a vertex with $d(S, S') = 1$.*
(1):(3) *S is an apartment and $S' = [[\Lambda'_0, \Lambda'_1]]$ with $[\Lambda'_0] \in S$ but $[\Lambda'_1] \notin S$.*
(1):(4) *S is an apartment and $S' = [[\Lambda'_0, \Lambda'_1]] \subset S$.*

This coincides with the classical configuration **(i):(iii)**.

(2):(2) *$S = [\Lambda_0], S' = [\Lambda'_0]$, where $d([\Lambda_0], [\Lambda'_0]) = 1$.*
(2):(3) *$S = [\Lambda_0]$ and $S' = [[\Lambda_0, \Lambda'_1]]$, where $[[\Lambda_0, \Lambda'_1]] = \mathcal{B}^{j'}$.*
(2):(4) *excluded.*
(3):(3) *excluded.*

(3):(4) *$S = S' = [[\Lambda_0, \Lambda_1]] = \mathcal{B}^j$ and $\mathcal{B}^{j'}$ the midpoint.*
(4):(4) *$S = S' = [[\Lambda_0, \Lambda_1]]$ with $\mathcal{B}^j = \mathcal{B}^{j'}$ the midpoint.*
These last two configurations coincide with the classical configuration for **(iii):(iii)**.

Finally, if $T = -2^\alpha \begin{pmatrix} 1 & \frac{1}{2} \\ \frac{1}{2} & 1 \end{pmatrix}$, *then α is even and S and S' are apartments with $d(S, S') = 2$.*

Proof. First note that, when T is diagonal so that j and j' anticommute, then conjugation by j preserves $\mathcal{O}'$ and acts by the Galois automorphism and similarly for j' and $\mathcal{O}$. In particular, the action of j' must preserve both fixed-point sets $\mathcal{B}^j$ and $\mathcal{B}^{\mathcal{O}^\times}$, and the action of j must preserve both fixed-point sets $\mathcal{B}^{j'}$ and $\mathcal{B}^{\mathcal{O}',\times}$ We now prove selected cases. For **(1):(1)**, j must act by a nontrivial automorphism of order 2 on the apartment $S' = \mathcal{B}^{\mathcal{O}',\times}$. It thus has a unique fixed point x_0 on the apartment S', which then has distance ≤ 1 from the apartment $S = \mathcal{B}^{\mathcal{O}^\times}$. But this point must have distance exactly 1 from S, since otherwise the intersection of S' with the tube of radius 1 around S, i.e., the fixed-point set $\mathcal{B}^j$, would contain another point. This confirms the configuration claimed for **(1):(1)**. Case **(1):(3)** is similar, since j' must have a unique fixed point on the apartment S. In Case **(1):(2)**, the fixed points of j' all have distance ≤ 1 from $S' = [\Lambda_0]$; since the apartment S contains exactly one of them, we must have $d(S, S') = 1$. The reasoning for Case **(1):(4)** is as in the classical situation. In Case **(2):(2)** we have $S = [\Lambda_0]$ and $\mathcal{B}^j = \{x \mid d(x, [\Lambda_0]) \leq 1\}$ and similarly for $S' = [\Lambda_0']$ and $\mathcal{B}^{j'}$. It follows that $[\Lambda_0] \in \mathcal{B}^{j'}$ and $[\Lambda_0'] \in \mathcal{B}^j$. On the other hand, we must have $[\Lambda_0] \neq [\Lambda_0']$. To see this, let us choose a basis of Λ_0 as in (A.10) of [4], so that j is given by the matrix $j = 2^{\frac{\alpha}{2}} \bar{j}$ with

$$\bar{j} = \begin{pmatrix} -1 & 2\lambda \\ 2 & 1 \end{pmatrix} \quad \text{with } \lambda \in \mathbb{Z}_2^\times. \tag{6A.3.1}$$

Let $j' = 2^{\frac{\beta}{2}} \bar{j}'$, where

$$\bar{j}' = \begin{pmatrix} a & b \\ c & -a \end{pmatrix}. \tag{6A.3.2}$$

The condition that j' anticommute with j is given by $b = a - \lambda c$. If $[\Lambda_0] = [\Lambda_0']$ then $\bar{j}'$ would have to be a scalar matrix modulo 2. But then $2 \mid c$ and $2 \mid a$ which contradicts $\det(\bar{j}') \in \mathbb{Z}_2^\times$.

In Case **(2):(3)**, $[\Lambda_0] = \mathcal{B}^{\mathcal{O}^\times}$ has to be fixed by j', which yields the assertion. For the same reason, Case **(2):(4)** is excluded since then j' fixes no vertex. This case could also have been excluded from the fact that j and

j' generate the matrix algebra $\mathrm{M}_2(\mathbb{Q}_p)$ which yields $(\varepsilon\, 2^\alpha, \varepsilon' 2^\beta)_2 = 1$. For the same reason, Case **(3):(3)** does not occur. The remaining diagonal cases are obvious.

For $T = -2^\alpha \begin{pmatrix} 1 & \frac{1}{2} \\ \frac{1}{2} & 1 \end{pmatrix}$, we note that T is representable by the quadratic space of the traceless matrices in $\mathrm{M}_2(\mathbb{Q}_2)$ (with q given by squaring) only when α is even. Let $j = 2^{\frac{\alpha}{2}} \bar{j}$ and $j' = 2^{\frac{\alpha}{2}}\, \bar{j}'$. Then $\bar{j}^2 = \bar{j}'^2 = 1$. We may choose a basis of $\mathbb{Q}_2^2$ such that

(6A.3.3)
$$\bar{j} = \begin{pmatrix} 1 & 0 \\ 0 & -1 \end{pmatrix}.$$

Then $\bar{j}' = g\bar{j}g^{-1}$ for a suitable $g \in \mathrm{GL}_2(\mathbb{Q}_p)$. By the Bruhat decomposition we may assume that $g \in NA$ or $g \in NwNA$, where N is the subgroup of unipotent upper triangular matrices, A the subgroup of diagonal matrices, and where $w = \begin{pmatrix} 0 & 1 \\ 1 & 0 \end{pmatrix}$. The condition is that

(6A.3.4)
$$1 = (\bar{j}, \bar{j}') = -\mathrm{tr}(\bar{j} \cdot \bar{j}'^{\iota}),$$

where $x \mapsto x^\iota$ denotes the main involution on $\mathrm{M}_2(\mathbb{Q}_2)$. It is easy to see that $g \in NA$ yields no solution. If $g \in NwNA$, we may assume that

(6A.3.5)
$$g = \begin{pmatrix} 1 & b_1 \\ 0 & 0 \end{pmatrix} \cdot \begin{pmatrix} 0 & 1 \\ 1 & 0 \end{pmatrix} \cdot \begin{pmatrix} 1 & b_2 \\ 0 & 1 \end{pmatrix}.$$

Then (6A.3.4) is equivalent to $b_1 b_2 = -\frac{3}{4}$, and hence we may take

(6A.3.6)
$$g = \begin{pmatrix} 1 & \frac{1}{4} \\ 1 & -\frac{3}{4} \end{pmatrix}.$$

Now S is the standard apartment $\mathcal{A} = \{[\Lambda_r] \mid r \in \mathbb{Z}\}$, with $\Lambda_r = [e_0, 2^r e_1]$, and $S' = gS$. But
(6A.3.7)
$$g[\Lambda_r] = [[e_0 + e_1, 2^{r-2}(e_0 - 3e_1)]] = \begin{cases} [[e_0 + e_1, 2^r e_1]] & \text{if } r \geq 2, \\ [[e_0 - 3e_1, 2^{4-r} e_1]] & \text{if } r < 2. \end{cases}$$

It is now easy to see that $S \cap S' = \emptyset$. It follows that $d(S, S') = 2$, the geodesic between S and S' being formed by the vertices
(6A.3.8)
$$S \ni [e_0, e_1] = [e_0 + e_1, e_1] \supset [e_0 + e_1, 2e_1] \supset [e_0 + e_1, 2^2 e_1] \in S'.$$

□

6A.4 CALCULATIONS

The principle of the proof of Theorem 6A.1.2 is the same as that of Theorem 6.1 of [3], which is as follows. We write $(Z(j), Z(j'))$ as the sum of two terms, namely of

$$(Z(j)^h, Z(j')^h) + (Z(j)^h, Z(j')^v) + (Z(j)^v, Z(j')^h) \tag{6A.4.1}$$

and of $(Z(j)^v, Z(j')^v)$. For the first term we use the Genestier equations and the information on $Z(j)^h$ and $Z(j')^h$ encoded in the cycle data of $Z(j)$ and $Z(j')$. The second term is purely combinatorial and only depends on the first two entries of the cycle data for $Z(j)$ and $Z(j')$. This allows us to dispose immediately of $(Z(j)^v, Z(j')^v)$ in the cases which are termed classical in the list, i.e., **(1):(4)**, **(3):(4)**, and **(4):(4)**. Indeed, these cases have already been dealt with in Section 6 of [3], except that there the multiplicities μ and μ' of the central components of $Z(j)^v$ and $Z(j')^v$ were equal to α and β with $\alpha \leq \beta$. Similarly, the calculation of the last two summands of (6A.4.1) is as in Section 6 of [3]. This gives the result claimed in Case **(1):(4)**. For the term $(Z(j)^h, Z(j')^h)$ in the Cases **(3):(4)** and **(4):(4)** we have to use the Genestier equation — and here the result is different from the case $p \neq 2$.

Case (3):(4). In this case $(Z^h(j), Z^h(j')) = 2$ (and not 1, as in the case $p \neq 2$). Indeed, let us choose coordinates as in (A.14) of the appendix to section 11 of [4] (relative to j) for

$$\Lambda_0 \underset{\neq}{\supset} \Lambda_1 \underset{\neq}{\supset} 2\Lambda_0. \tag{6A.4.2}$$

Then $j = p^{\frac{\alpha}{2}}\, \bar{j}$ with

$$\bar{j} = \begin{pmatrix} -1 & -2(1-2\lambda) \\ 1 & 1 \end{pmatrix}, \quad \text{where } \varepsilon_1 = 4\lambda - 1 \ . \tag{6A.4.3}$$

Furthermore $Z^h(j)$ equals, with $\mu = -(1-2\lambda)^{-1}$,
(6A.4.4)
$$\operatorname{Spec} W[X]/(X^2+2\mu X-2\mu) \simeq \operatorname{Spec} W[T_0, T_1]/(T_0T_1-2, T_0+2\mu-\mu T_1)$$

(where $T_0 \mapsto X$, $T_1 \mapsto \mu^{-1}\,(X + 2\mu)$). Now $j' = 2^{\frac{\beta-1}{2}}\, \bar{j}'$, where $\bar{j}'$ is given terms of the above coordinates as

$$\bar{j}' = \begin{pmatrix} 2a_0 & 2b_0 \\ c & -2a_0 \end{pmatrix}, \tag{6A.4.5}$$

where b_0 and c are units; cf. (A.17) of the Appendix to Section 11 of [3]. Then $Z^h(j')$ is given by the equation $b_0T_0 - 4a_0 - cT_1 = 0$ in

Spec $W[T_0, T_1]/(T_0T_1 - 2)$. Hence $(Z^h(j), Z^h(j'))$ equals the length of the Artin ring

$$\text{(6A.4.6)} \qquad W[X]/(X^2 + 2\mu X - 2\mu, b_0 X - 4a_0 - \frac{c}{\mu}(X + 2\mu)).$$

Now $W[X]/(X^2 + 2\mu X - 2\mu) = \tilde{W}$ is the ring of integers in a ramified quadratic extension, with the residue class $\tilde{\pi}$ of X as uniformizer. Hence the length is equal to the $\tilde{\pi}$-valuation of

$$\text{(6A.4.7)} \quad (b_0 - \frac{c}{\mu})\tilde{\pi} - 2(c + 2a_0) = (b_0 + c(1 + \lambda))\tilde{\pi} - 2(c + 2a_0).$$

The condition that j anticommute with j' is given as $-2a_0 + b_0 - c\,(1 - 2\lambda) = 0$. Hence $(b_0 + c(1 + \lambda))\tilde{\pi} = 2(b_0 - a_0)\tilde{\pi}$, so that the $\tilde{\pi}$-valuation of (6A.4.7) is equal to 2, as claimed.

Case (4):(4). In this case $(Z^h(j), Z^h(j')) = 3$ (and not 1 as in the case $p \neq 2$). In this case we may choose standard coordinates as in (A.17) in the Appendix to section 11 in [4] for $\Lambda_0 \underset{\neq}{\supset} \Lambda_1 \underset{\neq}{\supset} 2\Lambda_0$. Then $j = 2^{\frac{\alpha - 1}{2}} \bar{j}$ and $j' = 2^{\frac{\beta - 1}{2}} \bar{j}'$ where

$$\text{(6A.4.8)} \qquad \bar{j} = \begin{pmatrix} 2a_0 & 2b_0 \\ c & -2a_0 \end{pmatrix}, \quad \bar{j}' = \begin{pmatrix} 2a'_0 & 2b'_0 \\ c' & -2a'_0 \end{pmatrix},$$

where b_0, c, b'_0, c' are units. Then $Z^h(j)$ equals, with $\mu = c/b_0$ and $\nu = 2a_0/c$,

$$\begin{aligned} \text{(6A.4.9)} \quad & \text{Spec } W[X]/(X^2 - 2\mu\nu X - 2\mu) \\ & \simeq \text{Spec } W[T_0, T_1]/(T_0T_1 - 2, b_0T_0 - 4a_0 - cT_1) \end{aligned}$$

(where $T_0 \mapsto X$, $T_1 \mapsto \mu^{-1}X - 2\nu$). Furthermore $Z^h(j')$ is given by the equation $b'_0T_0 - 4a'_0 - c'T_1 = 0$ in Spec $W[T_0, T_1]/(T_0T_1 - p)$. Denoting by $\tilde{\pi}$ the residue class of X in $\tilde{W} = W[X]/(X^2 - 2\mu\nu X - 2\mu)$, we therefore have to determine the $\tilde{\pi}$-order of

$$\text{(6A.4.10)} \qquad \begin{aligned} & b'_0\tilde{\pi} - 4a'_0 - c'(\mu^{-1}\tilde{\pi} - 2\nu) \\ & = (b'_0 - c'/\mu)\tilde{\pi} + 2(\nu c' - 2a_0) \\ & = c^{-1}((b'_0c - b_0c')\tilde{\pi} + 4(a_0c' - a_0)). \end{aligned}$$

The condition that j and j' anticommute is given by $4a_0a'_0 + (b_0c' + b'_0c) = 0$. Hence $(b'_0c - b_0c')\tilde{\pi} = 2(b'_0c + 4a_0a'_0)\tilde{\pi}$ and the $\tilde{\pi}$-valuation of (6A.4.10) is 3, as claimed.

We now go through the remaining cases. In the combinatorial problem of calculating $(Z(j)^v, Z(j')^v)$, we use the following terminology. By the *contribution of a vertex* $[\Lambda]$ we mean the number

$$m_{[\Lambda]}(S, \mu) \cdot (\mathbb{P}_{[\Lambda]}, Z(j')^v). \tag{6A.4.11}$$

To calculate $(Z(j)^v, Z(j')^v)$, we have to add the contributions of all vertices.

It will be useful to note the following elementary results, valid for arbitrary p:

(I) A vertex $[\Lambda]$ of multiplicity $\nu \geq 1$ in $Z(j')$ will be called *regular in* $Z(j')$ if it has one neighbor with multiplicity $\nu + 1$ and p neighbors with multiplicity $\nu - 1$. Then $(\mathbb{P}_{[\Lambda]}, Z(j')^v) = 1 - p$. If a vertex $[\Lambda]$ has multiplicity 0 in $Z(j')^v$ and has a unique neighbor with multiplicity 1, then $(\mathbb{P}_{[\Lambda]}, Z(j')^v) = 1$.

(II) We have the following summation formulae, where $\nu \leq \mu$ in the second formula,

$$\mu + \sum_{t=1}^{\mu-1} (\mu - t)(p-1)p^{t-1} = \frac{p^\mu - 1}{p - 1}. \tag{6A.4.12}$$

$$\mu + \sum_{t=1}^{\nu-1} (\mu - t)(p-1)p^{t-1} - (\mu - \nu)p^{\nu-1} = \frac{p^\nu - 1}{p - 1}. \tag{6A.4.13}$$

Case (1):(1). In this case $Z(j)^h = Z(j')^h = 0$. It remains to compute $(Z(j)^v, Z(j')^v)$. We are assuming that $\mu \leq \mu'$. We divide the vertices $[\Lambda]$ into two groups:

Group 1: $d([\Lambda], S') < d([\Lambda], S)$.

Each vertex $[\Lambda]$ with $m_{[\Lambda]}(S, \mu) > 0$ is regular for $Z(j')$. Therefore this group contributes

$$-\sum_{t=1}^{\mu-1} (\mu - t)\, 2^{t-1} = -(2^\mu - 1) + \mu. \tag{6A.4.14}$$

Here we used (I) above.

Group 2: $d([\Lambda], S) \subset d([\Lambda], S')$.

The automorphism j' acts on this group, with a unique fixed point $[\Lambda_0]$, the vertex on S closest to S'. This vertex contributes $-\mu$. The contributions of the other vertices in this group are summed according to the closest vertex in S. Therefore this sum is twice (for the symmetry)

$$-\sum_{s=1}^{\mu'-2} (2^{\min(\mu, \mu'-1-s)} - 1) + \mu. \tag{6A.4.15}$$

Therefore, if $\mu \leq \mu' - 1$, Group 2 contributes

$$-\mu + 2(-\sum_{t=1}^{\mu-1}(2^t - 1) - (\mu' - \mu - 1)\,(2^\mu - 1) + \mu)$$
$$= -\mu + 2(\mu - 1) - 4(2^{\mu-1} - 1) - 2(\mu' - \mu - 1)\,(2^\mu - 1) + 2\mu$$
$$= \mu + 2\mu' - 2(\mu' - \mu)\,2^\mu.$$

Therefore, adding the contribution of Group 1, we obtain for $\mu \leq \mu' - 1$, i.e., $\alpha < \beta$,

$$2(\mu + \mu') + 1 - (2(\mu' - \mu) + 1)\,2^\mu, \tag{6A.4.16}$$

which confirms the formula in this case.

If $\mu = \mu'$, the contribution of Group 2 equals

$$-\mu + 2(-\sum_{s=1}^{\mu-2}(2^s - 1) + \mu) = 3\mu - 2^\mu. \tag{6A.4.17}$$

Therefore, adding the contribution of Group 1, we obtain for $\mu = \mu'$, i.e., $\alpha = \beta$,

$$4\mu + 1 - 2 \cdot 2^\mu, \tag{6A.4.18}$$

which confirms the formula in this case.

Case (1):(2). In this case $Z(j)^h = 0$ and $Z(j')^h$ meets the special fiber in the central component $\mathbb{P}_{[\Lambda_0']}$ of $Z(j')$ in two ordinary special points. Therefore, since $m_{[\Lambda_0']}(S, \mu) = \mu - 1$,

$$(Z(j)^h, Z(j')^h) + (Z(j)^h, Z(j')^v) + (Z(j')^h, Z(j)^v) = 2(\mu - 1). \tag{6A.4.19}$$

To calculate $(Z(j)^v, Z(j')^v)$, we divide the vertices into two groups:

Group 1: $d([\Lambda], [\Lambda_0']) < d([\Lambda], S)$.

The contribution of $[\Lambda_0']$ is $-3(\mu - 1)$. The other vertices in group 1 contribute

$$\begin{cases} -\sum_{t=1}^{\mu-2}(\mu - 1 - t)\,2^t & \text{if } \mu \leq \mu' + 1, \\ -\sum_{t=1}^{\mu'-1}(\mu - 1 - t)\,2^t + (\mu - \mu' - 1)\,2^{\mu'} & \text{if } \mu > \mu' + 1. \end{cases} \tag{6A.4.20}$$

Group 2: $d([\Lambda], [\Lambda_0']) > d([\Lambda], S)$.

The automorphism j' acts on S with a unique fixed vertex $[\Lambda_0]$ characterized by $d([\Lambda_0], [\Lambda_0']) = 1$. The contribution of $[\Lambda_0]$ is equal to $-\mu$. The contributions of the other vertices in this group are summed according to the vertex in S closest to the given vertex. Therefore this sum is twice (for the symmetry) the sum

$$-\sum_{s=1}^{\mu'-2}(2^{\min(\mu,\mu'-1-s)}-1)+\mu. \tag{6A.4.21}$$

This last sum is equal to
(6A.4.22)

$$\begin{cases} -(\mu'-\mu-1)\,(2^{\mu}-1)-\sum_{s=1}^{\mu-1}(2^{s}-1)+\mu & \text{if } \mu \le \mu'-1,\\ -\sum_{s=1}^{\mu'-2}(2^{s}-1)+\mu & \text{if } \mu > \mu'-1.\end{cases}$$

Using (II) above, this is equal to
(6A.4.23)

$$\begin{cases} -(\mu'-\mu-1)\,(2^{\mu}-1)+(\mu-1)-2(2^{\mu-1}-1)+\mu & \text{if } \mu \le \mu'-1,\\ (\mu'-2)-2(2^{\mu'-2}-1)+\mu & \text{if } \mu > \mu'-1.\end{cases}$$

The total for $(Z(j), Z(j'))$ is for $\mu \le \mu' - 1$, i.e., $\alpha < \beta$, equal to

$$\begin{aligned} 2(\mu{-}1)-3(\mu-1)-\sum_{t=2}^{\mu-1}(\mu-t)\,2^{t-1}-\mu& \\ &= -2(\mu'-\mu-1)\,(2^{\mu}-1)+2(\mu-1)-4\,(2^{\mu-1}-1)+2\mu\\ &= -(2^{\mu}-1)-2(\mu'-\mu-1)\,(2^{\mu}-1)+4\mu-2(2^{\mu}-1)\\ &= 4\mu-(3+2(\mu'-\mu-1))\,(2^{\mu}-1). \end{aligned}$$

This equals

$$2(\mu+\mu')+1-(2(\mu'-\mu)+1)\,2^{\mu}, \tag{6A.4.24}$$

which confirms the given formula in this case (recall $\mu = \frac{\alpha}{2}+1$, $\mu' = \frac{\beta}{2}+1$).

For $\mu = \mu'$, i.e., $\alpha = \beta$, we obtain

$$-(2^{\mu}-1)+2(\mu-2)-4(2^{\mu-2}-1)+2\mu = 4\mu+1-2\cdot 2^{\mu}, \tag{6A.4.25}$$

which confirms the formula in this case.

For $\mu = \mu' + 1$, i.e., $\alpha = \beta + 1$, we obtain

$$\begin{aligned} 2(\mu-1)-3(\mu-1)-\sum_{t=2}^{\mu-1}(\mu-t)\,2^{t-1}& \\ &-\mu+2(\mu-3)-4(2^{\mu-3}-1)+2\mu\\ &= -(2^{\mu}-1)+2(\mu-3)-4(2^{\mu-3}-1)+2\mu. \end{aligned}$$

This equals

$$4\mu - 1 - 2^{\mu-1} - 2^{\mu}, \tag{6A.4.26}$$

which confirms the formula in this case.

For $\mu > \mu' + 1$, i.e., $\alpha > \beta$, the total for $(Z(j), Z(j'))$ is equal to

$$\begin{aligned} 2(\mu-1) - 3(\mu-1) - \sum_{t=2}^{\mu'} (\mu - t)\, 2^{t-1} \\ + (\mu - \mu' - 1)\, 2^{\mu'} - \mu + 2(\mu' - 2) - 4(2^{\mu'-2} - 1) + 2\mu \\ = -(2^{\mu'+1} - 1) + 2(\mu' - 2) - 4(2^{\mu'-2} - 1) + 2\mu. \end{aligned}$$

This equals

$$2(\mu + \mu') + 1 - 2^{\mu'} - 2^{\mu'+1}, \tag{6A.4.27}$$

which confirms the formula in this case.

Case (1):(3). In this case $Z(j)^h = 0$ and $Z(j')$ is the spectrum of a ramified quadratic extension of W which passes through the double point $\mathrm{pt}_{[\Lambda_0'],[\Lambda_1']}$. Hence

$$(Z(h)^h, Z(j')^h) + (Z(j)^h, Z(j')^v) + (Z(j')^h, Z(j)^v) = 2\mu - 1 \ . \tag{6A.4.28}$$

To calculate $(Z(j)^v, Z(j')^v)$, we first note that the contribution of $[\Lambda_0'] \in S$ is equal to -2μ and the contribution of $[\Lambda_1']$ is equal to $-2(\mu - 1)$. The remaining vertices are divided into two groups.

Group 1: $0 < d([\Lambda], [\Lambda_1']) < d([\Lambda], [\Lambda_0'])$.

These vertices contribute

$$\begin{cases} -\sum_{t=2}^{\mu-1} (\mu - t)\, 2^{t-1} & \text{if } \mu \le \mu' + 1, \\ -\sum_{t=2}^{\mu'} (\mu - t)\, 2^{t-1} + (\mu - (\mu' + 1))\, 2^{\mu'} & \text{if } \mu > \mu' + 1. \end{cases} \tag{6A.4.29}$$

Group 2: $0 < d([\Lambda], [\Lambda_0']) < d([\Lambda], [\Lambda_1'])$.

Here we sum the contributions of the lattices according to the closest vertex on S. We obtain, taking into account the involution of S induced by j',

$$-2 \sum_{s=1}^{\mu'-1} (2^{\min(\mu, \mu' - s)} - 1) + 2\mu. \tag{6A.4.30}$$

This equals
(6A.4.31)

$$\begin{cases} -2\,(\mu' - \mu)\,(2^{\mu} - 1) - 2\,\sum_{s=1}^{\mu-1}(2^s - 1) + 2\mu & \text{if } \mu \leq \mu', \\ -2\,\sum_{s=1}^{\mu'-1}(2^s - 1) + 2\mu & \text{if } \mu \geq \mu' + 1. \end{cases}$$

For $\mu \leq \mu'$, the total for $(Z(j), Z(j'))$ is equal to

$$2\mu - 1 - 2\mu - 2(\mu - 1) - \sum_{t=2}^{\mu-1}(\mu - t)\,2^{t-1} - 2(\mu' - \mu)(2^{\mu} - 1) \\ - 2\sum_{s=1}^{\mu-1}(2^s - 1) + 2\mu.$$

This equals

$$2(\mu + \mu') + 3 - (2(\mu' - \mu) + 3)\,2^{\mu}, \tag{6A.4.32}$$

which confirms the formula in this case (note that $\mu = \frac{\alpha}{2} + 1$, $\mu' = \frac{\beta}{2}$ and that $\mu \leq \mu' \Leftrightarrow \alpha < \beta$).

If $\mu = \mu' + 1$, i.e., $\alpha = \beta$, the total is

$$2\mu - 1 - 2\mu - 2(\mu - 1) - \sum_{t=2}^{\mu-1}(\mu - t)\,2^{t-1} - 2\sum_{s=1}^{\mu-2}(2^s - 1) + 2\mu \\ = -(2^{\mu} - 1) + 2(\mu - 2) - 4(2^{\mu-2} - 1) + 2\mu.$$

This equals

$$4\mu + 1 - 2 \cdot 2^{\mu}, \tag{6A.4.33}$$

which confirms the formula in this case.

If $\mu > \mu' + 1$, i.e., $\alpha > \beta$, the total is

$$-2\mu + 1 - \sum_{t=2}^{\mu'}(\mu - t)\,2^{t-1} + (\mu - (\mu' + 1))\,2^{\mu'} - 2\sum_{s=1}^{\mu'-1}(2^s - 1) + 2\mu \\ = -(2^{\mu'+1} - 1) + 2(\mu' - 1) - 4(2^{\mu'-1} - 1) + 2\mu.$$

This equals

$$2(\mu + \mu') + 3 - 2 \cdot 2^{\mu'+1}, \tag{6A.4.34}$$

which confirms the formula in this case.

Case (2):(2). In this case $S = [\Lambda_0]$ and $S' = [\Lambda_0']$ with $d([\Lambda_0], [\Lambda_0']) = 1$. Now $Z(j)^h$ meets $Z(j')^v$ in the two ordinary special points of $\mathbb{P}_{[\Lambda_0]}$ and $Z(j')^h$ meets $Z(j)^v$ in the two ordinary special points of $\mathbb{P}_{[\Lambda_0']}$. Hence
(6A.4.35)

$$(Z(j)^h, Z(j')^h)+(Z(j)^h, Z(j')^v)+(Z(j)^v, Z(j')^h) = 2(\mu-1)+2(\mu'-1).$$

To calculate $(Z(j)^v, Z(j')^v)$ we divide the vertices into two groups. We may assume $\mu \leq \mu'$ by symmetry.

Group 1: $0 \leq d([\Lambda], [\Lambda_0]) < d([\Lambda], [\Lambda_0'])$.

This group contributes
(6A.4.36)

$$\begin{cases} -\mu - 2\sum_{t=1}^{\mu-1}(\mu - t)\, 2^{t-1} & \text{if } \mu \leq \mu' - 1 \\ -\mu - \sum_{t=1}^{\mu-2}(\mu - t)\, 2^{t} + (\mu - (\mu - 1))\, 2^{\mu-1} & \text{if } \mu = \mu'. \end{cases}$$

The quantity for $\mu = \mu'$ is equal to
(6A.4.37)

$$\mu - 2(\mu + \sum_{t=1}^{\mu-2}(\mu - t)\, 2^{t-1} - (\mu - (\mu - 1))\, 2^{\mu-2}) = \mu - 2(2^{\mu-1} - 1).$$

Hence the contribution of this group is equal to

$$\begin{cases} \mu - 2\,(2^{\mu} - 1) & \text{if } \mu \leq \mu' - 1, \\ \mu - 2(2^{\mu-1} - 1) & \text{if } \mu = \mu'. \end{cases} \tag{6A.4.38}$$

Group 2: $0 \leq d([\Lambda], [\Lambda_0'] < d([\Lambda], [\Lambda_0])$.

Here $[\Lambda_0']$ contributes $-3(\mu - 1)$. The remaining vertices in this group contribute

$$-\sum_{t=2}^{\mu-1}(\mu - t)\, 2^{t-1}. \tag{6A.4.39}$$

Hence this group contributes

$$-\mu + 2 - (2^{\mu} - 1). \tag{6A.4.40}$$

For the total, we obtain in case $\mu \leq \mu' - 1$, i.e., $\alpha < \beta$,

$$\begin{aligned} 2(\mu - 1) + 2(\mu' - 1) + \mu - 2(2^{\mu} - 1) - \mu + 2 - (2^{\mu} - 1) \\ = 2(\mu + \mu') + 1 - 3 \cdot 2^{\mu}, \end{aligned} \tag{6A.4.41}$$

which yields the formula in this case. When $\mu = \mu'$, i.e., $\alpha = \beta$, we get

$$\begin{aligned} 4(\mu - 1) + \mu - 2(2^{\mu-1} - 1) - \mu + 2 - (2^{\mu} - 1) \\ = 4\mu + 1 - 2 \cdot 2^{\mu}, \end{aligned} \tag{6A.4.42}$$

which confirms the formula in this case.

Case (2):(3). In this case $Z(j)^h$ is unramified and meets the special fiber in two ordinary special points of $\mathbb{P}_{[\Lambda_0]}$, and $Z(j')^h$ is ramified and meets the special fiber in the double point $\mathrm{pt}_{[\Lambda_0],[\Lambda_1']}$. Hence

$$(Z(j)^h, Z(j')^h) + (Z(j)^h, Z(j')^v) + (Z(j)^v, Z(j')^h) \\ = 2\mu' + 2\mu - 1 = 2(\mu + \mu') - 1. \tag{6A.4.43}$$

To calculate $(Z(j)^v, Z(j)^v)$, we note that the contribution of $[\Lambda_1']$ is $-2(\mu - 1)$ and that the contribution of $[\Lambda_0]$ is -2μ. The remaining vertices are divided into two groups.

Group 1: $0 < d([\Lambda], [\Lambda_1']) < d([\Lambda], [\Lambda_0])$.
These vertices contribute

$$\begin{cases} -\sum_{t=2}^{\mu-1} (\mu - t) 2^{t-1} & \text{if } \mu - 1 \leq \mu', \\ -\sum_{t=2}^{\mu'} (\mu - t)\, 2^{t-1} + (\mu - (\mu' + 1))\, 2^{\mu'} & \text{if } \mu - 1 > \mu'. \end{cases} \tag{6A.4.44}$$

Group 2: $0 < d([\Lambda], [\Lambda_0]) < d([\Lambda], [\Lambda_1'])$.
These vertices contribute

$$\begin{cases} -2\, \sum_{t=1}^{\mu-1} (\mu - t)\, 2^{t-1} & \text{if } \mu \leq \mu', \\ -2\, \sum_{t=1}^{\mu'-1} (\mu - t)\, 2^{t-1} + 2(\mu - \mu')\, 2^{\mu'-1} & \text{if } \mu > \mu'. \end{cases} \tag{6A.4.45}$$

The total for $(Z(j), Z(j'))$ for $\mu \leq \mu'$, i.e., $\alpha < \beta$, is equal to

$$2(\mu + \mu') - 1 - 4\mu + 2 - \sum_{t=2}^{\mu-1} (\mu - t)\, 2^{t-1} - 2 \sum_{t=1}^{\mu-1} (\mu - t)\, 2^{t-1} \\ = 2(\mu + \mu') - (2^\mu - 1) - 2(2^\mu - 1).$$

This equals

$$2(\mu + \mu') + 3 - 3 \cdot 2^\mu, \tag{6A.4.46}$$

which confirms the formula in this case.
For $\mu = \mu' + 1$, i.e., $\alpha = \beta$, the total is

$$2(\mu + \mu') - 1 - 4\mu + 2 - \sum_{t=2}^{\mu-1} (\mu - t)\, 2^{t-1} \\ - 2 \sum_{t=1}^{\mu'-1} (\mu - t)\, 2^{t-1} + 2(\mu - \mu')2^{\mu'-1} \\ = 2(\mu + \mu') - (2^\mu - 1) - 2(2^{\mu-1} - 1).$$

This equals

$$2(\mu+\mu')+3-2\cdot 2^{\mu}, \tag{6A.4.47}$$

which confirms the formula in this case.

For $\mu-1>\mu'$, i.e., $\alpha>\beta$, the total is

$$2(\mu+\mu')-1-4\mu+2-\sum_{t=2}^{\mu'}(\mu-t)\,2^{t-1}+(\mu-(\mu'+1))\,2^{\mu'} \\ -2\sum_{t=1}^{\mu'-1}(\mu-t)\,2^{t-1}+2(\mu-\mu')\,2^{\mu'-1}.$$

This equals

$$2(\mu+\mu')+3-4\cdot 2^{\mu'}, \tag{6A.4.48}$$

which confirms the formula in this case.

Case $T=-2^{\alpha}\begin{pmatrix}1 & \frac{1}{2}\\ \frac{1}{2} & 1\end{pmatrix}$**,** α **even**. In this case $Z(j)^h=Z(j')^h=0$. To calculate $(Z(j)^v, Z(j')^v)$, we divide the vertices into three groups.

Group1: $d([\Lambda],S)=d([\Lambda],S')$.

The contribution of these vertices is

$$-(\mu-1)-\sum_{t=1}^{\mu-2}(\mu-1-t)\,2^{t-1}=-(2^{\mu-1}-1). \tag{6A.4.49}$$

Group 2: $d([\Lambda],S')<d([\Lambda],S)$.

The vertex $[\Lambda_0']\in S'$ closest to S contributes $-(\mu-2)$. The remaining vertices are grouped according to the closest vertex on S'. If that vertex has distance $t\geq 1$ from $[\Lambda_0']$, the contribution is equal to

$$-(\mu-2-t)-\sum_{s=1}^{\mu-2-t-1}(\mu-2-t-s)\,2^{s-1}=-(2^{\mu-2-t}-1). \tag{6A.4.50}$$

We see that the group 2 contributes

$$-(\mu-2)-2\sum_{t=1}^{\mu-3}(2^{\mu-2-t}-1)=-(\mu-2)-2\sum_{t=1}^{\mu-3}(2^{t}-1). \tag{6A.4.51}$$

Group 3: $d([\Lambda],S)<d([\Lambda],S')$.

The vertex $[\Lambda_0] \in S$ closest to S' contributes $-\mu$. The remaining vertices are grouped according to the closest vertex on S. If that vertex on S has distance t with $1 \leq t \leq \mu - 3$ from $[\Lambda_0]$, the contribution is equal to

$$-\mu - \sum_{s=1}^{\mu-2-t-1} (\mu - s)\, 2^{s-1} + (\mu - (\mu - 2 - t))\, 2^{\mu-2-t} = -(2^{\mu-2-t} - 1). \tag{6A.4.52}$$

Finally a vertex on S with distance $\mu - 2$ from $[\Lambda_0]$ contributes μ. We see that Group 3 contributes

$$-\mu - 2\sum_{t=1}^{\mu-3}(2^{\mu-2-t} - 1) + 2\mu = \mu - 2\sum_{t=1}^{\mu-3}(2^t - 1). \tag{6A.4.53}$$

The total is

$$-(2^{\mu-1} - 1) - (\mu - 2) - 2\sum_{t=1}^{\mu-3}(2^t - 1) + \mu - 2\sum_{t=1}^{\mu-3}(2^t - 1)$$

which equals

$$4\mu - 1 - 3 \cdot 2^{\mu-1}, \tag{6A.4.54}$$

which confirms the formula in this case.

6A.5 THE FIRST NONDIAGONAL CASE

In this section we consider the case $T = -2^\alpha \begin{pmatrix} 0 & \frac{1}{2} \\ \frac{1}{2} & 0 \end{pmatrix}$. This case is different from the above cases in that the endomorphisms j and j' do not anticommute and, while being $\neq 0$, satisfy $q(j) = q(j') = 0$. The case where $j \in V'$ has $q(j) = 0$ was excluded from [3], therefore we have to describe first the special cycle $Z(j)$ in $\mathcal{M}$ in this case. The method is the same as in [3], therefore we will be brief.

Let, thus, $j \in V'$ be nonzero with $q(j) = 0$. The proof of Proposition 2.2 of [3] carries over to show that $Z(j) \cap \mathbb{P}_{[\Lambda]} \neq \emptyset$ iff $j(\Lambda) \subset \Lambda$, and in fact $\mathbb{P}_{[\Lambda]}$ then occurs in $Z(j)$ with multiplicity

$$\text{mult}_{[\Lambda]}(j) = \max\{r \geq 0 \mid j(\Lambda) \subset 2^r \Lambda\}. \tag{6A.5.1}$$

Indeed, this follows from the local equation for $Z(j)$ along $\hat{\Omega}_{[\Lambda]}$, (see [3], equation (3.13)): if $j = 2^r \begin{pmatrix} a & b \\ c & -a \end{pmatrix}$ with $a, b, c \in \mathbb{Z}_2$ not all simultaneously divisible by 2, then since $\det(j) = 0$, not both b and c are divisible by

2, so that the second factor in (3.13) of [3], which is $bT^2 - 2aT - c$, is not divisible by 2.

Furthermore, the horizontal component $Z(j)^h$ is trivial. This can be proved by considering the local equations as in section 3 of [3]; the global argument of Remark 8.2 of [3], that 0 is not represented by a division algebra, can also be applied.

We choose a basis e_1, e_2 of $\mathbb{Q}_2^2$ such that

$$j = \begin{pmatrix} 0 & 1 \\ 0 & 0 \end{pmatrix}. \tag{6A.5.2}$$

Then for $\Lambda_r = [e_1, 2^r e_2]$ on the standard apartment we have $\mathbb{P}_{[\Lambda_r]} \cap Z(j) \neq \emptyset$ iff $r \geq 0$ and then $\mathrm{mult}_{[\Lambda_r]}(j) = r$. Conversely, let Λ be such that $j(\Lambda) \subset \Lambda$, i.e., $\mathbb{P}_{[\Lambda]} \cap Z(j) \neq \emptyset$. Then $\Lambda \cap \mathrm{Ker}\, j$ is generated by a multiple $e_1' = \lambda e_1$ of e_1. Furthermore there is $e_2' \in \Lambda$ such that $j(e_2') = p^r e_1'$ and such that $\Lambda = [e_1', e_2']$. Then $r = \mathrm{mult}_{[\Lambda]}(j)$. In other words, Λ is the r-th vertex on the apartment corresponding to the basis e_1', e_2'; and this apartment shares a half-apartment with the standard apartment. For the multiplicity $\mathrm{mult}_{[\Lambda]}(j)$ there is the formula

$$\mathrm{mult}_{[\Lambda]}(j) = \mathrm{dist}([\Lambda], \partial S) + 1. \tag{6A.5.3}$$

Here $S = \{[\Lambda] \mid \mathrm{mult}_{[\Lambda]}(j) > 0\}$ and $[\Lambda] \in S$. We therefore obtain the picture for $Z(j)$ shown in the frontispiece of this book.

Now let $j, j' \in V'$ with matrix $T = -2^\alpha \begin{pmatrix} 0 & \frac{1}{2} \\ \frac{1}{2} & 0 \end{pmatrix}$, i.e., $q(j) = q(j') = 0$ and $jj' + j'j = 2^\alpha$. Let e_2 be a generator of $\ker(j')$ and let $e_1 = j(e_2)$. Then e_1 is a generator of $\ker(j)$. Thus e_1, e_2 is a basis of $\mathbb{Q}_2^2$ defining an apartment $\mathcal{A}$ with one half-apartment belonging to j (in the sense made clear above) and the opposite half-apartment belonging to j', i.e.,

$$j = \begin{pmatrix} 0 & 1 \\ 0 & 0 \end{pmatrix}, \quad j' = \begin{pmatrix} 0 & 0 \\ 2^\alpha & 0 \end{pmatrix}. \tag{6A.5.4}$$

It is immediate that

$$S \cap \mathcal{A} = \{\Lambda_r \mid r > 0\}, \quad S' \cap \mathcal{A} = \{\Lambda_r \mid r < \alpha\}. \tag{6A.5.5}$$

All vertices in $S' \setminus \partial S'$ are regular for $Z(j')$. Therefore the contribution of a vertex $[\Lambda]$ which is joined by a geodesic of length t to $\mathcal{A}$ and meets there the vertex $[\Lambda_r]$ is equal to

$$\begin{cases} -(r-t) & \text{if } t \leq r \text{ and } t \leq \alpha - r - 1, \\ r - t & \text{if } t \leq r \text{ and } t = \alpha - r, \\ 0 & \text{in all other cases.} \end{cases} \tag{6A.5.6}$$

Therefore the sum of the contributions of vertices with $[\Lambda_r]$ as the closest vertex on $\mathcal{A}$ is equal to

(6A.5.7)

$$\begin{cases} -r - \sum_{t=1}^{\alpha-r-1}(r-t)\,2^{t-1} + (r-(r-\alpha))\,2^{r-\alpha-1} & \text{if } \frac{\alpha}{2} \le r < \alpha, \\ -r - \sum_{t=1}^{r-1}(r-t)\,2^{t-1} & \text{if } 0 < r < \frac{\alpha}{2}, \\ \alpha & \text{if } r = \alpha. \end{cases}$$

This equals

$$\text{(6A.5.8)} \qquad \begin{cases} -(2^{\alpha-r}-1) & \text{if } \frac{\alpha}{2} \le r < \alpha, \\ -(2^r-1) & \text{if } 0 < r < \frac{\alpha}{2}, \\ \alpha & \text{if } r = \alpha. \end{cases}$$

If α is even, the total contribution is equal to

$$\begin{aligned} -\sum_{s=\frac{\alpha}{2}}^{\alpha-1}(2^{\alpha-s}-1) - \sum_{s=1}^{\frac{\alpha}{2}-1}(2^s-1) + \alpha \\ = -\sum_{s=1}^{\frac{\alpha}{2}}(2^s-1) - \sum_{s=1}^{\frac{\alpha}{2}-1}(2^s-1) + \alpha \\ = -2\sum_{s=1}^{\frac{\alpha}{2}-1}(2^s-1) - (2^{\frac{\alpha}{2}}-1) + \alpha. \end{aligned}$$

This equals

$$\text{(6A.5.9)} \qquad 2\alpha + 3 - 3\cdot 2^{\frac{\alpha}{2}},$$

which confirms the formula in this case.

If α is odd, the total contribution is equal to

$$\text{(6A.5.10)} \quad -\sum_{s=\frac{\alpha+1}{2}}^{\alpha-1}(2^{\alpha-s}-1) - \sum_{s=1}^{\frac{\alpha-1}{2}}(2^s-1) + \alpha = -2\sum_{s=1}^{\frac{\alpha-1}{2}}(2^s-1) + \alpha.$$

This equals

$$\text{(6A.5.11)} \qquad 2\alpha + 3 - 2\cdot 2^{\frac{\alpha+1}{2}},$$

which confirms the formula in this case.

Bibliography

[1] ARGOS (Arithmetische Geometrie Oberseminar), Proceedings of the Bonn seminar 2003/04, forthcoming.

[2] I. Bouw, *Invariants of ternary quadratic forms*, in [1].

[3] S. Kudla and M. Rapoport, *Height pairings on Shimura curves and p-adic uniformization*, Invent. math., **142** (2000), 153–223.

[4] S. Kudla, M. Rapoport, and T. Yang, *Derivatives of Eisenstein series and Faltings heights*, Compositio Math., **140** (2004), 887–951.

[5] T. Yang, *Local densities of 2-adic quadratic forms*, J. Number Theory, **108** (2004), 287–345.

Chapter Seven

An inner product formula

In Chapter 4 we defined the $\widehat{\mathrm{CH}}^1(\mathcal{M})$-valued generating function $\widehat{\phi}_1(\tau) = \sum_{t\in\mathbb{Z}} \widehat{\mathcal{Z}}(t,v)\, q^t$ and showed its modularity. It follows that the height pairing $\langle\, \widehat{\phi}_1(\tau_1), \widehat{\phi}_1(\tau_2)\,\rangle$ of two of these generating functions is then a series with coefficients in $\mathbb{R}$ which is a modular form of two variables, of weight $\frac{3}{2}$ in each. Here we are identifying, as in Chapter 2, $\widehat{\mathrm{CH}}^2(\mathcal{M})$ with $\mathbb{R}$ via the arithmetic degree map. On the other hand, in the previous chapter, we defined a generating series for 0-cycles with coefficients in $\widehat{\mathrm{CH}}^2(\mathcal{M})$ and proved that it is a Siegel modular form of genus two and weight $\frac{3}{2}$. This was done by calculating explicitly all coefficients of this generating series and showing that they coincide term by term with the corresponding coefficients of the Siegel Eisenstein series considered in Chapter 5.

In the present chapter we prove an inner product formula which asserts that after pulling back the generating series for 0-cycles under the diagonal map, it coincides with the inner product of the generating function for divisors with itself under the height pairing. Recall that there seemed to be some arbitrariness in which the coefficients of the generating series for 0-cycles was defined in Chapter 6, outside the case when the index $T \in \mathrm{Sym}_2(\mathbb{Z})^\vee$ is positive definite and good. Therefore we may view the inner product formula as an additional justification of these definitions, since this formula shows their coherence. It should be stressed that this result is purely geometric, and that formulas for the Fourier coefficients of Eisenstein series do not enter into its proof. In fact, it turns out that we do not need the explicit expressions that we obtained for the coefficients, but we do need the way in which these calculations were organized.

When forming the inner product of $\widehat{\phi}_1(\tau)$ with itself, there are the pairings of those special divisors $\mathcal{Z}(t,v)$ which have no common intersection in the generic fiber. For this part of the intersection product, the claimed formula turns out to be essentially equivalent to the decomposition of the intersection of two such special divisors according to their 'fundamental matrices'; see [3], (8.24). This part of the formula is therefore essentially tautological. For the special divisors which do have an intersection in the generic fiber, we use the adjunction formula in the form of Section 2.10. For this we construct in Section 7.3 a weakly biadmissible Green function on $\mathfrak{H} \times \mathfrak{H} \setminus \Delta$ and show that the induced metric on the sheaf of differentials on $\mathcal{M}(\mathbb{C})$ coincides exactly with the metric on ω defined in [4], including the mysterious constant $2C = \gamma + \log(4\pi)$. To apply then the adjunction formula, we decompose the cycles into irreducible components according to the conductor of the order generated by the special endomorphism (the 'type' of the special endomorphism). To calculate all the terms occurring in the adjunction formula, there are archimedean calculations similar to those in [4]. The calculations at the finite primes lead to Gross's theory of quasi-canonical liftings and the study of their ramification behavior.

7.1 STATEMENT OF THE MAIN RESULT

Let

$$\widehat{\phi}_1(\tau) = \sum_{t \in \mathbb{Z}} \widehat{\mathcal{Z}}(t, v)\, q^t \tag{7.1.1}$$

be the generating function for divisors on $\mathcal{M}$.[1] In Chapter 4, it was proved that this is a modular form of weight $\frac{3}{2}$, valued in $\widehat{\mathrm{CH}}^1(\mathcal{M})$.

On the other hand, in Chapter 6 we introduced the generating function for 0-cycles

$$\widehat{\phi}_2(\tau) = \sum_{T \in \mathrm{Sym}_2(\mathbb{Z})^\vee} \widehat{\mathcal{Z}}(T, v)\, q^T \tag{7.1.2}$$

and have proved that this is a Siegel modular form[2] of weight $\frac{3}{2}$. Therefore, the pullback of this function under the diagonal map

$$\mathfrak{H} \times \mathfrak{H} \longrightarrow \mathfrak{H}_2, \qquad (\tau_1, \tau_2) \mapsto \begin{pmatrix} \tau_1 & \\ & \tau_2 \end{pmatrix} \tag{7.1.3}$$

is a modular form of two variables, of weight $\frac{3}{2}$ in each of them. In this chapter we will prove that this pullback coincides with the inner product of $\widehat{\phi}_1$ with itself under the height pairing.

Theorem C. *For τ_1 and $\tau_2 \in \mathfrak{H}$,*

$$\langle\, \widehat{\phi}_1(\tau_1), \widehat{\phi}_1(\tau_2)\,\rangle = \widehat{\phi}_2\Big(\begin{pmatrix} \tau_1 & \\ & \tau_2 \end{pmatrix}\Big).$$

---

[1] We are using the convention that $\mathcal{Z}(t)$ denotes both the moduli stack of triples (A, ι, x) and its image in $\mathcal{M}$, viewed as a divisor.

[2] Recall that, as explained in Chapter 2, we identify $\widehat{\mathrm{CH}}^2(\mathcal{M})$ with $\mathbb{R}$ via $\widehat{\deg}$.

Equivalently, for all t_1, $t_2 \in \mathbb{Z}$, *and* v_1, $v_2 \in \mathbb{R}_{>0}$,

$$((\star)) \qquad \langle\, \widehat{\mathcal{Z}}(t_1, v_1), \widehat{\mathcal{Z}}(t_2, v_2)\,\rangle = \sum_{\substack{T \in \mathrm{Sym}_2(\mathbb{Z})^\vee \\ \mathrm{diag}(T)=(t_1,t_2)}} \widehat{\mathcal{Z}}(T, \begin{pmatrix} v_1 & \\ & v_2 \end{pmatrix}).$$

Here

$$(7.1.4) \quad \mathrm{Sym}_2(\mathbb{Z})^\vee = \{\, T \in \mathrm{Sym}_2(\mathbb{Q}) \mid \mathrm{tr}(Tb) \in \mathbb{Z}, \quad \forall b \in \mathrm{Sym}_2(\mathbb{Z})\,\}.$$

To prove the equality $((\star))$ of Theorem C, we first consider the case $t_1 t_2 \neq 0$. We write

$$(7.1.5) \qquad \mathcal{Z}(t_i) = \mathcal{Z}^{\mathrm{hor}}(t_i) + \mathcal{Z}^{\mathrm{ver}}(t_i),$$

where $\mathcal{Z}^{\mathrm{hor}}(t_i)$ is the closure in $\mathcal{M}$ of the generic fiber $\mathcal{Z}^{\mathrm{hor}}(t_i)_{\mathbb{Q}}$ and where $\mathcal{Z}^{\mathrm{ver}}(t_i)$ is a divisor supported in the fibers $\mathcal{M}_p$ for some $p \mid D(B)$.[3] By equipping these cycles with the 'standard' Green functions recalled in Section 3.5, we obtain classes

$$(7.1.6) \qquad \widehat{\mathcal{Z}}(t_i, v_i) = \widehat{\mathcal{Z}}^{\mathrm{hor}}(t_i, v_i) + \widehat{\mathcal{Z}}^{\mathrm{ver}}(t_i) \in \widehat{\mathrm{CH}}^1(\mathcal{M}),$$

where

$$(7.1.7) \qquad \widehat{\mathcal{Z}}^{\mathrm{hor}}(t_i, v_i) = (\mathcal{Z}^{\mathrm{hor}}(t_i), \Xi(t_i, v_i)) \in \widehat{\mathrm{CH}}^1(\mathcal{M}),$$

and

$$(7.1.8) \qquad \widehat{\mathcal{Z}}^{\mathrm{ver}}(t_i) = (\mathcal{Z}^{\mathrm{ver}}(t_i), 0) \in \widehat{\mathrm{CH}}^1(\mathcal{M}).$$

Notice that, when $t_i < 0$, then $\mathcal{Z}(t_i) = \phi$, and we have

$$\widehat{\mathcal{Z}}(t_i, v_i) = (0, \Xi(t_i, v_i)) \in \widehat{\mathrm{CH}}^1(\mathcal{M}).$$

Such classes might be thought of as being 'vertical at infinity'. We need to compute

$$(7.1.9) \qquad \begin{aligned} \langle\, \widehat{\mathcal{Z}}(t_1, v_1), \widehat{\mathcal{Z}}(t_2, v_2)\,\rangle = {} & \langle\, \widehat{\mathcal{Z}}^{\mathrm{hor}}(t_1, v_1), \widehat{\mathcal{Z}}^{\mathrm{hor}}(t_2, v_2)\,\rangle \\ & + \langle\, \widehat{\mathcal{Z}}^{\mathrm{hor}}(t_1, v_1), \widehat{\mathcal{Z}}^{\mathrm{ver}}(t_2)\,\rangle \\ & + \langle\, \widehat{\mathcal{Z}}^{\mathrm{ver}}(t_1), \widehat{\mathcal{Z}}^{\mathrm{hor}}(t_2, v_2)\,\rangle \\ & + \langle\, \widehat{\mathcal{Z}}^{\mathrm{ver}}(t_1), \widehat{\mathcal{Z}}^{\mathrm{ver}}(t_2)\,\rangle \end{aligned}$$

[3] Here it is essential to interpret the notation $\mathcal{Z}(t_i)$ as the image divisor in $\mathcal{M}$.

in terms of the $\widehat{\mathcal{Z}}(T,v)$'s with $\mathrm{diag}(T)=(t_1,t_2)$. When t_1t_2 is not a square, so that only nonsingular T's can arise, the required identity comes down essentially to the statement that the intersection can be written as a disjoint union according to the fundamental matrices in the form as, e.g., in [3]. We recall this in Section 7.2. Then, in Sections 7.3–7.9, we handle the cases in which $t_1t_2=m^2$ and $t_i>0$. In Section 7.10, we consider the case in which $t_1t_2=m^2$ but $t_1,\, t_2<0$.

Finally, the cases where $t_1t_2=0$ are considered in Section 7.11. Here we rely on arguments involving modularity of our various generating functions combined with the identities ((⋆)) for $t_1t_2\neq 0$ to conclude that the remaining terms must agree. As a consequence, we obtain the explicit value for $\langle\,\hat{\omega},\hat{\omega}\,\rangle$, which had been conjectured in [4].

Theorem 7.1.1.

$$\langle\,\hat{\omega},\hat{\omega}\,\rangle=\zeta_{D(B)}(-1)\left[2\frac{\zeta'(-1)}{\zeta(-1)}+1-2C-\frac{1}{2}\sum_{p|D(B)}\frac{p+1}{p-1}\cdot\log(p)\right],$$

where $2C=\log(4\pi)+\gamma$, *for Euler's constant* γ.

7.2 THE CASE t_1t_2 IS NOT A SQUARE

When t_1t_2 is not a square, the cycles $\mathcal{Z}(t_1)$ and $\mathcal{Z}(t_2)$ can have horizontal components, but these components are disjoint on the generic fiber. Thus, every height pairing on the right-hand side of (7.1.9) can be written as a sum over primes p, $p\leq\infty$, of local contributions, $\langle\,\widehat{\mathcal{Z}}^{\mathrm{hor}}(t_1,v_1),\widehat{\mathcal{Z}}^{\mathrm{hor}}(t_2,v_2)\,\rangle_p$, $\langle\,\widehat{\mathcal{Z}}^{\mathrm{hor}}(t_1,v_1),\widehat{\mathcal{Z}}^{\mathrm{ver}}(t_2)\,\rangle_p$, etc. For a finite prime p, these contributions come from intersections in the fiber at p. For $p=\infty$, they come from the Green functions.

Theorem 7.2.1. *Suppose that* t_1t_2 *is not a square.*
(i) For a finite prime p *with* $p\nmid D(B)$,

$$\begin{aligned}\langle\,\widehat{\mathcal{Z}}(t_1,v_1),\widehat{\mathcal{Z}}(t_2,v_2)\,\rangle_p&=\langle\,\widehat{\mathcal{Z}}^{\mathrm{hor}}(t_1,v_1),\widehat{\mathcal{Z}}^{\mathrm{hor}}(t_2,v_2)\,\rangle_p\\&=\left(\,\mathcal{Z}^{\mathrm{hor}}(t_1),\mathcal{Z}^{\mathrm{hor}}(t_2)\,\right)_p\cdot\log p\\&=\sum_{\substack{T>0\\ \mathrm{diag}(T)=(t_1,t_2)\\ \mathrm{Diff}(T,B)=\{p\}}}\widehat{\mathcal{Z}}(T).\end{aligned}$$

Here the quantity

$$\widehat{\mathcal{Z}}(T)=\widehat{\mathcal{Z}}(T,\begin{pmatrix}v_1&\\&v_2\end{pmatrix})$$

is independent of v_1 and v_2.
(ii) For a finite prime p with $p \mid D(B)$,

$$\langle\, \widehat{\mathcal{Z}}(t_1, v_1), \widehat{\mathcal{Z}}(t_2, v_2)\,\rangle_p = \sum_{\substack{T>0 \\ \mathrm{diag}(T)=(t_1,t_2) \\ \mathrm{Diff}(T,B)=\{p\}}} \widehat{\mathcal{Z}}(T).$$

and, again, $\widehat{\mathcal{Z}}(T)$ is independent of v.
(iii) For $p = \infty$,

$$\langle\, \widehat{\mathcal{Z}}(t_1, v_1), \widehat{\mathcal{Z}}(t_2, v_2)\,\rangle_\infty = \sum_{\substack{T \\ \mathrm{diag}(T)=(t_1,t_2) \\ \mathrm{Diff}(T,B)=\{\infty\}}} \widehat{\mathcal{Z}}(T, \begin{pmatrix} v_1 & \\ & v_2 \end{pmatrix}).$$

The statement in this theorem reflects the decomposition of the intersection pairing of the $\widehat{\mathcal{Z}}(t_i, v_i)$ according to 'fundamental matrices', as is implicit in [2], Theorem 14.11 (resp. Proposition 12.5) in the case $p \nmid D(B)$ (resp. the case $p = \infty$), and explicit in [3], Theorem 8.6 in the case $p \mid D(B)$.

Sketch of Proof. For a fixed prime $p < \infty$, let B' be the definite quaternion algebra with invariants $\mathrm{inv}_\ell(B') = \mathrm{inv}_\ell(B)$ for $\ell \neq p, \infty$ and $\mathrm{inv}_p(B') = -\mathrm{inv}_p(B)$. Let $H' = (B')^\times$, and let $V' = \{x \in B' \mid \mathrm{tr}(x) = 0\}$. For a fixed maximal order $O_{B'}$ in B', we fix an isomorphism $B'(\mathbb{A}_f^p) \simeq B(\mathbb{A}_f^p)$, so that $O_B \otimes_{\mathbb{Z}} \hat{\mathbb{Z}}^p$ is identified with $O_{B'} \otimes_{\mathbb{Z}} \hat{\mathbb{Z}}^p$. We then have identifications $H(\mathbb{A}_f^p) = H'(\mathbb{A}_f^p)$ and $V(\mathbb{A}_f^p) = V'(\mathbb{A}_f^p)$, and we set $K^p = (O_B \otimes \hat{\mathbb{Z}}^p)^\times$.

First suppose that $p \nmid D(B)$. In this case, B' is ramified at p. Since $t_1 t_2$ is not a square, the cycles $\mathcal{Z}(t_1)$ and $\mathcal{Z}(t_2)$ are disjoint on the generic fiber and only meet in $\mathcal{M}_p$ at supersingular points. We have

$$\langle \hat{\mathcal{Z}}(t_1, v_1), \hat{\mathcal{Z}}(t_2, v_2)\rangle_p = (\mathcal{Z}(t_1), \mathcal{Z}(t_2))_p \cdot \log(p), \tag{7.2.1}$$

where the geometric intersection number can be written as a sum of local intersection numbers

$$(\mathcal{Z}(t_1), \mathcal{Z}(t_2))_p = \sum_{x \in \mathcal{M}_p(\bar{\mathbb{F}}_p)^{\mathrm{ss}}} (\mathcal{Z}(t_1), \mathcal{Z}(t_2))_x \ . \tag{7.2.2}$$

If $x \in \mathcal{M}_p(\bar{\mathbb{F}})^{\mathrm{ss}}$ corresponds to (A, ι), then

$$(\mathcal{Z}(t_1), \mathcal{Z}(t_2))_x = \sum_{\substack{\mathbf{y}=(y_1,y_2) \in V(A,\iota)^2 \\ Q(y_i)=t_i}} \nu_p(\mathbf{y}). \tag{7.2.3}$$

Here $\nu_p(\mathbf{y})$ denotes the length of the deformation space of the pair of endomorphisms $\mathbf{y} = (y_1, y_2)$ of (A, ι), and each point is counted with multiplicity $e_{\mathbf{y}}^{-1}$. But $\nu_p(\mathbf{y}) = \nu_p(T)$ only depends on the *fundamental matrix* $T = Q(\mathbf{y})$ of $(A, \iota, \mathbf{y})$. Therefore, rearranging the sum according to the possible fundamental matrices we obtain, as in Section 6.1,

(7.2.4)
$$(\mathcal{Z}(t_1), \mathcal{Z}(t_2))_p = \sum_{\substack{T \in \mathrm{Sym}_2(\mathbb{Z})^\vee \\ \mathrm{diag}(T)=(t_1,t_2)}} \sum_{x=(A,\iota)\in \mathcal{M}_p(\bar{\mathbb{F}}_p)^{\mathrm{ss}}} \sum_{\substack{\mathbf{y}\in V(A,\iota)^2 \\ Q(\mathbf{y})=T}} e_{\mathbf{y}}^{-1}\, \nu_p(T)$$
$$= \sum_{\substack{T \in \mathrm{Sym}_2(\mathbb{Z})^\vee \\ \mathrm{diag}(T)=(t_1,t_2)}} |\Lambda_T| \cdot \nu_p(T),$$

with

$$|\Lambda_T| = \sum_{x\in \mathcal{M}_p(\bar{\mathbb{F}}_p)^{\mathrm{ss}}} |\{\mathbf{y} \in V(A,\iota)^2;\ Q(\mathbf{y}) = T\}|, \tag{7.2.5}$$

where again the cardinality is to be taken in the stack sense, i.e., $(A, \iota, \mathbf{y})$ counts with multiplicity $1/|\mathrm{Aut}\ (A, \iota, \mathbf{y})|$. Note that since $t_1 t_2$ is not a square, the sum (7.2.4) runs only over nonsingular matrices T.

Next, we consider the contributions of terms where $p \mid D(B)$. In this case, we also have to take into account the additional terms coming from vertical components. Let

(7.2.6) $O_{B'}[p^{-1}] = O_{B'} \otimes_{\mathbb{Z}} \mathbb{Z}[p^{-1}]$, (resp. $L'[p^{-1}] = L' \otimes_{\mathbb{Z}} \mathbb{Z}[p^{-1}]$),

and let $\Gamma' = (O_{B'}[p^{-1}])^\times$. Note that Γ' can be identified with an arithmetic subgroup of $B'(\mathbb{Q}_p)^\times \simeq \mathrm{GL}_2(\mathbb{Q}_p)$.

Recall that we are assuming that $t_1 t_2$ is not a square. Then we can write as in the development leading up to [3], Theorem 8.5, (cf. also the proof of Theorem 6.2.1)

$$\langle\, \widehat{\mathcal{Z}}(t_1, v_1), \widehat{\mathcal{Z}}(t_2, v_2)\,\rangle_p$$
$$= \langle\, \widehat{\mathcal{Z}}^{\mathrm{hor}}(t_1, v_1), \widehat{\mathcal{Z}}^{\mathrm{hor}}(t_2, v_2)\,\rangle_p + \langle\, \widehat{\mathcal{Z}}^{\mathrm{hor}}(t_1, v_1), \mathcal{Z}^{\mathrm{ver}}(t_2)\,\rangle_p$$
$$+ \langle\, \mathcal{Z}^{\mathrm{ver}}(t_1), \widehat{\mathcal{Z}}^{\mathrm{hor}}(t_2, v_2)\,\rangle_p + \langle\, \mathcal{Z}^{\mathrm{ver}}(t_1), \mathcal{Z}^{\mathrm{ver}}(t_2)\,\rangle_p$$

$$= \sum_{\substack{T \in \mathrm{Sym}_2(\mathbb{Z})^\vee \\ \mathrm{diag}(T)=(t_1,t_2)}} \sum_{\substack{\mathbf{y}=[y_1,y_2]\in (L'[p^{-1}])^2 \\ Q(\mathbf{y})=T \\ \mathrm{mod}\ \Gamma'}} e_{\mathbf{y}}^{-1} \cdot \nu_p(T) \log(p), \tag{7.2.7}$$

$$= \sum_{\substack{T \in \mathrm{Sym}_2(\mathbb{Z})^\vee \\ \mathrm{diag}(T)=(t_1,t_2)}} |\Lambda_T| \cdot \nu_p(T) \log(p),$$

where $e_{\mathbf{y}}$ is the order of the stabilizer of $\mathbf{y}$ in

$$\Gamma'^{,1} = \{ \gamma \in \Gamma' \mid \mathrm{ord}_p(\nu(\gamma)) = 0 \}, \tag{7.2.8}$$

and where

$$\Lambda_T = [\, \Gamma' \backslash \{ \mathbf{y} \in (L'[p^{-1}])^2 \mid Q(\mathbf{y}) = T \} \,], \tag{7.2.9}$$

and

$$|\Lambda_T| = \sum_{\mathbf{y} \in \Lambda_T} e_{\mathbf{y}}^{-1}. \tag{7.2.10}$$

The computation is done using the p-adic uniformization of the formal completion of the special fiber $\mathcal{M}_p$. The quantity

$$\frac{1}{2}\nu_p(T) = \chi(\mathcal{Z}(\mathbf{j}), \mathcal{O}_{\mathcal{Z}(j_1)} \otimes^{\mathbb{L}} \mathcal{O}_{\mathcal{Z}(j_2)}), \tag{7.2.11}$$

which depends only on the $\mathrm{GL}_2(\mathbb{Z}_p)$-equivalence class of $T \in \mathrm{Sym}_2(\mathbb{Z}_p)$, is 'global' on the Drinfeld space $\widehat{\Omega}$. Note that when the matrix T is changed by an element of $\mathrm{GL}_2(\mathbb{Z}_p)$, the geometry of the formal cycle $\mathcal{Z}(\mathbf{j})$ changes, although $\nu_p(T)$ does not. An explicit example is given in Example 5.5 at the end of Section 5 of [3]. This fact will be relevant to us here, since later, in the case in which t_1t_2 *is* a square, we will need to compute the individual terms $\langle \widehat{\mathcal{Z}}^{\mathrm{hor}}(t_1, v_1), \widehat{\mathcal{Z}}^{\mathrm{hor}}(t_2, v_2) \rangle_p$ and $\langle \mathcal{Z}^{\mathrm{ver}}(t_1), \mathcal{Z}^{\mathrm{ver}}(t_2) \rangle_p$.

Finally, we consider the archimedean part of the height pairing in the case in which t_1t_2 is not a square. This part of the pairing is given by the star product of the Green functions. Let D be the union of the upper and the lower half planes and let $M = \mathcal{M}_{\mathbb{C}} = [\Gamma \setminus D]$. Then, as in the proof of Proposition 12.5 of [2],

$$\begin{aligned}
\langle \widehat{\mathcal{Z}}(t_1, v_1), \widehat{\mathcal{Z}}(t_2, v_2) \rangle_\infty &= \frac{1}{2} \int_M \Xi(t_1, v_1) * \Xi(t_2, v_2) \\
&= \frac{1}{2} \int_{[\Gamma \setminus D]} \sum_{\substack{x_1 \in L \\ Q(x_1)=t_1}} \sum_{\substack{x_2 \in L \\ Q(x_2)=t_2}} \xi(v_1^{\frac{1}{2}} x_1) * \xi(v_2^{\frac{1}{2}} x_2) \\
&= \sum_{\substack{T \in \mathrm{Sym}_2(\mathbb{Z})^\vee \\ \mathrm{diag}(T)=(t_1,t_2)}} \sum_{\substack{\mathbf{x}=[x_1,x_2] \in L^2 \\ Q(\mathbf{x})=T \\ \mathrm{mod}\ \Gamma}} e_{\mathbf{x}}^{-1} \cdot \nu_\infty(T, v)
\end{aligned} \tag{7.2.12}$$

$$= \sum_{\substack{T \in \mathrm{Sym}_2(\mathbb{Z})^\vee \\ \mathrm{diag}(T)=(t_1,t_2)}} |\Lambda_T| \cdot \nu_\infty(T, v),$$

where

$$\nu_\infty(T, v) = \frac{1}{2} \int_D \xi(v_1^{\frac{1}{2}} x_1) * \xi(v_2^{\frac{1}{2}} x_2), \tag{7.2.13}$$

and

$$\Lambda_T = [\, \Gamma \backslash \{ \, \mathbf{x} \in (L)^2 \mid Q(\mathbf{x}) = T \, \} \,]. \tag{7.2.14}$$

As usual,

$$|\Lambda_T| = \sum_{\substack{\mathbf{x}=[x_1,x_2] \in L^2 \\ Q(\mathbf{x})=T \\ \mathrm{mod}\ \Gamma}} e_{\mathbf{x}}^{-1}. \tag{7.2.15}$$

Summarizing, we have, for any $T \in \mathrm{Sym}_2(\mathbb{Z})^\vee$ with $\det(T) \neq 0$ and $\mathrm{Diff}(T, B) = \{p\}$,

$$\widehat{\mathcal{Z}}(T, v) = \begin{cases} |\Lambda_T| \cdot \nu_p(T) \cdot \log(p) & \text{if } p < \infty, \\ |\Lambda_T| \cdot \nu_\infty(T, v) & \text{if } p = \infty, \end{cases} \tag{7.2.16}$$

where $|\Lambda_T|$ is given by (7.2.5), (7.2.10), and (7.2.15), respectively, $\nu_p(T)$ is given in (7.2.4) and (7.2.11), and $\nu_\infty(T, v)$ is given by (7.2.13). This yields the claimed result. □

7.3 A WEAKLY ADMISSIBLE GREEN FUNCTION

In this section, we construct a weakly μ-biadmissible Green function on $\mathfrak{H} \times \mathfrak{H} - \Delta$, as defined in Section 2.7, for the hyperbolic metric $\mu = y^{-2}\, dx \wedge dy$ on the upper half plane $\mathfrak{H}$. This Green function is rapidly decreasing away from the diagonal. The Green functions introduced in [2] (see also Section 3.5) can be recovered by restriction to the 'slices' of the form $z_1 \times \mathfrak{H}$.

For $t > 0$, let

$$\beta_1(t) = -\mathrm{Ei}(-t) = \int_1^\infty e^{-tr}\, r^{-1}\, dr \tag{7.3.1}$$

be the exponential integral.

Proposition 7.3.1. *For $t \in \mathbb{R}_{>0}$, the function*

$$g_t^0(z_1, z_2) = \beta_1(4\pi t \sinh^2(d(z_1, z_2))),$$

where $d(z_1, z_2)$ is the hyperbolic distance between z_1 and $z_2 \in \mathfrak{H}$, defines a weakly μ-biadmissible Green function on $\mathfrak{H} \times \mathfrak{H} - \Delta$. The associated function ϕ_t^0 is given by

$$\phi_t^0(z_1, z_2) = \left[\, 4t \cosh^2(d(z_1, z_2)) - \frac{1}{2\pi} \,\right] e^{-4\pi t \sinh^2(d(z_1,z_2))}.$$

For any $\gamma \in \mathrm{GL}_2(\mathbb{R})^+$,

$$g^0(\gamma z_1, \gamma z_2) = g^0(z_1, z_2) \qquad \textit{and} \qquad \phi^0(\gamma z_1, \gamma z_2) = \phi^0(z_1, z_2).$$

Proof. The symmetry (ii) in Definition 2.7.1 is obvious and condition (i) is immediate from the asymptotics of β_1 near zero; see Lemma 7.3.3. For the Green equation, we let $R = 2\sinh^2(d(z_1, z_2))$ and compute, first on all of $\mathfrak{H} \times \mathfrak{H} - \Delta$,

(7.3.2)

$$\begin{aligned} &dd^c\, g_t^0(z_1, z_2) \\ &= \frac{1}{2\pi i} \left[\, -2\pi t R^{-1}\, \partial R \wedge \bar{\partial} R - R^{-2}\, \partial R \wedge \bar{\partial} R + R^{-1}\, \partial\bar{\partial} R \,\right] e^{-2\pi t R} \\ &= \frac{1}{2\pi i} \left[\, -2\pi t R\, \partial \log(R) \wedge \bar{\partial} \log(R) + \partial\bar{\partial} \log(R) \,\right] e^{-2\pi t R}. \end{aligned}$$

Since

$$\cosh(d(z_1, z_2)) = \frac{|z_1 - z_2|^2}{2y_1 y_2} + 1, \tag{7.3.3}$$

we have

$$R = 2\, \frac{|z_1 - z_2|^2}{2y_1 y_2} \cdot \frac{|z_1 - \bar{z}_2|^2}{2y_1 y_2}, \tag{7.3.4}$$

$$\log(R) = -2\log(y_1 y_2) - \log(2) + \log|z_1 - z_2|^2 + \log|z_1 - \bar{z}_2|^2, \tag{7.3.5}$$

and

(7.3.6)

$$\begin{aligned} &\partial \log(R) \\ &= \left(i y_1^{-1} + \frac{1}{z_1 - z_2} + \frac{1}{z_1 - \bar{z}_2} \right) dz_1 + \left(i y_2^{-1} + \frac{1}{z_2 - z_1} + \frac{1}{z_2 - \bar{z}_1} \right) dz_2 \\ &= (\rho^{-1} + \sigma_1)\, dz_1 + (-\rho^{-1} + \sigma_2)\, dz_2, \end{aligned}$$

where, for convenience, we write $\rho = z_1 - z_2$, and

$$(7.3.7) \qquad \sigma_1 = iy_1^{-1} + \frac{1}{z_1 - \bar{z}_2}, \qquad \sigma_2 = iy_2^{-1} + \frac{1}{z_2 - \bar{z}_1}.$$

Then

$$(7.3.8) \qquad \bar{\partial}\log(R) = (\bar{\rho}^{-1} + \bar{\sigma}_1)\, d\bar{z}_1 + (-\bar{\rho}^{-1} + \bar{\sigma}_2)\, d\bar{z}_2$$

and

(7.3.9)

$$R\,\partial\log(R) \wedge \bar{\partial}\log(R)$$
$$= 2(2y_1y_2)^{-2}|z_1 - \bar{z}_2|^2\Big(|1+\rho\sigma_1|^2\, dz_1 \wedge d\bar{z}_1 + |-1+\rho\sigma_2|^2\, dz_2 \wedge d\bar{z}_2$$
$$+ (1+\rho\sigma_1)(-1+\bar{\rho}\bar{\sigma}_2)dz_1 \wedge d\bar{z}_2 + (-1+\rho\sigma_2)(1+\bar{\rho}\bar{\sigma}_1)dz_2 \wedge d\bar{z}_1\Big).$$

In addition,

$$(7.3.10) \quad \partial\bar{\partial}\log(R) = -\, iy_1^{-2}\, dx_1 \wedge dy_1 - iy_2^{-2}\, dx_2 \wedge dy_2 + (z_1 - \bar{z}_2)^{-2}\, dz_1 \wedge d\bar{z}_2 + (z_2 - \bar{z}_1)^{-2}\, dz_2 \wedge d\bar{z}_1.$$

Note that the 2-forms in (7.3.9) and (7.3.10) are both smooth on all of $\mathfrak{H} \times \mathfrak{H}$. Now, fixing z_1, we obtain

(7.3.11)

$$R\,\partial_2\log(R) \wedge \bar{\partial}_2\log(R) = 2\,(2y_1y_2)^{-2}|z_1 - \bar{z}_2|^2\,|-1+\rho\sigma_2|^2\, dz_2 \wedge d\bar{z}_2,$$

and

$$(7.3.12) \qquad \partial_2\bar{\partial}_2\log(R) = -iy_2^{-2}\, dx_2 \wedge dy_2,$$

so that

$$d_2d_2^c g_t^0(z_1, z_2) = \big[ty_1^{-2}|z_1 - \bar{z}_2|^2|1-\rho\sigma_2|^2 - \frac{1}{2\pi}\,\big]e^{-2\pi tR}y_2^{-2}dx_2 \wedge dy_2$$

$$(7.3.13) \qquad = \Big[\,4t\big(\frac{|z_1 - z_2|^2}{2y_1y_2} + 1\big)^2 - \frac{1}{2\pi}\,\Big]\, e^{-2\pi tR}\, y_2^{-2}\, dx_2 \wedge dy_2$$

$$= \Big[\,4t\cosh^2(d(z_1, z_2)) - \frac{1}{2\pi}\,\Big]\, e^{-2\pi tR}\, y_2^{-2}\, dx_2 \wedge dy_2.$$

In fact, to finish the proof of (iii) of Definition 2.7.1, we must still check the Green equation for currents, but this will follow from the calculation in [2], Proposition 11.1, once we have identified the restriction of g_t^0 to the 'slice' $z_1 \times \mathfrak{H}$. □

Remark 7.3.2. Away from the diagonal,

$$g_t^0(z_1, z_2) = O(\exp(-4\pi t \sinh^2(d(z_1, z_2)))), \tag{7.3.14}$$

and hence decays doubly exponentially with respect to the hyperbolic distance.

Next, we consider the pullback of the metric on $\mathcal{O}_{\mathfrak{H}\times\mathfrak{H}}(\Delta)$ determined, on $\mathfrak{H} \times \mathfrak{H} - \Delta$, by the identity

$$g_t^0(z_1, z_2) = -\log ||s_\Delta||^2, \tag{7.3.15}$$

where $s_\Delta = z_1 - z_2$. The point is to determine the behavior of this metric along Δ.

Lemma 7.3.3. *For the exponential integral β_1 of (7.3.1),*

$$\lim_{t\downarrow 0} \Big(\beta_1(t) + \log(t) \Big) = -\gamma,$$

where γ is Euler's constant. More precisely

$$\beta_1(t) = -\gamma - \log(t) + \beta_1^0(t)$$

where $\beta_1^0(t)$ is given by

$$\beta_1^0(t) = \int_0^t \frac{e^u - 1}{u}\, du.$$

In particular, $\lim_{t\downarrow 0} \beta_1^0(t) = 0$. □

It follows from this lemma together with (7.3.4) that

$$\begin{aligned} g_t^0(z_1, z_2) &= -\log|z_1 - z_2|^2 - \gamma - \log(4\pi t) \\ &\qquad + 2\log(2y_1y_2) - \log|z_1 - \bar{z}_2|^2 + \beta_1^0(2\pi t R). \end{aligned} \tag{7.3.16}$$

Recalling that $\rho = z_1 - z_2$ and using (7.3.6), we compute

$$\begin{aligned} -\log||1||^2 &= -\log||\rho^{-1}s_\Delta||^2 \\ &= \log|z_1 - z_2|^2 + g_t^0(z_1, z_2) \qquad (7.3.17) \\ &= \log|z_1 - z_2|^2 - \gamma - \log(2\pi t R) + \text{smooth} \\ &= -\gamma - \log(\pi t) + 2\log(y_1y_2) - \log|z_1 - \bar{z}_2|^2 + \text{smooth}. \end{aligned}$$

Restricting to the diagonal $z = z_1 = z_2$, we get

$$(-\log||1||^2)|_\Delta = -\gamma - \log(4\pi t) + 2\log(y). \tag{7.3.18}$$

Note that the corresponding Chern form is

$$-dd^c(\log ||1||^2)|_\Delta = dd^c(2\log(y)) = -\frac{1}{2\pi}\,\mu. \tag{7.3.19}$$

Formula (7.3.18) can be interpreted as follows. Let I be the ideal of holomorphic functions vanishing on the diagonal in $\mathfrak{H}\times\mathfrak{H}$. Taking $s_\Delta^\vee = (z_1-z_2)^{-1}$, as a section of $\mathcal{O}_{\mathfrak{H}\times\mathfrak{H}}(-\Delta)$, with divisor $-\Delta$, then for any open set U in $\mathfrak{H}\times\mathfrak{H}$, there is an isomorphism

$$\Gamma(U,I) \xrightarrow{\sim} \Gamma(U,\mathcal{O}_{\mathfrak{H}\times\mathfrak{H}}(-\Delta)), \qquad f \mapsto f\cdot s_\Delta^\vee. \tag{7.3.20}$$

Then the canonical isomorphism

$$i_\Delta^*(\mathcal{O}_{\mathfrak{H}\times\mathfrak{H}}(-\Delta)) \xrightarrow{\sim} \Omega^1_{\mathfrak{H}} \tag{7.3.21}$$

is given explicitly by

$$f\cdot s_\Delta^\vee \mapsto i_\Delta^*(h)\cdot dz, \tag{7.3.22}$$

where

$$f(z_1,z_2) = h(z_1,z_2)\,(z_1-z_2). \tag{7.3.23}$$

In particular, $\rho\cdot s_\Delta^\vee \mapsto dz$ in (7.3.21). The metric on $\Omega^1_{\mathfrak{H}}$ which makes (7.3.21) an isometry is then determined by

$$-\log ||dz||^2 = -\log ||\rho\cdot s_\Delta^\vee||^2 = \gamma + \log(4\pi t) - 2\log(y), \tag{7.3.24}$$

where the sign has changed compared to (7.3.18), since we are considering $\mathcal{O}_{\mathfrak{H}\times\mathfrak{H}}(-\Delta)$, rather than $\mathcal{O}_{\mathfrak{H}\times\mathfrak{H}}(\Delta)$.

Suppose that $\Gamma\subset \mathrm{GL}_2^+(\mathbb{R})$ is a discrete group acting properly discontinuously on $\mathfrak{H}$ with compact quotient $\Gamma\backslash\mathfrak{H}$. Let $M = [\Gamma\backslash\mathfrak{H}]$ be the corresponding compact orbifold, and let

$$\mathrm{pr} : \mathfrak{H} \longrightarrow [\Gamma\backslash\mathfrak{H}] = M \tag{7.3.25}$$

be the natural projection. We also suppose that there is a normal subgroup $\Gamma'\subset\Gamma$ of finite index which acts without fixed points in $\mathfrak{H}$. Thus, there is a compact Riemann surface $X = \Gamma'\backslash\mathfrak{H}$ with a finite group of automorphisms $\Gamma^0 = \Gamma/\Gamma'$ so that $M = [\Gamma^0\backslash X]$. Thus, we are in the situation of Sections 2.3 and 2.7. The function

$$g_t(P,Q) = \sum_{\gamma\in\Gamma'} g_t^0(z_1,\gamma z_2), \tag{7.3.26}$$

for $P = \mathrm{pr}(z_1)$ and $Q = \mathrm{pr}(z_2) \in X$, is then a weakly μ-biadmissible Green function on $X \times X - \Delta_X$ with associated function

$$\phi_t(P,Q) = \sum_{\gamma \in \Gamma'} \phi_t^0(z_1, \gamma z_2). \tag{7.3.27}$$

Since Γ' is normal in Γ, these functions are invariant with respect to the diagonal action of $\Gamma^0 = \Gamma/\Gamma'$.

In the arithmetic adjunction formula of Section 2.7, we need to compute the metric on $\Omega^1_X = i^*_\Delta \mathcal{O}_{X\times X}(-\Delta)$ determined by the function $-g_t$. Since Γ' has no elliptic fixed points on $\mathfrak{H}$, we have[4]

$$-\log ||dz||^2 = \gamma + \log(4\pi t) - 2\log(y) - \psi_t(z), \tag{7.3.28}$$

where

$$\psi_t(z) = \sum_{\substack{\gamma \in \Gamma' \\ \gamma \neq 1}} g_t^0(z, \gamma z). \tag{7.3.29}$$

Here, the first three terms on the right-hand side are the contribution for $\gamma = 1$, as computed in (7.3.24) above. Since Γ' is normal in Γ, the function $\psi_t(z)$ is actually Γ-invariant, as is the quantity $-\log ||dz||^2 + 2\log(y)$. In particular, ψ_t defines a function on M. If f is a holomorphic automorphic form of weight 2 with respect to Γ, then the corresponding section $f(z)\,dz$ of Ω^1_M has norm

$$||f(z)dz||^2 = |f(z)|^2\, y^2 \cdot e^{-2C - \log(t) + \psi_t(z)}, \tag{7.3.30}$$

where

$$2C = \gamma + \log(4\pi). \tag{7.3.31}$$

Note that the constant C is exactly the quantity occurring in Definition 3.4 of [4], so that its, seemingly mysterious, appearance there is explained in a natural way by (7.3.30). Also note that the Chern form of the metric on Ω^1_M determined by $-g_t$ is

$$\frac{1}{2\pi}\,\mu - dd^c\psi_t. \tag{7.3.32}$$

Of course, X, ψ_t, etc. depend on the choice of the subgroup Γ'.

---

[4]Of course, the reader will not confuse the γ here, which is Euler's constant, with an element of Γ!

Next, we return to the orbifold $M = [\Gamma\backslash D] = \mathcal{M}(\mathbb{C})$, where $\Gamma = O_B^\times$, so that we now have a uniformization by $D = \mathfrak{H}^+ \cup \mathfrak{H}^-$,

$$\mathrm{pr} : D \longrightarrow [\Gamma\backslash D] = M, \tag{7.3.33}$$

where $\Gamma \subset \mathrm{GL}_2(\mathbb{R})$. We want to determine the relation between the Green functions g_t^0 just constructed and the Green functions introduced in Section 3.5. As in [2], let

$$V = \{\mathbf{x} \in \mathrm{M}_2(\mathbb{R}) \mid \mathrm{tr}(\mathbf{x}) = 0\}, \tag{7.3.34}$$

with quadratic form $Q(\mathbf{x}) = \det(\mathbf{x}) = \mathbf{x}^\iota \mathbf{x}$ and associated bilinear form $(\mathbf{x}, \mathbf{y}) = \mathrm{tr}(\mathbf{x}^\iota \mathbf{y})$, where ι is the main involution on $\mathrm{M}_2(\mathbb{R})$.[5] The quadratic space (V, Q) has signature $(1, 2)$. Let

$$\mathbb{H} = \{\mathbf{x} \in V \mid Q(\mathbf{x}) = 1\} \tag{7.3.35}$$

be the hyperboloid of two sheets. We embed $D = \mathfrak{H}^+ \cup \mathfrak{H}^-$ into V by the map

$$z = x + iy \mapsto \mathbf{x}(z) = y^{-1} \begin{pmatrix} -x & |z|^2 \\ -1 & x \end{pmatrix}. \tag{7.3.36}$$

Note that $Q(\mathbf{x}(z)) = 1$ and that $\mathbf{x}(\bar{z}) = -\mathbf{x}(z)$. This embedding identifies D with $\mathbb{H}$ and is equivariant for the action of $\mathrm{GL}_2(\mathbb{R})$:

$$g \cdot \mathbf{x}(z) = g\,\mathbf{x}(z) g^{-1} = \mathbf{x}(g(z)). \tag{7.3.37}$$

Note that every $\mathbf{x} \in V$ with $t = Q(\mathbf{x}) > 0$, i.e., every vector inside the light cone, can be written uniquely in the form $\mathbf{x} = \sqrt{t} \cdot \mathbf{x}(z)$ for $z \in D$.

Remark 7.3.4. Note that this parametrization is different from the one introduced in Section 3.2. Here it is more convenient to identify D with the oriented positive lines in V. Thus we use $z \mapsto \mathbb{R} \cdot \mathbf{x}(z)$ rather than the map $z \mapsto \mathbb{C}^\times \cdot w(z)$ of (3.2.4).

Given $z \in D$, there is an orthogonal decomposition

$$\begin{aligned} V &= \mathbb{R}\,\mathbf{x}(z) \oplus \mathbf{x}(z)^\perp, \\ \mathbf{x} &= \frac{(\mathbf{x}, \mathbf{x}(z))}{(\mathbf{x}(z), \mathbf{x}(z))}\,\mathbf{x}(z) + \mathbf{x}', \end{aligned} \tag{7.3.38}$$

[5] We apologize for the awkward notation with $\mathbf{x} \in V$ and $z = x + iy \in \mathfrak{H}$. Soon the real part of z will no longer be mentioned, and we will revert to $x \in V$ and $z \in \mathfrak{H}$.

and the restriction of Q to $\mathbf{x}(z)^\perp$ is negative definite. In these coordinates, the quantity $R(\mathbf{x}, z)$, defined in (3.5.1) can be written as

$$R(\mathbf{x}, z) := -(\mathbf{x}', \mathbf{x}') = \frac{1}{2}(\mathbf{x}, \mathbf{x}(z))^2 - (\mathbf{x}, \mathbf{x}). \tag{7.3.39}$$

Recall that $R(\mathbf{x}, z) \geq 0$ and that $R(\mathbf{x}, z) = 0$ if and only if $\mathbf{x} \in \mathbb{R}\,\mathbf{x}(z)$. Also, $R(-\mathbf{x}, z) = R(\mathbf{x}, \bar{z}) = R(\mathbf{x}, z)$. Similarly, in these coordinates, the functions φ and ξ of (3.5.4) and (3.5.6) become

$$\varphi(\mathbf{x}, z) = \left((\mathbf{x}, \mathbf{x}(z))^2 - \frac{1}{2\pi}\right) e^{-2\pi R(\mathbf{x}, z)}, \tag{7.3.40}$$

and

$$\xi(\mathbf{x}, z) = \beta_1(2\pi R(\mathbf{x}, z)), \qquad R(\mathbf{x}, z) > 0. \tag{7.3.41}$$

The following result shows that these Green functions are, indeed, the Green functions determined by g_t^0 via restriction to slices. First, we extend g_t^0 and ϕ_t^0, defined in Proposition 7.3.1, to $D \times D$ by setting:

$$g_t^0(z_1, z_2) = \begin{cases} g_t^0(z_1, z_2), & \text{if } z_1 \text{ and } z_2 \in \mathfrak{H}^+, \\ g_t^0(\bar{z}_1, \bar{z}_2), & \text{if } z_1 \text{ and } z_2 \in \mathfrak{H}^-, \\ 0, & \text{otherwise,} \end{cases} \tag{7.3.42}$$

and similarly for ϕ_t^0. These extensions are invariant under the diagonal action of $\mathrm{GL}_2(\mathbb{R})$.

Lemma 7.3.5. *If* $\mathbf{x} = \pm\sqrt{t}\,\mathbf{x}(z_0)$, *with* $z_0 \in \mathfrak{H}$, *then,*

$$\xi(\mathbf{x}, z) = g_t^0(z_0, z) + g_t^0(\bar{z}_0, z),$$
$$\varphi(\mathbf{x}, z) = \phi_t^0(z_0, z) + \phi_t^0(\bar{z}_0, z).$$

Proof. For z_1 and z_2 in $\mathfrak{H} = \mathfrak{H}^+$,

$$(\mathbf{x}(z_1), \mathbf{x}(z_2)) = 2\cosh(d(z_1, z_2)). \tag{7.3.43}$$

Also note that, if $t = Q(\mathbf{x}) > 0$ and $\mathbf{x} = \pm\sqrt{t} \cdot \mathbf{x}(z_0)$, for $z_0 \in \mathfrak{H}$, then, for $z \in \mathfrak{H}$,

$$R(\mathbf{x}, z) = t \cdot 2\sinh^2(d(z_0, z)). \tag{7.3.44}$$

The claimed identities are then immediate. □

For $\mathbf{x} \in V$ with $Q(\mathbf{x}) = t > 0$, and $\mathbf{x} = \pm\sqrt{t}\,\mathbf{x}(z_0)$, we have $D_{\mathbf{x}} = \{z_0, \bar{z}_0\}$. Recall from Propsition 3.5.1 that

$$dd^c\xi(\mathbf{x}) + \delta_{D_{\mathbf{x}}} = \varphi(\mathbf{x}) \cdot \mu, \tag{7.3.45}$$

as currents on D, and hence that

$$\int_D \varphi(\mathbf{x}, z)\, d\mu(z) = 2. \tag{7.3.46}$$

The first of these identities completes the proof for g_t^0 of property (iii) of Definition 2.7.1. Note that $\xi(\mathbf{x})$ is a Green function for the cycle $D_{\mathbf{x}}$ consisting of 2 points.

From the definition of the $*$-product and Lemma 7.3.5, we obtain the following fact will be useful in Section 7.5 below.

Lemma 7.3.6. *Write* $\mathbf{x}_1 = \pm\sqrt{t_1}\,\mathbf{x}(z_1)$ *and* $\mathbf{x}_2 = \pm\sqrt{t_2}\,\mathbf{x}(z_2)$ *with* z_1 *and* $z_2 \in \mathfrak{H}$. *Then*

$$\begin{aligned}\int_D \xi(\mathbf{x}_1) * \xi(\mathbf{x}_2) \;=\;& g_{t_1}^0(z_1, z_2) + \int_D g_{t_2}^0(z_2, \zeta)\, \phi_{t_1}^0(z_1, \zeta)\, d\mu(\zeta) \\ &+ g_{t_1}^0(\bar{z}_1, \bar{z}_2) + \int_D g_{t_2}^0(\bar{z}_2, \zeta)\, \phi_{t_1}^0(\bar{z}_1, \zeta)\, d\mu(\zeta).\end{aligned}$$

For $\mathbf{x} \in V$, with $Q(\mathbf{x}) = t > 0$, and $z \in D$, we set

$$\Xi(\mathbf{x}, z) = \sum_{\gamma \in \Gamma_{\mathbf{x}} \backslash \Gamma} \xi(\mathbf{x}, \gamma z) = \bar{e}_{\mathbf{x}}^{-1} \sum_{\gamma \in \bar{\Gamma}} \xi(\mathbf{x}, \gamma z), \tag{7.3.47}$$

and

$$\Phi(\mathbf{x}, z) = \sum_{\gamma \in \Gamma_{\mathbf{x}} \backslash \Gamma} \varphi(\mathbf{x}, \gamma z) = \bar{e}_{\mathbf{x}}^{-1} \sum_{\gamma \in \bar{\Gamma}} \varphi(\mathbf{x}, \gamma z). \tag{7.3.48}$$

Here $\bar{\Gamma} = \Gamma/\Gamma_D$, where Γ_D is the subgroup of Γ acting trivially on D. Note that the function $\Xi(t, v)$ in Corollary 3.5.2 is then obtained by summing $\Xi(\mathbf{x}, v)$ over the Γ orbits in the set $\{\mathbf{x} \in O_B \cap V \mid Q(x) = t\}$.

Lemma 7.3.7. $\Xi(\mathbf{x})$ *is a Green function on the stack* M *for the divisor* $\mathrm{pr}(D_{\mathbf{x}})$; *see Section 2.3. Explicitly,*

$$dd^c\Xi(\mathbf{x}) + \delta_{\mathrm{pr}(D_{\mathbf{x}})} = \Phi(\mathbf{x}) \cdot \mu.$$

Proof. In fact, if $\phi \in C^\infty(M)$, we have

$$\begin{aligned}\langle\, dd^c\Xi(\mathbf{x}), \phi\,\rangle_M &= \int_M \Xi(\mathbf{x}) \cdot dd^c\phi \\ &= \int_{[\Gamma\backslash D]} \Xi(\mathbf{x}) \cdot dd^c\phi \\ &= e_{\mathbf{x}}^{-1} \int_D \xi(\mathbf{x}, z) \cdot \mathrm{pr}^*(dd^c\phi)(z) \\ &= e_{\mathbf{x}}^{-1} \langle\, dd^c\xi(\mathbf{x}), \mathrm{pr}^*\phi\,\rangle \\ &= e_{\mathbf{x}}^{-1} \Big(-\langle\, \delta_{D_{\mathbf{x}}}, \mathrm{pr}^*\phi\,\rangle + \int_D \varphi(\mathbf{x}, z) \cdot \mathrm{pr}^*\phi(z) \cdot d\mu(z) \Big) \\ &= -\langle\, \delta_{\mathrm{pr}(D_{\mathbf{x}})}, \phi\,\rangle_M + \int_M \Phi(\mathbf{x}) \cdot \phi \cdot \mu.\end{aligned}$$

Here the points in $\mathrm{pr}(D_{\mathbf{x}})$ on the orbifold M are counted with multiplicity $e_{\mathbf{x}}^{-1} = |\Gamma_{\mathbf{x}}|^{-1}$. Note, for example, that, if 1_M is the constant function on M, then

$$\langle\, \delta_{\mathrm{pr}(D_{\mathbf{x}})}, 1_M\,\rangle_M = 2\, e_{\mathbf{x}}^{-1}.$$

□

7.4 A FINER DECOMPOSITION OF SPECIAL CYCLES

There is a decomposition of the cycle $\mathcal{Z}^{\mathrm{hor}}(t)$ via the 'conductor', defined as follows. For $t \in \mathbb{Z}_{>0}$, write $4t = n^2 d$ where $-d$ is a fundamental discriminant. Let $\boldsymbol{k}_t = \mathbb{Q}(\sqrt{-t}) = \mathbb{Q}(\sqrt{-d})$. Then, on the generic fiber, there is a decomposition

(7.4.1)
$$\mathcal{Z}^{\mathrm{hor}}(t)_{\mathbb{Q}} = \coprod_{\substack{c|n \\ (c,D(B))=1}} \mathcal{Z}^{\mathrm{hor}}(t:c)_{\mathbb{Q}},$$

where $\mathcal{Z}^{\mathrm{hor}}(t:c)_{\mathbb{Q}}$ is the locus of triples (A, ι, x) in $\mathcal{Z}^{\mathrm{hor}}(t)_{\mathbb{Q}}$ such that, for the embedding $\phi_x : \boldsymbol{k}_t \hookrightarrow \mathrm{End}^0(A, \iota)$ determined by x,

(7.4.2)
$$\phi_x^{-1}(\,\mathbb{Q}[x] \cap \mathrm{End}(A, \iota)\,) = O_{c^2 d}.$$

Here $O_{c^2 d} \subset O_d$ is the order of conductor c; see Section 10 of [4]. Recall that the order on the left-hand side of (7.4.2) is maximal at all primes dividing $D(B)$, hence the restriction $(c, D(B)) = 1$ in the decomposition. Note that, for $r \geq 1$,

(7.4.3)
$$\mathcal{Z}^{\mathrm{hor}}(r^2 t : c)_{\mathbb{Q}} = \mathcal{Z}^{\mathrm{hor}}(t:c)_{\mathbb{Q}},$$

since both are the locus of triples (A, ι, ι'), where ι' is an embedding

$$\iota' : O_{c^2 d} \hookrightarrow \mathrm{End}(A, \iota) \tag{7.4.4}$$

which is optimal, in the obvious sense. Let[6]

$$\mathcal{Z}^{\mathrm{hor}}(t : c) = \overline{\mathcal{Z}^{\mathrm{hor}}(t : c)_{\mathbb{Q}}}, \tag{7.4.5}$$

Note that, by (7.4.3),

$$\mathcal{Z}^{\mathrm{hor}}(r^2 t : c) = \mathcal{Z}^{\mathrm{hor}}(t : c). \tag{7.4.6}$$

Remark 7.4.1. The conductor condition (7.4.2) does not lead to a good moduli problem over $\mathbb{Z}_{(p)}$, so that we are forced to use the closure in our global definition of $\mathcal{Z}^{\mathrm{hor}}(t : c)$.

Lemma 7.4.2. $\mathcal{Z}^{\mathrm{hor}}(t : c)_{\mathbb{Q}}$ *is irreducible. More precisely, its coarse moduli space is isomorphic to* Spec $\boldsymbol{k}_{t,c}$, *where* $\boldsymbol{k}_{t,c}$ *is the ray class field of* $\boldsymbol{k}_t$ *with norm subgroup* $\boldsymbol{k}_t^{\times} \cdot \mathbb{C}^{\times} \cdot \hat{O}_{c^2 d}^{\times}$ *in* $\mathbb{A}_{\boldsymbol{k}_t}^{\times}$.

Proof. Let (A, ι, x) be a $\mathbb{C}$-valued point of $\mathcal{Z}^{\mathrm{hor}}(t : c)$. Then A is of the form $\mathrm{Lie}(A)/\Gamma$, where Γ is an O_B-module from the left and where the order $O_{c^2 d}$ acts from the right on Γ through holomorphic endomorphisms, and is maximal with this property. Now the finite ideles of $\boldsymbol{k}_t$ act in the usual way transitively on the set of such Γ's with stabilizer equal to $\boldsymbol{k}_t^{\times} \cdot \hat{O}_{c^2 d}^{\times}$. Since this action commutes with the action of the Galois group of $\boldsymbol{k}_t$, the action of an element $\sigma \in \mathrm{Gal}_{\boldsymbol{k}_t}$ is given by translation by an element of $\mathbb{A}_{\boldsymbol{k}_t, f}^{\times}$. Now the theory of complex multiplication shows that this element is the image of σ under the class field isomorphism. The assertion concerning the coarse moduli space of $\mathcal{Z}^{\mathrm{hor}}(t : c)$ follows, and this also implies the first claim. $\square$

We have the decomposition

$$\mathcal{Z}(t) = \bigcup_{\substack{c \mid n \\ (c, D(B))=1}} \mathcal{Z}^{\mathrm{hor}}(t : c) \cup \mathcal{Z}^{\mathrm{ver}}(t). \tag{7.4.7}$$

To obtain the decomposition of the class $\widehat{\mathcal{Z}}(t, v) \in \widehat{\mathrm{CH}}^1(\mathcal{M})$ corresponding to (7.4.7), we must decompose the Green function. As in Section 5 of [4], let $L = O_B \cap V$,

$$L(t) = \{\, x \in L \mid Q(x) = t \,\} \tag{7.4.8}$$

and

[6] Here again, as in the definition of $\mathcal{Z}^{\mathrm{hor}}(t)$, we use the notation $\mathcal{Z}^{\mathrm{hor}}(t : c)_{\mathbb{Q}}$ to denote the corresponding image divisor in $\mathcal{M}_{\mathbb{Q}}$.

(7.4.9) $$L(t:c) = \{\, x \in L \mid Q(x) = t,\ \text{type}\,(x) = c \,\},$$

where $\text{type}(x) = c$ means that

(7.4.10) $$\mathbb{Q}[x] \cap O_B \;=\; \text{an order of conductor } c \text{ in } \mathbb{Q}[x].$$

Here $\mathbb{Q}[x] \simeq \mathbb{Q}[\sqrt{-Q(x)}]$. Green functions for $\mathcal{Z}^{\text{hor}}(t)$ and $\mathcal{Z}^{\text{hor}}(t:c)$ are given by

(7.4.11) $$\Xi(t,v)(z) = \sum_{x \in L(t)} \xi(v^{\frac{1}{2}}x, z),$$

and

(7.4.12) $$\Xi(t:c,v)(z) = \sum_{x \in L(t:c)} \xi(v^{\frac{1}{2}}x, z).$$

By the constructions explained in Section 7.3, if $Q = \text{pr}(z)$, then[7]

(7.4.13) $$\Xi(t,v)(z) = \sum_{P \in \mathcal{Z}^{\text{hor}}(t)(\mathbb{C})} g_{tv}(P,Q),$$

and

(7.4.14)
$$\begin{aligned}\Xi(t:c,v)(z) &= \sum_{P \in \mathcal{Z}^{\text{hor}}(t:c)(\mathbb{C})} g_{tv}(P,Q)\\ &= \sum_{P \in \mathcal{Z}^{\text{hor}}(r^2t:c)(\mathbb{C})} g_{tv}(P,Q)\\ &= \Xi(r^2t:c, r^{-2}v).\end{aligned}$$

In the last steps we used (7.4.6). Let

(7.4.15) $$\widehat{\mathcal{Z}}^{\text{hor}}(t:c,v) = (\mathcal{Z}^{\text{hor}}(t:c), \Xi(t:c,v)) \;\in\; \widehat{\text{CH}}^1(\mathcal{M}).$$

Thus we obtain the decomposition

(7.4.16) $$\widehat{\mathcal{Z}}^{\text{hor}}(t,v) \;=\; \sum_{c|n} \widehat{\mathcal{Z}}^{\text{hor}}(t:c,v).$$

Note that, by (7.4.6) and (7.4.14),

(7.4.17) $$\widehat{\mathcal{Z}}^{\text{hor}}(r^2t:c,v) = \widehat{\mathcal{Z}}^{\text{hor}}(t:c,r^2v).$$

We now consider the pairing $\langle\, \widehat{\mathcal{Z}}(t_1,v_1), \widehat{\mathcal{Z}}(t_2,v_2)\,\rangle$ in the case where t_1t_2 is a square. We begin by determining the common components of $\mathcal{Z}^{\text{hor}}(t_1)$ and $\mathcal{Z}^{\text{hor}}(t_2)$.

---

[7]Here the points in $\mathcal{Z}^{\text{hor}}(t)(\mathbb{C})$ are counted with their fractional multiplicities.

Lemma 7.4.3. *Write* $t_1 = n_1^2 t$ *and* $t_2 = n_2^2 t$ *with* $(n_1, n_2) = 1$*, and, as above, write* $4t = n^2 d$*. Then the greatest common divisor of the divisors*[8] $\mathcal{Z}^{\text{hor}}(t_1)$ *and* $\mathcal{Z}^{\text{hor}}(t_2)$ *is equal to*

$$\mathcal{Z}^{\text{hor}}(t) = \sum_{c \mid n} \mathcal{Z}^{\text{hor}}(t : c).$$

Proof. We have

$$\mathcal{Z}^{\text{hor}}(t_i) = \sum_{\substack{c \mid n_i n \\ (c, D(B))=1}} \mathcal{Z}^{\text{hor}}(n_i^2 t : c). \tag{7.4.18}$$

Thus, the common part is the sum of the $\mathcal{Z}^{\text{hor}}(n_i^2 t : c)$'s where c divides both $n_1 n$ and $n_2 n$, hence $c \mid n$. Since, for such c, $\mathcal{Z}^{\text{hor}}(n_i^2 t : c) = \mathcal{Z}^{\text{hor}}(t : c)$, this sum is just $\mathcal{Z}^{\text{hor}}(t)$. □

Then, we can write $\langle\, \widehat{\mathcal{Z}}(t_1, v_1), \widehat{\mathcal{Z}}(t_2, v_2)\,\rangle = A + B + C$, where

$$A = \sum_{\substack{c_1, c_2 \\ c_1 \mid n_1 n,\ c_2 \mid n_2 n \\ c_1 \neq c_2}} \langle\, \widehat{\mathcal{Z}}^{\text{hor}}(t_1 : c_1, v_1), \widehat{\mathcal{Z}}^{\text{hor}}(t_2 : c_2, v_2)\,\rangle, \tag{7.4.19}$$

$$B = \sum_{c \mid n} \langle\, \widehat{\mathcal{Z}}^{\text{hor}}(t_1 : c, v_1), \widehat{\mathcal{Z}}^{\text{hor}}(t_2 : c, v_2)\,\rangle, \tag{7.4.20}$$

and

$$C = \langle\, \widehat{\mathcal{Z}}^{\text{hor}}(t_1), \widehat{\mathcal{Z}}^{\text{ver}}(t_2)\,\rangle + \langle\, \widehat{\mathcal{Z}}^{\text{ver}}(t_1), \widehat{\mathcal{Z}}^{\text{hor}}(t_2)\,\rangle + \langle\, \widehat{\mathcal{Z}}^{\text{ver}}(t_1), \widehat{\mathcal{Z}}^{\text{ver}}(t_2)\,\rangle. \tag{7.4.21}$$

There are, thus, several types of intersections to be computed. First, in the terms in (7.4.19), since $c_1 \neq c_2$, the cycles are disjoint on the generic fiber, and hence

$$\begin{aligned} &\langle\, \widehat{\mathcal{Z}}^{\text{hor}}(t_1 : c_1, v_1), \widehat{\mathcal{Z}}^{\text{hor}}(t_2 : c_2, v_2)\,\rangle \\ &\qquad = \sum_{p \leq \infty} \langle\, \widehat{\mathcal{Z}}^{\text{hor}}(t_1 : c_1, v_1), \widehat{\mathcal{Z}}^{\text{hor}}(t_2 : c_2, v_2)\,\rangle_p. \end{aligned} \tag{7.4.22}$$

We let $A = \sum_{p \leq \infty} A_p$, where

$$A_p = \sum_{\substack{c_1, c_2 \\ c_1 \mid n_1 n,\ c_2 \mid n_2 n \\ c_1 \neq c_2}} \langle\, \widehat{\mathcal{Z}}^{\text{hor}}(t_1 : c_1, v_1), \widehat{\mathcal{Z}}^{\text{hor}}(t_2 : c_2, v_2)\,\rangle_p. \tag{7.4.23}$$

[8]Here the $\mathcal{Z}^{\text{hor}}(t_i)$'s denote the image divisors on $\mathcal{M}$.

Similarly, let $C = \sum_p C_p$, where

$$\text{(7.4.24)} \quad C_p = \langle\, \widehat{\mathcal{Z}}^{\text{hor}}(t_1, v_1), \widehat{\mathcal{Z}}^{\text{ver}}(t_2)\,\rangle_p + \langle\, \widehat{\mathcal{Z}}^{\text{ver}}(t_1), \widehat{\mathcal{Z}}^{\text{hor}}(t_2, v_2)\,\rangle_p + \langle\, \widehat{\mathcal{Z}}^{\text{ver}}(t_1), \widehat{\mathcal{Z}}^{\text{ver}}(t_2)\,\rangle_p.$$

Here only primes p with $p \mid D(B)$ occur. Finally, by (7.4.17), the terms in B have the form
(7.4.25)

$$\langle\, \widehat{\mathcal{Z}}^{\text{hor}}(t_1 : c, v_1), \widehat{\mathcal{Z}}^{\text{hor}}(t_2 : c, v_2)\,\rangle = \langle \widehat{\mathcal{Z}}^{\text{hor}}(t : c, n_1^2\, v_1), \widehat{\mathcal{Z}}^{\text{hor}}(t : c, n_2^2\, v_2)\rangle.$$

These terms are not a sum of local contributions and will be computed in the next section.

7.5 APPLICATION OF ADJUNCTION

In this section, we apply the arithmetic adjunction formula, as formulated in Section 2.7, to compute the arithmetic intersection numbers (7.4.25) of common components of $\widehat{\mathcal{Z}}^{\text{hor}}(t_1, v_1)$ and $\widehat{\mathcal{Z}}^{\text{hor}}(t_2, v_2)$. Recall that, as in (7.3.33), $M = \mathcal{M}(\mathbb{C}) = [\Gamma\backslash D]$.

We begin by making explicit what the adjunction formula of Section 2.7 gives for the self-intersection number of an irreducible horizontal divisor $\mathcal{Z}$ on $\mathcal{M}$. We write

$$\mathcal{Z}(\mathbb{C}) = \sum_i P_i.$$

Taking $z_i \in D$ with $\text{pr}(z_i) = P_i$, the Green function for $\mathcal{Z}$ determined by g_v^0, for a parameter $v > 0$, is given by

$$\Xi_{\mathcal{Z}}(v) = \sum_i e_{z_i}^{-1} \sum_{\gamma\in\Gamma} g_v^0(z_i, \gamma z),$$

with associated function

$$\Phi_{\mathcal{Z}}(v) = \sum_i e_{z_i}^{-1} \sum_{\gamma\in\Gamma} \phi_v^0(z_i, \gamma z).$$

The adjunction formula of Theorem 2.7.2 can be applied here for the infinite presentation $M = [\Gamma\backslash D]$.

Proposition 7.5.1. *Let $\mathcal{Z} = \mathcal{Z}^{\text{hor}}$ be an irreducible, reduced, horizontal divisor on $\mathcal{M}$. For v_1 and $v_2 \in \mathbb{R}_{>0}$, let $\Xi(v_1)$ and $\Xi(v_2)$ be the Green* *functions on M for $\mathcal{Z}$ with parameters v_1 and v_2. Let $\widehat{\mathcal{Z}}(v_1) = (\mathcal{Z}, \Xi_{\mathcal{Z}}(v_1))$ and $\widehat{\mathcal{Z}}(v_2) = (\mathcal{Z}, \Xi_{\mathcal{Z}}(v_2)) \in \widehat{\mathrm{CH}}^1(\mathcal{M})$ be the corresponding classes. Then,*

$$
\begin{aligned}
&\langle\, \widehat{\mathcal{Z}}(v_1), \widehat{\mathcal{Z}}(v_2)\, \rangle_{\mathcal{M}} \\
&= -h_{\hat{\omega}}(\mathcal{Z}) - \frac{1}{2}\, \deg_{\mathbb{Q}}(\mathcal{Z}) \cdot \log(4D(B)\, v_1) + \mathfrak{d}_{\mathcal{Z}} \\
&\quad + \sum_{P,P' \in \mathcal{Z}(\mathbb{C})} e_P^{-1}\, e_{P'}^{-1} \sum_{\substack{\gamma\in\Gamma \\ z \neq \gamma z'}} \frac{1}{2} \Big(g^0_{v_1}(z, \gamma z') \\
&\qquad\qquad + \int_D \phi^0_{v_1}(z,\zeta)\, g^0_{v_2}(\gamma z', \zeta)\, d\mu(\zeta) \Big) \\
&\quad + \sum_{P \in \mathcal{Z}(\mathbb{C})} e_P^{-1}\, \frac{1}{2} \int_D \phi^0_{v_1}(z,\zeta)\, g^0_{v_2}(z,\zeta)\, d\mu(\zeta),
\end{aligned}
$$

where $\hat{\omega}$ is the Hodge bundle ω on $\mathcal{M}$ with the metric defined by (3.3.4), and where $\mathfrak{d}_{\mathcal{Z}}$ is the discriminant degree of $\mathcal{Z}$.

Proof. The formula of Theorem 2.7.2 involves the height of $\mathcal{Z}$ with respect to $\widehat{\boldsymbol{\omega}}_1 = \widehat{\boldsymbol{\omega}}_{\mathcal{M}/S}$, where $\boldsymbol{\omega}_{\mathcal{M}/S}$ is the relative dualizing sheaf and the metric is determined by $-g^0_{v_1}$ on $D \times D - \Delta_D$, as in Section 7.3 above. By Proposition 3.3.1, as sheaves $\boldsymbol{\omega}_{\mathcal{M}/S} = \omega$, where ω is the Hodge bundle on $\mathcal{M}$. The following lemma compares the two metrics on this sheaf.

Lemma 7.5.2. *As classes in $\widehat{\mathrm{CH}}^1(\mathcal{M})$,*

$$\widehat{\boldsymbol{\omega}}_1 = \hat{\omega} + (\, 0, \log(v_1) + \log(4\, D(B))\,).$$

Proof. This follows by comparison of the metric on $\hat{\boldsymbol{\omega}}_1$ determined by formula (7.3.24) above with that on $\hat{\omega}$ given in (3.3.4). To compare them, we use the fact that, under the Kodaira-Spencer isomorphism

$$\omega = \wedge^2(\Omega^1_{\mathcal{A}/\mathcal{M}}) \xrightarrow{\sim} \boldsymbol{\omega}_1, \tag{7.5.1}$$

we have

$$2\pi i\, f(z)\, \alpha_z \mapsto f(z)\, dz, \tag{7.5.2}$$

where $f(z)\,\alpha_z$ is as in the discussion after (3.13) of [4]. Now the claimed identity is evident. □

Using the expression of the lemma, we obtain

$$h_{\widehat{\boldsymbol{\omega}}_1}(\mathcal{Z}) = h_{\hat{\omega}}(\mathcal{Z}) + \frac{1}{2}\, \deg_{\mathbb{Q}}(\mathcal{Z}) \Big(\, \log(v_1) + \log(4D(B)) \Big) \tag{7.5.3}$$

as claimed. □

It will be important to note that the adjunction formula is not 'homogeneous'; some of the terms involved are linear in the cycle $\mathcal{Z}$, while others are quadratic. For example, if $\mathcal{Z} = 2\,\mathcal{Z}_0$, with $\mathcal{Z}_0$ irreducible, reduced, and horizontal, then

$$\begin{aligned}
&\langle\, \widehat{\mathcal{Z}}(v_1), \widehat{\mathcal{Z}}(v_2)\,\rangle_{\mathcal{M}} \\
&\quad = 4\,\langle\, \widehat{\mathcal{Z}}_0(v_1), \widehat{\mathcal{Z}}_0(v_2)\,\rangle_{\mathcal{M}} \\
&\quad = -2\,h_{\hat{\omega}}(\mathcal{Z}) - \deg_{\mathbb{Q}}(\mathcal{Z})\cdot\log(4D(B)\,v_1) + 2\,\mathfrak{d}_{\mathcal{Z}} \\
&\qquad + \sum_{P,P'\in\mathcal{Z}(\mathbb{C})} e_P^{-1}\,e_{P'}^{-1} \sum_{\substack{\gamma\in\Gamma \\ z\neq\gamma z'}} \frac{1}{2}\Big(\, g^0_{v_1}(z,\gamma z') \\
&\qquad\qquad + \int_D \phi^0_{v_1}(z,\zeta)\, g^0_{v_2}(\gamma z',\zeta)\,d\mu(\zeta)\Big) \\
&\qquad + \sum_{P\in\mathcal{Z}(\mathbb{C})} e_P^{-1} \int_D \phi^0_{v_1}(z,\zeta)\, g^0_{v_2}(z,\zeta)\,d\mu(\zeta).
\end{aligned} \tag{7.5.4}$$

Here, $h_{\hat{\omega}}(\mathcal{Z}) = 2\,h_{\hat{\omega}}(\mathcal{Z}_0)$, $\deg(\mathcal{Z}) = 2\,\deg(\mathcal{Z}_0)$, and $\mathfrak{d}_{\mathcal{Z}} = 2\,\mathfrak{d}_{\mathcal{Z}_0}$, so that these terms occur multiplied by an additional factor of 2 in the pairing $\langle\, \widehat{\mathcal{Z}}(v_1), \widehat{\mathcal{Z}}(v_2)\,\rangle_{\mathcal{M}} = 4\,\langle\, \widehat{\mathcal{Z}}_0(v_1), \widehat{\mathcal{Z}}_0(v_2)\,\rangle_{\mathcal{M}}$. The same is true for the last sum, over points of $\mathcal{Z}(\mathbb{C})$. On the other hand, the sum over pairs of points of $\mathcal{Z}(\mathbb{C})$ amounts to 4 times the corresponding sum over pairs of points of $\mathcal{Z}_0(\mathbb{C})$, and hence has no additional factor.

Since (A,ι,y) and $(A,\iota,-y)$ occur together in $\mathcal{Z}^{\mathrm{hor}}(t:c)$, all components of $\mathcal{Z}^{\mathrm{hor}}(t)$ occur with multiplicity 2. Thus, applying (7.5.4) and the irreducibility of $\mathcal{Z}^{\mathrm{hor}}(t:c)$, Lemma 7.4.2, we get

Corollary 7.5.3. *Let t_1t_2 be a square, and define t as in Lemma 4.2. Then, for $c \mid n$,*

$$\begin{aligned}
&\langle\, \widehat{\mathcal{Z}}^{\mathrm{hor}}(t_1:c,v_1),\ \widehat{\mathcal{Z}}^{\mathrm{hor}}(t_2:c,v_2)\,\rangle \\
&= -2\,h_{\hat{\omega}}(\mathcal{Z}^{\mathrm{hor}}(t:c)) - \deg_{\mathbb{Q}}(\mathcal{Z}^{\mathrm{hor}}(t:c))\cdot\log(4D(B)\,t_1v_1) + 2\,\mathfrak{d}_{\mathcal{Z}^{\mathrm{hor}}(t:c)} \\
&\quad + \sum_{P,P'\in\mathcal{Z}^{\mathrm{hor}}(t:c)(\mathbb{C})} e_P^{-1}\,e_{P'}^{-1} \sum_{\substack{\gamma\in\Gamma \\ z\neq\gamma z'}} \frac{1}{2}\Big(\, g^0_{t_1v_1}(z,\gamma z') \\
&\qquad\qquad + \int_D \phi^0_{t_1v_1}(z,\zeta)\, g^0_{t_2v_2}(\gamma z',\zeta)\,d\mu(\zeta)\Big) \\
&\quad + \sum_{P\in\mathcal{Z}^{\mathrm{hor}}(t:c)(\mathbb{C})} e_P^{-1} \int_D \phi^0_{t_1v_1}(z,\zeta)\, g^0_{t_2v_2}(z,\zeta)\,d\mu(\zeta).
\end{aligned}$$

Using Lemma 7.3.6, we obtain the identity

(7.5.5)

$$\begin{aligned}
&\sum_{P,P'\in\mathcal{Z}^{\text{hor}}(t:c)(\mathbb{C})} e_P^{-1}\, e_{P'}^{-1} \sum_{\substack{\gamma\in\Gamma\\ z\neq\gamma z'}} \frac{1}{2}\Big(g^0_{t_1 v_1}(z,\gamma z') \\
&\qquad\qquad + \int_D \phi^0_{t_1 v_1}(z,\zeta)\, g^0_{t_2 v_2}(\gamma z',\zeta)\, d\mu(\zeta)\Big) \\
&= \sum_{\substack{x_1\in L(t_1:c)\\ \text{mod }\Gamma}} \sum_{\substack{x_2\in L(t_2:c)\\ \text{mod }\Gamma}} e_{x_1}^{-1}\, e_{x_2}^{-1} \sum_{\substack{\gamma\in\Gamma\\ \gamma D_{x_2}\neq D_{x_1}}} \frac{1}{2}\int_D \xi(v_1^{\frac{1}{2}} x_1) * \xi(v_2^{\frac{1}{2}}\gamma x_2). \\
&= \sum_{\substack{T\in\text{Sym}_2(\mathbb{Z})^\vee\\ \text{diag}(T)=(t_1,t_2)\\ \det(T)\neq 0}} \sum_{\substack{\mathbf{x}=[x_1,x_2]\in L^2\\ Q(\mathbf{x})=T\\ \text{type}(\mathbf{x})=(c,c)\\ \text{mod }\Gamma}} e_{\mathbf{x}}^{-1}\, \frac{1}{2}\int_D \xi(v_1^{\frac{1}{2}} x_1) * \xi(v_2^{\frac{1}{2}} x_2) \\
&= \sum_{\substack{T\in\text{Sym}_2(\mathbb{Z})^\vee\\ \text{diag}(T)=(t_1,t_2)\\ \det(T)\neq 0}} |\Lambda_T(c,c)|\cdot \nu_\infty(T,v),
\end{aligned}$$

where $\nu_\infty(T,v)$ is given by (7.2.13) and

(7.5.6)

$$\Lambda_T(c_1,c_2) = \big[\,\Gamma\backslash\{\,\mathbf{x}=[x_1,x_2]\in L^2 \mid Q(\mathbf{x})=T,\ \text{type}(\mathbf{x})=(c_1,c_2)\,\}\,\big].$$

It is interesting to compare this expression with the 'nonsingular' part of the archimedean height pairing from Definition 12.3 of [2]. Recall that, for $x\in L$, the type of x is defined by (7.4.10).

We need to evaluate the integrals in the last term of Corollary 7.5.3.

Lemma 7.5.4.

$$\int_D \phi^0_{v_1}(z,\zeta)\, g^0_{v_2}(z,\zeta)\, d\mu(\zeta) = \log\Big(\frac{v_1+v_2}{v_2}\Big) - J(4\pi(v_1+v_2)),$$

where

$$J(t) = \int_0^\infty e^{-tw}\big[\,(w+1)^{\frac{1}{2}} - 1\,\big]\, w^{-1}\, dw.$$

Proof. Note that we may as well assume that $z\in\mathfrak{H}$, so that the contribution of $\mathfrak{H}^-$ is zero. Let $\rho = d(z,\zeta)$ be the hyperbolic distance from z to ζ. Then

$$\phi^0_{v_1}(z,\zeta) = \big[\,4v_1\cosh^2(\rho) - \frac{1}{2\pi}\,\big]\, e^{-4\pi v_1 \sinh^2(\rho)} \tag{7.5.7}$$

and

(7.5.8) $$g^0_{v_2}(z,\zeta) = \beta_1(4\pi v_2 \sinh^2(\rho)).$$

As in the proof of Proposition 12.1 of [4],

(7.5.9)
$$\begin{aligned}
&\int_D \phi^0_{v_1}(z,\zeta)\, g^0_{v_2}(z,\zeta)\, d\mu(\zeta) \\
&= \int_0^\pi \int_0^\infty \left[\, 4v_1 \cosh^2(\rho) - \frac{1}{2\pi}\,\right] e^{-4\pi v_1 \sinh^2(\rho)} \\
&\qquad\qquad \times \beta_1(4\pi v_2 \sinh^2(\rho))\, 2\sinh(\rho)\, d\rho\, d\theta \\
&= \pi \int_0^\infty \int_1^\infty \left[\, 4v_1(w+1) - \frac{1}{2\pi}\,\right] e^{-4\pi(v_1+v_2 r)w}\, r^{-1}\, dr\, (w+1)^{-\frac12}\, dw.
\end{aligned}$$

By integration by parts, we have

(7.5.10)
$$\begin{aligned}
&\int_0^\infty e^{-4\pi(v_1+v_2 r)w}\,(w+1)^{-\frac12}\, dw \\
&\qquad = 8\pi(v_1+v_2 r) \int_0^\infty e^{-4\pi(v_1+v_2 r)w}\left[\,(w+1)^{\frac12} - 1\,\right] dw.
\end{aligned}$$

Multiplying this by $-(2\pi)^{-1}$ and substituting back into the main integral, we obtain

(7.5.11)
$$\begin{aligned}
&\int_0^\infty \int_1^\infty \left[\, -4\pi v_2 r\left[\,(w+1)^{\frac12} - 1\,\right] + 4\pi v_1\,\right] e^{-4\pi(v_1+v_2 r)w}\, r^{-1}\, dr\, dw \\
&\qquad = -\int_0^\infty e^{-4\pi(v_1+v_2)w}\left[\,(w+1)^{\frac12} - 1\,\right] w^{-1}\, dw \\
&\qquad\qquad + \int_1^\infty v_1(v_1+v_2 r)^{-1}\, r^{-1}\, dr \\
&\qquad = -J(4\pi(v_1+v_2)) + \log(v_1+v_2) - \log(v_2).
\end{aligned}$$

□

Thus, the last term in Corollary 7.5.3 becomes

(7.5.12)
$$\begin{aligned}
&\sum_{P \in \mathcal{Z}^{\text{hor}}(t:c)(\mathbb{C})} e_P^{-1} \int_D \phi^0_{t_1 v_1}(z,\zeta)\, g^0_{t_2 v_2}(z,\zeta)\, d\mu(\zeta) \\
&= \deg_{\mathbb{Q}} \mathcal{Z}^{\text{hor}}(t:c)\Big[-J(4\pi(t_1 v_1 + t_2 v_2)) + \log(t_1 v_1 + t_2 v_2) - \log(t_2 v_2)\Big].
\end{aligned}$$

Thus, we obtain the following pretty result:

Proposition 7.5.5.

$$\begin{aligned}
&\langle\, \widehat{\mathcal{Z}}^{\mathrm{hor}}(t_1 : c, v_1), \widehat{\mathcal{Z}}^{\mathrm{hor}}(t_2 : c, v_2)\,\rangle \\
&= -2\, h_{\hat{\omega}}(\mathcal{Z}^{\mathrm{hor}}(t : c)) + 2\, \mathfrak{d}_{\mathcal{Z}^{\mathrm{hor}}(t:c)} - \deg_{\mathbb{Q}}(\mathcal{Z}^{\mathrm{hor}}(t : c)) \cdot \log(4D(B)) \\
&\quad + \deg_{\mathbb{Q}}(\mathcal{Z}^{\mathrm{hor}}(t : c)) \left[\, \log(t_1 v_1 + t_2 v_2) - \log(t_1 v_1) - \log(t_2 v_2)\, \right] \\
&\quad - \deg_{\mathbb{Q}}(\mathcal{Z}^{\mathrm{hor}}(t : c))\, J(4\pi(t_1 v_1 + t_2 v_2)) \\
&\quad + \sum_{\substack{T \in \mathrm{Sym}_2(\mathbb{Z})^\vee \\ \mathrm{diag}(T) = (t_1, t_2) \\ \det(T) \neq 0}} |\Lambda_T(c, c)| \cdot \nu_\infty(T, v).
\end{aligned}$$

Summing on c, this yields the following expression for B from (7.4.20).

Corollary 7.5.6. *The quantity*

$$B = \sum_{c|n} \langle\, \widehat{\mathcal{Z}}^{\mathrm{hor}}(t_1 : c, v_1), \widehat{\mathcal{Z}}^{\mathrm{hor}}(t_2 : c, v_2)\,\rangle$$

is given by

$$\begin{aligned}
B =\ & -2\, h_{\hat{\omega}}(\mathcal{Z}^{\mathrm{hor}}(t)) + 2 \sum_{c|n} \mathfrak{d}_{\mathcal{Z}^{\mathrm{hor}}(t:c)} - \deg_{\mathbb{Q}}(\mathcal{Z}(t)) \cdot \log(4D(B)) \\
& + \deg_{\mathbb{Q}}(\mathcal{Z}(t)) \left[\, \log(t_1 v_1 + t_2 v_2) - \log(t_1 v_1) - \log(t_2 v_2)\, \right] \\
& - \deg_{\mathbb{Q}}(\mathcal{Z}(t))\, J(4\pi(t_1 v_1 + t_2 v_2)) + B_\infty,
\end{aligned}$$

where

$$B_\infty = \sum_{\substack{T \in \mathrm{Sym}_2(\mathbb{Z})^\vee \\ \mathrm{diag}(T) = (t_1, t_2) \\ \det(T) \neq 0}} \sum_{c|n} |\Lambda_T(c, c)| \cdot \nu_\infty(T, v).$$

Similarly, when $t_1 t_2$ is a square, by the same calculations as in Section 2, we have

Proposition 7.5.7. *The quantity*

$$A_\infty = \sum_{c_1 \neq c_2} \langle\, \widehat{\mathcal{Z}}^{\mathrm{hor}}(t_1 : c_1, v_1), \widehat{\mathcal{Z}}^{\mathrm{hor}}(t_2 : c_2, v_2)\,\rangle_\infty$$

is given by

$$A_\infty = \sum_{\substack{T \in \mathrm{Sym}_2(\mathbb{Z})^\vee \\ \mathrm{diag}(T) = (t_1, t_2)}} \sum_{c_1 \neq c_2} |\Lambda_T(c_1, c_2)| \cdot \nu_\infty(T, v).$$

Here note that the condition $c_1 \neq c_2$ does not allow x_1 and x_2 to be colinear, so that only T's with $\det(T) \neq 0$ contribute to the sum.

7.6 CONTRIBUTIONS FOR $p \mid D(B)$

In this section, we consider the contributions of terms A_p and C_p for fixed p where $p \mid D(B)$. To do this, we use the analysis on pp. 214–215 of [3]. For convenience, let $\hat{\mathcal{C}}_i$ denote the base change to $W = W(\bar{\mathbb{F}}_p)$ of the formal completion of $\mathcal{Z}(t_i)$ for $i = 1, 2$, along its special fiber, and write

$$\hat{\mathcal{C}} = \hat{\mathcal{C}}_1 \times_{\hat{\mathcal{A}}} \hat{\mathcal{C}}_2 = \coprod_T \hat{\mathcal{C}}_T, \tag{7.6.1}$$

as in (8.27) and (8.28) of [3]. Here T ranges over all $T \in \mathrm{Sym}_2(\mathbb{Z})^\vee_{\geq 0}$ with $\mathrm{diag}(T) = (t_1, t_2)$.

Now [3], (8.30), gives, as already used in Section 6.2, a description of $\hat{\mathcal{C}}_T$ as a subset of a quotient space,

$$\hat{\mathcal{C}}_T \hookrightarrow H'(\mathbb{Q})\backslash\Big(V'(\mathbb{Q})^2_T \times \hat{\Omega}^\bullet \times H(\mathbb{A}^p_f)/K^p\Big). \tag{7.6.2}$$

By strong approximation, we have

$$H'(\mathbb{A}_f) = H'(\mathbb{Q})H'(\mathbb{Q}_p)K^p, \tag{7.6.3}$$

so that, since $\Gamma' = H'(\mathbb{Q}) \cap H'(\mathbb{Q}_p)K^p$, the quotient on the right side in (7.6.2) can be written as

$$\Gamma'\backslash\Big(V'(\mathbb{Q})^2_T \times \hat{\Omega}^\bullet\Big). \tag{7.6.4}$$

Recalling the incidence relations given after (8.30) in [3], which describe the image of $\hat{\mathcal{C}}_T$ in the right side of (7.6.2), we have

$$\hat{\mathcal{C}}_T \simeq \sum_{\substack{\mathbf{y}=[y_1,y_2]\in(L'[p^{-1}])^2 \\ Q(\mathbf{y})=T \\ \bmod \Gamma'}} \Gamma'_{\mathbf{y}}\backslash\mathcal{Z}^\bullet(\mathbf{j}). \tag{7.6.5}$$

Here, as in [3], $\mathbf{j}$ is the $\mathbb{Z}_p$-span of the pair (j_1, j_2) of endomorphisms associated to $\mathbf{y} = (y_1, y_2)$, of the model p-divisible group $\mathbb{X}$.

Recall from [3] that the quantity

$$\nu_p(T) = 2\,\chi(\mathcal{Z}(\mathbf{j}), \mathcal{O}_{\mathcal{Z}(j_1)} \otimes^{\mathbb{L}} \mathcal{O}_{\mathcal{Z}(j_2)}), \tag{7.6.6}$$

(see (7.2.11) above) is finite provided $\det(T) \neq 0$ and gives twice the full intersection number $(\mathcal{Z}(j_1), \mathcal{Z}(j_2))$ of the cycles $\mathcal{Z}(j_1)$ and $\mathcal{Z}(j_2)$ in the Drinfeld space $\hat{\Omega}$, [3], Theorem 6.1. We can decompose this quantity further by writing

$$\mathcal{Z}(j_i) = \mathcal{Z}^{\mathrm{hor}}(j_i) + \mathcal{Z}^{\mathrm{ver}}(j_i) \tag{7.6.7}$$

for the vertical and horizontal components. Then, let

$$\mathcal{G} := \mathcal{O}_{\mathcal{Z}^{\mathrm{hor}}(j_1)} \otimes^{\mathbb{L}} \mathcal{O}_{\mathcal{Z}^{\mathrm{hor}}(j_2)}. \tag{7.6.8}$$

For $\det(T) \neq 0$, the quantity

$$\nu_p^{\mathrm{hor}}(T) := 2\,\chi(\mathcal{Z}(\mathbf{j}), \mathcal{G}) \tag{7.6.9}$$

is finite and gives twice the intersection number $(\mathcal{Z}^{\mathrm{hor}}(j_1), \mathcal{Z}^{\mathrm{hor}}(j_2))$ of the horizontal parts.

Remark 7.6.1. It is asserted here implicitly that the right-hand side of (7.6.9) only depends on the matrix T. To see this, note that if a pair of special endomorphisms (j_1', j_2') has the same matrix of inner products as (j_1, j_2), then it follows from Witt's theorem that there is an element $g \in \mathrm{GL}_2(\mathbb{Q}_p)$ which carries the pair (j_1, j_2) into the pair (j_1', j_2'). But then g induces an automorphism of $\hat{\Omega}$ which carries the pair of special cycles $[\mathcal{Z}(j_1), \mathcal{Z}(j_2)]$ into $[\mathcal{Z}(j_1'), \mathcal{Z}(j_2')]$, and this automorphism preserves the intersection numbers $\chi(\mathcal{Z}(\mathbf{j}), \mathcal{O}_{\mathcal{Z}^{\mathrm{hor}}(j_1)} \otimes^{\mathbb{L}} \mathcal{O}_{\mathcal{Z}^{\mathrm{hor}}(j_2)})$, resp. $\chi(\mathcal{Z}(\mathbf{j}), \mathcal{O}_{\mathcal{Z}^{\mathrm{hor}}(j_1')} \otimes^{\mathbb{L}} \mathcal{O}_{\mathcal{Z}^{\mathrm{hor}}(j_2')})$.

We now consider

$$A_p = \sum_{c_1 \neq c_2} \langle\, \widehat{\mathcal{Z}}^{\mathrm{hor}}(t_1 : c_1, v_1), \widehat{\mathcal{Z}}^{\mathrm{hor}}(t_2 : c_2, v_2)\,\rangle_p.$$

By the same method as in the case where $t_1 t_2$ is not a square (see (7.2.7)), i.e., decomposing according to fundamental matrices, we have,

Proposition 7.6.2.

$$A_p = \sum_{\substack{T \in \mathrm{Sym}_2(\mathbb{Z})^\vee \\ \mathrm{diag}(T) = (t_1, t_2)}} A_p(T),$$

where only T with $\det(T) \neq 0$ occur, and where

$$A_p(T) = \sum_{c_1 \neq c_2} |\Lambda_T(c_1, c_2)| \cdot \nu_p^{\mathrm{hor}}(T)\, \log(p),$$

Here, for any $T \in \mathrm{Sym}_2(\mathbb{Z})^\vee$,

$$(7.6.10)\quad \Lambda_T(c_1,c_2) = [\,\Gamma'\backslash\{\,\mathbf{y}=[y_1,y_2]\in(L'[p^{-1}])^2 \mid Q(\mathbf{y})=T,\ \mathrm{type}(y_1,y_2)=(c_1,c_2)\,\}\,].$$

For $y \in L'[p^{-1}]$, type$(y) = c$ means

$$(7.6.11)\qquad \mathbb{Q}[y]\cap L'[p^{-1}] = \text{ a } \mathbb{Z}[\frac{1}{p}]\text{-order of conductor } c \text{ in } \mathbb{Q}[y].$$

Note that c is prime to p. In particular, the condition $c_1 \neq c_2$ implies that y_1 and y_2 are not colinear, and hence $\det(T) \neq 0$. Also, the factor 2 occurs in the definition (7.6.9) of $\nu_p^{\mathrm{hor}}(T)$ since two 'sheets' $\mathcal{Z}(\mathbf{j})$ occur in the quotient $\Gamma'_{\mathbf{y}}\backslash\mathcal{Z}^\bullet(\mathbf{j})$; see [3], (8.37) and (8.40).

Next we compute

$$(7.6.12)\quad C_p = \langle\, \widehat{\mathcal{Z}}^{\mathrm{hor}}(t_1,v_1), \mathcal{Z}^{\mathrm{ver}}(t_2)\,\rangle_p + \langle\, \mathcal{Z}^{\mathrm{ver}}(t_1), \widehat{\mathcal{Z}}^{\mathrm{hor}}(t_2,v_2)\,\rangle_p + \langle\, \mathcal{Z}^{\mathrm{ver}}(t_1), \mathcal{Z}^{\mathrm{ver}}(t_2)\,\rangle_p\,.$$

To do this, note that the analysis on pp. 214–215 of [3] leading up to formula (8.30) makes no use of the fact that the given pair of special endomorphisms span a rank 2 module over $\mathbb{Z}_p$. Thus, the same argument applies in the present situation. We write

$$(7.6.13)\qquad \hat{\mathcal{C}}_i = \hat{\mathcal{C}}_i^h + \hat{\mathcal{C}}_i^v$$

for the vertical and horizontal components. Then, using the intersection calculus of Section 4 of [3], we have

$$(7.6.14)\quad \frac{C_p}{\log(p)} = (\hat{\mathcal{C}}_1^h,\hat{\mathcal{C}}_2^v)_p + (\hat{\mathcal{C}}_1^v,\hat{\mathcal{C}}_2^h)_p + (\hat{\mathcal{C}}_1^v,\hat{\mathcal{C}}_2^v)_p = \sum_T \chi(\hat{\mathcal{C}}_T,\mathcal{F}),$$

where

$$(7.6.15)\qquad \mathcal{F} = \mathcal{O}_{\hat{\mathcal{C}}_1^h}\otimes^{\mathbb{L}}\mathcal{O}_{\hat{\mathcal{C}}_2^v} + \mathcal{O}_{\hat{\mathcal{C}}_1^v}\otimes^{\mathbb{L}}\mathcal{O}_{\hat{\mathcal{C}}_2^h} + \mathcal{O}_{\hat{\mathcal{C}}_1^v}\otimes^{\mathbb{L}}\mathcal{O}_{\hat{\mathcal{C}}_2^v}.$$

The quantity

$$(7.6.16)\qquad C_p(T) := \chi(\hat{\mathcal{C}}_T,\mathcal{F})\cdot\log(p),$$

is finite for any T with diag$(T) = (t_1,t_2)$ and depends only on T; see Remark 7.6.1.

Proposition 7.6.3. *For* $\det(T)\neq 0$ *with* diag$(T) = (t_1,t_2)$,

$$C_p(T) = |\Lambda_T|\cdot(\nu_p(T) - \nu_p^{\mathrm{hor}}(T))\,\log(p),$$

where $|\Lambda_T|$ *is as in (7.2.10).*

Proof. The condition $\det(T) \neq 0$ allows us the make essentially the same analysis as on pp. 215–216 of [3]. Using (7.6.5), we have

$$\frac{C_p(T)}{\log(p)} = \sum_{\substack{\mathbf{y}\in(L'[p^{-1}])^2 \\ Q(\mathbf{y})=T \\ \bmod \Gamma'}} \chi(\Gamma'_{\mathbf{y}} \setminus \mathcal{Z}^{\bullet}(\mathbf{j}), \mathcal{F}). \tag{7.6.17}$$

The stabilizer of $\mathbf{y}$ in Γ' is $\Gamma' \cap Z'(\mathbb{Q})$, where $Z'(\mathbb{Q})$ is the center of $H'(\mathbb{Q})$. Thus $\Gamma' \cap Z'(\mathbb{Q}) \simeq \mathbb{Z}[p^{-1}]^{\times} \simeq \{\pm 1\} \times \mathbb{Z}$, and the generator of the infinite factor of this group acts on $\mathcal{Z}^{\bullet}(\mathbf{j})$ by translating the 'sheet' by 2. The contribution of $\mathbf{y}$ to (7.6.17) is then

$$\begin{aligned} e_{\mathbf{y}}^{-1} \cdot 2 \cdot \big(\, \chi(\mathcal{Z}(\mathbf{j}), \mathcal{O}_{\hat{\mathcal{C}}_1} \otimes^{\mathbb{L}} \mathcal{O}_{\hat{\mathcal{C}}_2}) - \chi(\mathcal{Z}(\mathbf{j}), \mathcal{G}) \,\big) \\ = e_{\mathbf{y}}^{-1} \cdot (\nu_p(T) - \nu_p^{\mathrm{hor}}(T)). \end{aligned} \tag{7.6.18}$$

Here the 2 arises from the two sheets, as in (8.40) of [3], and $e_{\mathbf{y}} = 2$ is the order of the remaining part of the stabilizer of $\mathbf{y}$. □

Finally, we determine the contribution of the two singular T's which occur in the sum (7.6.14).

Proposition 7.6.4. *For a matrix* $T \in \mathrm{Sym}_2(\mathbb{Z})^{\vee}$ *with* $\mathrm{diag}(T) = (t_1, t_2)$ *and* $\det(T) = 0$, *i.e., for*

$$T = \begin{pmatrix} t_1 & m \\ m & t_2 \end{pmatrix}, \qquad m^2 = t_1 t_2,$$

write $t_1 = n_1 t$ *and* $t_2 = n_2 t$ *with* $(n_1, n_2) = 1$, *and let* $4t = n^2 d$, *where* $-d$ *is the discriminant of the quadratic extension* $\boldsymbol{k} = \mathbb{Q}(\sqrt{-t})$. *Let* $k = \mathrm{ord}_p(n)$ *and let* $\chi = \chi_d(p)$.
(i) If p is ramified or inert in $\mathbb{Q}(\sqrt{-t})$, *then*

$$C_p(T) = \frac{1}{2} \deg \mathcal{Z}(t)_{\mathbb{Q}} \cdot \tilde{\nu}_p(T) \cdot \log(p).$$

Here,

$$\tilde{\nu}_p(T) = \mathrm{ord}_p(t_1 t_2) + 2(1+\chi-\mathrm{ord}_p(d/4)) - \begin{cases} \frac{(p+1)(p^k-1)}{p-1} & \text{if } p \text{ is inert,} \\ 2\,\frac{p^{k+1}-1}{p-1} & \text{if } p \text{ is ramified.} \end{cases}$$

(ii) If p is split in $\mathbb{Q}(\sqrt{-t})$, *then*

$$C_p(T) = \delta(d; D(B)/p) \cdot H_0(t; D(B)) \cdot \tilde{\nu}_p(T) \cdot \log(p),$$

with $\delta(d; D(B)/p)$ and $H_0(t; D(B))$ given by (7.6.21) and (7.6.22), and

$$\tilde{\nu}_p(T) = -2 \cdot (p^k - 1).$$

Proof. Since in this case the components y_1 and y_2 of $\mathbf{y}$ with $Q(\mathbf{y}) = T$, and hence j_1 and j_2, are colinear, the analysis is nearly the same as that dealt with in Section 11 of [4] for the cycle given by a single special endomorphism. More precisely, writing $t_1 = n_1^2 t$ and $t_2 = n_2^2 t$, with n_1 and n_2 relatively prime, we have $n_2 y_1 = n_1 y_2$ and we let $y = n_1^{-1} y_1 = n_2^{-1} y_2$, and similarly for the corresponding j_1, j_2 we let $j = n_1^{-1} j_1 = n_2^{-1} j_2$. There will be three cases depending on whether p is inert, ramified, or split in the field $\boldsymbol{k} = \mathbb{Q}(\sqrt{-t})$. Let

(7.6.19)

$$\alpha = \min\{\text{ord}_p(t_1), \text{ord}_p(t_2)\} \qquad \text{and} \qquad \beta = \max\{\text{ord}_p(t_1), \text{ord}_p(t_2)\}.$$

In particular, $\alpha = \text{ord}_p(t) = 2k + \text{ord}_p(d/4)$ and $\alpha + \beta = \text{ord}_p(t_1 t_2)$.

If p is either inert or ramified in $\boldsymbol{k} = \mathbb{Q}(\sqrt{-t})$, then the discussion of the first part of Section 11 of [4] can be applied, so that the intersection number attached to $\hat{\mathcal{C}}_T$ will be given by

$$\text{(7.6.20)} \quad \delta(d; D(B)) \cdot H_0(t; D(B)) \cdot \big[\, (\mathcal{Z}^{\text{hor}}(j_1), \mathcal{Z}^{\text{ver}}(j_2)) + (\mathcal{Z}^{\text{ver}}(j_1), \mathcal{Z}^{\text{hor}}(j_2)) + (\mathcal{Z}^{\text{ver}}(j_1), \mathcal{Z}^{\text{ver}}(j_2)) \,\big],$$

where the effect of the 'sheets' is taken into account; see Lemma 11.4 of [4]. Here

$$\text{(7.6.21)} \qquad \delta(d; D) = \prod_{\ell | D} (1 - \chi_d(\ell))$$

and

$$\text{(7.6.22)} \qquad H_0(t; D) = \frac{h(d)}{w(d)} \sum_{\substack{c | n \\ (c,D)=1}} c \prod_{\ell | c} (1 - \chi_d(\ell)\ell^{-1}).$$

Recall that by [4], Proposition 9.1,

$$\text{(7.6.23)} \qquad \deg \mathcal{Z}(t)_{\mathbb{Q}} = 2\,\delta(d; D(B))\, H_0(t; D(B)).$$

The sum of intersection numbers on the right-hand side of (7.6.20) can be computed using the results of [3], for $p \neq 2$, and of the Appendix to Section 11 of [4].

Recall from the appendix to Chapter 6 that, for any j, the vertical cycle $\mathcal{Z}^{\text{ver}}(j)$ is determined by data (S, μ), where $S = S(j)$ is a subset of the building $\mathcal{B}$ and $\mu = \mu(j) \geq 0$, by the formula

$$\mathcal{Z}^{\text{ver}}(j) = \sum_{[\Lambda]} \text{mult}_{[\Lambda]}(S, \mu)\, \mathbb{P}_{[\Lambda]}, \tag{7.6.24}$$

where the multiplicity function is given by

$$\text{mult}_{[\Lambda]}(S, \mu) = \max\{\mu - d([\Lambda], S), 0\}. \tag{7.6.25}$$

Write $q(j) = j^2 = \epsilon\, p^{\alpha}$, with $\alpha = \text{ord}_p(q(j))$, and let $\boldsymbol{k}_p = \mathbb{Q}_p(\sqrt{q(j)})$.[9] For p inert in $\boldsymbol{k}_p$, we have $S = [\Lambda_0]$ and,

$$\mu = \begin{cases} \frac{\alpha}{2} & \text{if } p \neq 2, \\ \frac{\alpha}{2} + 1 & \text{if } p = 2. \end{cases} \tag{7.6.26}$$

For p ramified in $\boldsymbol{k}_p$, we have $S = [[\Lambda_0, \Lambda_1]]$, the closure of an edge, and

$$\mu = \begin{cases} \frac{\alpha-1}{2} & \text{if } p \neq 2, \\ \frac{\alpha}{2} & \text{if } p = 2, \text{ and } \alpha \text{ is even}, \\ \frac{\alpha-1}{2} & \text{if } p = 2, \text{ and } \alpha \text{ is odd}. \end{cases} \tag{7.6.27}$$

For p split in $\boldsymbol{k}_p$, we have $S = \mathcal{A}$, an apartment in $\mathcal{B}$, and

$$\mu = \begin{cases} \frac{\alpha}{2} & \text{if } p \neq 2, \\ \frac{\alpha}{2} + 1 & \text{if } p = 2. \end{cases} \tag{7.6.28}$$

Now suppose that j_1 and j_2 are colinear special endomorphisms, such that

$$\{\text{ord}_p(q(j_1)), \text{ord}_p(q(j_2))\} = \{\alpha, \beta\},$$

with $\alpha \leq \beta$. The resulting pair of vertical cycles $\mathcal{Z}^{\text{ver}}(j_1)$, $\mathcal{Z}^{\text{ver}}(j_2)$ is determined by the collection of data (S, μ_1, μ_2).

If p is either inert or ramified, the data (S, μ_1, μ_2) coincides with one considered in [3] for a pair of *anticommuting* special endomorphisms j_1' and j_2', and so, by the chart (6.15) in [3],

$$(\mathcal{Z}^{\text{ver}}(j_1), \mathcal{Z}^{\text{ver}}(j_2)) = \begin{cases} -(p+1)\frac{p^{\mu}-1}{p-1} & \text{if } \chi_d(p) = -1, \\ 2 - 2\,\frac{p^{\mu+1}-1}{p-1} & \text{if } \chi_d(p) = 0, \end{cases}$$

---

[9]Note that, if j arises from a global special endomorphism y with $Q(y) = -y^2 = t$, then $q(j) = -t$. The same change of sign occurs in [3] and in the appendix to Chapter 6.

where $\mu = \min\{\mu_1, \mu_2\}$. Adding the quantity

$$(\mathcal{Z}^{\text{hor}}(j_1), \mathcal{Z}^{\text{ver}}(j_2)) + (\mathcal{Z}^{\text{ver}}(j_1), \mathcal{Z}^{\text{hor}}(j_2)) = 2\mu_1 + 2\mu_2, \tag{7.6.29}$$

and recalling (7.6.17), (7.6.26), and (7.6.27), we get the claimed value.

Finally, we consider the case in which p splits in $\boldsymbol{k}$, so that there are no horizontal components. The situation is now like the one considered after Lemma 11.4 in [4]. By (11.19) and (11.20) of [4], the intersection number attached to $\hat{C}_T$ will be given by

$$\delta(d; D(B)/p) \cdot H_0(t; D(B)) \tag{7.6.30}$$

times the intersection multiplicity of $\langle \epsilon(y)^{\mathbb{Z}} \rangle \backslash \mathcal{Z}(\mathbf{j})$. To compute this multiplicity, recall that the configuration $\mathcal{Z}(\mathbf{j})$ in question is now determined by data (S, μ_1, μ_2), where $S = \mathcal{A}$ is an apartment. We use the following easy projection formula:

Lemma 7.6.5. *Let* pr $: \hat{\Omega} \to \Delta\backslash\hat{\Omega}$ *be the natural projection, where* $\Delta = \langle \epsilon(y)^{\mathbb{Z}} \rangle$. *Let* $\mathcal{Z}_1 \subset \Delta\backslash\hat{\Omega}$ *and* $\mathcal{Z}_2 \subset \hat{\Omega}$ *be divisors such that the intersections of the supports* $|\mathcal{Z}_1| \cap |\text{pr}_*(\mathcal{Z}_2)|$ *and* $|\text{pr}^*(\mathcal{Z}_1)| \cap |\mathcal{Z}_2|$ *are subsets of the special fibers proper over the base. Then*

$$(\mathcal{Z}_1, \text{pr}_*(\mathcal{Z}_2))_{\Delta\backslash\hat{\Omega}} = (\text{pr}^*(\mathcal{Z}_1), \mathcal{Z}_2)_{\hat{\Omega}}.$$

Now take $\mathcal{Z}_1$ to be the image of $\mathcal{Z}(j_1)$ in $\Delta\backslash\hat{\Omega}$ and let $\mathcal{Z}_2$ be a fundamental domain for the action of Δ on $\mathcal{Z}(j_2)$. Since Δ acts by translation by 2 along the fixed apartment, we can take $\mathcal{Z}_2$ to be the collection of $\mathbb{P}_{[\Lambda]}$'s associated to two neighboring vertices on the apartment and to the set of all vertices joined to them by geodesics running away from the apartment and having a distance at most μ_2 from them. The $\mathbb{P}_{[\Lambda]}$'s occur with the multiplicity associated to $\mathcal{Z}(j_2)$, i.e.,

$$\text{mult}(\mathbb{P}_{[\Lambda]}) = \text{mult}_{[\Lambda]}(j_2) = \mu_2 - d([\Lambda], \mathcal{B}^j). \tag{7.6.31}$$

We then compute the sum of the intersection numbers of components in $\mathcal{Z}(j_1)$ and $\mathcal{Z}_2$, as in the proof of Theorem 6.1 of [3]. The total is

$$-2(p^{\mu} - 1), \tag{7.6.32}$$

where $\mu = \min\{\mu_1, \mu_2\}$, and we get the claimed result from (7.6.28). □

7.7 CONTRIBUTIONS FOR $p \nmid D(B)$

In this section we compute the quantity

$$(7.7.1) \qquad A_p = \sum_{\substack{c_1, c_2 \\ c_1 | n_1 n,\ c_2 | n_2 n \\ c_1 \neq c_2}} \langle\, \widehat{\mathcal{Z}}^{\text{hor}}(t_1 : c_1, v_1), \widehat{\mathcal{Z}}^{\text{hor}}(t_2 : c_2, v_2)\,\rangle_p$$

of (7.4.23) in the case $p \nmid D(B)$. Note that

$$(7.7.2) \quad \langle\, \widehat{\mathcal{Z}}^{\text{hor}}(t_1 : c_1, v_1), \widehat{\mathcal{Z}}^{\text{hor}}(t_2 : c_2, v_2)\,\rangle_p \\ = (\, \mathcal{Z}^{\text{hor}}(t_1 : c_1), \mathcal{Z}^{\text{hor}}(t_2 : c_2)\,)_p \cdot \log(p),$$

where $(\ ,\ )_p$ is the geometric intersection number at points in the fiber $\mathcal{M}_p$.

Lemma 7.7.1. *(i) If $p \nmid D(B)$ is not split (resp. split) in $\boldsymbol{k}_t = \mathbb{Q}(\sqrt{-t})$, then all points of $\mathcal{Z}^{\text{hor}}(t : c)(\bar{\mathbb{F}}_p)$ are supersingular (resp. ordinary).*
(ii) Suppose that $\tilde{x} \in \mathcal{Z}^{\text{hor}}(t : c)(\bar{\mathbb{F}}_p)$ corresponds to (A, ι, y), and let

$$\psi_y : \boldsymbol{k}_t \hookrightarrow \text{End}(A, \iota) \otimes_{\mathbb{Z}} \mathbb{Q}, \qquad \sqrt{-t} \mapsto y.$$

Then, the order $\psi_y^{-1}(\text{End}(A, \iota))$ in $\boldsymbol{k}_t$ has conductor c^0, where $c = c^0\, p^s$ with $p \nmid c^0$.
(iii) Suppose that p is split in $\boldsymbol{k}_t$. Then the embedding ψ_y in (ii) is an isomorphism, and

$$V(A, \iota) \otimes_{\mathbb{Z}} \mathbb{Q} \ \simeq\ \{\, x \in \boldsymbol{k}_t \mid \text{tr}(x) = 0\,\}$$

is one-dimensional. For $c_1 = c_1^0\, p^{s_1}$ and $c_2 = c_2^0\, p^{s_2}$ with $p \nmid c_i^0$,

$$c_1^0 \neq c_2^0 \quad \Longrightarrow \quad (\, \mathcal{Z}^{\text{hor}}(t_1 : c_1), \mathcal{Z}^{\text{hor}}(t_2 : c_2)\,)_p = 0.$$

Proof. (i) Let $x \in \mathcal{M}(\bar{\mathbb{F}}_p)$ corresponding to (A, ι) be a point in the image of $\mathcal{Z}^{\text{hor}}(t : c)(\bar{\mathbb{F}}_p)$. Then the action of $O_B \otimes \mathbb{Z}_p \simeq M_2(\mathbb{Z}_p)$ shows that the p-divisible group of (A, ι) is isomorphic to X^2, where X is a p-divisible group of dimension 1 and height 2. If p splits in $\boldsymbol{k}_t$, then $X \sim \hat{\mathbb{G}}_m \times \mathbb{Q}_p/\mathbb{Z}_p$ and conversely, which proves (i).

To prove (ii), let $\tilde{x}$ be the specialization of a point $\underline{\tilde{x}}$ of $\mathcal{Z}^{\text{hor}}(t : c)$ with values in a complete discrete valuation ring with residue field $\bar{\mathbb{F}}_p$ and fraction field of characteristic 0. If $\underline{\tilde{x}}$ corresponds to $(\underline{A}, \underline{\iota}, \underline{y})$ then the order

$\psi_{\underline{y}}^{-1}(\mathrm{End}(\underline{A}\,,\underline{\iota}))$ in $\boldsymbol{k}_t$ has conductor c. We have the commutative diagram

$$\begin{array}{ccc} \boldsymbol{k}_t & \overset{\psi_{\underline{y}}}{\hookrightarrow} & \mathrm{End}(\underline{A}\,,\underline{\iota})\otimes\mathbb{Q} \\ \| & & \cap\!\downarrow \\ \boldsymbol{k}_t & \overset{\psi_y}{\hookrightarrow} & \mathrm{End}(A,\iota)\otimes\mathbb{Q} \end{array} \quad . \tag{7.7.3}$$

Hence it suffices to show that the inclusion $\mathrm{End}(\underline{A}\,,\underline{\iota}) \hookrightarrow \mathrm{End}(A,\iota)$ induces an isomorphism

$$\psi_{\underline{y}}^{-1}(\mathrm{End}(\underline{A}\,,\underline{\iota})\otimes\mathbb{Z}[p^{-1}]) = \psi_y^{-1}(\mathrm{End}(A,\iota)\otimes\mathbb{Z}[p^{-1}]). \tag{7.7.4}$$

This comes down to showing that any endomorphism α of $(\underline{A}\,,\underline{\iota})$ whose reduction modulo p is divisible by a prime number $\ell \neq p$ in $\mathrm{End}(A,\iota)$ is itself divisible by ℓ. But α is divisible by ℓ if and only if α annihilates the group scheme $\underline{A}\,[\ell]$. Since this group scheme is étale, the assertion follows.

(iii) The first assertion of (iii) is trivial. To prove the second, suppose that (A,ι) is a point of intersection of $\mathcal{Z}(t_1:c_1)$ and $\mathcal{Z}(t_2:c_2)$ in the fiber $\mathcal{M}_p$ for a prime p split in $\boldsymbol{k}_t$. Thus, (A,ι) has special endomorphisms y_1 and y_2 with $Q(y_1)=t_1$ and $Q(y_2)=t_2$. By (ii), the orders $\psi_{y_1}^{-1}(\mathrm{End}(A,\iota))$ and $\psi_{y_2}^{-1}(\mathrm{End}(A,\iota))$ have conductors c_1^0 and c_2^0 respectively. On the other hand, by the first part of (iii), $\psi_{y_2}=\psi_{\pm y_1}$, so that $c_1^0=c_2^0$. □

Remark 7.7.2. If p is not split in $\boldsymbol{k}_t$, and if (A,ι) corresponds to a point $x\in\mathcal{M}(\bar{\mathbb{F}}_p)^{\mathrm{ss}}$, then the space $V(A,\iota)\otimes\mathbb{Q}$ has dimension 3, so that there can be pairs $\mathbf{y}=[y_1,y_2]$ of special endomorphisms with nonsingular fundamental matrix $T=Q(\mathbf{y})$. The corresponding orders $\psi_{y_1}^{-1}(\mathrm{End}(A,\iota))$ and $\psi_{y_2}^{-1}(\mathrm{End}(A,\iota))$ can have conductors c_1^0 and c_2^0 with $c_1^0\neq c_2^0$, where $c_1\mid n_1 n$ and $c_2\mid n_2 n$. In particular, $(\,\mathcal{Z}(t_1:c_1),\mathcal{Z}(t_2:c_2)\,)_p$ can be nonzero in this case.

By (i) of the lemma, we can write

$$(\,\mathcal{Z}(t_1:c_1),\mathcal{Z}(t_2:c_2)\,)_p = \sum_{x\in\mathcal{M}_p(\bar{\mathbb{F}}_p)^{\bullet}} (\,\mathcal{Z}(t_1:c_1),\mathcal{Z}(t_2:c_2)\,)_x, \tag{7.7.5}$$

where

$$\bullet = \begin{cases} \text{s.s.} & \text{if } p \text{ is not split in } \boldsymbol{k}, \\ \text{ord} & \text{if } p \text{ is split in } \boldsymbol{k}. \end{cases} \tag{7.7.6}$$

For a point $x \in \mathcal{M}_p(\bar{\mathbb{F}}_p)^\bullet$ corresponding to (A, ι), we now describe the local structure of a cycle $\mathcal{Z}^{\text{hor}}(t : c)$ in the formal neighborhood $(\mathcal{M})_x^\wedge$ of x. Let $y \in V(A, \iota)$ be a special endomorphism with $Q(y) = t = -y^2$. The corresponding embedding $\boldsymbol{k}_t = \mathbb{Q}(\sqrt{-t}) \to \text{End}(A, \iota)_{\mathbb{Q}}$ defines an embedding

$$\psi = \psi_y : \mathbb{Q}_p(\sqrt{-t}) \longrightarrow \text{End}(A(p), \iota) \otimes_{\mathbb{Z}_p} \mathbb{Q}_p, \tag{7.7.7}$$

where $A(p)$ is the p-divisible group of A. Note that the action of the idempotents in $O_B \otimes \mathbb{Z}_p \simeq M_p(\mathbb{Z}_p)$ gives a decomposition $A(p) \simeq X^2$, where X is the p-divisible group of a supersingular (resp. ordinary) elliptic curve over $\bar{\mathbb{F}}_p$ when p is not split (resp. split) in $\boldsymbol{k}_t$. Since ψ embeds $\mathbb{Q}_p(\sqrt{-t})$ into $\text{End}(X)_{\mathbb{Q}_p}$, we may apply the theory of quasicanonical liftings of Gross [1]. In the case in which p is split in $\boldsymbol{k}_t$, Gross's theory amounts to the classical Serre-Tate theory for an ordinary elliptic curve over $\bar{\mathbb{F}}_p$, as is pointed out at the end of [1], so we will use the same terminology in the two cases.

For $s \in \mathbb{Z}_{\geq 0}$, let $\mathcal{W}_s(\psi)$ be the quasi-canonical divisor of level s associated to ψ. Recall[10] that $\mathcal{W}_s(\psi)$ is a reduced, irreducible, regular divisor in the spectrum of the formal completion of $\mathcal{M}$ at x, such that the pullback of the universal p-divisible group on $\mathcal{M}$ to $\mathcal{W}_s(\psi)$ has as its endomorphism algebra the order $\mathbb{Z}_p + p^s O_{\boldsymbol{k}_p}$ of conductor p^s in $\boldsymbol{k}_p$. For example, when p splits in $\boldsymbol{k}$, $O_{\boldsymbol{k}_p} \simeq \mathbb{Z}_p \oplus \mathbb{Z}_p$, the order of conductor p^s is $\{(a, b) \in \mathbb{Z}_p^2 \mid a \equiv b \mod p^s\}$, and

$$\mathcal{W}_s(\psi) \simeq \text{Spf}\, W[T]/\Phi_s(T), \tag{7.7.8}$$

where $\Phi_s(T)$ is the cyclotomic polynomial whose roots are the primitive p^s-th roots of 1. Let M be the quotient field of $W = W(\bar{\mathbb{F}}_p)$, and let M_s be the Galois extension of M over which the quasi-canonical liftings of level s are defined. Recall that the extension M_s/M is totally ramified, and, in the case when p is split in $\boldsymbol{k}_t$, M_s is the extension generated by a primitive p^sth root of unity. Also note that

$$m_0(p) := |M_0 : M| = \begin{cases} 2 & \text{if } p \text{ is ramified in } \boldsymbol{k}_t, \\ 1 & \text{otherwise.} \end{cases} \tag{7.7.9}$$

We write

$$\mathcal{W}_s(\psi) = \text{Spf}\, W_s, \tag{7.7.10}$$

where W_s is the integral closure of W in M_s.

[10] Note that the fact that the canonical morphism from $\mathcal{W}_s(\psi)$ to $(\mathcal{M})_x^\wedge$ is a closed immersion follows from [1], Proposition 5.3, (3).

Lemma 7.7.3. *Suppose that $\tilde{x} \in \mathcal{Z}(t)(\bar{\mathbb{F}}_p)$ corresponding to (A, ι, y). Let $\mathcal{Z}(t)^{\wedge}_{\tilde{x}}$ be the formal completion of $\mathcal{Z}(t)$ at $\tilde{x}$. Then, for $4t = n^2 d$, as usual, there is an equality of formal schemes*

$$\mathcal{Z}(t)^{\wedge}_{\tilde{x}} = \bigcup_{s=0}^{\operatorname{ord}_p(n)} \mathcal{W}_s(\psi),$$

where $\psi = \psi_y : \boldsymbol{k}_{t,p} \hookrightarrow \operatorname{End}(A(p), \iota) \otimes_{\mathbb{Z}_p} \mathbb{Q}_p$ is the embedding corresponding to y.

Proposition 7.7.4. *Let $i : \mathcal{Z}(t) \to \mathcal{M}$ be the natural morphism, defined by $(A, \iota, y) \mapsto (A, \iota)$. Then i is unramified, and, for $x \in \mathcal{M}_p$ corresponding to (A, ι), there is an equality of formal divisors*[11]

$$(\, i_* \mathcal{Z}(t)\,)^{\wedge}_{x} = \sum_{\substack{y \in V(A,\iota) \\ Q(y)=t}} \sum_{s=0}^{\operatorname{ord}_p(n)} \mathcal{W}_s(\psi).$$

Similarly, if $c = c^0 p^s$ with $p \nmid c^0$, then

$$(\, i_* \mathcal{Z}^{\text{hor}}(t : c)\,)^{\wedge}_{x} = \sum_{\substack{y \in V(A,\iota) \\ Q(y)=t \\ \text{type}(y)=c^0}} \mathcal{W}_s(\psi).$$

Here, for $y \in V(A, \iota)$, $\text{type}(y) = c^0$ if the order $\psi_y^{-1}(\operatorname{End}(A, \iota))$ in $\boldsymbol{k}_t$ has conductor c^0.

Since the intersection number $(\, \mathcal{W}_{s_1}(\psi_{y_1}), \mathcal{W}_{s_2}(\psi_{y_2})\,)$ of two distinct quasi-canonical divisors is always finite, we obtain

Corollary 7.7.5. *Let $x \in \mathcal{M}_p(\bar{\mathbb{F}}_p)^{\bullet}$ be a point corresponding to (A, ι). For $c_1 \neq c_2$, write $c_1 = c_1^0 p^{s_1}$ and $c_2 = c_2^0 p^{s_2}$, where $p \nmid c_1^0$ and $p \nmid c_2^0$, as above. Then*

$$(\, \mathcal{Z}^{\text{hor}}(t_1 : c_1), \mathcal{Z}^{\text{hor}}(t_2 : c_2)\,)_x = \sum_{\substack{T \\ \text{diag}(T)=(t_1,t_2)}} \sum_{\substack{\mathbf{y}=[y_1,y_2] \in V(A,\iota)^2 \\ Q(\mathbf{y})=T \\ \text{type}(\mathbf{y})=(c_1^0,c_2^0)}} e_{\mathbf{y}}^{-1} (\, \mathcal{W}_{s_1}(\psi_{y_1}), \mathcal{W}_{s_2}(\psi_{y_2})\,).$$

[11]Note that, elsewhere in this paper, we have been omitting i_* from the notation, often writing $\mathcal{Z}(t)$ for $i_* \mathcal{Z}(t)$.

Taking the sum on c_1, c_2, and x and regrouping according to the fundamental matrices T, we obtain

Corollary 7.7.6.

$$A_p = \sum_{\substack{T \\ \mathrm{diag}(T)=(t_1,t_2)}} A_p(T),$$

where

$$A_p(T) = \sum_{x \in \mathcal{M}(\bar{\mathbb{F}}_p)^\bullet} \sum_{\substack{c_1, c_2 \\ c_1 | n_1 n,\ c_2 | n_2 n \\ c_1 \neq c_2}} \sum_{\substack{\mathbf{y}=[y_1,y_2] \in V(A,\iota)^2 \\ Q(\mathbf{y})=T \\ \mathrm{type}(\mathbf{y})=(c_1^0,c_2^0)}} e_{\mathbf{y}}^{-1} \big(\mathcal{W}_{s_1}(\psi_{y_1}), \mathcal{W}_{s_2}(\psi_{y_2}) \big) \cdot \log(p).$$

For p inert or ramified in $\boldsymbol{k}$, both singular and nonsingular matrices T can make a nonzero contribution to A_p. For nonsingular T's, the expression for $A_p(T)$ given in Corollary 7.7.6 will be just what is needed for the comparison to be made in Section 7.9, so it only remains to determine the contribution of singular T's. On the other hand, for p split in $\boldsymbol{k}_t$, *only* singular T's can contribute to A_p, due to (iii) of Lemma 7.7.1. Thus, for the remainder of this section, we consider the quantity $A_p(T)$ for $T \in \mathrm{Sym}_2(\mathbb{Z})^\vee$ with $\mathrm{diag}(T) = (t_1, t_2)$ and $\det(T) = 0$. Note that there are precisely two such T's.

When $\det(T) = 0$, a pair of special endomorphisms $\mathbf{y} \in V(A,\iota)^2$ with fundamental matrix $Q(\mathbf{y}) = T$ has colinear components y_1 and y_2. By (ii) of Lemma 7.7.1, it follows that $c_1^0 = \mathrm{type}(y_1) = \mathrm{type}(y_2) = c_2^0$. Since $c_1 \mid n_1 n$ and $c_2 \mid n_2 n$, and $(n_1, n_2) = 1$, we must have $c_1^0 = c_2^0 = c^0$ with $c^0 \mid n$. Thus, we obtain,

(7.7.11)

$$A_p(T) = \sum_{x \in \mathcal{M}(\bar{\mathbb{F}}_p)^\bullet} \sum_{c^0} \sum_{\substack{\mathbf{y}=[y_1,y_2] \in V(A,\iota)^2 \\ Q(\mathbf{y})=T \\ \mathrm{type}(\mathbf{y})=(c^0,c^0)}} e_{\mathbf{y}}^{-1} \sum_{s_1=0}^{\mathrm{ord}_p(n_1 n)} \sum_{\substack{s_2=0 \\ s_1 \neq s_2}}^{\mathrm{ord}_p(n_2 n)} \big(\mathcal{W}_{s_1}(\psi_{y_1}), \mathcal{W}_{s_2}(\psi_{y_2}) \big) \cdot \log(p)$$

$$= |\Lambda_T| \cdot \sum_{s_1=0}^{\mathrm{ord}_p(n_1 n)} \sum_{\substack{s_2=0 \\ s_1 \neq s_2}}^{\mathrm{ord}_p(n_2 n)} (\, \mathcal{W}_{s_1}(\psi), \mathcal{W}_{s_2}(\psi) \,) \cdot \log(p).$$

Here c^0 runs over divisors of n which are prime to $p\,D(B)$, and

$$\begin{aligned} |\Lambda_T| &= \sum_{x \in \mathcal{M}(\bar{\mathbb{F}}_p)^\bullet} \sum_{\substack{\mathbf{y} \in V(A,\iota)^2 \\ Q(\mathbf{y})=T}} e_{\mathbf{y}}^{-1} \\ &= \sum_{x \in \mathcal{M}(\bar{\mathbb{F}}_p)^\bullet} \sum_{\substack{y \in V(A,\iota) \\ Q(y)=t}} e_y^{-1} \\ &=: |\Lambda_t|. \end{aligned} \tag{7.7.12}$$

Here, we have used the bijection given by the map $y \mapsto [n_1 y, n_2 y] = \mathbf{y}$ and that $e_y = e_{\mathbf{y}}$ when y is mapped to $\mathbf{y}$.

Proposition 7.7.7. *(i) For an embedding $\psi : \mathbb{Q}_p(\sqrt{-t}) \to \mathrm{End}^0(X)$, and for $s = \min\{s_1, s_2\}$,*

$$(\mathcal{W}_{s_1}(\psi), \mathcal{W}_{s_2}(\psi)) = |M_s : M| = m_0(p) \cdot \begin{cases} 1 & \text{if } s = 0, \\ p^{s-1}(p - \chi_d(p)) & \text{if } s \geq 1. \end{cases}$$

(ii)

$$\begin{aligned} \deg \mathcal{Z}(t)_{\mathbb{Q}} &= |\Lambda_t| \cdot \sum_{s=0}^{\mathrm{ord}_p(n)} |M_s : M| \\ &= |\Lambda_t| \cdot m_0(p) \cdot \left(1 + (p - \chi_d(p)) \frac{p^{\mathrm{ord}_p(n)} - 1}{p-1} \right), \end{aligned}$$

where $|\Lambda_t|$ is given by (7.7.12) and $m_0(p) = |M_0 : M|$, as in (7.7.9).

Proof. The closed embedding of $\mathcal{W}_s(\psi)$ into $\mathcal{M}_x^\wedge$ is given by a homomorphism of complete W-algebras,

$$W[[T]] \longrightarrow W_s, \ \ T \longmapsto \pi_s \tag{7.7.13}$$

where π_s is a uniformizer of W_s, provided that $|M_s : M| > 1$. In this case, since M_s/M is totally ramified we have

$$W_s = W[T]/(P_s(T)), \tag{7.7.14}$$

where $P_s(T)$ is the minimal polynomial of π_s over M and is an Eisenstein polynomial of degree $|M_s : M|$. Now for $s_1 < s_2$, and if $|M_{s_1} : M| > 1$, we obtain

$$\begin{aligned}(\mathcal{W}_{s_1}(\psi), \mathcal{W}_{s_2}(\psi)) &= \text{length } W_{s_1} \otimes_{W[[T]]} W_{s_2} \\ &= \text{length } (W[T]/(P_{s_1}(T))) \otimes_{W[T]} (W[T]/(P_{s_2}(T))) \qquad (7.7.15)\\ &= \text{length } W_{s_2}/(P_{s_1}(\pi_{s_2})) \\ &= |M_{s_1} : M|.\end{aligned}$$

The situation is a little different when $M_{s_1} = M$. In this case, in (7.7.13), $T \in W[[T]]$ is sent to some element $a \in W$, and the same argument as before works provided $|M_{s_2} : M| > 1$. There is one exceptional case which occurs when $p = 2$ is split in $\boldsymbol{k}$, $s_1 = 0$ and $s_2 = 1$. In this case, $M_0 = M_1 = M$. Then one checks that the maps $W[[T]] \to W_0$ and $W[[T]] \to W_1$ send T to 0 and -2, respectively, and hence $W_0 \otimes_{W[[T]]} W_1 \simeq W/2W$, so that the formula in the proposition is again true.

Part (ii) follows from Lemma 7.7.3 and (7.7.12), since

$$\deg \mathcal{Z}(t)_{\mathbb{Q}} = \sum_{x \in \mathcal{M}(\bar{\mathbb{F}}_p)^{\bullet}} \sum_{\substack{y \in V(A,\iota) \\ Q(y)=t}} e_y^{-1} \sum_{s=0}^{\text{ord}_p(n)} |M_s : M|, \qquad (7.7.16)$$

where the inner sum is independent of y. □

Corollary 7.7.8. *For $T \in \text{Sym}_2(\mathbb{Z})^\vee$ with $\text{diag}(T) = (t_1, t_2)$ and $\det(T) = 0$,*

$$\begin{aligned}\frac{A_p(T)}{\log(p)} &= \deg \mathcal{Z}(t)_{\mathbb{Q}} \cdot \left[\text{ord}_p(n_1 n_2) + 2\text{ord}_p(n)\right] \\ &\quad - 2\,|\Lambda_t| \cdot m_0(p) \cdot \frac{p-\chi}{p-1}\left(kp^k - \frac{p^k-1}{p-1}\right),\end{aligned}$$

where $\chi = \chi_d(p)$ and $k = \text{ord}_p(n)$.

Proof. Setting

$$k = \min\{\, \text{ord}_p(n_1 n), \text{ord}_p(n_2 n)\,\}$$

and

$$\ell = \max\{\, \text{ord}_p(n_1 n), \text{ord}_p(n_2 n)\,\},$$

we have $k = \mathrm{ord}_p(n)$ and $\ell - k = \mathrm{ord}_p(n_1 n_2)$. Then, by (7.7.11) and Proposition 7.7.7,

$$A_p(T) = |\Lambda_t| \cdot \mu_p(k, \ell) \log(p), \tag{7.7.17}$$

where

$$\begin{aligned} \mu_p(k,\ell) &= \sum_{\substack{s_1=0 \\ s_1 \neq s_2}}^{k} \sum_{s_2=0}^{\ell} |M_{\min\{s_1,s_2\}} : M| \\ &= (\ell - k) \sum_{s=0}^{k} |M_s : M| + 2 \sum_{s=0}^{k-1} |M_s : M| \, (k-s) \\ &= [\,(\ell - k) + 2k\,] \sum_{s=0}^{k} |M_s : M| - 2 \sum_{s=0}^{k} |M_s : M| \, s. \quad \square \end{aligned} \tag{7.7.18}$$

7.8 COMPUTATION OF THE DISCRIMINANT TERMS

In this section we will calculate the discriminant term occurring in Corollary 7.5.3:

$$\mathfrak{d}_{\mathcal{Z}^{\mathrm{hor}}(t)} = \sum_{c|n} \mathfrak{d}_{\mathcal{Z}^{\mathrm{hor}}(t:c)} = \sum_{c|n} \log |\mathcal{D}(c)^{-1} : O(c)|. \tag{7.8.1}$$

Here $O(c)$ is the ring of regular functions on $\mathcal{Z}^{\mathrm{hor}}(t : c)$ and

$$\mathcal{D}(c)^{-1} = \{ \alpha \in O(c) \otimes_{\mathbb{Z}} \mathbb{Q} \mid \mathrm{tr}(\alpha \cdot O(c)) \subset \mathbb{Z} \}. \tag{7.8.2}$$

Let $\tilde{\mathcal{Z}}^{\mathrm{hor}}(t : c)$ be the normalization of $\mathcal{Z}^{\mathrm{hor}}(t : c)$ and let $\tilde{O}(c)$ be its ring of regular functions. The inclusion $O(c) \subset \tilde{O}(c)$ corresponds to the normalization morphism

$$\pi : \tilde{\mathcal{Z}}^{\mathrm{hor}}(t : c) \longrightarrow \mathcal{Z}^{\mathrm{hor}}(t : c). \tag{7.8.3}$$

Let $\tilde{\mathcal{D}}(c)^{-1} = \{\alpha \in O(c) \otimes_{\mathbb{Z}} \mathbb{Q} \mid \mathrm{tr}(\alpha \cdot \tilde{O}(c)) \subset \mathbb{Z}\}$ be the dual of $\tilde{O}(c)$ with respect to the trace form. Then

$$O(c) \subset \tilde{O}(c) \subset \tilde{\mathcal{D}}(c)^{-1} \subset \mathcal{D}(c)^{-1} \tag{7.8.4}$$

and $|\tilde{O}(c) : O(c)| = |\mathcal{D}(c)^{-1} : \tilde{\mathcal{D}}(c)^{-1}|$, so that

$$\mathfrak{d}_{\mathcal{Z}^{\mathrm{hor}}(t:c)} = 2 \cdot \log |\tilde{O}(c) : O(c)| + \log |\tilde{\mathcal{D}}(c)^{-1} : \tilde{O}(c)|. \tag{7.8.5}$$

We then write

$$\mathfrak{d}_{\mathcal{Z}^{\mathrm{hor}}(t)} = \sum_{p} (\, 2\delta_p(t) + \partial_p(t) \,) \log(p), \tag{7.8.6}$$

where

$$\delta_p(t) = \sum_c \operatorname{ord}_p|\tilde{O}(c) : O(c)|, \tag{7.8.7}$$

and

$$\partial_p(t) = \sum_c \operatorname{ord}_p|\tilde{\mathcal{D}}(c)^{-1} : \tilde{O}(c)|. \tag{7.8.8}$$

To calculate $\delta_p(t)$, recall Serre's δ-invariant [7] associated to a point $\tilde{x} \in \mathcal{Z}^{\text{hor}}(t : c)(\bar{\mathbb{F}}_p)$, given by

$$\delta_{\tilde{x}} = \text{length } (\, \pi_* \mathcal{O}_{\tilde{\mathcal{Z}}^{\text{hor}}(t:c)} / \mathcal{O}_{\mathcal{Z}^{\text{hor}}(t:c)}\,)_{\tilde{x}}. \tag{7.8.9}$$

Then, $\delta_{\tilde{x}} = 0$ unless $\tilde{x}$ is a singular point of $\mathcal{Z}^{\text{hor}}(t : c)$. Furthermore

$$\delta_p(t) = \sum_{c|n} \operatorname{ord}_p|\tilde{O}(c) : O(c)| = \sum_{c|n} \sum_{\tilde{x} \in \mathcal{Z}^{\text{hor}}(t:c)(\bar{\mathbb{F}}_p)} \delta_{\tilde{x}}. \tag{7.8.10}$$

To calculate $\delta_{\tilde{x}}$, we use the following elementary fact.

Lemma 7.8.1. *Let (X, x) be the spectrum of a complete regular local ring R of dimension 2. Let C be a reduced divisor on X through x and let*

$$C = \sum_{i=1}^{n} C_i$$

be the decomposition of C into its formal branches through x, i.e., C_i is irreducible, for all i. We assume that C_i is regular at x, $\forall i$. Let

$$\pi : \tilde{C} \longrightarrow C$$

be the normalization morphism, i.e., $\tilde{C} = \coprod_{i=1}^{n} C_i$. Let

$$\delta_x = \text{length } (\pi_* \mathcal{O}_{\tilde{C}} / \mathcal{O}_C)_x$$

be the Serre invariant of C at x. Then

$$2\delta_x = \sum_{i \neq j} (C_i, C_j)_x.$$

Proof. We proceed by induction on n, the case $n = 1$ being trivial since then $\delta = 0$. Let $C' = \sum_{i=1}^{n-1} C_i$, so that $C = C' + C_n$. Then the normalization morphism factors as

$$\tilde{C} \xrightarrow{\pi'} C' \textstyle\coprod C_n \xrightarrow{\pi_1} C. \tag{7.8.11}$$

Let $f' = 0$, resp. $f_n = 0$, be the equations of C', resp. C_n. Then f' and f_n are relatively prime elements of R. We have the following diagram of local rings at x with exact rows and columns:

$$\begin{array}{ccccccccc} & & 0 & & 0 & & 0 & & \\ & & \uparrow & & \uparrow & & \uparrow & & \\ 0 & \to & \mathcal{O}_C & \to & \mathcal{O}_{C'} \oplus \mathcal{O}_{C_n} & \to & \Delta & \to & 0 \\ & & \uparrow & & \uparrow & & \uparrow & & \\ 0 & \to & R & \to & R \oplus R & \to & R & \to & 0 \\ & & \uparrow & & \uparrow & & \uparrow & & \\ 0 & \to & (f' f_n) & \to & (f') \oplus (f_n) & \to & (f', f_n) & \to & 0 \\ & & \uparrow & & \uparrow & & \uparrow & & \\ & & 0 & & 0 & & 0 & & \end{array} \tag{7.8.12}$$

The first row shows that length $\Delta =$ length $(\pi_{1*}(\mathcal{O}_{C' \sqcup C_n})/\mathcal{O}_C)_x$. The last column shows that length $\Delta = (C', C_n)_x$. By induction hypothesis,

$$\text{length}\,(\pi'_*(\mathcal{O}_{\tilde{C}})/\mathcal{O}_{C'})_x = \sum_{1 \le i < j \le n-1} (C_i, C_j)_x. \tag{7.8.13}$$

Hence

$$\begin{aligned} \delta_x &= \text{length}\,(\pi_*(\mathcal{O}_{\tilde{C}})/\mathcal{O}_C)_x \\ &= \text{length}\,(\pi'_*(\mathcal{O}_{\tilde{C}})/\mathcal{O}_{C' \sqcup C_n})_{(x,x)} \\ &\qquad\qquad + \text{length}\,(\pi'_{1*}(\mathcal{O}_{C' \sqcup C_n})/\mathcal{O}_C)_x \\ &= \sum_{1 \le i < j \le n-1} (C_i, C_j)_x + \sum_{i=1}^{n-1} (C_i, C_n)_x \\ &= \sum_{1 \le i < j \le n} (C_i, C_j)_x. \end{aligned} \tag{7.8.14}$$

□

From these calculations we obtain the following formula.

Proposition 7.8.2. *Let $t_1 t_2$ be a square and define t as in Lemma 7.4.3. Then*

$$2\,\delta_p(t) = \sum_{\substack{T \in \mathrm{Sym}_2(\mathbb{Z})^\vee \\ \mathrm{diag}(T) = (t_1, t_2) \\ \det(T) \neq 0}} 2\,\delta_p(T),$$

where $2\,\delta_p(T)$ is given as follows:
(i) If $p \nmid D(B)$, then, for p split in $\boldsymbol{k}_t$, $2\delta_p(T) = 0$, while for p inert or

ramified,

$$2\,\delta_p(T) \;=\; \sum_{x\in\mathcal{M}(\bar{\mathbb{F}}_p)^{ss}} \sum_{c^0} \sum_{\substack{\mathbf{y}=[y_1,y_2]\in V(A,\iota)^2\\ Q(\mathbf{y})=T\\ \text{type}(\mathbf{y})=(c^0,c^0)}} e_{\mathbf{y}}^{-1} \sum_{s=0}^{\text{ord}_p(n)} (\mathcal{W}_s(\psi_{y_1}),\mathcal{W}_s(\psi_{y_2})),$$

where c^0 runs over divisors of n which are prime to $pD(B)$.
(ii) If $p \mid D(B)$, then

$$\begin{aligned} 2\,\delta_p(T) &= \sum_{c} \sum_{\mathbf{y}=[y_1,y_2]\in\Lambda_T(c,c)} e_{\mathbf{y}}^{-1}\; 2\,(\,\mathcal{Z}^{\text{hor}}(j_1),\mathcal{Z}^{\text{hor}}(j_2)\,)\\ &= \sum_{c} |\Lambda_T(c,c)|\cdot\nu_p^{\text{hor}}(T), \end{aligned}$$

where c runs over divisors of n which are prime to $D(B)$.

Note that $\delta_p(T)$ can only be nonzero when $\text{Diff}(T,B)=\{p\}$, and that, in (ii), the conductor c is prime to p.

Proof. By (7.8.10), we must sum the intersection number expression of Lemma 7.8.1 over the (singular) points $x\in i_*\mathcal{Z}^{\text{hor}}(t:c)(\bar{\mathbb{F}}_p)$.

First suppose that $p\nmid D(B)$. Using the expression for $(\,i_*\mathcal{Z}^{\text{hor}}(t:c)\,)^{\wedge}_x$ given by Proposition 7.7.4, we have

$$(7.8.15)\quad 2\,\delta_x = \sum_{\substack{T'\in\text{Sym}_2(\mathbb{Z})^\vee\\ \text{diag}(T')=(t,t)\\ \det(T')\neq 0}} \sum_{\substack{\mathbf{y}'=[y_1',y_2']\in V(A,\iota)^2\\ \mathbb{Q}(\mathbf{y}')=T'\\ \text{type}(\mathbf{y}')=(c^0,c^0)}} e_{\mathbf{y}'}^{-1}\,(\,\mathcal{W}_s(\psi_{y_1'}),\mathcal{W}_s(\psi_{y_2'})\,),$$

where $c=c^0\,p^s$ with $s=\text{ord}_p(c)$. If p is split in $\boldsymbol{k}_t$, (iii) of Lemma 7.8.1 implies that the sum here is empty so that $\delta_p(t)=0$. If p is inert or ramified, we use the following fact.

Lemma 7.8.3. *There is a bijection between the sets*

$$\{\,\mathbf{y}'\in V(A,\iota)^2 \mid Q(\mathbf{y}')=T',\ \text{type}(\mathbf{y}')=(c^0,c^0)\,\}$$

and

$$\{\,\mathbf{y}\in V(A,\iota)^2 \mid Q(\mathbf{y})=T,\ \text{type}(\mathbf{y})=(c^0,c^0)\,\}$$

given by $\mathbf{y}'=[y_1',y_2']\mapsto[n_1y_1',n_2y_2']=\mathbf{y}$. Furthermore, $\text{Aut}(A,\iota,\mathbf{y}')=\text{Aut}(A,\iota,\mathbf{y})$. *Here*

$$T=[n_1,n_2]\,T'\begin{bmatrix}n_1\\n_2\end{bmatrix}.$$

Proof. We must show that the map is surjective. It suffices to show that, if $y \in V(A,\iota)$ is of type c with $Q(y) = m^2 t$, where $c \mid n$ and $p \nmid c$, then $m^{-1}y \in \mathrm{End}(A,\iota)$. By the definition of the type, this is equivalent to showing that $\psi_y^{-1}(m^{-1}y) = \frac{n}{2}\sqrt{-d}$ is in the order O_{c^2d} of conductor c. If $d = 4d_0$, then $O_{c^2d} = \mathbb{Z}[c\sqrt{-d_0}] \ni n\sqrt{-d_0}$. If d is odd, we have $4t = n^2 d$, so that $2 \mid n$. But then, for c even, we have $\frac{c}{2}\sqrt{-d} \in O_{c^2d}$ and $\frac{c}{2} \mid \frac{n}{2}$, while, for c odd, $c\sqrt{-d} \in O_{c^2d}$ and $c \mid \frac{n}{2}$. □

Since $\mathcal{W}_s(\psi_{n_i y_i'}) = \mathcal{W}_s(\psi_{y_i'})$, we obtain the claimed expression by summing (7.8.15) over x and c^0 and collecting the terms with a fixed T.

The proof in the case $p \mid D(B)$ is similar. Again, by the argument in Remark 7.6.1, the term in between the two equality signs in (ii) only depends on T. □

Next, we calculate the quantity $\partial_p(t)$.

Proposition 7.8.4. *(i) For $p \nmid D(B)$,*

$$\partial_p(t) = |\Lambda_t| \cdot m_0(p) \cdot \left[\frac{p-\chi}{p-1}\left(k\,p^k - 2\,\frac{p^k-1}{p-1}\right) + \frac{1-\chi}{p-1}\,k \right] + \frac{1}{2}\deg \mathcal{Z}(t)_{\mathbb{Q}} \cdot \mathrm{ord}_p(d),$$

where $\chi = \chi_d(p)$, $k = \mathrm{ord}_p(n)$, and $m_0(p) = 2$ if p is ramified in $\boldsymbol{k}_t$ and 1 otherwise.
(ii) For $p \mid D(B)$,

$$\partial_p(t) = \frac{1}{2}\deg \mathcal{Z}(t)_{\mathbb{Q}} \cdot \mathrm{ord}_p(d).$$

Proof. Using elementary properties of the different, we have

$$\begin{aligned} &\mathrm{ord}_p|\mathcal{O}_{\tilde{\mathcal{Z}}^{\mathrm{hor}}(t:c)}/\tilde{\mathcal{D}}(c)| \\ &= \mathrm{ord}_p|\mathcal{O}_{\tilde{\mathcal{Z}}^{\mathrm{hor}}(t:c)_{\mathbb{Z}_p}}/\tilde{\mathcal{D}}(c)_{\mathbb{Z}_p}| \qquad (7.8.16)\\ &= \sum_{\tilde{x}\in\tilde{\mathcal{Z}}^{\mathrm{hor}}(t:c)(\bar{\mathbb{F}}_p)} e_{\tilde{x}}^{-1}\ \mathrm{length}\ (\mathcal{O}_{\tilde{\mathcal{Z}}^{\mathrm{hor}}(t:c)_{(\tilde{x})}}/\tilde{\mathcal{D}}(c)_{(\tilde{x})}). \end{aligned}$$

Here $\tilde{\mathcal{D}}(c)$ denotes the sheaf associated to the module above with the same name.

If $p \nmid D(B)$, and summing over all c, we obtain

$$\partial_p(t) = \sum_{x \in \mathcal{M}(\bar{\mathbb{F}}_p)^\bullet} \sum_{\substack{y \in V(A,\iota) \\ Q(y)=t}} e_y^{-1} \sum_{s=0}^{\mathrm{ord}_p(n)} \partial(s) \tag{7.8.17}$$

$$= |\Lambda_t| \cdot \sum_{s=0}^{\mathrm{ord}_p(n)} \partial(s),$$

where

$$\partial(s) = \text{length } W_s : \mathcal{D}(W_s/W) = \mathrm{ord}_{M_s}(\mathcal{D}(M_s/M)), \tag{7.8.18}$$

for $\mathcal{D}(W_s/W) = \mathcal{D}(M_s/M)$, the different of W_s over W, hence of M_s over M.

Proposition 7.8.5. *For* $\chi = \chi_p(d)$,

$$\partial(s) = |M_0 : M| \cdot \left(p^{s-1}(p-\chi)\, s - \frac{p^s - \chi p^{s-1} + \chi - 1}{p-1} \right) + p^s \,\mathrm{ord}_p(d).$$

Proof. First suppose that p is split in $\boldsymbol{k}$. Then, $W_0 = W = W(\bar{\mathbb{F}}_p)$, and M_s is the totally ramified extension of $M_0 = M = W \otimes_{\mathbb{Z}_p} \mathbb{Q}_p$ generated by the p^sth roots of unity. Under the reciprocity isomorphism

$$(\mathbb{Z}_p/p^s\mathbb{Z}_p)^\times \xrightarrow{\sim} \mathrm{Gal}(M_s/M), \tag{7.8.19}$$

the ramification function i_G, [6], p. 70, pulls back to the function $i_G(a) = p^r$ if $\mathrm{ord}_p(a-1) = r$, for $r < s$. Then, by [6], Chapter 4, Proposition 4, we have, for $s \geq 1$, the well-known formula
(7.8.20)

$$\mathrm{ord}_{M_s}(\mathcal{D}(M_s/M)) = \sum_{\substack{a \in (\mathbb{Z}_p/p^s\mathbb{Z}_p)^\times \\ a \neq 1}} p^{\mathrm{ord}_p(a-1)} = p^{s-1}(p-1)s - p^{s-1}.$$

If p is inert or ramified in $\boldsymbol{k}$, we follow [1], but with slightly modified notation. Let $O = O_{\boldsymbol{k}_p}$ be the ring of integers of $\boldsymbol{k}_p$, and let π be a uniformizer. Let $q = |O/\pi O|$ be the order of the residue field. We identify M_0 with the completion of the maximal unramified extension of $\boldsymbol{k}_p$ and W_0 with its ring of integers. Recall that, if p is inert, $M_0 = M$, while, if p is ramified, then M_0 is a ramified quadratic extension of M. Let G be the base change to W_0 of the Lubin-Tate formal group over O for which

$$[\pi](x) = \pi x + x^q. \tag{7.8.21}$$

Let $L = M_0(G[\pi^n])$ be the extension of M_0 which is generated by all of the π^n-division points of G. Then, under the reciprocity law for Lubin-Tate groups, there is an isomorphism

$$(O/\pi^n O)^\times \xrightarrow{\sim} \mathrm{Gal}(L/M_0). \tag{7.8.22}$$

Moreover, the pullback under this isomorphism of the ramification function i_G on $\mathrm{Gal}(L/M_0)$ is given by

$$i_G(\alpha) = q^{\mathrm{ord}_O(\alpha-1)}, \tag{7.8.23}$$

for $\mathrm{ord}_O(\alpha - 1) < n$, [5], p. 371. Thus, we have

(7.8.24)
$$\mathrm{ord}_L(\mathcal{D}(L/M_0)) = \sum_{\substack{\alpha\in(O/\pi^s O)^\times \\ \alpha\neq 1}} q^{\mathrm{ord}_O(\alpha-1)} = q^{n-1}(q-1)\,n - q^{n-1}.$$

Now let L_s be the extension of M_0 generated by the p^s-division points of G, so that $L_s = M_0(G[\pi^s])$, if p is inert in $\boldsymbol{k}$, and $L_s = M_0(G[\pi^{2s}])$, if p is ramified in $\boldsymbol{k}$. Then, by [1], the field M_s is the fixed field in L_s of the subgroup $(\mathbb{Z}_p/p^s\mathbb{Z}_p)^\times \subset (O/p^s O)^\times$. Note that the extension M_s is totally ramified over M and that

$$|M_s : M| = |M_0 : M| \cdot p^{s-1}(p-\chi), \tag{7.8.25}$$

for $\chi = \chi_d(p)$. Again by the formula from [6], we have

(7.8.26) $\mathrm{ord}_{L_s}(\mathcal{D}(L_s/M_s))$
$$\begin{aligned}
&= \sum_{\substack{a\in(\mathbb{Z}_p/p^s\mathbb{Z}_p)^\times \\ a\neq 1}} q^{\mathrm{ord}_O(a-1)} \\
&= (p-2)p^{s-1} + (p-1)p^{s-2}\cdot p^2 \\
&\qquad + \ldots (p-1)p^{s-r}\cdot p^{2r-2} + \cdots + (p-1)p^{2s-2} \\
&= p^{2s-1} - 2p^{s-1}.
\end{aligned}$$

Note that here, in the inert case, $q = p^2$ and $\mathrm{ord}_O(a-1) = \mathrm{ord}_p(a-1)$ and, in the ramified case, $q = p$ and $\mathrm{ord}_O(a-1) = 2\,\mathrm{ord}_p(a-1)$, so that the result is the same in both cases. Finally, we use the relation

(7.8.27)
$$\mathrm{ord}_{L_s}(\mathcal{D}(L_s/M_0)) = |L_s : M_s|\cdot\mathrm{ord}_{M_s}(\mathcal{D}(M_s/M_0)) + \mathrm{ord}_{L_s}(\mathcal{D}(L_s/M_s)).$$

Solving for $\mathrm{ord}_{M_s}(\mathcal{D}(M_s/M_0))$, we obtain

(7.8.28)
$$\mathrm{ord}_{M_s}(\mathcal{D}(M_s/M_0)) = |M_0 : M|\left(p^{s-1}(p-\chi)s - \frac{p^s - \chi p^{s-1} + \chi - 1}{p-1}\right).$$

Finally, to get the formulas stated, we must add the term

$$|M_s : M_0| \cdot \mathrm{ord}_{M_0}(\mathcal{D}(M_0/M)) = p^s \,\mathrm{ord}_p(d), \tag{7.8.29}$$

when p is ramified in $\boldsymbol{k}$. □

To finish the proof of (i) of Proposition 7.8.4, we sum the expression for $\partial(s)$ over s and multiply by $|\Lambda_t|$. For the $\mathrm{ord}_p(d)$ term, we use (ii) of Proposition 7.7, so that

$$\deg \mathcal{Z}(t)_{\mathbb{Q}} = 2\,|\Lambda_t|\,\frac{p^{k+1}-1}{p-1}, \tag{7.8.30}$$

when p is ramified.

To prove (ii), we need to evaluate the last expression in (7.8.16). This sum is empty if p is split in $\boldsymbol{k}_t$ and is zero if p is inert since then $\mathcal{Z}^{\mathrm{hor}}(t)$ is unramified over $\mathrm{Spec}\,\mathbb{Z}_p$. If p is ramified, then $\tilde{\mathcal{Z}}^{\mathrm{hor}}(t : c)_{(\tilde{x})}$ is the spectrum of a ramified quadratic extension of $W(\bar{\mathbb{F}}_p)$. If $p \neq 2$, then $\tilde{\mathcal{D}}_{(\tilde{x})}$ generates the maximal ideal $\mathfrak{m}_{\tilde{x}}$ in $\mathcal{O}_{\tilde{x}}$, whereas if $p = 2$, then $\tilde{\mathcal{D}}_{(\tilde{x})}$ is equal to $\mathfrak{m}_{\tilde{x}}^{\delta}$ where $\delta = \mathrm{ord}_p(d)$, as one calculates using the equations (A.16) resp. (A.18) of [4]. Hence the last expression of (7.8.16) is equal to

$$\begin{aligned} &\mathrm{ord}_p(d) \cdot \sum_{\tilde{x} \in \tilde{\mathcal{Z}}^{\mathrm{hor}}(t:c)(\bar{\mathbb{F}}_p)} e_{\tilde{x}}^{-1} \\ &= \frac{1}{2}\,\mathrm{ord}_p(d) \cdot \sum_{\tilde{x} \in \tilde{\mathcal{Z}}^{\mathrm{hor}}(t)(\bar{\mathbb{F}}_p)} e_{\tilde{x}}^{-1} \cdot \mathrm{ram}_{\tilde{x}} \\ &= \frac{1}{2}\,\mathrm{ord}_p(d) \cdot \deg \mathcal{Z}^{\mathrm{hor}}(t : c)_{\mathbb{Q}}, \end{aligned} \tag{7.8.31}$$

where $\mathrm{ram}_{\tilde{x}} = 2$ denotes the ramification index of $\tilde{\mathcal{Z}}^{\mathrm{hor}}(t : c)_{(\tilde{x})}$ over W. Summing over $c|n$ yields now (ii). □

7.9 COMPARISON FOR THE CASE $t_1, t_2 > 0$, AND $t_1 t_2 = m^2$

In this section, we complete the computation of the arithmetic intersection number

$$\langle\, \widehat{\mathcal{Z}}(t_1, v_1), \widehat{\mathcal{Z}}(t_2, v_2)\,\rangle \;=\; A + B + C \tag{7.9.1}$$

in the case $t_1 t_2 = m^2$, with t_1 and $t_2 > 0$, by assembling the quantities computed above and comparing with the right hand side of the identity $((\star))$ of Theorem C.

First, recall that the term B, computed via adjunction, can be written as

$$B = -2\,h_{\hat{\omega}}(\mathcal{Z}^{\text{hor}}(t)) + 2\sum_p \partial_p(t)\,\log(p) + \sum_p\sum_T 2\delta_p(T)\log(p)$$

$$+\deg_{\mathbb{Q}}(\mathcal{Z}(t))\left[\log(t_1v_1+t_2v_2)-\log(t_1v_1)-\log(t_2v_2)\right] \tag{7.9.2}$$

$$-\deg_{\mathbb{Q}}(\mathcal{Z}(t))\,J(4\pi(t_1v_1+t_2v_2))$$

$$+B_\infty - \deg_{\mathbb{Q}}(\mathcal{Z}(t))\cdot\log(4D(B)).$$

We also have

$$A = \sum_{p\leq\infty}\sum_{\substack{T\in\text{Sym}_2(\mathbb{Z})^\vee\\ \text{diag}(T)=(t_1,t_2)\\ \det(T)\neq 0}} A_p(T) + \sum_{\substack{p<\infty\\ p\nmid D(B)}}\sum_{\substack{T\in\text{Sym}_2(\mathbb{Z})^\vee\\ \text{diag}(T)=(t_1,t_2)\\ \det(T)=0}} A_p(T), \tag{7.9.3}$$

and

$$C = \sum_{p\mid D(B)}\sum_{\substack{T\in\text{Sym}_2(\mathbb{Z})^\vee\\ \text{diag}(T)=(t_1,t_2)\\ \det(T)\neq 0}} C_p(T) + \sum_{p\mid D(B)}\sum_{\substack{T\in\text{Sym}_2(\mathbb{Z})^\vee\\ \text{diag}(T)=(t_1,t_2)\\ \det(T)=0}} C_p(T). \tag{7.9.4}$$

Of course, when $\det(T)\neq 0$, $A_p(T)$ and $C_p(T)$ can only be nonzero when $\text{Diff}(T,B)=\{p\}$. There is no such restriction on p for the two singular T's in the sums.

We must compare the quantity in (7.9.1) with the sum

(7.9.5)

$$\sum_{p\leq\infty}\sum_{\substack{T\in\text{Sym}_2(\mathbb{Z})^\vee\\ \text{diag}(T)=(t_1,t_2)\\ \det(T)\neq 0\\ \text{Diff}(T,B)=\{p\}}} \widehat{\mathcal{Z}}\left(T,\begin{pmatrix}v_1&\\&v_2\end{pmatrix}\right) + \sum_{\substack{T\in\text{Sym}_2(\mathbb{Z})^\vee\\ \text{diag}(T)=(t_1,t_2)\\ \det(T)=0}} \widehat{\mathcal{Z}}\left(T,\begin{pmatrix}v_1&\\&v_2\end{pmatrix}\right).$$

Here, for any $T\in\text{Sym}_2(\mathbb{Z})^\vee$ with $\det(T)\neq 0$ and $\text{Diff}(T,B)=\{p\}$,

$$\widehat{\mathcal{Z}}(T,v) = \begin{cases} |\Lambda_T|\cdot\nu_p(T)\cdot\log(p) & \text{if } p<\infty,\, p\nmid D(B),\\ |\Lambda_T|\cdot\nu_p(T)\cdot\log(p) & \text{if } p<\infty,\, p\mid D(B),\\ |\Lambda_T|\cdot\nu_\infty(T,v) & \text{if } p=\infty,\end{cases} \tag{7.9.6}$$

where $|\Lambda_T|$ is given by (7.2.5), (7.2.10), and (7.2.14), respectively, $\nu_p(T)$ is given in (7.2.4) and (7.2.11), and $\nu_\infty(T,v)$ is given by (7.2.13). In the rank

1 case we defined in Section 6.4,

(7.9.7)
$$\widehat{\mathcal{Z}}(T,v) = -\langle\, \widehat{\mathcal{Z}}(t, t^{-1}\mathrm{tr}(Tv))\, ,\, \hat{\omega}\,\rangle + \frac{1}{2}\, \deg_{\mathbb{Q}}(\mathcal{Z}(t))\cdot\log\left(\frac{\mathrm{tr}(Tv)}{t\det(v)D(B)}\right).$$

This can be made more explicit.

Lemma 7.9.1. *Suppose that $T \in \mathrm{Sym}_2(\mathbb{Z})^\vee$ with $\det(T) = 0$ has rank 1 with t_1 and $t_2 > 0$, and that $v = \mathrm{diag}(v_1, v_2)$. Then*

$$\begin{aligned}
&\widehat{\mathcal{Z}}(T,v) \\
&= -h_{\hat{\omega}}(\mathcal{Z}^{\mathrm{hor}}(t)) - h_{\hat{\omega}}(\mathcal{Z}^{\mathrm{ver}}(t)) - \frac{1}{2}\, \deg_{\mathbb{Q}}(\mathcal{Z}(t)) \cdot J(4\pi(t_1v_1 + t_2v_2)) \\
&- \frac{1}{2}\deg_{\mathbb{Q}}(\mathcal{Z}(t))\Big(\log(v_1v_2) - \log(t_1v_1 + t_2v_2) + \log(t) + \log(D(B))\Big).
\end{aligned}$$

Here,

$$h_{\hat{\omega}}(\mathcal{Z}^{\mathrm{ver}}(t)) = \sum_{p|D(B)} h_{\hat{\omega}}(\mathcal{Z}^{\mathrm{ver}}(t)_p),$$

and, by Theorem 11.5 of [4], for $p \mid D(B)$ and $4t = n^2 d$,

$$\frac{h_{\hat{\omega}}(\mathcal{Z}^{\mathrm{ver}}(t)_p)}{\log p} = -\deg_{\mathbb{Q}}(\mathcal{Z}(t))\cdot \begin{cases} k - \frac{(p+1)(p^k-1)}{2(p-1)} & \text{if } \chi_d(p) = -1, \\ k + 1 - \frac{p^{k+1}-1}{p-1} & \text{if } \chi_d(p) = 0, \end{cases}$$

where $k = \mathrm{ord}_p(n)$. If $\chi_d(p) = 1$, i.e., if p splits in $\boldsymbol{k}_d$, then

$$\frac{h_{\hat{\omega}}(\mathcal{Z}^{\mathrm{ver}}(t)_p)}{\log p} = 2\, H_0(t; D(B))\, \delta(d; D(B)/p)\cdot(p^k - 1),$$

where $H_0(t; D(B))$ and $\delta(d; D(B)/p)$ are given by (7.6.21) and (7.6.22), respectively.

Proof. For $v = \mathrm{diag}(v_1, v_2)$, we have

$$\begin{aligned}
&\widehat{\mathcal{Z}}(T,v) \\
&= -h_{\hat{\omega}}(\mathcal{Z}^{\mathrm{hor}}(t)) - h_{\hat{\omega}}(\mathcal{Z}^{\mathrm{ver}}(t)) - \frac{1}{2}\int_{\mathcal{M}(\mathbb{C})} \Xi(t, t^{-1}\mathrm{tr}(Tv))\, d\mu \\
&- \frac{1}{2}\deg_{\mathbb{Q}}(\mathcal{Z}(t))\Big(\log(v_1v_2) - \log(t_1v_1 + t_2v_2) + \log(t) + \log(D(B))\Big).
\end{aligned}$$

By Proposition 12.1 of [4], for $t > 0$, we have

(7.9.8)
$$\frac{1}{2}\int_{\mathcal{M}(\mathbb{C})} \Xi(t, t^{-1}\mathrm{tr}(Tv))\, d\mu = \frac{1}{2}\deg \mathcal{Z}(t)_{\mathbb{Q}} \cdot J(4\pi(t_1v_1 + t_2v_2)).$$

□

We can now begin the comparison. First, comparing Corollary 7.5.6 and Proposition 7.5.7 with (7.2.16) for $p = \infty$, it is immediate that

(7.9.9)
$$\begin{aligned} &A_\infty + B_\infty \\ &= \sum_{\substack{T\in\mathrm{Sym}_2(\mathbb{Z})^\vee \\ \mathrm{diag}(T)=(t_1,t_2) \\ \det(T)\neq 0 \\ \mathrm{Diff}(T,B)=\{\infty\}}} \Big(\sum_{c_1\neq c_2} |\Lambda_T(c_1, c_2)| + \sum_{c} |\Lambda_T(c, c)| \Big) \cdot \nu_\infty(T, v) \\ &= \sum_{\substack{T\in\mathrm{Sym}_2(\mathbb{Z})^\vee \\ \mathrm{diag}(T)=(t_1,t_2) \\ \det(T)\neq 0 \\ \mathrm{Diff}(T,B)=\{\infty\}}} \widehat{\mathcal{Z}}(T, v), \end{aligned}$$

as required.

Next suppose that $p \mid D(B)$, and that $T \in \mathrm{Sym}_2(\mathbb{Z})^\vee$, with $\mathrm{diag}(T) = (t_1, t_2)$, $\det(T) \neq 0$, and $\mathrm{Diff}(T, B) = \{p\}$. Then, Propositions 7.6.2 and 7.6.3 together with (ii) of Proposition 7.8.2 yield

(7.9.10)
$$\begin{aligned} &A_p(T) + 2\,\delta_p(T)\,\log(p) + C_p(T) \\ &= \sum_{c_1\neq c_2} |\Lambda_T(c_1, c_2)| \cdot \nu_p^{\mathrm{hor}}(T)\,\log(p) + \sum_{c} |\Lambda_T(c, c)| \cdot \nu_p^{\mathrm{hor}}(T)\,\log(p) \\ &\qquad + |\Lambda_T| \cdot (\nu_p(T) - \nu_p^{\mathrm{hor}}(T))\,\log(p) \\ &= |\Lambda_T| \cdot \nu_p(T)\,\log(p) \\ &= \widehat{\mathcal{Z}}(T, v). \end{aligned}$$

Finally, suppose that $p \nmid D(B)$, and that $T \in \mathrm{Sym}_2(\mathbb{Z})^\vee$, with $\mathrm{diag}(T) =$

(t_1, t_2), $\det(T) \neq 0$ and $\mathrm{Diff}(T, B) = \{p\}$. Then, by Corollary 7.7.6

(7.9.11)

$$\begin{aligned} &A_p(T) \\ &= \sum_{c_1^0 \neq c_2^0} |\Lambda_T(c_1^0, c_2^0)| \cdot \sum_{s_1=0}^{\mathrm{ord}_p(n_1 n)} \sum_{s_2=0}^{\mathrm{ord}_p(n_2 n)} (\,\mathcal{W}_{s_1}(\psi_1), \mathcal{W}_{s_2}(\psi_2)\,) \cdot \log(p) \\ &\quad + \sum_{c^0} |\Lambda_T(c^0, c^0)| \cdot \sum_{\substack{s_1=0 \\ s_1 \neq s_2}}^{\mathrm{ord}_p(n_1 n)} \sum_{s_2=0}^{\mathrm{ord}_p(n_2 n)} (\,\mathcal{W}_{s_1}(\psi_1), \mathcal{W}_{s_2}(\psi_2)\,) \cdot \log(p), \end{aligned}$$

whereas, by (i) of Proposition 7.8.2,
(7.9.12)

$$2\,\delta_p(T)\,\log(p) \;=\; \sum_{c^0} |\Lambda_T(c^0, c^0)| \cdot \sum_{s=0}^{\mathrm{ord}_p(n)} (\mathcal{W}_s(\psi_1), \mathcal{W}_s(\psi_2))\,\log(p).$$

Thus, we have

$$A_p(T) + 2\delta_p(T)\,\log(p) \;=\; \widehat{\mathcal{Z}}(T, v). \tag{7.9.13}$$

Note that identities (7.9.9), (7.9.10), and (7.9.13) account for all terms associated to nonsingular T's on the two sides.

It remains to compare the global terms and terms associated to the two singular T's with diagonal (t_1, t_2). First, there is the remaining part of B:

(7.9.14)

$$\begin{aligned} B^{\mathrm{global}} &= -2\,h_{\hat{\omega}}(\mathcal{Z}^{\mathrm{hor}}(t)) + 2\sum_p \partial_p(t)\,\log(p) \\ &\quad - \deg_{\mathbb{Q}}(\mathcal{Z}(t)) \cdot \log(4D(B)) \\ &\quad + \deg_{\mathbb{Q}}(\mathcal{Z}(t))\,\big[\,\log(t_1 v_1 + t_2 v_2) - \log(t_1 t_2) - \log(v_1 v_2)\,\big] \\ &\quad - \deg_{\mathbb{Q}}(\mathcal{Z}(t))\,J(4\pi(t_1 v_1 + t_2 v_2)). \end{aligned}$$

Next, there is the contribution of each of the two singular T's. For any $p \nmid D(B)$, Corollary 7.7.8 gives

$$\begin{aligned} \frac{A_p(T)}{\log(p)} &= \deg \mathcal{Z}(t)_{\mathbb{Q}} \cdot \big[\,\mathrm{ord}_p(n_1 n_2) + 2\mathrm{ord}_p(n)\,\big] \\ &\qquad - 2\,|\Lambda_t|\, m_0(p) \cdot \frac{p - \chi}{p - 1}\left(k p^k - \frac{p^k - 1}{p - 1} \right), \end{aligned} \tag{7.9.15}$$

where $\chi = \chi_d(p)$ and $k = \mathrm{ord}_p(n)$. For $p \mid D(B)$, recalling Proposition 7.6.4 and the expression for $h_{\hat{\omega}}(\mathcal{Z}^{\mathrm{ver}}(t)_p)$ in Lemma 7.9.1, we have for p inert or ramified in $\boldsymbol{k}_t$,

(7.9.16)

$$\begin{aligned} C_p(T) & \\ = \deg \mathcal{Z}(t)_{\mathbb{Q}} \log(p) \cdot & \begin{cases} \frac{1}{2}\,\mathrm{ord}_p(t_1t_2) - \frac{(p+1)(p^k-1)}{2(p-1)} & \text{if } p \text{ is inert,} \\ \frac{1}{2}\,\mathrm{ord}_p(t_1t_2) - \frac{p^{k+1}-1}{p-1} & \text{if } p \text{ is ramified,} \end{cases} \\ & + \deg \mathcal{Z}(t)_{\mathbb{Q}} \log(p) \cdot (1 + \chi - \mathrm{ord}_p(d/4)) \\ = -h_{\hat{\omega}}(\mathcal{Z}^{\mathrm{ver}}(t)_p) & + \deg \mathcal{Z}(t)_{\mathbb{Q}} \left[\, \frac{1}{2}\,\mathrm{ord}_p(t_1t_2) - k - \mathrm{ord}_p(d/4) \,\right] \log(p). \end{aligned}$$

If p is split in $\boldsymbol{k}_t$, then the comparison of Proposition 7.6.4 and Lemma 7.9.1 shows that

$$C_p(T) = -h_{\hat{\omega}}(\mathcal{Z}^{\mathrm{ver}}(t)_p). \tag{7.9.17}$$

Taking into account that there are two singular T's and cancelling the terms which the expressions $B^{\mathrm{global}} + 2\,A_p(T) + 2\,C_p(T)$ and $2\,\mathcal{Z}(T, v)$ have in common, we are left with

(7.9.18)

$$\begin{aligned} 2 \sum_p \partial_p(t) \log(p) & - \deg_{\mathbb{Q}}(\mathcal{Z}(t)) \log(4\,t_1t_2) \\ & + 2 \sum_{p \mid D(B)} \deg \mathcal{Z}(t)_{\mathbb{Q}} \left[\, \frac{1}{2}\,\mathrm{ord}_p(t_1t_2) - k - \mathrm{ord}_p(d/4) \,\right] \log(p) \\ & + 2 \sum_{p \nmid D(B)} \Bigg(\deg \mathcal{Z}(t)_{\mathbb{Q}} \cdot \left[\, \mathrm{ord}_p(n_1n_2) + 2\,\mathrm{ord}_p(n) \,\right] \\ & \qquad - 2\,|\Lambda_t|\, m_0(p) \cdot \frac{p-\chi}{p-1} \left(kp^k - \frac{p^k-1}{p-1} \right) \Bigg), \end{aligned}$$

on the intersection pairing side, and

$$-\deg \mathcal{Z}(t)_{\mathbb{Q}} \log(t), \tag{7.9.19}$$

on the genus two generating function side.

Finally, we use the values for the discriminant terms $\partial_p(t)$ and compute the coefficient of each $\log(p)$ in (7.9.18).

First suppose that $p \nmid D(B)$. We then have terms

$$\partial_p(t) = |\Lambda_t|\, m_0(p) \Big[\frac{p-\chi}{p-1} \left(k\, p^k - 2\, \frac{p^k-1}{p-1} \right) + \frac{1-\chi}{p-1}\, k \Big] + \frac{1}{2} \deg \mathcal{Z}(t)_{\mathbb{Q}} \cdot \mathrm{ord}_p(d), \tag{7.9.20}$$

$$\frac{A_p(T)}{\log(p)} = \deg \mathcal{Z}(t)_{\mathbb{Q}} \cdot \big[\, \mathrm{ord}_p(n_1 n_2) + 2\, \mathrm{ord}_p(n) \,\big] - 2\, |\Lambda_t|\, m_0(p) \cdot \frac{p-\chi}{p-1} \left(kp^k - \frac{p^k-1}{p-1} \right), \tag{7.9.21}$$

and

$$-\frac{1}{2} \deg \mathcal{Z}(t)_{\mathbb{Q}}\, \mathrm{ord}_p(4 t_1 t_2) = -\deg \mathcal{Z}(t)_{\mathbb{Q}} \,\big[\, \mathrm{ord}_p(n_1 n_2) + 2\, \mathrm{ord}_p(n) + \mathrm{ord}_p(d/2) \,\big], \tag{7.9.22}$$

whose sum is

$$\begin{aligned} &\deg \mathcal{Z}(t)_{\mathbb{Q}} \left[\mathrm{ord}_p(2) - \frac{1}{2}\, \mathrm{ord}_p(d)\right] - k |\Lambda_t|\, m_0(p) \left(\frac{p^{k+1} - \chi\, p^k + \chi - 1}{p-1} \right) \\ &\quad = \deg \mathcal{Z}(t)_{\mathbb{Q}} \left[\, \mathrm{ord}_p(2) - \frac{1}{2}\, \mathrm{ord}_p(d) - \mathrm{ord}_p(n) \right] \\ &\quad = -\frac{1}{2} \deg \mathcal{Z}(t)_{\mathbb{Q}} \cdot \mathrm{ord}_p(t). \end{aligned} \tag{7.9.23}$$

Here we have used (ii) of Proposition 7.7.7,

$$\deg \mathcal{Z}(t)_{\mathbb{Q}} = |\Lambda_t|\, m_0(p) \left(1 + (p-\chi)\, \frac{p^k-1}{p-1} \right). \tag{7.9.24}$$

Since there are two such terms in (7.9.18), we get the required coincidence with the $\log(p)$ part of (7.9.19).

Next suppose that $p \mid D(B)$. We then have terms

$$\frac{1}{2} \deg \mathcal{Z}(t)_{\mathbb{Q}} \cdot \mathrm{ord}_p(d), \tag{7.9.25}$$

$$\deg \mathcal{Z}(t)_{\mathbb{Q}} \cdot \big[\, \frac{1}{2}\, \mathrm{ord}_p(t_1 t_2) - \mathrm{ord}_p(n) - \mathrm{ord}_p(d/4) \,\big], \tag{7.9.26}$$

and

$$-\frac{1}{2}\deg \mathcal{Z}(t)_{\mathbb{Q}} \cdot \mathrm{ord}_p(4t_2t_2), \tag{7.9.27}$$

In this case, the sum is

$$-\frac{1}{2}\deg \mathcal{Z}(t)_{\mathbb{Q}} \cdot \mathrm{ord}_p(t),$$

so that we again have the claimed agreement. This finishes the proof of the identity (($\star$)) of Theorem C in the case in which t_1t_2 is a square and t_1, $t_2 > 0$.

7.10 THE CASE $t_1, t_2 < 0$ WITH $t_1t_2 = m^2$

In this case, the classes

$$\widehat{\mathcal{Z}}(t_1, v_1) = (0, \Xi(t_1, v_1)), \ \widehat{\mathcal{Z}}(t_2, v_2) = (0, \Xi(t_2, v_2)) \in \widehat{\mathrm{CH}}^1(\mathcal{M})$$

only involve the Green functions. More precisely, the cycles $\mathcal{Z}(t_i)$ are empty and the functions $\Xi(t_i, v_i)$ are smooth on $\mathcal{M}(\mathbb{C})$. We then have

$$\begin{aligned}
\langle\, \widehat{\mathcal{Z}}(t_1, v_1), \widehat{\mathcal{Z}}(t_2, v_2)\,\rangle &= \frac{1}{2}\int_{\mathcal{M}(\mathbb{C})} \Xi(t_1, v_1) * \Xi(t_2, v_2) \\
&= \frac{1}{2}\int_{[\Gamma\backslash D]} \sum_{\substack{x_1\in L \\ Q(x_1)=t_1}} \sum_{\substack{x_2\in L \\ Q(x_2)=t_2}} \xi(v_1^{\frac{1}{2}}x_1) * \xi(v_2^{\frac{1}{2}}x_2) \\
&= \sum_{\substack{T\in \mathrm{Sym}_2(\mathbb{Z})^\vee \\ \mathrm{diag}(T)=(t_1,t_2)}} \frac{1}{2}\int_{[\Gamma\backslash D]} \sum_{\substack{\mathbf{x}\in L^2 \\ Q(\mathbf{x})=T}} \xi(v_1^{\frac{1}{2}}x_1) * \xi(v_2^{\frac{1}{2}}x_2) \\
&= \sum_{\substack{T\in \mathrm{Sym}_2(\mathbb{Z})^\vee \\ \mathrm{diag}(T)=(t_1,t_2) \\ \det T\neq 0}} |\Lambda_T| \cdot \nu_\infty(T, v) \qquad (7.10.1)\\
&\quad + \int_{[\Gamma\backslash D]} \sum_{\substack{\mathbf{x}\in L^2 \\ Q(\mathbf{x})=T_0}} \xi(v_1^{\frac{1}{2}}x_1) * \xi(v_2^{\frac{1}{2}}x_2)
\end{aligned}$$

$$= \sum_{\substack{T \in \mathrm{Sym}_2(\mathbb{Z})^\vee \\ \mathrm{diag}(T)=(t_1,t_2) \\ \det T \neq 0}} \widehat{\mathcal{Z}}(T,v)$$

$$+ \int_{[\Gamma\backslash D]} \sum_{\substack{\mathbf{x}\in L^2 \\ Q(\mathbf{x})=T_0}} \xi(v_1^{\frac{1}{2}}x_1) * \xi(v_2^{\frac{1}{2}}x_2),$$

where, just as in (7.2.12), $\nu_\infty(T,v)$ is given by (7.2.13) and $|\Lambda_T|$ is given by (7.2.15). Here the matrix T_0 is one of the two singular matrices with $\mathrm{diag}(T_0) = (t_1, t_2)$. It remains to prove the following identity.

Proposition 7.10.1. *For a singular matrix* T_0 *with* $\mathrm{diag}(T_0) = (t_1, t_2)$, *and for* $v = \mathrm{diag}(v_1, v_2)$,

$$\widehat{\mathcal{Z}}(T_0,v) = -\langle\, \widehat{\mathcal{Z}}(t, t^{-1}\mathrm{tr}(Tv)), \hat{\omega}\,\rangle = \frac{1}{2}\int_{[\Gamma\backslash D]} \sum_{\substack{\mathbf{x}\in L^2 \\ Q(\mathbf{x})=T_0}} \xi(v_1^{\frac{1}{2}}x_1)*\xi(v_2^{\frac{1}{2}}x_2).$$

Here recall that $\widehat{\mathcal{Z}}(T_0,v)$ is defined as in Section 6.4.

Proof. As in Section 6.4, we write

$$T_0 = \begin{pmatrix} t_1 & m \\ m & t_2 \end{pmatrix},$$

and choose n_1 and n_2 relatively prime with $t_1 = n_1^2 t$, $t_2 = n_2^2 t$, $m = n_1 n_2 t$. Note that we have $t^{-1}\mathrm{tr}(Tv) = v_1 n_1^2 + v_2 n_2^2$. Thus, by the results of [4] (see (iii) of Theorem 8.8),

(7.10.2)

$$-\langle\, \widehat{\mathcal{Z}}(t,t^{-1}\mathrm{tr}(Tv)), \hat{\omega}\,\rangle = -2\,\delta(d; D(B))\, H_0(t; D(B))$$

$$\times \frac{1}{4\pi}|t|^{-\frac{1}{2}}(v_1 n_1^2 + v_2 n_2^2)^{-\frac{1}{2}} \int_1^\infty e^{-4\pi|t|(v_1 n_1^2 + v_2 n_2^2)\,u}\, u^{-\frac{3}{2}}\, du.$$

Now for any $\mathbf{x}$ in the sum in the last expression in the proposition, we can write $x_1 = n_1 y$ and $x_2 = n_2 y$, for a unique vector $y \in L$ with $Q(y) = t$. Using Lemma 11.4 of [2] and unwinding, we have

(7.10.3)
$$\frac{1}{2}\int_{[\Gamma\backslash D]} \sum_{\substack{\mathbf{x}\in L^2 \\ Q(\mathbf{x})=T_0}} \xi(v_1^{\frac{1}{2}}x_1) * \xi(v_2^{\frac{1}{2}}x_2)$$

$$= \sum_{\substack{y\in L(t) \\ \mathrm{mod}\ \Gamma}} \frac{1}{4}\int_{\Gamma_y\backslash D} \xi(v_1^{\frac{1}{2}}n_1 y, z)\cdot \varphi(v_2^{\frac{1}{2}}n_2 y, z)\, d\mu(z),$$

where φ is as in (3.5.5). The extra factor of $\frac{1}{2}$ in the second line comes from the stack $[\Gamma\backslash D]$ in the first line. The integral here can be computed by the method used in the second part of the proof of Proposition 12.1 of [4]. To clean up the notation, we temporarily write v_1 for $v_1 n_1^2$ and v_2 for $v_2 n_2^2$. As in (12.9) of [4], we shift y to a standard vector and choose polar coordinates $z = re^{i\theta}$ in $\mathfrak{H}$. By Lemma 11.5 of [2], we can write the integrand as

$$-\mathrm{Ei}(-2\pi R_1) \cdot (2\,R_2 - 4v_2|t| - \frac{1}{2\pi}) \cdot e^{-2\pi R_2},$$

where

$$R_1 = \frac{2v_1|t|}{\sin^2(\theta)} \qquad \text{and} \qquad R_2 = \frac{2v_2|t|}{\sin^2(\theta)}.$$

Then, the integral becomes[12]

$$\begin{aligned}
&2\,\delta_y^{-1} \int_{\Gamma_y^+\backslash D^+} \xi(v_1^{\frac{1}{2}} y, z) \cdot \varphi(v_2^{\frac{1}{2}} y, z)\, d\mu(z) \\
&= 8\,\delta_y^{-1} \log|\epsilon(y)| \cdot \int_0^{\frac{\pi}{2}} -\mathrm{Ei}\left(-\frac{4\pi v_1|t|}{\sin^2(\theta)}\right) \cdot \left(\frac{4v_2|t|\cos^2(\theta)}{\sin^2(\theta)} - \frac{1}{2\pi}\right) \\
&\qquad\qquad \times \exp\left(-\frac{4\pi v_2|t|}{\sin^2(\theta)}\right) \cdot (\sin^2(\theta))^{-1}\, d\theta \\
&= 4\,\delta_y^{-1} \log|\epsilon(y)| \cdot \int_1^\infty \left(\int_1^\infty e^{-4\pi v_1|t|rw}\, w^{-1}\, dw\right) \\
&\qquad\qquad \times \left(4v_2|t|(r-1) - \frac{1}{2\pi}\right) e^{-4\pi v_2|t|r}\, (r-1)^{-\frac{1}{2}}\, dr \\
&= 4\,\delta_y^{-1} \log|\epsilon(y)| \cdot \int_1^\infty e^{-4\pi|t|(v_1 w + v_2)} \int_0^\infty e^{-4\pi|t|r(v_1 w+v_2)} \\
&\qquad\qquad \times \left(4v_2|t|r - \frac{1}{2\pi}\right) r^{-\frac{1}{2}}\, dr\, w^{-1}\, dw.
\end{aligned}$$

The integral with respect to r here is

$$-\frac{1}{4\pi} \cdot |t|^{-\frac{1}{2}}\, (v_1 w + v_2)^{-\frac{3}{2}}\, v_1 w,$$

[12]Note that in [4], $d\mu(z) = \frac{1}{2\pi}\, y^{-2}\, dx\, dy$, whereas here we omit the $\frac{1}{2\pi}$; see Section 3.5.

so we get the expression

$$- 4\,\delta_y^{-1}\,\log|\epsilon(y)|\cdot|t|^{-\frac{1}{2}}\int_1^\infty e^{-4\pi|t|(v_1 w+v_2)}\,\frac{v_1}{4\pi}\,(v_1 w+v_2)^{-\frac{3}{2}}\,dw$$

$$= -4\,\delta_y^{-1}\,\log|\epsilon(y)|\cdot|t|^{-\frac{1}{2}}\,\frac{1}{4\pi}\,(v_1+v_2)^{-\frac{1}{2}}\int_1^\infty e^{-4\pi|t|(v_1+v_2)u}\,u^{-\frac{3}{2}}\,du,$$

where we have used the substitution $u = (v_1 w + v_2)/(v_1 + v_2)$ in the last step. Returning to (7.10.3), we obtain

$$\sum_{\substack{y\in L(t)\\ \bmod \Gamma}} -\delta_y^{-1}\,\log|\epsilon(y)|\cdot|t|^{-\frac{1}{2}} \times \frac{1}{4\pi}\,(v_1 n_1^2 + v_2 n_2^2)^{-\frac{1}{2}}\int_1^\infty e^{-4\pi|t|(v_1 n_1^2+v_2 n_2^2)u}\,u^{-\frac{3}{2}}\,du$$

$$= -2\,\delta(d;D(B))\,H_0(t;D(B))\cdot|t|^{-\frac{1}{2}} \times \frac{1}{4\pi}\,(v_1 n_1^2 + v_2 n_2^2)^{-\frac{1}{2}}\int_1^\infty e^{-4\pi|t|(v_1 n_1^2+v_2 n_2^2)u}\,u^{-\frac{3}{2}}\,du,$$

by Lemma 12.2 of [4]. □

7.11 THE CONSTANT TERMS

In this section we prove the remaining cases of identity (($\star$)) of Theorem C. For clarity, we will write $\widehat{\mathcal{Z}}_2(T, v)$ for a coefficient of the genus two generating function.

Theorem 7.11.1. *(i) For $t_1 \neq 0$,*

$$\langle\,\widehat{\mathcal{Z}}(t_1, v_1), \widehat{\mathcal{Z}}(0, v_2)\,\rangle = \widehat{\mathcal{Z}}_2(T, v),$$

where $T = \mathrm{diag}(t_1, 0)$ and $v = \mathrm{diag}(v_1, v_2)$.
(ii)

$$\langle\,\widehat{\mathcal{Z}}(0, v_1), \widehat{\mathcal{Z}}(0, v_2)\,\rangle = \widehat{\mathcal{Z}}_2(0, v),$$

where $v = \mathrm{diag}(v_1, v_2)$.

Corollary 7.11.2. *The constant* $\mathbf{c} = -\log(D(B))$ *and so*

$$\langle\,\hat{\omega}, \hat{\omega}\,\rangle = \zeta_{D(B)}(-1)\left[2\frac{\zeta'(-1)}{\zeta(-1)} + 1 - 2C - \frac{1}{2}\sum_{p|D(B)}\frac{p+1}{p-1}\cdot\log(p)\right].$$

Proof. Recall that, by (5.13) of [4],

$$\widehat{\mathcal{Z}}(0,v) = -\hat{\omega} - (0, \log(v) - \mathbf{c}), \tag{7.11.1}$$

for the constant $\mathbf{c}$ given, as in (0.15) of [4], by the relation

(7.11.2)
$$\frac{1}{2}\deg(\hat{\omega})\cdot\mathbf{c} = \langle\, \hat{\omega}, \hat{\omega}\,\rangle - \zeta_{D(B)}(-1)\left[2\frac{\zeta'(-1)}{\zeta(-1)} + 1 - 2C - \sum_{p|D(B)} \frac{p\log(p)}{p-1}\right].$$

Therefore,

(7.11.3)
$$\langle\, \widehat{\mathcal{Z}}(t_1, v_1), \widehat{\mathcal{Z}}(0, v_2)\,\rangle = -\langle\, \widehat{\mathcal{Z}}(t_1, v_1), \hat{\omega}\,\rangle - \frac{1}{2}\deg_{\mathbb{Q}}(\mathcal{Z}(t_1))(\log(v_2) - \mathbf{c})$$

and

(7.11.4)
$$\langle\, \widehat{\mathcal{Z}}(0, v_1), \widehat{\mathcal{Z}}(0, v_2)\,\rangle = \langle\, \hat{\omega}, \hat{\omega}\,\rangle + \frac{1}{2}\deg_{\mathbb{Q}}(\hat{\omega})\cdot(\,\log(v_1 v_2) - 2\,\mathbf{c}\,).$$

On the other hand, in the sum on the right-hand side of (($\star$)), all contributions of nonsingular matrices T are zero by Remark 6.3.2. Therefore the only matrix T which contributes is the matrix $T = \mathrm{diag}(t_1, 0)$. But for T and v as in (i) of the theorem, we have $t = t_1$, $n_1 = 1$, $n_2 = 0$, and $t^{-1}\mathrm{tr}(Tv) = v_1$, so that, by (7.9.7),

(7.11.5)
$$\widehat{\mathcal{Z}}_2(T, v) = -\langle\, \widehat{\mathcal{Z}}(t_1, v_1), \hat{\omega}\,\rangle - \frac{1}{2}\deg_{\mathbb{Q}}(\mathcal{Z}(t_1))\cdot(\,\log(v_2) + \log(D(B))).$$

Finally, for $v = \mathrm{diag}(v_1, v_2)$,

$$\widehat{\mathcal{Z}}_2(0, v) = \langle\, \hat{\omega}, \hat{\omega}\,\rangle + \frac{1}{2}\deg_{\mathbb{Q}}(\hat{\omega})\cdot(\,\log(v_1 v_2) - \mathbf{c} + \log(D(B))). \tag{7.11.6}$$

Now, for a fixed cycle $\widehat{\mathcal{Z}}(t_1, v_1)$, with $t_1 > 0$, the function

$$\langle\, \widehat{\mathcal{Z}}(t_1, v_1), \hat{\phi}_1(\tau_2)\,\rangle = \sum_{t_2\in\mathbb{Z}} \langle\, \widehat{\mathcal{Z}}(t_1, v_1), \widehat{\mathcal{Z}}(t_2, v_2)\,\rangle\, q_2^{t_2} \tag{7.11.7}$$

is a modular form of weight $\frac{3}{2}$, as is the function

$$\hat{\phi}_2(\mathrm{diag}(\tau_1, \tau_2))_{t_1}, \tag{7.11.8}$$

obtained by taking the t_1 Fourier coefficient of $\hat{\phi}_2(\mathrm{diag}(\tau_1, \tau_2))$. We have proved that, for any $t_2 \neq 0$,

$$\langle\, \widehat{\mathcal{Z}}(t_1, v_1), \hat{\phi}_1(\tau_2)\,\rangle_{t_2} = \hat{\phi}_2(\mathrm{diag}(\tau_1, \tau_2))_{t_1, t_2}, \tag{7.11.9}$$

and hence, it follows that the constant terms must agree as well (!), i.e.,

$$\langle\, \widehat{\mathcal{Z}}(t_1, v_1), \widehat{\mathcal{Z}}(0, v_2)\,\rangle = \hat{\phi}_2(\mathrm{diag}(\tau_1, \tau_2))_{t_1,0} = \widehat{\mathcal{Z}}_2(T, v) \tag{7.11.10}$$

for $T = \mathrm{diag}(t_1, 0)$ and $v = \mathrm{diag}(v_1, v_2)$. Comparing (7.11.3) and (7.11.5), we conclude that

$$\mathbf{c} = -\log(D(B)).$$

Inserting this value into (7.11.4) and (7.11.6), we obtain (ii) of the Theorem. Since

$$\deg(\hat{\omega}) = \mathrm{vol}(\mathcal{M}(\mathbb{C})) = -\zeta_{D(B)}(-1),$$

we obtain the corollary. □

The proof of Theorem C is now complete.

Bibliography

[1] B. Gross, *On canonical and quasi-canonical liftings*, Invent. math. **84** (1986), 321–326.

[2] S. Kudla, *Central derivatives of Eisenstein series and height pairings*, Annals of Math., **146** (1997), 545–646.

[3] S. Kudla and M. Rapoport, *Height pairings on Shimura curves and p-adic uniformization*, Invent. math., **142** (2000), 153–223.

[4] S. Kudla, M. Rapoport, and T. Yang, *Derivatives of Eisenstein series and Faltings heights*, Compositio Math. **140**, 887–951.

[5] J. Neukirch, Algebraische Zahlentheorie, Springer-Verlag, Berlin, 1992.

[6] J.-P. Serre, Corps locaux, Hermann, Paris, 1968.

[7] ______, Groupes algébriques et corps de classes, Hermann, Paris, 1959.

Chapter Eight

On the doubling integral

In this chapter, we obtain some information about the doubling zeta integral for the metaplectic cover $G'_{\mathbb{A}}$ of $\mathrm{SL}_2(\mathbb{A})$. In the standard doubling integral, as considered in [5], one integrates the pullback to $\mathrm{Sp}_n(\mathbb{A}) \times \mathrm{Sp}_n(\mathbb{A})$ of a Siegel Eisenstein series on $\mathrm{Sp}_{2n}(\mathbb{A})$ against a pair of cusp forms f_1 and f_2 on $\mathrm{Sp}_n(\mathbb{A})$. If f_1 and f_2 lie in an irreducible cuspidal automorphic representation σ, the result is a product of a partial L-function $L^S(s+\frac{1}{2},\sigma)$ of σ associated to the standard degree $2n+1$ representation of the L-group ${}^L\mathrm{Sp}_n = SO_{2n+1}$ and certain 'bad' local zeta integrals at the places in the finite set S where the data is ramified. If f_1 and f_2 lie in different irreducible cuspidal automorphic representations, then the global doubling integral vanishes. A similar construction can be made for cusp forms on the metaplectic group [15].

Here, we consider a variant of this procedure, which is closer to what is done in [3] and [2]. The pullback $E(\iota(g_1,g_2),s,\Phi)$ of the Siegel Eisenstein series determines a kernel function on $\mathrm{Sp}_n(\mathbb{A}) \times \mathrm{Sp}_n(\mathbb{A})$, and the associated integral operator on the space of cusp forms preserves each irreducible cuspidal automorphic representation σ. If the Eisenstein series is defined by a factorizable section, the resulting endomorphism of $\sigma \simeq \otimes_p \sigma_p$ is given as a product of endomorphisms of the local components σ_p, and, when all the data is unramified, the unramified vector in σ_p is an eigenvector with eigenvalue $L_p(s+\frac{1}{2},\sigma_p)$ (up to a standard normalizing factor independent of σ_p). The difficulty is to determine what happens at the ramified places, including the archimedean places.

In this chapter, we restrict ourselves to the case $n=1$, and we consider genuine irreducible cuspidal automorphic representations σ of the metaplectic extension $G'_{\mathbb{A}}$ of $\mathrm{Sp}_1(\mathbb{A}) = \mathrm{SL}_2(\mathbb{A})$. In this case, the doubling integral represents the degree 2 standard L-function of the representation $\pi = \mathrm{Wald}(\sigma,\psi)$ associated to σ by the Shimura-Waldspurger lift. Our main result is an explicit construction of 'good test vectors', $\boldsymbol{f}_p \in \sigma_p$ and $\Phi_p(s) \in I_p(s,\chi_p)$, the local induced representation, such that $\boldsymbol{f}_p$ is an eigenvector of the local zeta integral operator defined by $\Phi_p(s)$. We do this only for unramified principal series and special representations, for p odd, and for mildly ramified principal series and special representations for $p=2$. The main local result is Theorem 8.3.1. As a consequence, we can construct certain global eigenfunctions $\boldsymbol{f} \in \sigma$ for the doubling integral operator described above. The resulting explicit doubling formulas are given in Theorem 8.3.2 and Theorem 8.3.3. The latter of these formulas will be used in Chapter 9 to prove, in certain cases, a nonvanishing criterion for the arithmetic theta lift.

This chapter is quite long for two reasons. First, in Section 8.2, we have provided a sketch, from our point of view, of Waldspurger's results [23], [28], which we need in order to best formulate our results. The most important of these are the relations among central signs, local dichotomy invariants, and local root numbers. Second, we have provided a substantial amount of background material concerning coordinates on the metaplectic groups, Weil representations, etc., which is not readily accessible in the literature in the form we need. We hope that the resulting precision will justify the additional space required.

8.1 THE GLOBAL DOUBLING INTEGRAL

As in Sections 5.1 and 5.5, we let $G'_{\mathbb{A}}$ be the metaplectic extension of $\mathrm{Sp}_1(\mathbb{A}) = \mathrm{SL}_2(\mathbb{A})$ and $G_{\mathbb{A}}$ be the metaplectic extension of $\mathrm{Sp}_2(\mathbb{A})$. Let $\underline{P}' \subset \mathrm{Sp}_1$ (resp. $\underline{P} \subset \mathrm{Sp}_2$) be the upper triangular Borel subgroup (resp. the Siegel parabolic subgroup), and let $P'_{\mathbb{A}}$ (resp. $P_{\mathbb{A}}$) be the full inverse image of $\underline{P}'(\mathbb{A})$ (resp. $\underline{P}(\mathbb{A})$) in $G'_{\mathbb{A}}$ (resp. $G_{\mathbb{A}}$).

Let $\underline{i}_0 : \mathrm{Sp}_1 \times \mathrm{Sp}_1 \to \mathrm{Sp}_2$ be the embedding

$$\underline{i}_0 : \begin{pmatrix} a_1 & b_1 \\ c_1 & d_1 \end{pmatrix} \times \begin{pmatrix} a_2 & b_2 \\ c_2 & d_2 \end{pmatrix} \longmapsto \begin{pmatrix} a_1 & & b_1 & \\ & a_2 & & b_2 \\ c_1 & & d_1 & \\ & c_2 & & d_2 \end{pmatrix}. \tag{8.1.1}$$

Also let

$$g^{\vee} = \mathrm{Ad}\begin{pmatrix} 1 & \\ & -1 \end{pmatrix} \cdot g, \tag{8.1.2}$$

and put

$$\underline{i}(g_1, g_2) = \underline{i}_0(g_1, g_2^{\vee}). \tag{8.1.3}$$

Recall that there are two orbits of $\mathrm{Sp}_1(\mathbb{Q}) \times \mathrm{Sp}_1(\mathbb{Q})$ on the coset space $\underline{P}(\mathbb{Q})\backslash\mathrm{Sp}_2(\mathbb{Q})$, so that

$$\mathrm{Sp}_2(\mathbb{Q}) = \underline{P}(\mathbb{Q})\,\underline{i}(\mathrm{Sp}_1(\mathbb{Q}) \times \mathrm{Sp}_1(\mathbb{Q})) \;\cup\; \underline{P}(\mathbb{Q})\underline{\delta}\,\underline{i}(\mathrm{Sp}_1(\mathbb{Q}) \times \mathrm{Sp}_1(\mathbb{Q})),$$

where

$$\underline{\delta} = \begin{pmatrix} & & 1 & \\ & 1 & & \\ -1 & 1 & & \\ & & 1 & 1 \end{pmatrix} = w_1\, m(\begin{pmatrix} 1 & -1 \\ & 1 \end{pmatrix}). \tag{8.1.4}$$

Here $w_1 = \underline{i}_0(w_1, 1)$, in the notation of Section 8.5.1. The stabilizers of the chosen orbit representatives are

$$\underline{P}(\mathbb{Q}) \cap \underline{i}(\mathrm{Sp}_1(\mathbb{Q}) \times \mathrm{Sp}_1(\mathbb{Q})) = \underline{i}(\underline{P}'(\mathbb{Q}) \times \underline{P}'(\mathbb{Q})),$$

and

$$\underline{\delta}^{-1}\underline{P}(\mathbb{Q})\underline{\delta} \cap \underline{i}(\mathrm{Sp}_1(\mathbb{Q}) \times \mathrm{Sp}_1(\mathbb{Q})) = \underline{i} \circ \Delta(\mathrm{Sp}_1(\mathbb{Q})),$$

where Δ is the diagonal embedding. In fact, we have

$$\underline{\delta}\, \underline{i}(g, g) = p(g)\, \underline{\delta},$$

where

$$p(g) = \begin{pmatrix} d & c & -c & \\ b & a & & -b \\ & & a & -b \\ & & -c & d \end{pmatrix}.$$

The map $\underline{i}$ has a unique lift to a homomorphism

$$i : G'_{\mathbb{A}} \times G'_{\mathbb{A}} \longrightarrow G_{\mathbb{A}},$$

whose restriction to $\mathbb{C}^1 \times \mathbb{C}^1$ is given by $i(z_1, z_2) = z_1 z_2^{-1}$.

Let ψ be the standard character of $\mathbb{A}/\mathbb{Q}$ which is unramified and such that $\psi_\infty(x) = e(x) = e^{2\pi i x}$. For $\xi \in \mathbb{Q}^\times$, let ψ_ξ be the character $\psi_\xi(x) = \psi(\xi x)$. As explained in Section 8.5.5, there is then a group isomorphism

$$\underline{P}(\mathbb{A}) \times \mathbb{C}^1 \xrightarrow{\sim} P_{\mathbb{A}}, \qquad (p, z) \mapsto [p, z]_{L,\psi} = [p, z]_{\mathrm{L}}.$$

As in Chapter 5, for a character χ of $\mathbb{A}^\times/\mathbb{Q}^\times$, we also write χ for the character of $P_{\mathbb{A}}$ defined by

$$\chi([n(b)m(a), z]_{\mathrm{L}}) = z\, \chi(\det a),$$

and, for $s \in \mathbb{C}$, we let $I(s, \chi)$ be the global degenerate principal series representation of $G_{\mathbb{A}}$ on smooth functions $\Phi(s)$ satisfying

$$\Phi([nm(a), z]_{\mathrm{L}} g, s) = z\, \chi(\det a)\, |\det a|^{s+\frac{3}{2}}\, \Phi(g, s), \tag{8.1.5}$$

where we require that $\Phi(s)$ be K_∞-finite. A section $\Phi(s) \in I(s,\chi)$ is called standard if its restriction to K is independent of s. For a standard section $\Phi(s) \in I(s,\chi)$, the corresponding Siegel Eisenstein series

$$(8.1.6) \qquad E(g,s,\Phi) = \sum_{\gamma \in \underline{P}(\mathbb{Q})\backslash \underline{G}(\mathbb{Q})} \Phi(\gamma g, s)$$

converges for $\mathrm{Re}(s) > \frac{3}{2}$.

Let $\sigma \simeq \otimes_{p\leq\infty}\sigma_p$ be an irreducible cuspidal automorphic representation of $G'_{\mathbb{A}}$ lying in the space $\mathcal{A}_{00}(G')$ of cusp forms orthogonal to all $\mathrm{O}(1)$ theta functions. For a function $f \in \sigma$ and a section $\Phi(s) \in I(s,\chi)$, we consider the global doubling integral defined by

$$(8.1.7) \qquad Z(s,\Phi,f)(g_1') = \int_{\mathrm{Sp}_1(\mathbb{Q})\backslash \mathrm{Sp}_1(\mathbb{A})} E(i(g_1',g_2'),s,\Phi)\, f(g_2')\, dg_2.$$

Here $g_2' \in G'_{\mathbb{A}}$ is any element with image $g_2 \in \mathrm{Sp}_1(\mathbb{A})$, and we take Tamagama measure dg_2 on $\mathrm{Sp}_1(\mathbb{Q})\backslash\mathrm{Sp}_1(\mathbb{A})$. Note that this is not quite the standard doubling integral of Piatetski-Shapiro and Rallis [5], since we are integrating against only one cusp form, but is of the type considered by Böcherer [2] in classical language.

Unwinding in the usual way, using the coset information above, we get

$$\begin{aligned} Z(s,\Phi,f)(g_1') &= \int_{\mathrm{Sp}_1(\mathbb{A})} \Phi(\underline{\delta}\, i(g_1',g_2'),s)\, f(g_2')\, dg_2 \\ &\quad + \int_{\underline{P}'(\mathbb{Q})\backslash \mathrm{Sp}_1(\mathbb{A})} \sum_{\gamma_1 \in \underline{P}'(\mathbb{Q})\backslash \mathrm{Sp}_1(\mathbb{Q})} \Phi(i(\gamma_1 g_1', g_2'),s)\, f(g_2')\, dg_2'. \end{aligned}$$

The second term here vanishes identically, since it can be expressed in terms of the constant term of f. By (i) of Lemma 8.4.1, below, we have

$$\Phi(\underline{\delta}\, i(g_1',g_2'),s) = \Phi(\underline{\delta}\, i(1,(g_1')^{-1}g_2'),s),$$

so that, for $\mathrm{Re}(s) > \frac{3}{2}$,

$$Z(s,\Phi,f)(g_1') = \int_{\mathrm{Sp}_1(\mathbb{A})} \Phi(\underline{\delta}\, i(1,g_2'),s)\, f(g_1'g_2')\, dg_2.$$

Thus, the function $Z(s,\Phi,f)$ again lies in the space of σ, and the doubling integral gives the operation of the function $g' \mapsto \Phi(\underline{\delta}\, i'(1,g'),s)$ on $G'_{\mathbb{A}}$ in the representation σ.

Recall that

$$I(s,\chi) = \otimes_{p\leq\infty} I_p(s,\chi_p),$$

and suppose that $\Phi(s) = \otimes_{p\le\infty}\Phi_p(s)$ and that $f \simeq \otimes_{p\le\infty} f_p$ under the isomorphism $\sigma \simeq \otimes_{p\le\infty}\sigma_p$. Then, under the latter isomorphism,

$$Z(s,\Phi,f) \simeq \bigotimes_{p\le\infty} Z_p(s,\Phi_p,f_p), \tag{8.1.8}$$

where

$$Z_p(s,\Phi_p,f_p) = \int_{\mathrm{Sp}_1(\mathbb{Q}_p)} \Phi_p(\delta_p\, i(1,g'),s)\,\sigma_p(g')f_p\,dg \quad \in \sigma_p \tag{8.1.9}$$

is the local doubling integral. Here $\delta_p \in G_p$ is an element projecting to $\underline{\delta} \in \mathrm{Sp}_2(\mathbb{Q}_p)$, with $\delta_p \in K_p$ for almost all p, and such that $\underline{\delta} = \prod_p \delta_p \in G_{\mathbb{Q}}$. For $p<\infty$ we choose the measure dg_p on $\mathrm{SL}_2(\mathbb{Q}_p)$ for which $\mathrm{vol}(\mathrm{SL}_2(\mathbb{Z}_p)) = 1$. The measure dg_∞ on $\mathrm{SL}_2(\mathbb{R})$ is then determined by the requirement that the product measure $\prod_p dg_p$ is Tamagawa measure. See Lemma 8.4.29 for an explicit description of dg_∞.

In Section 8.4 below, we will consider the case in which σ has 'square free' level and will calculate the local integrals $Z_p(s,\Phi_p,f_p) \in \sigma_p$ explicitly for a certain good choice of f_p and $\Phi_p(s)$. To describe the results, it will be helpful to first review Waldspurger's theory of the Shimura-Shintani correspondence.

8.2 REVIEW OF WALDSPURGER'S THEORY

8.2.1 Local theory

In this section, we review local theta dichotomy as proved by Waldspurger in [28], in particular the important relations among Whittaker models, the central sign, and the representation[1] $\mathrm{Wald}(\sigma,\psi)$ and its root number. In addition, we recall the definition of the Waldspurger involution. Since our formulation differs slightly from that given in [28], we explain in some detail how to derive our statements from those of Waldspurger.

Let F be a local field and let G' be the metaplectic extension of $\mathrm{Sp}_1(F) = \mathrm{SL}_2(F)$. As described in the Section 8.5, for a nondegenerate additive character ψ of F, there is an isomorphism $[\ ,\ ]_R : \mathrm{Sp}_1(F)\times\mathbb{C}^1 \xrightarrow{\sim} G'$ giving the Rao coordinates $g' = [g,z]_R$ of an element $g' \in G'$. We will sometimes write N' for the group of F-points of the unipotent radical $\underline{N}'$ of the standard Borel subgroup $\underline{P}' \subset \mathrm{Sp}_1$, and we identify this group with its image under the unique splitting homomorphism $N' \to G'$, $n \mapsto [n,1]_R$.

The *central sign* of an irreducible admissible genuine representation σ of G' is defined as follows ([28], p. 225):

[1] This notation was introduced in [7].

We have

$$\sigma([m(-1),1]_R)^2 = \sigma([m(-1),1]_R^2) = \sigma([1,(-1,-1)_F]_R) = (-1,-1)_F,$$

where $(\cdot,\cdot)_F$ is the quadratic Hilbert symbol for F. Since the Weil index $\gamma_F(-1,\psi)$ satisfies

$$\gamma_F(-1,\psi)^2 = (-1,-1)_F,$$

there is a sign $z(\sigma,\psi)$ defined by the relation

$$\sigma([m(-1),1]_R) = z(\sigma,\psi)\,\gamma_F(-1,\psi)^{-1}. \tag{8.2.1}$$

Since

$$\begin{aligned} \sigma([m(-1),1]_{\mathrm{L}}) &= \gamma_F(-1,\psi_{\frac{1}{2}})\,\sigma([m(-1),1]_{\mathrm{R}}) \\ &= (-1,2)_F\,\gamma_F(-1,\psi)\,\sigma([m(-1),1]_{\mathrm{R}}), \end{aligned} \tag{8.2.2}$$

in Leray coordinates we have

$$z(\sigma,\psi) = \sigma([m(-1),1]_{\mathrm{L}})\,\chi_2(-1), \tag{8.2.3}$$

where $\chi_2(x) = (x,2)_F$.

For an irreducible admissible genuine representation σ of G', let

$$\hat{F}(\sigma) = \{\, \psi \in \hat{F} \mid W(\sigma,\psi) \neq 0 \,\}, \tag{8.2.4}$$

where $W(\sigma,\psi)$ is the ψ-Whittaker model of σ.

For a quadratic space (V,Q) over F and a nondegenerate additive character ψ of F, let $\omega_{\psi,V}$ be the Weil representation of G' on the space $S(V)$; see Section 8.5.3 for the conventions used. We assume that $\dim_F(V)$ is odd.

Suppose that $\xi \in F^\times$ is such that
(i) $\psi_\xi \in \hat{F}(\sigma)$, and
(ii) there is a vector $x_\xi \in V$ such that $Q(x_\xi) = \xi$.
Then, for $W \in W(\sigma,\psi_\xi)$, $\varphi \in S(V)$, and $h \in \mathrm{SO}(V)$, we can consider the integral

$$u_\psi(h;W,\varphi;\xi,V) = \int_{N'\backslash \mathrm{Sp}_1(F)} W(g')\,\overline{\omega_{\psi,V}(g')\varphi(h^{-1}x_\xi)}\,dg, \tag{8.2.5}$$

as in [28], IV, p. 238, up to a change in notation. Here $g' \in G'$ is any element which projects to $g \in \mathrm{Sp}_1(F)$. Note that a central element $[1,z]_R$ acts by multiplication by z in both σ and the Weil representation so that the integrand is invariant under $\mathbb{C}^1$. Formally, we may view this integral as defining a map[2]

$$(8.2.6) \qquad u_\psi(\xi, V) : S(V) \longrightarrow W(\sigma, \psi_\xi)^\vee \otimes C^\infty(\mathrm{SO}(V)_{x_\xi} \backslash \mathrm{SO}(V)).$$

This map is $G' \times \mathrm{SO}(V)$-equivariant. Let

$$(8.2.7) \qquad \theta_\psi(\sigma; \xi, V) \subset C^\infty(\mathrm{SO}(V)_{x_\xi} \backslash \mathrm{SO}(V))$$

be the subspace spanned by the functions $u_\psi(h; W, \varphi; \xi, V)$ as W and φ vary; this submodule is stable under the action of $\mathrm{SO}(V)$.

For a quaternion algebra B over F, let

$$\mathbf{V}^B = \{ x \in B \mid \mathrm{tr}(x) = 0 \},$$

with quadratic form[3] $q(x) = -\nu(x) = x^2$, and let $H^B = B^\times$. Recall that the action of H^B on $\mathbf{V}^B$ by conjugation gives an isomorphism $H^B \simeq \mathrm{GSpin}(\mathbf{V}^B)$. We write

$$\mathbf{V}^B = \mathbf{V}^\pm$$

and $H^B = H^\pm$, if $\mathrm{inv}(B) = \pm 1$.

Let $\widetilde{\mathcal{P}}$ be the set of (isomorphism classes of) irreducible admissible, genuine unitary representations of G' which are not of the form $\theta_\psi(\mathbb{1}, U_\alpha)$, where $\theta_\psi(\mathbb{1}, U_\alpha)$ is the Weil representation ω_{ψ, U_α} of G' on the space of even functions in the Schwartz space $S(U_\alpha)$, where $U_\alpha = F$ with quadratic form $Q(x) = \alpha x^2$.

Waldspurger proves the following result in [28].

Theorem 8.2.1. (Waldspurger). *Assume that $\sigma \in \widetilde{\mathcal{P}}$. Suppose that $\xi \in F^\times$ with $\psi_\xi \in \hat{F}(\sigma)$.*
(i) If $V = \mathbf{V}^+$, then

$$\theta_\psi(\sigma; \xi, \mathbf{V}^+) \neq 0 \iff \psi \in \hat{F}(\sigma).$$

If $\psi \in \hat{F}(\sigma)$, then $\theta_\psi(\sigma; \xi, \mathbf{V}^+)$ is an irreducible representation of H^+. Moreover, the isomorphism class of $\theta_\psi(\sigma; \xi, \mathbf{V}^+)$ is independent of ξ.

---

[2]We get a $\mathbb{C}$-linear map by composing with complex conjugation $\varphi \mapsto \bar{\varphi}$ on $S(V)$. The resulting map $S(V) \longrightarrow W(\sigma, \psi_\xi)^\vee \otimes C^\infty(\mathrm{SO}(V)_{x_\xi} \backslash \mathrm{SO}(V))$ is equivariant for $G' \times SO(V)$, where G' acts on $S(V)$ by $g \mapsto \omega_{V,\psi}(g^\vee)$; see Lemma 8.5.8. Since the representation $\sigma(g^\vee)$ is isomorphic to the contragradient $\sigma^\vee(g)$, [17], Chapitre 4, we obtain an intertwining map $S(V) \longrightarrow \sigma \otimes C^\infty(\mathrm{SO}(V)_{x_\xi} \backslash \mathrm{SO}(V))$. Thus, the representation $\theta_\psi(\sigma; V)$ defined below corresponds to σ under local Howe duality [6].

[3]Here we follow the convention in [28], and hence use the notation $\mathbf{V}^B$ to distinguish this space from the space V^B where the quadratic form $Q(x) = \nu(x) = -x^2$ is used.

(ii) If $V = \mathbf{V}^-$, there is an $x_\xi \in \mathbf{V}^-$ with $Q(x_\xi) = \xi$ if and only if $\xi \notin F^{\times,2}$. Then

$$\theta_\psi(\sigma; \xi, \mathbf{V}^-) \neq 0 \iff \psi \notin \hat{F}(\sigma).$$

If $\psi \notin \hat{F}(\sigma)$, then $\theta_\psi(\sigma; \xi, \mathbf{V}^-)$ is an irreducible representation of H^- and the isomorphism class of $\theta_\psi(\sigma; \xi, \mathbf{V}^-)$ is independent of ξ.

In case (i) of the theorem, let

$$\theta_\psi(\sigma, \mathbf{V}^+) \simeq \theta_\psi(\sigma; \xi, \mathbf{V}^+),$$

and let $\theta_\psi(\sigma, \mathbf{V}^-) = 0$. In case (ii) of the theorem, let

$$\theta_\psi(\sigma, \mathbf{V}^-) \simeq \theta_\psi(\sigma; \xi, \mathbf{V}^-),$$

and let $\theta_\psi(\sigma, \mathbf{V}^+) = 0$. Define the *dichotomy sign* to be

$$\delta(\sigma, \psi) = \begin{cases} +1 & \text{if } \theta_\psi(\sigma, \mathbf{V}^+) \neq 0, \\ -1 & \text{if } \theta_\psi(\sigma, \mathbf{V}^-) \neq 0. \end{cases} \tag{8.2.8}$$

It follows that, if σ is as in Theorem 8.2.1, then

$$\delta(\sigma, \psi) = +1 \iff \psi \in \hat{F}(\sigma). \tag{8.2.9}$$

For example, if σ is an irreducible principal series representation, then, by [28], Lemma 3, p. 227, $\hat{F}(\sigma) = \hat{F}$, and so $\delta(\sigma, \psi) = +1$ for all ψ.

For $\sigma \in \widetilde{\mathcal{P}}$, take ξ so that $\psi_\xi \in \hat{F}(\sigma)$, and let

$$\mathrm{Wald}(\sigma, \psi) := \theta_{\psi_\xi}(\sigma, \mathbf{V}^+) \otimes \chi_\xi, \tag{8.2.10}$$

where $\chi_\xi(x) = (x, \xi)_F$ is the quadratic character attached to ξ.[4] In [28], Proposition 15, p. 266, Waldspurger proves that this representation is independent of the choice of ξ, for $\sigma \in \widetilde{\mathcal{P}}$. Note that the relation

$$\mathrm{Wald}(\sigma, \psi_\alpha) = \mathrm{Wald}(\sigma, \psi) \otimes \chi_\alpha \tag{8.2.11}$$

is immediate from the definition.

The following result, which is implicit in [28], relates the dichotomy sign, the central sign, and the root number of $\mathrm{Wald}(\sigma, \psi)$.

Theorem 8.2.2. (Waldspurger). *For $\sigma \in \widetilde{\mathcal{P}}$,*

$$\delta(\sigma, \psi) = \epsilon(\frac{1}{2}, \mathrm{Wald}(\sigma, \psi)) \cdot z(\sigma, \psi).$$

[4] This space is called $S_\psi(T)$ in [28], p. 280.

Proof. First recall that Waldspurger defines Shintani liftings taking irreducible admissible unitary representations of $H^{\pm} = \mathrm{SO}(\mathbf{V}^{\pm})$ to irreducible admissible genuine representations of G': $\pi \mapsto \theta(\pi, \psi)$, p. 228 (resp. $\pi' \mapsto \theta(\pi', \psi)$, p. 235). For the first of these, he proves in [28], Theorem 1, p.249, that the maps $\pi \mapsto \theta(\pi, \psi)$ and $\sigma \mapsto \theta_{\psi}(\sigma, \mathbf{V}^+)$ define reciprocal bijections:

$$(8.2.12) \quad \left\{ \begin{array}{c} \text{irreducible, admissible,} \\ \text{unitary, generic} \\ \text{rep's } \pi \text{ of PGL}(2) \end{array} \right\} \quad \leftrightarrow \quad \left\{ \begin{array}{c} \text{irreducible, admissible,} \\ \text{unitary, genuine} \\ \text{rep's } \sigma \text{ of } G' \\ \sigma \not\simeq \theta_{\psi}(\mathbb{1}, U_{\alpha}) \\ \text{with } \psi \in \hat{F}(\sigma) \end{array} \right\}.$$

On the left-hand side here, generic just means infinite dimensional. For the second of these, he proves in [28], Proposition 14, p. 266, that the maps $\pi' \mapsto \theta(\pi', \psi)$ and $\sigma \mapsto \theta_{\psi}(\sigma, \mathbf{V}^-)$ define reciprocal bijections:

$$(8.2.13) \quad \left\{ \begin{array}{c} \text{irreducible, spherical} \\ \text{rep's } \pi' \text{ of SO}(\mathbf{V}^-) \end{array} \right\} \quad \leftrightarrow \quad \left\{ \begin{array}{c} \text{special or supercuspidal} \\ \text{(resp., discrete series)} \\ \text{rep's } \sigma \text{ of } G' \\ \text{with } \psi \notin \hat{F}(\sigma) \end{array} \right\}.$$

On the left-hand side here, spherical means that the representation π' occurs in the space $C^{\infty}(H^-_x \backslash H^-)$ for some nonzero $x \in V^-$. On the right-hand side here, σ is special or supercuspidal in the nonarchimedean case and is a discrete series representation in the archimedean case. In the nonarchimedean case, π' has dimension > 1 if and only if σ is supercuspidal and is not an odd Weil representation. Note that, in fact, all irreducible representations of $\mathrm{SO}(\mathbf{V}^-)$ are spherical [28], Proposition 18, p. 277.

Also recall that for an irreducible admissible unitary generic representation π of $\mathrm{PGL}(2) = \mathrm{SO}(\mathbf{V}^+)$,

$$(8.2.14) \qquad F(\pi) := \{ \xi \in F^{\times} \mid \mathcal{U}(\pi, \xi) \neq 0 \},$$

where $\mathcal{U}(\pi, \xi)$ is the 'hyperboloid model' of π, [28], p. 226, i.e., a realization of π in the space of functions $C^{\infty}(H_{x_{\xi}} \backslash H)$ for $H = \mathrm{GL}_2(F)$.

The following fundamental relation is then Lemma 6, p. 234 in [28].

Lemma 8.2.3. *(i)*

$$\hat{F}(\theta(\pi, \psi)) = \{ \psi_{\xi} \mid \xi \in F(\pi) \}.$$

(ii)

$$z(\theta(\pi, \psi), \psi) = \epsilon(\frac{1}{2}, \pi).$$

This second relation says that the central sign of $\theta(\pi, \psi)$ coincides with the root number of π.

Suppose that σ lies in the set on the right side of (8.2.12). Then, on the one hand, $\theta_\psi(\sigma, \mathbf{V}^+) \neq 0$, so that $\delta(\sigma, \psi) = +1$. On the other hand, $\sigma = \theta(\pi, \psi)$ for some π on the left side of (8.2.12), and, by (ii) of the fundamental relation,

$$z(\sigma, \psi) = \epsilon(\frac{1}{2}, \pi). \tag{8.2.15}$$

Moreover, $\psi \in \hat{F}(\sigma)$, so that

$$\pi = \theta_\psi(\sigma, \mathbf{V}^+) = \text{Wald}(\sigma, \psi), \tag{8.2.16}$$

where the first equality is given by the bijection (8.2.12). This gives the identity of Theorem 8.2.2 in this case.

Next suppose that σ is as on the right side of (8.2.12), except that $\psi \notin \hat{F}(\sigma)$. This condition implies that σ is not an irreducible principal series, and hence lies in the set on the right side of (8.2.13), since the even Weil representations have been excluded. Hence $\sigma = \theta(\pi', \psi)$ for some π' in the set of representations on the left side of (8.2.13). Let π be the representation of $SO(\mathbf{V}^+) = PGL(2)$ associated to π' under the Jacquet-Langlands correspondence, and let $\sigma' = \theta(\pi, \psi)$. Here is the picture:

$$\begin{array}{ccccccc} G' & & \sigma & \leftrightarrow & \pi' & & SO(\mathbf{V}^-) \\ & \text{W.inv.} & \updownarrow & & \updownarrow & \text{JL} & \\ G' & & \sigma' & \leftrightarrow & \pi & & SO(\mathbf{V}^+), \end{array} \tag{8.2.17}$$

where the vertical arrow on the right is the Jacquet-Langlands correspondence and the vertical arrow on the left is, by definition, the Waldspurger involution. On the one hand, since $\theta_\psi(\sigma, \mathbf{V}^-) \neq 0$, we have $\delta(\sigma, \psi) = -1$. On the other hand, by (3) of Theorem 2, p. 277 of [28] combined with (ii) of Lemma 8.2.3, we have

$$z(\sigma, \psi) = -z(\sigma', \psi) = -\epsilon(\frac{1}{2}, \pi). \tag{8.2.18}$$

It then remains to relate π and $\text{Wald}(\sigma, \psi)$. First suppose that σ is not an odd Weil representation, so that σ lies in Waldspurger's set $\widetilde{\mathcal{P}}_1$. Then Proposition 15, p. 266 of Waldspurger implies that

$$\pi = \text{Wald}(\sigma, \psi).$$

This proves Theorem 8.2.2 in this case.

Finally, we need to consider the case in which σ is an odd Weil representation, i.e., $\sigma = \theta_\psi(\mathrm{sgn}, U_\alpha)$, where $U_\alpha = (F, \alpha x^2)$. The condition

$$\psi \notin \hat{F}(\sigma) = \{\psi_\xi \mid \xi \in \alpha \cdot F^{\times,2}\} \tag{8.2.19}$$

implies that $\alpha \notin F^{\times,2}$. By Lemma 20, p. 249 of [28], we have

$$\theta_{\psi_\alpha}(\sigma, \mathbf{V}^+) = \sigma(|\ |^{\frac{1}{2}}, |\ |^{-\frac{1}{2}}), \tag{8.2.20}$$

and hence

$$\mathrm{Wald}(\sigma, \psi) = \sigma(|\ |^{\frac{1}{2}}, |\ |^{-\frac{1}{2}}) \otimes \chi_\alpha. \tag{8.2.21}$$

Since $\chi_\alpha \neq 1$, we have

$$\begin{aligned} \epsilon(\frac{1}{2}, \mathrm{Wald}(\sigma, \psi)) &= \epsilon(\frac{1}{2}, \sigma(|\ |^{\frac{1}{2}}, |\ |^{-\frac{1}{2}}) \otimes \chi_\alpha) \\ &= \chi_\alpha(-1), \end{aligned} \tag{8.2.22}$$

by the relation given on p. 274 of [28]. On the other hand, the central sign can be computed from the formulas for the Weil representation:

$$\omega_{\psi,U}([m(a), 1]_R)\varphi(x) = \gamma_F(a, \psi)^{-1}\, |a|^{\frac{1}{2}}\, \chi_\alpha(a)\, \varphi(xa). \tag{8.2.23}$$

Taking $a = -1$ and assuming that φ is an odd function, we have, via (8.2.1),

$$z(\sigma, \psi) = -\chi_\alpha(-1), \tag{8.2.24}$$

so that the relation of Theorem 8.2.2 holds. □

Finally, we give a small table of the corresponding representations. Recall that, in [28], for a character μ of $F^\times$, the principal series representation is defined on the space of functions $\mathcal{B}(\mu)$ on G' satisfying

$$f([nm(a), z]_R g') = z\, \mu(a)\, \gamma_F(a, \psi)^{-1}\, |a|\, f(g'). \tag{8.2.25}$$

Note that this parametrization depends on ψ. Notice also that $\mathcal{B}(\mu) = I(\mu\chi_2)$ in our notation for induced representations of G'; see (8.4.3) below. If the action of G' on this space is irreducible, the resulting principal series representation is denoted by $\tilde{\pi}(\mu)$. In the nonarchimedean case, if $\mu = \chi_\alpha|\ |^{\frac{1}{2}}$, the action of G' is not irreducible. The irreducible subrepresentation of $\mathcal{B}(\mu)$ will be the special representation $\tilde{\sigma}(\chi_\alpha|\ |^{\frac{1}{2}})$ and the irreducible quotient will be $\theta_\psi(1\!\!1, U_\alpha)$, the even Weil representation for the one-dimensional quadratic space $U_\alpha = (F, \alpha x^2)$.

In this table, the additive character ψ is fixed and the parameterization of the representations in the first column depends on ψ, as just explained.

Table 2. The Local Theta Correspondence

σ	$\mathrm{Wald}(\sigma, \psi_\beta)$	$\epsilon(\frac{1}{2}, \mathrm{Wald}(\sigma, \psi_\beta))$	$\delta(\sigma, \psi_\beta)$	$z(\sigma, \psi_\beta)$
$\tilde{\pi}(\mu)$	$\pi(\chi_\beta\mu, \chi_\beta\mu^{-1})$	$\chi_\beta\mu(-1)$	$+1$	$\chi_\beta\mu(-1)$
$\tilde{\sigma}(\chi_\alpha\lvert\,\rvert^{\frac{1}{2}})$	$\sigma(\chi_{\alpha\beta}\lvert\,\rvert^{\frac{1}{2}}, \chi_{\alpha\beta}\lvert\,\rvert^{-\frac{1}{2}})$	$\chi_{\alpha\beta}(-1)$	$+1$	$\chi_{\alpha\beta}(-1)$
$\tilde{\sigma}(\chi_\beta\lvert\,\rvert^{\frac{1}{2}})$	$\sigma(\lvert\,\rvert^{\frac{1}{2}}, \lvert\,\rvert^{-\frac{1}{2}})$	-1	-1	1
$\theta_\psi(\mathrm{sgn}, U_\alpha)$	$\sigma(\chi_{\alpha\beta}\lvert\,\rvert^{\frac{1}{2}}, \chi_{\alpha\beta}\lvert\,\rvert^{-\frac{1}{2}})$	$\chi_{\alpha\beta}(-1)$	-1	$-\chi_{\alpha\beta}(-1)$
$\theta_\psi(\mathrm{sgn}, U_\beta)$	$\sigma(\lvert\,\rvert^{\frac{1}{2}}, \lvert\,\rvert^{-\frac{1}{2}})$	-1	$+1$	-1
other s.c.	s.c.	...	...	...
$\tilde{\pi}_\ell^\pm$	$\mathrm{DS}_{2\ell-1}$	$(-1)^{\ell-\frac{1}{2}}$	$\pm\chi_\beta(-1)$	$\pm\chi_\beta(-1)\,(-1)^{\ell-\frac{1}{2}}$

Here, in the second and fourth rows, $\chi_{\alpha\beta} \neq 1$. In the next to last row, 's.c.' stands for supercuspidal representations other than the odd Weil representations. For these, we do not record information about the root number, dichotomy invariant, and central sign, since these quantities depend on the detailed parametrization of such representations [16], and we will not need this information.

In the last row, for the archimedean case, $\ell \geq \frac{3}{2}$, and we suppose that $\psi(x) = e(x)$. Here $\tilde{\pi}_\ell^+ = \mathrm{HDS}_\ell$ is the holomorphic discrete series representation with lowest weight ℓ, while $\tilde{\pi}_\ell^-$ is the antiholomorphic discrete series representation with highest weight $-\ell$. Note that the signatures of the quadratic spaces in [28] are $\mathrm{sig}(\mathbf{V}^+) = (2, 1)$ and $\mathrm{sig}(\mathbf{V}^-) = (0, 3)$. Also $\mathrm{DS}_{2\ell-1}$ denotes the discrete series representation of $\mathrm{PGL}_2(\mathbb{R})$ of weight $2\ell - 1$.

The action of the Waldspurger involution can be seen in the table.[5] The pairs of representations which are switched by the involution are

$$\{\, \tilde{\sigma}(\chi_\alpha\lvert\,\rvert^{\frac{1}{2}}), \theta_\psi(\mathrm{sgn}, U_\alpha) \,\} \quad \text{and} \quad \{\, \tilde{\pi}_\ell^+, \tilde{\pi}_\ell^- \,\}. \tag{8.2.26}$$

8.2.2 Global theory

We now review parts of Waldspurger's global theory. For simplicity, we work over $\mathbb{Q}$. Let $\mathcal{A}_0(G')$ be the space of genuine cusp forms on $G'_{\mathbb{A}}$ and let $\mathcal{A}_{00}(G')$ be the orthogonal complement in $\mathcal{A}_0(G')$ of the subspace spanned by the theta functions coming from one-dimensional quadratic spaces.

For an irreducible genuine cuspidal representation $\sigma \simeq \otimes_{p\leq\infty}\sigma_p$ of $G'_{\mathbb{A}}$ occuring in $\mathcal{A}_{00}(G')$, the representation

$$\mathrm{Wald}(\sigma, \psi) = \otimes_{p\leq\infty}\mathrm{Wald}(\sigma_p, \psi_p) \tag{8.2.27}$$

[5]except for the supercuspidals, of course, whose data we have not recorded.

is, in fact, a cuspidal automorphic representation of $\mathrm{PGL}_2(\mathbb{A})$. It can also be defined by using global theta integrals; see [23] and [28], p. 280.[6] The central signs of the components σ_p of σ satisfy the product formula, see [28], p. 280 (1),

$$\prod_{p\leq\infty} z(\sigma_p, \psi_p) = 1, \tag{8.2.28}$$

and the fibers of the map $\sigma \mapsto \mathrm{Wald}(\sigma, \psi)$ have the following description, see [28], Corollaire 2, p. 286:

Proposition 8.2.4. *Let Σ be the set of all $p \leq \infty$ at which $\mathrm{Wald}(\sigma_p, \psi_p)$ is not an irreducible principal series representation. Then the set of all irreducible cuspidal genuine σ''s such that $\mathrm{Wald}(\sigma', \psi) = \mathrm{Wald}(\sigma, \psi)$ has cardinality 1 if $|\Sigma| = 0$ and $2^{|\Sigma|-1}$ if $|\Sigma| > 0$. In the latter case, the representations σ' have the form*

$$\sigma' \simeq (\otimes_{p\in\Sigma'}\sigma_p^W) \otimes (\otimes_{p\notin\Sigma'}\sigma_p),$$

where $\Sigma' \subset \Sigma$ is any subset with even cardinality. Here $\sigma_p \mapsto \sigma_p^W$ is the Waldspurger involution.

Note that, since the Waldspurger involution switches the central sign, the condition on the cardinality of Σ' is necessary in order to preserve the product formula for the central signs of the local components of σ'.

Because of the product formula for central signs, Theorem 8.2.2 implies the relation

$$\prod_{p\leq\infty} \delta(\sigma_p, \psi_p) = \prod_{p\leq\infty} \epsilon(\frac{1}{2}, \mathrm{Wald}(\sigma_p, \psi_p)) = \epsilon(\frac{1}{2}, \mathrm{Wald}(\sigma, \psi)) \tag{8.2.29}$$

between the product of the local dichotomy signs and the global root number of $\mathrm{Wald}(\sigma, \psi)$.

As usual, for any quaternion algebra B over $\mathbb{Q}$, let $\mathbf{V}^B$ be the $\mathbb{Q}$-vector space of trace zero elements in B with quadratic form $q(x) = -\nu(x) = x^2$, and let $H^B = B^\times \simeq \mathrm{GSpin}(\mathbf{V}^B)$. For $g' \in G'_{\mathbb{A}}$, $h \in H^B(\mathbb{A})$ and $\varphi \in S(\mathbf{V}^B(\mathbb{A}))$, there is a theta function

$$\theta(g', h; \varphi) = \sum_{x\in\mathbf{V}^B(\mathbb{Q})} \omega(g')\varphi(h^{-1}x), \tag{8.2.30}$$

where $\omega = \omega_{\mathbf{V}^B,\psi}$ is the Weil representation of $G'_{\mathbb{A}}$ in $S(\mathbf{V}^B(\mathbb{A}))$. For a cusp form $\mathbf{f} \in \sigma$, where σ is as above, and for $\varphi \in S(\mathbf{V}^B(\mathbb{A}))$, the global

[6] where $\mathrm{Wald}(\sigma, \psi)$ is denoted by $S_\psi(T)$.

theta integral is given by

$$\theta(\mathbf{f},\varphi)(h) = \int_{\mathrm{SL}_2(\mathbb{Q})\backslash \mathrm{SL}_2(\mathbb{A})} \mathbf{f}(g')\,\overline{\theta(g',h;\varphi)}\,dg, \tag{8.2.31}$$

where $g' \in G'_{\mathbb{A}}$ is any element mapping to $g \in \mathrm{SL}_2(\mathbb{A})$. This function lies in the space $\mathcal{A}_0(H^B)$ of cusp forms on $H^B(\mathbb{A})$ with trivial central character. Let

$$\theta_\psi(\sigma, \mathbf{V}^B) \quad \subset \quad \mathcal{A}_0(H^B) \tag{8.2.32}$$

be the subspace spanned by the $\theta(\mathbf{f},\varphi)$'s as $\mathbf{f}$ varies in σ and φ varies in $S(\mathbf{V}^B(\mathbb{A}))$. This global construction is compatible with the local construction of Section 8.2.1, and we have

$$\theta_\psi(\sigma, \mathbf{V}^B) \simeq \begin{cases} \otimes_{p\le\infty}\theta_{\psi_p}(\sigma_p, \mathbf{V}^B_p), \\ 0, \end{cases} \tag{8.2.33}$$

so that $\theta_\psi(\sigma, \mathbf{V}^B)$ is either an irreducible cuspidal automorphic representation or zero. Now local theta dichotomy comes into play, and

$$\theta_\psi(\sigma, \mathbf{V}^B) \neq 0 \quad \Longrightarrow \quad \mathrm{inv}_p(B) = \delta(\sigma_p, \psi_p), \quad \forall p \le \infty. \tag{8.2.34}$$

Waldspurger's beautiful result is then the following.

Theorem 8.2.5. *(i) If* $\epsilon(\frac{1}{2}, \mathrm{Wald}(\sigma,\psi)) = -1$, *then* $\theta_\psi(\sigma, \mathbf{V}^B) = 0$ *for all* B.
(ii) If $\epsilon(\frac{1}{2}, \mathrm{Wald}(\sigma,\psi)) = 1$, *then there is a unique quaternion algebra* B *over* $\mathbb{Q}$ *such that*

$$\otimes_{p\le\infty}\theta_{\psi_p}(\sigma_p, \mathbf{V}^B_p) \neq 0.$$

In this case,

$$\theta_\psi(\sigma, \mathbf{V}^B) \neq 0 \quad \Longleftrightarrow \quad L(\frac{1}{2}, \mathrm{Wald}(\sigma,\psi)) \neq 0.$$

Note: Elsewhere in this chapter and, indeed, throughout the book, we work with the quadratic spaces V^B with quadratic form $Q(x) = \nu(x) = -x^2$. As explained in Section 8.5, the associated Weil representation can be written as

$$\omega_{V^B,\psi} = \omega_{\mathbf{V}^B,\psi_{-1}} \tag{8.2.35}$$

in both the local and global situations. This implies that

$$\theta_\psi(\sigma, V^B) = \theta_{\psi_{-1}}(\sigma, \mathbf{V}^B). \tag{8.2.36}$$

For convenient later reference, we state the corresponding version of Theorem 8.2.5.

Corollary 8.2.6. *(i) If* $\epsilon(\frac{1}{2}, \text{Wald}(\sigma, \psi_{-1})) = -1$, *then* $\theta_\psi(\sigma, V^B) = 0$ *for all* B.
(ii) If $\epsilon(\frac{1}{2}, \text{Wald}(\sigma, \psi_{-1})) = 1$, *then there is a unique quaternion algebra* B *over* $\mathbb{Q}$ *such that*

$$\otimes_{p\leq\infty}\theta_{\psi_p}(\sigma_p, V_p^B) \neq 0.$$

In this case,

$$\theta_\psi(\sigma, V^B) \neq 0 \quad \Longleftrightarrow \quad L(\frac{1}{2}, \text{Wald}(\sigma, \psi_{-1})) \neq 0.$$

In the case $\epsilon(\frac{1}{2}, \text{Wald}(\sigma, \psi_{-1})) = -1$, a result analogous to (ii) on the nonvanishing of the *arithmetic* theta lifting in terms of the central *derivative* $L'(\frac{1}{2}, \text{Wald}(\sigma, \psi_{-1}))$ is proved in Chapter 9 in some special cases.

8.3 AN EXPLICIT DOUBLING FORMULA

We now return to the doubling integral of Section 8.1. The additive character ψ is now fixed to be unique unramified character with $\psi_\infty(x) = e(x)$. Let $\chi = \chi_{2\kappa}$, where $\kappa = \pm1$, be a global quadratic character. For convenience, we will write ψ_p^κ for the local component at p of ψ_κ.

Let $\sigma \simeq \otimes_{p\leq\infty}\sigma_p$ be an irreducible genuine cuspidal automorphic representation in the space $\mathcal{A}_{00}(G')$. We will determine good 'test vectors' $\boldsymbol{f}_p \in \sigma_p$ and $\Phi_p(s) \in I_p(s, \chi)$ such that $\boldsymbol{f}_p$ is an eigenvector for the operator $Z(s, \Phi_p, \cdot)$. We limit ourselves to those σ_p's which occur in our arithmetic application, although our method can be extended to all σ_p's. Specifically, we will prove the following result, which collects together from Section 8.4 the results of the calculations of local doubling integrals. Recall that we normalize the Haar measure on $\text{SL}_2(\mathbb{Q}_p)$ used in the definition of the local doubling integral by $\text{vol}(\text{SL}_2(\mathbb{Z}_p)) = 1$, for $p < \infty$, and we take $dg = \frac{12}{\pi} \cdot a^{-3}\, da\, db\, d\theta$, for $g = n(b)m(a)k_\theta \in \text{SL}_2(\mathbb{R})$ with $a > 0$, so that the product of these local measures gives Tamagawa measure on $\text{SL}_2(\mathbb{A})$; see Lemma 8.4.29. Also, recall that the parametrization of local components σ_p is as in Table 2 of Section 8.2 with respect to the local additive character ψ_p determined by ψ.

Theorem 8.3.1. *For a finite prime* p, *and for* $\kappa \in \mathbb{Z}_p^\times$, *let* $\Phi_p^0(s)$, $\Phi_p^1(s)$, *and* $\Phi_p^{\text{ra}}(s) \in I_p(s, \chi_{2\kappa})$ *be the standard sections defined in (8.4.13) below.*
(i) For $p \neq 2$, *suppose that* $\sigma_p = I(\mu_p) = \tilde{\pi}(\chi_2\mu_p)$ *is an unramified principal series representation. Let* $\boldsymbol{f}_p \in \sigma_p$ *be the unramified vector. Then*

$$Z(s, \Phi_p^0, \boldsymbol{f}_p) = \frac{L(s+\frac{1}{2}, \text{Wald}(\sigma_p, \psi_p^\kappa))}{\zeta_p(2s+2)} \cdot \boldsymbol{f}_p.$$

(ii) For $p \neq 2$, *suppose that* $\sigma_p = \tilde{\sigma}(\chi_\alpha|\ |^{\frac{1}{2}})$, *with* $\alpha \in \mathbb{Z}_p^\times$, *is an unramified special representation, and let* $\boldsymbol{f}_p = \boldsymbol{f}_{\rm sp} \in \sigma_p$ *be the Iwahori fixed vector. Then,* $Z(s, \Phi_p^0, \boldsymbol{f}_p) = 0$,

$$Z(s, \Phi_p^1, \boldsymbol{f}_p) = p^{-2}\,\frac{p-1}{p+1}(1 + \delta(\sigma_p, \psi_p^\kappa)\, p^{-s})L(s + \frac{1}{2}, \mathrm{Wald}(\sigma_p, \psi_p^\kappa))\boldsymbol{f}_p$$

and

$$Z(s, \Phi_p^{\rm ra}, \boldsymbol{f}_p) = -p^{-2}(1 - \delta(\sigma_p, \psi_p^\kappa)\, p^{-s})L(s + \frac{1}{2}, \mathrm{Wald}(\sigma_p, \psi_p^\kappa))\boldsymbol{f}_p.$$

Here $\delta(\sigma_p, \psi_p^\kappa)$ *is the local dichotomy invariant defined in (8.2.8); see Theorem 8.2.2.*

(iii) For $p = 2$, *suppose that* $\sigma_p = I(\mu_p) = \tilde{\pi}(\chi_\alpha|\ |^t)$, *with* $\mu_p = \chi_{2\alpha}|\ |^t$, $\alpha \in \mathbb{Z}_p^\times$ *and* $t \neq \pm\frac{1}{2}$, *is an irreducible principal series representation. Let* $\boldsymbol{f}_p = \boldsymbol{f}_{\rm ev} \in \sigma_p^{(\mathbf{J}', \chi_\alpha)}$ *be the 'good newvector' defined in part (i) of Theorem 8.4.26. Assume that* $\alpha \equiv \kappa \mod 4$. *Then*

$$Z(s, \Phi_p^0, \boldsymbol{f}_p) = \delta_\kappa\,\frac{1}{2\sqrt{2}} \cdot \frac{L(s + \frac{1}{2}, \mathrm{Wald}(\sigma_p, \psi_p^\kappa))}{\zeta_p(2s+2)} \cdot \boldsymbol{f}_p,$$

where $\delta_\kappa = 1$ *if* $\kappa \equiv 1 \mod 4$ *and* $\delta_\kappa = i$ *if* $\kappa \equiv 3 \mod 4$.

(iv) For $p = 2$, *suppose that* $\sigma_p = \tilde{\sigma}(\chi_\alpha|\ |^{\frac{1}{2}})$ *with* $\alpha \in \mathbb{Z}_p^\times$ *is a special representation, and let* $\boldsymbol{f}_p = \boldsymbol{f}_{\rm sp} \in \sigma_p^{(\mathbf{J}', \chi_\alpha)}$ *be the 'good newvector' defined in part (ii) of Theorem 8.4.26. Assume that* $\alpha \equiv \kappa \mod 4$. *Then,* $Z(s, \Phi_p^0, \boldsymbol{f}_p) = 0$,

$$Z(s, \Phi_p^1, \boldsymbol{f}_p) = \star\, p^{-2}\,\frac{p-1}{p+1}(1 + \delta(\sigma_p, \psi_p^\kappa)\, p^{-s})L(s + \frac{1}{2}, \mathrm{Wald}(\sigma_p, \psi_p^\kappa))\boldsymbol{f}_p$$

and

$$Z(s, \Phi_p^{\rm ra}, \boldsymbol{f}_p) = -\star\, p^{-2}(1 - \delta(\sigma_p, \psi_p^\kappa)\, p^{-s})L(s + \frac{1}{2}, \mathrm{Wald}(\sigma_p, \psi_p^\kappa))\boldsymbol{f}_p,$$

where $\star = \delta_\kappa\,\frac{1}{2\sqrt{2}}$.

(v) Suppose that $p = \infty$ *and that* $\sigma_\infty = \tilde{\pi}_\ell^+$ *is the holomorphic discrete series representation of weight* $\ell \in \frac{1}{2} + \mathbb{Z}_{>0}$. *Let* $\boldsymbol{f}_\infty \in \sigma_\infty$ *be the weight* ℓ *vector. Let* $\Phi_\infty^\ell(s) \in I_\infty(s, \chi)$ *be the standard normalized section of weight* ℓ. *Here* $\chi(-1) = \chi_{2\kappa}(-1) = (-1)^{\ell - \frac{1}{2}}$. *Then*

$$Z(s, \Phi_\infty^\ell, \boldsymbol{f}_\infty) = 24 \cdot e(-\tfrac{1}{4}\,(\ell - \tfrac{1}{2})) \cdot \frac{1}{2^{s+\frac{1}{2}}\,(s + \ell - \frac{1}{2})} \cdot \boldsymbol{f}_\infty.$$

Returning to the global representation $\sigma \simeq \otimes_p \sigma_p$, we make the following assumptions about the local components:

(i) For $p = \infty$

$$\sigma_\infty \simeq \tilde{\pi}_\ell^+,$$

for some $\ell \in \frac{1}{2} + \mathbb{Z}_{>0}$ with $\kappa = (-1)^{\ell - \frac{1}{2}}$.

(ii) For $p = 2$,

$$\sigma_p \simeq \begin{cases} \tilde{\pi}(\chi_\alpha |\ |^t) & \text{with } \alpha \in \mathbb{Z}_p^\times \text{ and } t \neq \frac{1}{2}, \text{ or} \\ \tilde{\sigma}(\chi_\alpha |\ |^{\frac{1}{2}}) & \text{with } \alpha \in \mathbb{Z}_p^\times. \end{cases}$$

(iii) For a square free odd integer N_o,

$$\sigma_p \simeq \begin{cases} \tilde{\pi}(\chi_\alpha |\ |^t) & \text{with } \alpha \in \mathbb{Z}_p^\times \text{ and } t \neq \frac{1}{2}, \text{ if } p \nmid 2N_o, \\ \tilde{\sigma}(\chi_\alpha |\ |^{\frac{1}{2}}) & \text{with } \alpha \in \mathbb{Z}_p^\times, \text{ if } p \mid N_o. \end{cases}$$

Note that in (ii) we are allowing a small amount of ramification.

Recall that, by (8.2.1) and (8.2.28), the product over all p of the central signs $z(\sigma_p, \psi_p)$ must be 1. Note that $z(\sigma_p, \psi_p)$ is given in the last column of Table 2. Under assumption (iii), $z(\sigma_p, \psi_p) = 1$ for $p \neq 2, \infty$, since $\chi_\alpha(-1) = 1$ for such p's. Thus we have

$$1 = \prod_{p \leq \infty} z(\sigma_p, \psi_p) = (-1)^{\ell - \frac{1}{2}} \cdot \chi_{\alpha,2}(-1), \tag{8.3.1}$$

where $\chi_\alpha = \chi_{\alpha,2}$ is the character occurring in σ_2. From this relation and (i), it follows that $\chi_{\alpha,2}(-1) = \kappa$, and so

$$\alpha \equiv \kappa \mod 4. \tag{8.3.2}$$

Let $N = N_o$ if σ_2 is an irreducible principal series representation and $N = 2N_o$ if σ_2 is a special representation. The level of $\mathrm{Wald}(\sigma, \psi_\kappa)$ is then N. For p odd, this is clear, while for $p = 2$, it follows from the fact that

$$\mathrm{Wald}(\sigma_2, \psi_2^\kappa) = \mathrm{Wald}(\sigma_2, \psi_2) \otimes \chi_\kappa = \begin{cases} \sigma(\chi_{\alpha\kappa} |\ |^{\frac{1}{2}}, \chi_{\alpha\kappa} |\ |^{-\frac{1}{2}}) & \text{if } 2 \mid N, \\ \pi(\chi_{\alpha\kappa} |\ |^t, \chi_{\alpha\kappa} |\ |^{-t}) & \text{if } 2 \nmid N, \end{cases}$$

where, by (8.3.2), the character $\chi_{\alpha\kappa}$ is unramified. Note that

$$\delta(\sigma_p, \psi_p^\kappa) = \begin{cases} -1 & \text{if } p \mid N \text{ and } \chi_{\alpha\kappa,p} = 1, \\ +1 & \text{otherwise.} \end{cases}$$

We write $N = N^+ N^-$, where

$$N^{\pm} = \prod_{\substack{p|N \\ \delta(\sigma_p,\psi_p^\kappa)=\pm 1}} p.$$

We then take $\boldsymbol{f} \in \sigma$, with $\boldsymbol{f} \simeq \boldsymbol{f}_\infty \otimes (\otimes_{p<\infty} \boldsymbol{f}_p)$, and

$$\Phi(s) = \Phi_\infty^\ell(s) \otimes (\otimes_{p|N^+} \Phi_p^1(s)) \otimes (\otimes_{p|N^-} \Phi_p^{\text{ra}}(s)) \otimes (\otimes_{p\nmid N} \Phi_p^0(s)),$$

for the test vectors described in Theorem 8.3.1. Here, of course, for almost all p, $\boldsymbol{f}_p$ is normalized to be the distinguished unramified vector in σ_p used in the definition of the restricted tensor product. We will refer to $\boldsymbol{f}$ as the good newvector in σ; it is unique up to scaling. Note that, by (8.4.17), the finite part $\Phi_f(s)$ of the section $\Phi(s) = \Phi_\infty^\ell(s) \otimes \Phi_f(s)$ is a standard Siegel-Weil section associated to the quadratic space V_κ^B of trace zero elements in the quaternion algebra B with invariants at the finite primes given by

$$\text{inv}_p(B) = \begin{cases} +1 & \text{if } \delta(\sigma_p, \psi_p^\kappa) = +1, \\ -1 & \text{if } \delta(\sigma_p, \psi_p^\kappa) = -1. \end{cases}$$

Thus $D(B) = N^-$. By the relation (8.2.29),

$$\text{inv}_\infty(B) = \epsilon(\frac{1}{2}, \text{Wald}(\sigma, \psi_\kappa)) \cdot \delta(\sigma_\infty, \psi_\infty^\kappa).$$

The quadratic form on V_κ^B is given by $Q(x) = -\kappa\, \nu(x)$. Then

$$\Phi_f(0) = \lambda_V(\varphi \otimes \varphi),$$

where $\varphi \in S(V_\kappa^B(\mathbb{A}_f))$ is the characteristic function of the set $\hat{O} \cap V_\kappa^B(\mathbb{A}_f)$, where O is an Eichler order in B of level N^+.

As a consequence of Theorem 8.3.1, we obtain the following explicit doubling formula, which is the main result of this chapter.

Theorem 8.3.2. *Suppose that σ satisfies conditions (i), (ii), and (iii) stated after Theorem 8.3.1. For the choice of $\boldsymbol{f} \in \sigma$ and $\Phi(s) \in I(s, \chi_{2\kappa})$ made above*

$$\begin{aligned} Z(s, \Phi, \boldsymbol{f})(g_1') &= \int_{\text{Sp}_1(\mathbb{Q})\backslash \text{Sp}_1(\mathbb{A})} E(i(g_1', g_2'), s, \Phi)\, \boldsymbol{f}(g_2')\, dg_2 \\ &= C(s, N^+, N^-) \cdot \frac{L(s+\frac{1}{2}, \text{Wald}(\sigma, \psi^\kappa))}{(s+\ell-\frac{1}{2})\, \zeta_N(2s+2)} \cdot \boldsymbol{f}(g_1'), \end{aligned}$$

where

$$C(s,N^+,N^-) = (-1)^* \cdot \frac{6}{2^s N^2} \cdot \prod_{p|N} C_p(s,N^+,N^-),$$

with

$$C_p(s,N^+,N^-) = (1+p^{-s}) \cdot \begin{cases} \frac{p-1}{p+1} & \text{if } p \mid N^+, \\ -1 & \text{if } p \mid N^-, \end{cases}$$

and

$$(-1)^* = \begin{cases} 1 & \text{if } \ell \equiv \frac{1}{2}, \frac{3}{2} \mod (4), \\ -1 & \text{if } \ell \equiv \frac{5}{2}, \frac{7}{2} \mod (4). \end{cases}$$

Recall that dg is Tamagawa measure on $\mathrm{Sp}_1(\mathbb{Q})\backslash\mathrm{Sp}_1(\mathbb{A})$, and note that

$$(-1)^* = e(-\tfrac{1}{4}(\ell-\tfrac{1}{2}))\, \delta_\kappa.$$

In Chapter 9, we will need a variant of Theorem 8.3.2 that involves the normalized Eisenstein series studied in Chapter 5. Let B be an indefinite quaternion algebra over $\mathbb{Q}$ ramified precisely at the primes dividing the square free positive integer $D(B)$. For $\chi = \chi_{2\kappa}$ with $\kappa = -1$, let $\tilde{\Phi}^B(s)$ be the (nonstandard) section given by (5.1.36):

$$\tilde{\Phi}^B(s) = \Phi_\infty^{\frac{3}{2}}(s) \otimes \big(\otimes_{p|D(B)} \tilde{\Phi}_p(s)\big) \otimes \big(\otimes_{p\nmid D(B)} \Phi_p^0(s)\big),$$

where, as in (5.1.33),

$$\tilde{\Phi}_p(s) = \Phi_p^-(s) + A_p(s)\,\Phi_p^0(s) + B_p(s)\,\Phi_p^1(s),$$

with rational functions $A_p(s)$ and $B_p(s)$ of p^{-s} satisfying conditions (5.1.34) and (5.1.35). Then, the normalized Eisenstein series is given by

$$\mathcal{E}_2(g,s,B) = -\frac{1}{2}\, c(D)(s+1)\Lambda_D(2s+2) \cdot E(g,s,\Phi), \tag{8.3.3}$$

where $D = D(B)$,

$$\Lambda_D(2s+2) = \big(\frac{D}{\pi}\big)^{s+1}\Gamma(s+1)\,\zeta_D(2s+2),$$

and

$$c(D) = -\frac{D}{2\pi}\prod_{p|D}(p+1)^{-1}.$$

Let $D(B)_o$ be the odd part of $D(B)$. Let $\sigma \simeq \otimes_{p\le\infty}\sigma_p$ be a genuine cuspidal automorphic representation of weight $\frac{3}{2}$, satisfying the conditions above with $N = D(B)$. Let $\boldsymbol{f} \in \sigma$ be the good newvector. Then the same argument using Corollary 8.4.9 instead of Corollary 8.4.8 gives the following:

Theorem 8.3.3. *Let*

$$\mathcal{Z}(s,\boldsymbol{f},B)(g_1') = \int_{\mathrm{Sp}_2(\mathbb{Q})\backslash \mathrm{Sp}_1(\mathbb{A})} \mathcal{E}_2(i(g_1',g_2'),s,B)\,\boldsymbol{f}(g_2')\,dg_2.$$

Then,

$$\mathcal{Z}(s,\boldsymbol{f},B) = \mathcal{C}(s)\cdot L(s+\frac{1}{2},\mathrm{Wald}(\sigma,\psi_{-1}))\cdot \boldsymbol{f},$$

where

$$\mathcal{C}(s) = \frac{3}{2\pi^2}\left(\frac{D}{2\pi}\right)^s \Gamma(s+1)\cdot\prod_{p|D}(p+1)^{-1}\,\mathcal{C}_p(s),$$

with

$$\mathcal{C}_p(s) = (1-\delta(\sigma_p,\psi_p^-)\,p^{-s}) - \frac{p-1}{p+1}(1+\delta(\sigma_p,\psi_p^-)\,p^{-s})\,B_p(s).$$

Here we note that local theta dichotomy is evident in the factors $\mathcal{C}_p(s)$, where

$$\mathcal{C}_p(0) = \begin{cases} 2 & \text{if } \delta(\sigma_p,\psi_p^-) = -1,\\ 0 & \text{if } \delta(\sigma_p,\psi_p^-) = +1.\end{cases}$$

Also note that, if $\delta(\sigma_p,\psi_p^-) = +1$, then

$$\mathcal{C}_p'(0) \;=\; \log(p) - \frac{p-1}{p+1}2\cdot B_p'(0) \;=\; 0. \tag{8.3.4}$$

Thus, the seemingly strange section $\widetilde{\Phi}_p(s)$, which arises from geometric considerations in the fibers of bad reduction of the arithmetic surface $\mathcal{M}$ ([13] and Chapter 5), also gives very clean constants in the local doubling formula for the special representation and the somewhat remarkable cancellation (8.3.4).[7]

Recall that

$$\Lambda(s+\frac{1}{2},\pi) = 2\,(\frac{D}{2\pi})^{s+1}\Gamma(s+1)\cdot L(s+\frac{1}{2},\pi)$$

[7] When writing [11], the author had the wrong sign in the second term in $\mathcal{C}_p(s)$. Thus the remarks made there about the case $\epsilon(\frac{1}{2},\mathrm{Wald}(\sigma,\psi^\kappa)) = +1$ are incorrect.

is the complete L-function of $\pi = \mathrm{Wald}(\sigma, \psi_{-1})$. Thus

$$\mathcal{Z}(s, \boldsymbol{f}, B) = \mathcal{C}_0(s) \cdot \Lambda(s + \frac{1}{2}, \mathrm{Wald}(\sigma, \psi_{-1})) \cdot \boldsymbol{f},$$

where

$$\mathcal{C}_0(s) = \frac{3}{2\pi D} \cdot \prod_{p|D} (p+1)^{-1}\, \mathcal{C}_p(s).$$

8.4 LOCAL DOUBLING INTEGRALS

In this section, we derive the local doubling formulas that are stated in Theorem 8.3.1. After reviewing some notation, we compute the local zeta integrals in the case of a nonarchimedean local field F of residue characteristic p, first for p odd and then for $p = 2$. Then we consider the case where $F = \mathbb{R}$. In the nonarchimedean case, F has ring of integers $\mathcal{O}$, uniformizer ϖ, residue field $\mathbb{F}_q$, and a fixed nontrivial unramified additive character ψ.

8.4.1 Coordinates and local doubling integrals

We use the following notation, which is explained in more detail in Section 8.5.[8] Let G (resp. G') be the metaplectic extension of $\mathrm{Sp}_2(F)$ (resp. $\mathrm{Sp}_1(F)$). Let $\underline{P}'$ be the upper triangular Borel subgroup of $\mathrm{Sp}_1 = \mathrm{SL}_2$ and let P' be the full inverse image of $\underline{P}'(F)$ in G'. Similarly, let $\underline{P}$ be the Siegel parabolic subgroup of Sp_2 and let P be the full inverse image of $\underline{P}(F)$ in G. We let $\underline{K}' = \mathrm{Sp}_1(\mathcal{O})$ and $\underline{K} = \mathrm{Sp}_2(\mathcal{O})$, and we write K' and K for the full inverse images of these groups in the metaplectic extensions G' and G respectively.

We will frequently use the notation

$$\mathbf{m}(a) = [m(a), 1]_{\mathrm{L}} \qquad \text{and} \qquad \mathbf{n}(b) = [n(b), 1]_{\mathrm{L}}$$

for the images of the elements $m(a)$ and $n(b)$ under the splitting homomorphism $\underline{P}(F) \longrightarrow G$, $p \mapsto [p, 1]_{\mathrm{L}}$ in Leray coordinates; see Section 8.5.1.

When the residue characteristic p of F is odd, there are normalized coordinates

$$\mathrm{Sp}_2(F) \times \mathbb{C}^1 \xrightarrow{\sim} G, \qquad (g, z) \mapsto [g, z] = [g, z\,\lambda(g)]_{\mathrm{L}},$$

where λ is given by (8.5.10), whose cocycle is trivial on $\underline{K} \times \underline{K}$. Thus, there is a splitting homomorphism

$$\underline{K} \longrightarrow G, \qquad k \mapsto \mathbf{k} = [k, 1].$$

[8]Note that there is a slight shift in notation here from that used in the Section 8.5.

The two splittings agree on $\underline{P}(\mathcal{O}) = \underline{P}(F) \cap \underline{K}$. The situation for G', $\underline{P}'(F)$, and $\underline{K}'$ is the same.

The embedding $\underline{i} : \mathrm{Sp}_1(F) \times \mathrm{Sp}_1(F) \to \mathrm{Sp}_2(F)$, defined as in (8.1.1)–(8.1.3), lifts to an embedding $i : G' \times G' \to G$ given, in Leray coordinates, by

$$i : [g_1, z_1]_{\mathrm{L}} \times [g_2, z_2]_{\mathrm{L}} \longmapsto [\underline{i}(g_1, g_2), z_1 z_2^{-1}]_{\mathrm{L}}. \tag{8.4.1}$$

In the case of odd residue characteristic and for ψ unramified, the same formula holds for the normalized coordinates.

As in Section 8.1, if χ is a character of $F^\times$, we define a character χ of P by

$$\chi([n(b)m(a), z]_{\mathrm{L}}) = z\, \chi(\det a).$$

For $s \in \mathbb{C}$, the induced representation $I(s, \chi)$ of G is realized on the space of smooth functions $\Phi(s)$ on G such that, for $g \in G$ and $p \in P$,

$$\Phi(p\, g, s) = \chi(p)\, |\det a\,|^{s+\frac{3}{2}}\, \Phi(g, s).$$

For convenience, we will sometimes write $\Phi_s(g)$ instead of $\Phi(g, s)$.

Let $(\sigma, \mathcal{V}_\sigma)$ be an irreducible, admissible, genuine representation of G'. Then, for a fixed preimage[9] $\delta \in G$ of $\underline{\delta}$, given by (8.1.4), and for $f \in \mathcal{V}_\sigma$ and $\Phi(s) \in I(s, \chi)$, the local doubling integral is given by

$$Z(s, \Phi, f) = \int_{\mathrm{Sp}_1(F)} \Phi_s(\delta\, i(1, g'))\, \sigma(g') f\, dg \quad \in \mathcal{V}_\sigma, \tag{8.4.2}$$

where $g' \in G'$ is any element projecting to $g \in \mathrm{Sp}_1(F)$; note that the integrand depends only on g, via (8.4.1). The integral (8.4.2) is absolutely convergent for $\mathrm{Re}(s)$ sufficiently large [5] and can be viewed as giving the action in the representation $(\sigma, \mathcal{V}_\sigma)$ of the function $g' \mapsto \Phi_s(\delta\, i(1, g'))$ on G'.

The proof of (i) of the following result will be given in the Section 8.5; see Lemma 8.5.5. The other statements are easy consequences of the definition.

Lemma 8.4.1. *(i) For any choice of $\delta \in G$ with image $\underline{\delta} \in \mathrm{Sp}_2(F)$ and for any $g' \in G'$,*

$$\delta\, i(g', g') = p(g')\, \delta,$$

where the element $p(g') \in P$ satisfies $\chi(p(g')) = 1$. In particular, for g'_0, g'_1, and $g'_2 \in G'$,

$$\Phi_s(\delta i(g'_0 g'_1, g'_0 g'_2)) = \Phi_s(\delta i(g'_1, g'_2)).$$

[9] A specific choice will be made in Lemma 8.4.3 below.

(ii) For $g' \in G'$,

$$Z(s, r(i(1, g'))\Phi, \sigma(g')f) = Z(s, \Phi, f).$$

and

$$\sigma(g')\, Z(s, \Phi, f) = Z(s, r(i(g', 1))\Phi, f).$$

Here r is the right multiplication.

Part (ii) says that the zeta integral gives a $G' \times G'$-intertwining map

$$Z(s) : I(s, \chi) \longrightarrow \mathrm{Hom}(\sigma, \sigma), \qquad \Phi \mapsto Z(s, \Phi, \cdot),$$

where the action on $\mathrm{Hom}(\sigma, \sigma)$ is given by pre- and post-multiplication, i.e.,

$$Z(s, r(i(g_1', g_2'))\Phi, f) = \sigma(g_1')\, Z(s, \Phi, \sigma(g_2')^{-1} f).$$

Corollary 8.4.2. *Suppose that there exists a subgroup $A \subset G'$ such that Φ is invariant under $i(A, 1)$. Then*

$$Z(s, \Phi, f) \in \sigma^A.$$

In particular, if $\sigma^A = 0$, then $Z(s, \Phi, f) = 0$ for all $f \in \sigma$.

8.4.2 Induced representations

Suppose that μ is a character of $F^\times$ and let $I(\mu)$ be the genuine principal series representation of G' on the space of smooth functions f on G' such that

$$f([n(b)m(a), z]_{\mathrm{L}}\, g') = z\, \mu(a)|a| f(g').$$

Note that this definition depends on the fixed additive character ψ used to define the Leray coordinates. In particular, since the Rao and Leray coordinates are related by

$$[m(a), z]_R = [m(a), z\beta(m(a))]_{\mathrm{L}}$$

with

$$\beta(m(a)) = \gamma(a, \psi_{\frac{1}{2}})^{-1} = (a, 2)_F\, \gamma(a, \psi)^{-1},$$

(see (8.5.17)), we have[10]

$$f([n(b)m(a), z]_{\mathrm{R}}\, g') = z\, \chi_2(a)\mu(a)\, \gamma(a, \psi)^{-1}\, |a| f(g'),$$

[10] In [28], Waldspurger uses the restriction to $\mathrm{SL}_2(F)$ of the cocycle from [4], p. 19. The cocycle α^* there restricts to c_{R} on $\mathrm{SL}_2(F)$. Thus, the corresponding 'coordinates' used in

where $\chi_2(a) = (a, 2)_F$. Hence, comparing with p. 225 of [28], we find that

$$I(\mu) = \mathcal{B}(\mu\chi_2) \tag{8.4.3}$$

in the parametrization of Waldspurger used in Table 2 in Section 8.2.

Suppose that $\sigma \subset I(\mu)$ is an irreducible submodule. Then, for $f \in \sigma$, $Z(s, \Phi, f) \in \sigma \subset I(\mu)$, and its value at the point $g'_0 \in G'$ is

(8.4.4)

$$\begin{aligned} Z(s, \Phi, f)(g'_0) &= \int_{\mathrm{Sp}_1(F)} \Phi_s(\delta i(1, g'))\, f(g'_0 g')\, dg \\ &= \int_{\mathrm{Sp}_1(F)} \Phi_s(\delta i(g'_0, g'))\, f(g')\, dg \\ &= \int_{\underline{K}'} \int_{F^\times} \int_F \Phi_s(\delta i(g'_0, \mathbf{n}(b)\, \mathbf{m}(a) k'))\, \mu(a)|a|^{-1}\, db\, d^*a\, f(k')\, dk \end{aligned}$$

We now use an interpolation method. Recall that, for a quadratic space (V, Q) over F, G acts on $S(V^2)$ via the Weil representation determined by ψ and there is a G-intertwining map

$$\lambda_V : S(V^2) \longrightarrow I(s_0, \chi_V), \qquad \varphi \mapsto \lambda_V(\varphi)(g) := \omega_V(g)\varphi(0),$$

with $s_0 = \frac{1}{2}\dim(V) - \frac{3}{2}$. The following basic observation can be found in [15].

Lemma 8.4.3. *Suppose that* $\Phi_{s_0} = \lambda_V(\varphi_1 \otimes \overline{\varphi}_2) \in I(s_0, \chi_V)$, *with* $\varphi_i \in S(V)$. *Then*

$$\Phi_{s_0}(\delta i(g'_0, g')) = \gamma(V) \int_V \omega_V(g'_0)\varphi_1(x) \cdot \overline{\omega_V(g')\varphi_2(-x)}\, dx,$$

where

$$\delta = [w_1, 1]_{\mathrm{L}}\, \mathbf{m}(\begin{pmatrix} 1 & -1 \\ & 1 \end{pmatrix}),$$

---

[28] are given by

$$[g, z]_{\mathrm{Wald}} = [g, z\, s(g)]_{\mathrm{R}},$$

where

$$s(g) = \begin{cases} (c, d) & \text{if } cd \neq 0 \text{ and } \mathrm{ord}(c) \text{ is odd}, \\ 1 & \text{otherwise}. \end{cases}$$

Since s is identically 1 on $\underline{P}'(F)$, we can use $[m(a), z]_{\mathrm{Wald}} = [m(a), z]_{\mathrm{R}}$ in the calculation above.

and

$$\gamma(V) = \chi_V(-1)\,\gamma(\eta)\,\gamma(\eta\circ V)^{-1},$$

for $\eta = \psi_{\frac{1}{2}}$.

Proof. Writing $\alpha = \begin{pmatrix} 1 & -1 \\ & 1 \end{pmatrix}$ and using the properties of the Weil representation reviewed in Section 8.5.2, we have

$$\begin{aligned}\Phi_{s_0}(\delta i(g_0', g')) &= \omega_V([w_1', 1]_{\mathrm{L}}\, \mathbf{m}(\alpha)\, i(g_0', g'))(\varphi_1 \otimes \overline{\varphi}_2)(0)\\ &= \omega_V([w_1', 1]_{\mathrm{L}}\, \mathbf{m}(\alpha))(\omega_V(g_0')\varphi_1 \otimes \overline{\omega_V(g')\varphi_2})(0).\end{aligned}$$

Absorbing g_0' and g' into φ_1 and φ_2, for convenience, we have

$$\begin{aligned}&\omega_V([w_1', 1]_{\mathrm{L}}\, \mathbf{m}(\alpha))(\varphi_1 \otimes \overline{\varphi}_2)(0)\\ &\qquad = \gamma(V) \int_V \omega_V(\mathbf{m}(\alpha))(\varphi_1 \otimes \overline{\varphi}_2)(x, 0)\, dx\\ &\qquad = \gamma(V) \int_V \varphi_1(x)\, \overline{\varphi_2(-x)}\, dx,\end{aligned}$$

where $\gamma(V) = \chi_V(-1)\,\gamma(\eta)\,\gamma(\eta\circ V)^{-1}$; see Lemma 8.5.6. □

Using this lemma and taking $g' = \mathbf{n}(b)\mathbf{m}(a)k'$, we obtain the following expression:

$$\begin{aligned}(8.4.5)\quad Z(s_0, \Phi, f)(g_0') &= \gamma(V) \int_{F^\times} \chi_V \mu(a) |a|^{\frac{m}{2}-1}\\ &\times \int_F \int_V \omega_V(g_0')\varphi_1(x) \cdot I(f, \varphi_2)(-xa) \cdot \psi(-b\, Q(x))\, dx\, db\, d^\times a,\end{aligned}$$

where $I(f, \varphi_2) \in S(V)$ is given by

$$(8.4.6)\qquad I(f, \varphi_2)(x) = \int_{\underline{K}'} f(k')\, \overline{\omega_V(k')\varphi_2(x)}\, dk.$$

Here, as usual, k' is any element of K' with image k in $\underline{K}'$.

We now assume that ψ *is unramified.*

Let $V_r = V + V_{r,r}$, where $V_{r,r} = F^{2r}$ with bilinear form with matrix $\begin{pmatrix} & 1_r \\ 1_r & \end{pmatrix}$. For any $\varphi \in S(V)$, let $\varphi^{(r)} = \varphi \otimes \varphi_r^0$, where φ_r^0 is the characteristic function of the lattice $\mathcal{O}^{2r}$ in $V_{r,r}$; this lattice is self-dual, and φ_r^0

is invariant under K' for the Weil representation $\omega_{V_{r,r}}$. Then we obtain the useful formula

$$(8.4.7)\quad Z(s_0+r,\,\Phi,\,f)(g_0') = \gamma(V)\int_{F^\times}\chi_V\mu(a)|a|^{\frac{m}{2}+r-1}\,d^\times a \\ \times\int_F\int_{V_r}\omega_{V_r}(g_0')\varphi_1^{(r)}(x)\cdot I(f,\varphi_2)^{(r)}(-ax)\cdot\psi(-b\,Q(x))\,dx\,db,$$

where

$$I(f,\varphi_2)^{(r)} = I(f,\varphi_2)\otimes\varphi_r^0.$$

We define a family of bilinear pairings on $S(V)$ by

$$(8.4.8)\quad B(r,a;\varphi_1,\varphi_2) = \int_F\int_{V_r}\varphi_1^{(r)}(x)\,\varphi_2^{(r)}(-ax)\,\psi(-b\,Q(x))\,dx\,db,$$

where φ_1 and $\varphi_2\in S(V)$. Note that

$$(8.4.9)\qquad B(r,a;\varphi_1,\varphi_2) = |a|^{2-m-2r}\,B(r,a^{-1};\varphi_2,\varphi_1).$$

Then, since, for $k'\in K'$, $\omega_{V_r}(k')\varphi_1^{(r)} = (\omega_V(k')\varphi_1)^{(r)}$, we may write

$$(8.4.10)\quad Z(s_0+r,\Phi,f)(k') \\ = \gamma(V)\int_{F^\times}\chi_V\mu(a)\,|a|^{\frac{m}{2}+r-1}B(r,a;\omega_V(k')\varphi_1,I(f,\varphi_2))\,d^\times a.$$

Also, for $\varphi_L\in S(V)$ the characteristic function of a lattice $L\subset V$, we let

$$(8.4.11)\qquad W_0(r,L) := \int_F\int_{V_r}\varphi_L^{(r)}(x)\,\psi(-b\,Q(x))\,dx\,db.$$

Note that (8.4.11) is precisely the type of integral appearing in (13.2) of [14], except that we require dx to be the Haar measure on V_r which is self-dual with respect to $\psi\circ(\ ,\ )$, where $(\ ,\ )$ is the bilinear form associated to Q. It will be useful to record the following fact for future use.

Lemma 8.4.4. *Let L and L' be lattices in V such that*

$$\varpi L'\subset L\subset L'.$$

Let φ_L, $\varphi_{L'}\in S(V)$ be their characteristic functions. Then

$$B(r,a;\varphi_L,\varphi_{L'}) = \begin{cases} W_0(r,L) & \text{if } \operatorname{ord}(a)\ge 0,\\ |a|^{-2s_0-2r}\,W_0(r,L') & \text{if } \operatorname{ord}(a)<0,\end{cases}$$

where $s_0 = \frac{m}{2} - 1$. *If* χ *is unramified, then*

$$\int_{F^\times} \chi(a)|a|^{s_0+r}\, B(r, a; \varphi_L, \varphi_{L'})\, d^\times a = L(s_0 + r, \chi) \cdot W_0(r, L) \\ + \chi(\varpi)^{-1} q^{-s_0-r}\, L(s_0 + r, \chi^{-1}) \cdot W_0(r, L').$$

If χ *is ramified, this integral is identically zero.*

Proof. If $\mathrm{ord}(a) \geq 0$, then $\varphi_L^{(r)}(x)\, \varphi_{L'}^{(r)}(-ax) = \varphi_L^{(r)}(x)$, and the contribution to the integral becomes

$$\begin{aligned} Z_1 &= \int_{\mathrm{ord}(a)\geq 0} \chi(a)|a|^{s_0+r}\, d^\times a \cdot \int_F \int_{V_r} \varphi_L^{(r)}(x)\, \psi(-b\, Q(x))\, dx\, db \\ &= L(s_0 + r, \chi) \cdot W_0(r, L). \end{aligned}$$

If $\mathrm{ord}(a) < 0$, then $\varphi_L^{(r)}(x)\, \varphi_{L'}^{(r)}(-ax) = \varphi_{L'}^{(r)}(ax)$ and the integral becomes

$$\begin{aligned} Z_2 &= \int_{\mathrm{ord}(a)<0} \chi(a)|a|^{s_0+r} \cdot \int_F \int_{V_r} \varphi_{L'}^{(r)}(ax)\, \psi(-b\, Q(x))\, dx\, db\, d^\times a \\ &= \int_{\mathrm{ord}(a)<0} \chi(a)|a|^{-s_0-r}\, d^\times a \cdot \int_F \int_{V_r} \varphi_{L'}^{(r)}(x)\, \psi(-b\, Q(x))\, dx\, db \\ &= \chi(\varpi)^{-1} q^{-s_0-r}\, L(s_0 + r, \chi^{-1}) \cdot W_0(r, L'). \end{aligned}$$

Together these give the claimed expression. □

8.4.3 Unramified representations

Suppose that the residue characteristic of F is odd and that the additive character ψ and the characters χ, with $\chi^2 = 1$, and μ are unramified. Let $\mathbf{K}'$ be the image of $\underline{K}' = \mathrm{Sp}_1(\mathcal{O})$ in G' under the splitting homomorphism, and let f^0 be the unique right $\mathbf{K}'$-invariant function in $I(\mu)$ with $f^0(1) = 1$. Similarly, let $\mathbf{K}$ be the image of $\underline{K} = \mathrm{Sp}_2(\mathcal{O})$ in G under the splitting homomorphism, and let $\Phi^0(s) \in I(s, \chi)$ be the unique right $\mathbf{K}$-invariant function with $\Phi^0(1, s) = 1$. Since $\Phi^0(s)$ is $i(\mathbf{K}' \times 1)$-invariant and since $I(\mu)^{\mathbf{K}'}$ is one-dimensional with basis vector f^0, it follows from Corollary 8.4.2 that $Z(s, \Phi^0, f^0)$ is a multiple of f^0.

Proposition 8.4.5.

$$Z(s, \Phi^0, f^0) = \frac{L(s + \frac{1}{2}, \chi\mu)\, L(s + \frac{1}{2}, \chi\mu^{-1})}{\zeta(2s + 2)} \cdot f^0.$$

Here $\zeta(s) = (1 - q^{-s})^{-1}$.

Proof. We evaluate $Z(r, \Phi^0, f^0)(1)$ by using (8.4.10) and Lemma 8.4.4. Write $\chi(x) = (x, 2\kappa)_F$ for $\kappa \in F^\times$ with $\mathrm{ord}(\kappa) = 0$. Let $V = V_\kappa^B$ with $B = M_2(\mathbb{Q}_p)$, and with quadratic form $Q(x) = -\kappa\,\nu(x) = \kappa x^2$. Note that this normalization gives $\chi_V = \chi$. Let φ_L be the characteristic function of $L = V \cap M_2(\mathbb{Z}_p)$, so that φ_L is $\mathbf{K}'$-invariant. Also note that, setting $\varphi_L^{(r)} = \varphi_L \otimes \varphi_r^0$, we have

$$\lambda_{V+V_{r,r}}(\varphi_L^{(r)} \otimes \varphi_L^{(r)}) = \Phi^0(r) \in I(r, \chi),$$

where $\Phi^0(r)$ is the unique normalized $\mathbf{K}$-fixed vector. Note that $\gamma(\eta \circ V) = \gamma(\kappa, \eta)\,\gamma(\eta) = 1$, so that the constant $\gamma(V) = 1$ in (8.4.10).

In addition, since $f^0(\mathbf{k}') = 1$ for $\mathbf{k}' \in \mathbf{K}'$, and φ_L is $\mathbf{K}'$ fixed,

$$I(f^0, \varphi_L)(x) = \overline{\varphi_L(x)} = \varphi_L(x) \qquad \text{and} \qquad I(f^0, \varphi_L)^{(r)}(x) = \varphi_L^{(r)}(x).$$

Thus,

$$\begin{aligned}
&Z(r, \Phi^0, f^0)(1)\\
&= \int_{F^\times} \chi\mu(a)|a|^{r+\frac{1}{2}}\, B(r, a; \varphi_L, \varphi_L)\, d^\times a\\
&= \left(L(r + \frac{1}{2}, \chi\mu) + \chi\mu(\varpi)^{-1}\, q^{-\frac{1}{2}-r}\, L(r + \frac{1}{2}, \chi^{-1}\mu^{-1}) \right) \cdot W_0(r, L)\\
&= \frac{L(r + \frac{1}{2}, \chi\mu)\, L(r + \frac{1}{2}, \chi\mu^{-1})}{\zeta(2r + 1)} \cdot W_0(r, L).
\end{aligned}$$

Finally, we use the formula

$$W_0(s, L) = \frac{\zeta(2s + 1)}{\zeta(2s + 2)},$$

which is given in [14], Section 13. □

In the notation of Table 2 in Section 8.2, we are considering the representation $\sigma = \tilde{\pi}(\mu\chi_2)$, so that $\mathrm{Wald}(\sigma, \psi) = \pi(\mu\chi_2, \mu^{-1}\chi_2)$, whereas $\chi = \chi_{2\kappa}$. Thus, since $\mathrm{Wald}(\sigma, \psi_\kappa) = \mathrm{Wald}(\sigma, \psi) \otimes \chi_\kappa$, the L-function in the numerator is $L(s + \frac{1}{2}, \mathrm{Wald}(\sigma, \psi_\kappa))$ and we have the following expression for the unramified local zeta integral.

Corollary 8.4.6. *Suppose that $p \neq 2$ and that $\chi = \chi_{2\kappa}$ is unramified. Then for $f^0 \in \sigma^{\mathbf{K}'}$ and for $\Phi^0(s) \in I(s, \chi)$, the normalized unramified section,*

$$Z(s, \Phi^0, f^0) = \frac{L(s + \frac{1}{2}, \mathrm{Wald}(\sigma, \psi_\kappa))}{\zeta(2s + 2)} \cdot f^0.$$

8.4.4 Special representations, $p \neq 2$

In this section, we assume that the residue characteristic of F is odd and consider the case in which σ is an unramified special representation. We also assume that $\chi(x) = \chi_{2\kappa}(x) = (x, 2\kappa)_F$ is unramified, so that κ is a unit. Let $\mu = |\ |^{\frac{1}{2}}\chi_{2\alpha}$ with $\alpha \in \mathbb{Z}_p^\times$, and let $\sigma = \sigma(\mu) \subset I(\mu)$ be the unique irreducible G' submodule. Note that $\sigma = \tilde{\sigma}(\mu\chi_2) = \tilde{\sigma}(\chi_\alpha|\ |^{\frac{1}{2}})$ in the notation of the Table 2 in Section 8.2.

Let $\underline{J}' \subset \underline{K}' = \mathrm{Sp}_1(\mathcal{O})$ be the Iwahori subgroup and let $\mathbf{J}' \subset \mathbf{K}'$ be its image in G' under the splitting homomorphism $\underline{K}' \to G'$. Then $\dim \sigma^{\mathbf{J}'} = 1$, and we take $\boldsymbol{f}_{\mathrm{sp}} \in \sigma$ to be the unique $\mathbf{J}'$ fixed vector with $\boldsymbol{f}_{\mathrm{sp}}(1) = 1$.

Similarly, we let $\underline{J}$ be the Siegel parahoric subgroup of $\underline{K} = \mathrm{Sp}_2(\mathcal{O})$ and let $\mathbf{J} \subset \mathbf{K}$ be its image in G under the splitting homomorphism $\underline{K} \to G$. The space $I(s,\chi)^{\mathbf{J}}$ has dimension 3 and is spanned by the functions $\Phi^0(s)$, $\Phi^1(s)$, and $\Phi^{\mathrm{ra}}(s)$ whose restrictions to $\mathbf{K}$, which are independent of s, are defined as follows. Let $V_\kappa^\pm$ be the space of trace zero elements in the quaternion algebra $B^\pm$ over F with $\mathrm{inv}_F(B^\pm) = \pm 1$ with quadratic from $Q_\kappa(x) = -\kappa\nu(x) = \kappa x^2$.

$$(8.4.12)\qquad \begin{aligned} L^0 &= M_2(\mathcal{O}) \cap V_\kappa^+, \\ L^1 &= \{\begin{pmatrix} a & b \\ \varpi c & d \end{pmatrix} \mid a,\, b,\, c,\, d \in \mathcal{O}\} \cap V_\kappa^+, \\ L^{\mathrm{ra}} &= O_{B^-} \cap V_\kappa^-, \end{aligned}$$

where O_{B^-} is the maximal order in B^-. Let $\varphi_0, \varphi_1 \in S(V_\kappa^+)$ and $\varphi_{\mathrm{ra}} \in S(V_\kappa^-)$ be their characteristic functions. Then, for $r \in \mathbb{Z}_{\geq 0}$,

$$(8.4.13)\qquad \begin{aligned} \Phi^0(r) &= \lambda(\varphi_0^{(r)} \otimes \varphi_0^{(r)}), \\ \Phi^1(r) &= \lambda(\varphi_1^{(r)} \otimes \varphi_1^{(r)}), \\ \Phi^{\mathrm{ra}}(r) &= \lambda(\varphi_{\mathrm{ra}}^{(r)} \otimes \varphi_{\mathrm{ra}}^{(r)}), \end{aligned}$$

for standard sections $\Phi^0(s)$, $\Phi^1(s)$ and $\Phi^{\mathrm{ra}}(s)$ of $I(s,\chi_{2\kappa})$. By Lemma 8.5.10, these sections are right $\mathbf{J}$-invariant. Since $i(\mathbf{J}' \times \mathbf{J}') \subset \mathbf{J}$, for $\Phi(s) = \Phi^0(s)$, $\Phi^1(s)$, or $\Phi^{\mathrm{ra}}(s)$, $Z(s,\Phi,f)$ is a $\mathbf{J}'$-invariant vector in σ and hence is a multiple of $\boldsymbol{f}_{\mathrm{sp}}$. We want to calculate $Z(s,\Phi,\boldsymbol{f}_{\mathrm{sp}})$ in each case.

Theorem 8.4.7. *For odd residue characteristic and for $\boldsymbol{f}_{\mathrm{sp}}$ the $\mathbf{J}'$-fixed vector in the unramified special representation $\sigma(\mu)$, where $\mu = \chi_{2\alpha}|\ |^{\frac{1}{2}}$,*

$$Z(s,\Phi^0,\boldsymbol{f}_{\mathrm{sp}}) = 0,$$

$$Z(s, \Phi^1, \boldsymbol{f}_{\rm sp}) = q^{-2}\frac{q-1}{q+1} \cdot (1 - \chi_{\alpha\kappa}(\varpi)\, q^{-s}) \cdot L(s + \frac{1}{2}, \chi_{2\kappa}\mu) \cdot \boldsymbol{f}_{\rm sp},$$

and

$$Z(s, \Phi^{\rm ra}, \boldsymbol{f}_{\rm sp}) = -q^{-2} \cdot (1 + \chi_{\alpha\kappa}(\varpi)\, q^{-s}) \cdot L(s + \frac{1}{2}, \chi_{2\kappa}\mu) \cdot \boldsymbol{f}_{\rm sp}.$$

Again, in the notation of Table 2 in Section 8.2, we are considering the special representation $\sigma = \tilde{\sigma}(\mu\chi_2) = \tilde{\sigma}(\chi_\alpha|\ |^{\frac{1}{2}})$, so that $\text{Wald}(\sigma, \psi) = \sigma(\chi_\alpha|\ |^{\frac{1}{2}}, \chi_\alpha|\ |^{-\frac{1}{2}})$. Thus,

$$L(s + \frac{1}{2}, \chi_{2\kappa}\mu) = L(s + \frac{1}{2}, \text{Wald}(\sigma, \psi) \otimes \chi_\kappa) = L(s + \frac{1}{2}, \text{Wald}(\sigma, \psi_\kappa)).$$

Also note that, since $\text{Wald}(\sigma, \psi_\kappa) = \sigma(\chi_{\alpha\kappa}|\ |^{\frac{1}{2}}, \chi_{\alpha\kappa}|\ |^{-\frac{1}{2}})$, Table 2 in Section 8.2 gives

$$-\chi_{\alpha\kappa}(\varpi) = \delta(\sigma, \psi_\kappa).$$

Thus, the factor $(1 + \chi_{\alpha\kappa}(\varpi)\, q^{-s})$ occurring for the section $\Phi^1(s)$ associated to V^+ is nonvanishing when $\delta(\sigma, \psi_\kappa) = +1$, and the analogous factor $(1 - \chi_{\alpha\kappa}(\varpi)\, q^{-s})$ occurring for the section $\Phi^{\rm ra}(s)$ associated to V^- is nonvanishing when $\delta(\sigma, \psi_\kappa) = -1$.

Corollary 8.4.8. *Let the notation be as in Theorem 8.4.7. Then*

$$Z(s, \Phi^1, \boldsymbol{f}_{\rm sp}) = q^{-2}\frac{q-1}{q+1}(1 + \delta(\sigma, \psi_\kappa)q^{-s}) \cdot L(s + \frac{1}{2}, \text{Wald}(\sigma, \psi_\kappa)) \cdot \boldsymbol{f}_{\rm sp},$$

and

$$Z(s, \Phi^{\rm ra}, \boldsymbol{f}_{\rm sp}) = -q^{-2}(1 - \delta(\sigma, \psi_\kappa)q^{-s}) \cdot L(s + \frac{1}{2}, \text{Wald}(\sigma, \psi_\kappa)) \cdot \boldsymbol{f}_{\rm sp}.$$

In later applications, we will be interested in the section $\widetilde{\Phi}(s)$ defined as follows[11] (see (5.1.33) and [13]):

$$\widetilde{\Phi}(s) = \Phi^{\rm ra}(s) + A(s)\, \Phi^0(s) + B(s)\, \Phi^1(s), \tag{8.4.14}$$

where $A(s)$ and $B(s)$ are rational functions of q^{-s} with the property that

$$A(0) = B(0) = 0, \quad A'(0) = \frac{-2}{q^2-1}\log(q), \quad B'(0) = \frac{1}{2}\frac{q+1}{q-1}\log(q). \tag{8.4.15}$$

[11] The section here is a little more general, since we allow $\chi = \chi_{2\kappa}$ for any unit κ, whereas, in Chapter 5, $\kappa = -1$.

Corollary 8.4.9. *For $f_{\rm sp}$ as in Theorem 8.4.7,*

$$Z(s,\widetilde{\Phi},f_{\rm sp}) = \Big[-(1-\delta(\sigma,\psi_\kappa)\,q^{-s}) + \frac{q-1}{q+1}\,(1+\delta(\sigma,\psi_\kappa)\,q^{-s})\,B(s)\Big] \times q^{-2}\cdot L(s+\frac{1}{2},\mathrm{Wald}(\sigma,\psi_\kappa))\cdot f_{\rm sp}.$$

Proof of Theorem 8.4.7. To calculate the expression on the right side of (8.4.10), for $\varphi=\varphi_0,\varphi_1$, or $\varphi_{\rm ra}$, we need to calculate

$$I(f_{\rm sp},\varphi)(x) = \int_{\underline{K}'} f_{\rm sp}(\mathbf{k}')\,\overline{\omega_V(\mathbf{k}')\varphi(x)}\,dk',$$

where $\mathbf{k}' = [k',1]$ is the image of k' under the splitting homomorphism. Using the coset decomposition

$$\underline{K}' = \underline{J}' \cup (N'\cap\underline{K}')w\underline{J}',$$

with $w = \begin{pmatrix} & 1\\ -1 & \end{pmatrix}$, and the $\mathbf{J}'$-invariance of φ, we have

$$\text{(8.4.16)}\quad I(f_{\rm sp},\varphi)(x) = f_{\rm sp}(1)\,\mathrm{vol}(\underline{J}')\,\overline{\varphi(x)} + f_{\rm sp}(\mathbf{w})\,\mathrm{vol}(\underline{J}')\sum_{b\in O/\varpi O}\psi(-b\,Q(x))\,\overline{\omega_V(\mathbf{w})\varphi(x)},$$

where $\mathbf{w} = [w,1] = [w,1]_L$. Note that $f_{\rm sp}$ is determined by its values $f_{\rm sp}(1)=1$ and $f_{\rm sp}(\mathbf{w})$.

Lemma 8.4.10. $f_{\rm sp}(\mathbf{w}) = -q^{-1}$.

Proof. Since $\Phi^0(s)$ is right invariant under $i(\mathbf{K}'\times\mathbf{K}')$ and $\sigma^{\mathbf{K}'}=0$, Lemma 8.4.2 implies that $Z(s,\Phi^0,f_{\rm sp})=0$. Since $\varphi=\varphi_0$ is $\mathbf{K}'$-invariant, we get

$$I(f_{\rm sp},\varphi_0)(x) = \mathrm{vol}(\underline{J}')\,(f_{\rm sp}(1)+q\,f_{\rm sp}(\mathbf{w}))\,\varphi_0(x),$$

and the same calculation as in the unramified case yields

$$Z(s,\Phi^0,f_{\rm sp})(1) = \mathrm{vol}(\underline{J}')\,[\,f_{\rm sp}(1)+q\,f_{\rm sp}(\mathbf{w})\,] \times \frac{L(s+\frac{1}{2},\chi\mu)\,L(s+\frac{1}{2},\chi\mu^{-1})}{\zeta(2s+1)}\cdot W_0(s,L).$$

Since this function of s must vanish identically, we obtain the relation

$$f_{\rm sp}(\mathbf{w}) = -q^{-1}\,f_{\rm sp}(1).$$

□

To calculate the zeta integrals $Z(s, \Phi, \boldsymbol{f}_{\text{sp}})$ in the remaining two cases, using (8.4.10) and (8.4.16), we note that

$$\omega_V(\mathbf{w})\varphi(x) = \gamma(V)\,\widehat{\varphi}(x) = \gamma(V)\int_V \psi((x,y))\,\varphi(y)\,dy,$$

where dy is self-dual with respect to $\psi \circ (\ ,\ )$ and

$$\gamma(V) = \chi_V(-1)\,\gamma_F(\eta)\,\gamma_F(\eta \circ V)^{-1}.$$

Lemma 8.4.11. $\gamma(V) = \gamma(V_\kappa^{\pm}) = \pm 1.$

Proof. Note that $\chi_V(-1) = \chi_{2\kappa}(-1) = 1$. Now $\gamma_F(\eta \circ V) = \gamma_F(\psi \circ Q)$, and we have

$$\gamma_F(\eta) = \gamma_F(\psi_{\frac{1}{2}}) = \gamma_F(\frac{1}{2}, \psi)\,\gamma_F(\psi) = \gamma_F(\psi) = 1,$$

since the residue characteristic is odd and ψ is unramified. Also,

$$Q_1 = \kappa \begin{pmatrix} 1 & & \\ & \varpi & \\ & & -\varpi \end{pmatrix} \qquad \text{and} \qquad Q_{\text{ra}} = \kappa \begin{pmatrix} \beta & & \\ & \varpi & \\ & & -\varpi\beta \end{pmatrix},$$

so that, since κ is a unit, $\epsilon(Q_1) = 1$, $\epsilon(Q_{\text{ra}}) = -1$. Thus

$$\gamma(\psi \circ Q_0) = \gamma(-\kappa, \psi)\gamma(\psi)^3\,\epsilon(Q_0) = 1,$$

and

$$\gamma(\psi \circ Q_{\text{ra}}) = \gamma(-\kappa, \psi)\gamma(\psi)^3\,\epsilon(Q_{\text{ra}}) = -1,$$

so that $\gamma(V) = \gamma(V_\kappa^{\pm})$ has the claimed value. □

Using these facts and (8.4.16) , we get

$$\begin{aligned}
&I(\boldsymbol{f}_{\text{sp}}, \varphi)(x)\\
&= \boldsymbol{f}_{\text{sp}}(1)\,\text{vol}(\underline{J}')\,\varphi(x) + \gamma(V)\,\boldsymbol{f}_{\text{sp}}(\mathbf{w})\,\text{vol}(\underline{J}') \sum_{b \in \mathcal{O}/\varpi\mathcal{O}} \psi(-b\,Q(x))\widehat{\varphi}(x)\\
&= \boldsymbol{f}_{\text{sp}}(1)\,\text{vol}(\underline{J}')\,\varphi(x) + \gamma(V)\,\boldsymbol{f}_{\text{sp}}(\mathbf{w})\,\text{vol}(\underline{J}') \cdot q\,\text{char}_{\mathcal{O}}(Q(x)) \cdot \widehat{\varphi}(x)\\
&= \text{vol}(\underline{J}')\,[\,\varphi(x) - \gamma(V) \cdot \text{char}_{\mathcal{O}}(Q(x)) \cdot \widehat{\varphi}(x)\,].
\end{aligned}$$

Consider the case of $L^1 = \mathcal{O}^3$, where the dual lattice is $L^{1,\sharp} = \mathcal{O} \oplus \varpi^{-1}\mathcal{O}^2$. Let

$$L^{1,\pm} = \{\,[x_0, x_1, x_2] \in L^{1,\sharp} \mid x_1 \equiv \pm x_2 \mod \mathcal{O}\,\},$$

and note that the quadratic form $Q_1^{\pm}$ on this lattice is $\mathrm{GL}_3(\mathcal{O})$ equivalent to

$$Q_0 = \kappa \begin{pmatrix} 1 & & \\ & 1 & \\ & & -1 \end{pmatrix}.$$

An easy calculation yields the following:

Lemma 8.4.12. *(i) Let φ_1 (resp. $\varphi_{1,\pm}$) be the characteristic function of L^1 (resp. $L^{1,\pm}$). Then*

$$(\mathrm{char}(\mathcal{O}) \circ Q\,) \cdot \widehat{\varphi}_1 = q^{-1}\left[\varphi_{1,+} + \varphi_{1,-} - \varphi_1\right].$$

(ii)

$$(\mathrm{char}(\mathcal{O}) \circ Q\,) \cdot \widehat{\varphi}_{\mathrm{ra}} = q^{-1}\varphi_{\mathrm{ra}}.$$

Here note that $\mathrm{vol}(L^1) = q^{-1}$ with respect to the measure on V which is self-dual with respect to $\psi \circ (\ ,\ )$. Using this expression and noting that $\mathrm{vol}(\underline{J}') = (q+1)^{-1}$, we obtain

$$q(q+1)\, I(f, \varphi_1) = (q+1)\,\varphi_1 - \varphi_{1,+} - \varphi_{1,-}, \tag{8.4.17}$$

and

$$q(q+1)\, I(f, \varphi_{\mathrm{ra}}) = (q+1)\varphi_{\mathrm{ra}}. \tag{8.4.18}$$

In the case of φ_{ra}, this gives immediately, using (8.4.10) and Lemma 8.4.4,

$$\begin{aligned} Z(s, \Phi^{\mathrm{ra}}, \boldsymbol{f}_{\mathrm{sp}})(1) &= \gamma(V^-) \int_{F^\times} \chi\mu(a)\, |a|^{\frac{m}{2}+r-1} B(r, a; \varphi_{\mathrm{ra}}, I(f, \varphi_{\mathrm{ra}}))\, d^\times a \\ &= -\frac{L(s+\frac{1}{2}, \chi\mu)\, L(s+\frac{1}{2}, \chi\mu^{-1})}{\zeta(2s+1)} \cdot q^{-1}\, W_0(s, L_{\mathrm{ra}}). \end{aligned}$$

In the remaining case, we observe that $\varpi L^{1,\pm} \subset L^1$ and use Lemma 8.4.4 again to obtain

(8.4.19)

$$\begin{aligned} &q(q+1)\, Z(s, \Phi^1, \boldsymbol{f}_{\mathrm{sp}})(1) \\ &= (q+1)\Big[\, L(s+\frac{1}{2}, \chi\mu) + \chi\mu(\varpi^{-1})\, q^{-s-\frac{1}{2}}\, L(s+\frac{1}{2}, \chi\mu^{-1})\Big]\, W_0(s, L^1) \\ &\quad - 2\Big[\, L(s+\frac{1}{2}, \chi\mu)\, W_0(s, L^1) \\ &\qquad\qquad + \chi\mu(\varpi^{-1})\, q^{-s-\frac{1}{2}}\, L(s+\frac{1}{2}, \chi\mu^{-1})\, W_0(s, L^0)\Big]. \end{aligned}$$

Next, we recall the following result.

Proposition 8.4.13. ([30], Section 8) *Let the notation be as above. Then*

$$W_0(s, L^0) = \frac{\zeta(2s+1)}{\zeta(2s+2)},$$

$$W_0(s, L^1) = q^{-1}\Big(2 - \frac{\zeta(2s+1)}{\zeta(2s)}\Big),$$

and

$$W_0(s, L^{\mathrm{ra}}) = q^{-1}\frac{\zeta(2s+1)}{\zeta(2s)}.$$

Writing these as

$$W_0(s, L^0) = \zeta(2s+1)\,(1 - q^{-2s-2})$$

and

$$W_0(s, L^1) = q^{-1}\,\zeta(2s+1)\Big(1 + q^{-2s} - 2\,q^{-2s-1}\Big)$$

and substituting into (8.4.19), we obtain, after a short manipulation,

$$\begin{aligned} &q(q+1)\,Z(s, \Phi^1, \boldsymbol{f}_{\mathrm{sp}}) \\ &= q^{-1}(q-1)\,\zeta(2s+1)\,L(s+\tfrac{1}{2}, \chi\mu)\,L(s+\tfrac{1}{2}, \chi\mu^{-1}) \\ &\qquad\qquad\qquad \times (1 - q^{-2s-1})(1 - \chi\mu^{-1}q^{-s-\frac{1}{2}})^2 \\ &= q^{-1}(q-1)\,L(s+\tfrac{1}{2}, \chi\mu)\,(1 - \chi\mu^{-1}q^{-s-\frac{1}{2}}). \end{aligned}$$ □

8.4.5 The case when $p = 2$

In this section we compute the local doubling integrals for certain principal series and special representations with mild ramification in the case $F = \mathbb{Q}_2$. We use the results of the Section 8.5.4, with a slight shift in notation. In particular, we let

$$\underline{J}' = \{\begin{pmatrix} a & b \\ c & d \end{pmatrix} \in \mathrm{Sp}_1(\mathbb{Z}_2) \mid \mathrm{ord}(c) \geq 2\}.$$

Recall that there is a splitting homomorphism $\underline{J}' \to G'$ given by $k \mapsto \mathbf{k} = [k, \lambda(k)]_{\mathrm{L}}$, and let $\mathbf{J}'$ be the image of $\underline{J}'$ in G'. Let K' be the full inverse image of $\underline{K}'$ in G'.

We let $I(\mu)$ be the induced representation of G', defined as in Section 8.4.2 above, for a character μ of $\mathbb{Q}_2^\times$. We begin by determining the characters ξ of $\mathbf{J}'$ for which the ξ- eigenspace $I(\mu)^{(\mathbf{J}',\xi)}$ is nonzero.

Notice that every element of $\underline{J}'$ can be uniquely written as

$$k = n(b)m(a)n_-(4c), \quad a \in \mathbb{Z}_2^\times, b, c \in \mathbb{Z}_2, \tag{8.4.20}$$

so that every character of $\underline{J}'$ has the form

$$\xi = (\chi, \psi^+, \psi^-) : \xi(n(b)m(a)n_-(4c)) = \chi(a)\psi^+(b)\psi^-(c), \tag{8.4.21}$$

where χ is a character of $\mathbb{Z}_2^\times$, and $\psi^\pm$ are additive characters of $\mathbb{Z}_2$.

Proposition 8.4.14. *Let the notation be as above. Then* $\xi = (\chi, \psi^+, \psi^-)$ *gives a character of* $\underline{J}'$ *if and only if the following two conditions hold.*

(i) The conductors of χ *and* $\psi^\pm$ *are all less than or equal to* 3, *i.e.,* $\psi^\pm(8\mathbb{Z}_2) = 1$ *and* $\chi(1 + 8\mathbb{Z}_2) = 1$. *In particular,* $\chi^2 = 1$ *on* $\mathbb{Z}_2^\times$.

(ii) $\chi(5) = \psi^+(4)\psi^-(4)$.

Proof. It is easy to check the identity

$$\begin{aligned} & n(b_1)m(a_1)n_-(c_1)n(b_2)m(a_2)n_-(c_2) \\ &= n\left(b_1 + \frac{b_2 a_1^2}{1 + b_2 c_1}\right) m\left(\frac{a_1 a_2}{1 + b_2 c_1}\right) n_-\left(c_2 + \frac{c_1 a_2^2}{1 + b_2 c_1}\right). \end{aligned} \tag{8.4.22}$$

Then, ξ is a character if and only if, for all $a_i \in \mathbb{Z}_2^\times, b_i, c_i \in \mathbb{Z}_2$,
(8.4.23)

$$\chi(1 + 4b_2c_1)^{-1}\psi^+\left(\frac{b_2 a_1^2}{1 + 4b_2 c_1}\right)\psi^-\left(\frac{c_1 a_2^2}{1 + 4b_2 c_1}\right) = \psi^+(b_2)\psi^-(c_1).$$

Setting $b_2 = 1$ and $c_1 = 0$, one gets $\psi^+(a_1^2 - 1) = 1$, for all $a_1 \in \mathbb{Z}_2^\times$. This implies that ψ^+ has conductor ≤ 3, i.e., $\psi^+(8\mathbb{Z}_2) = 1$. The same argument with $b_2 = 0$ and $c_1 = 1$ shows that ψ^- has conductor ≤ 3. Setting $a_1 = a_2 = 1$, one sees from (8.4.23) that

$$\chi(1 + 4bc) = \psi^+\left(\frac{-4b^2c}{1 + 4bc}\right)\psi^-\left(\frac{-4bc^2}{1 + 4bc}\right) \tag{8.4.24}$$

for all $b, c \in \mathbb{Z}_2$. Since $\psi^\pm$ has conductor ≤ 3, this identity implies that χ has conductor ≤ 3 as well. This proves that condition (i) holds if ξ is a character. When $b = c = 1$, the identity (8.4.24) becomes condition (ii). Conversely, if $\xi = (\chi, \psi^+, \chi^-)$ satisfies conditions (i) and (ii), condition (ii) implies (8.4.24) and then (8.4.23) via condition (i). □

Proposition 8.4.15. *Let μ be a character of $\mathbb{Q}_2^\times$ and let $\xi = (\chi, \psi^+, \chi^-)$ be a character of $\underline{J}' \simeq \mathbf{J}'$. Let $I(\mu)^{(\mathbf{J}',\xi)}$ be the ξ-eigenspace for the right action of $\mathbf{J}'$ on $I(\mu)$.*
(i) When $\psi^\pm = 1$, and $\chi = \mu\chi_2$ has conductor ≤ 2, i.e., $\chi(x) = (\alpha, x)_2$ for some $\alpha \in \mathbb{Z}_2^\times$, then

$$I(\mu)^{(\mathbf{J}',\xi)} = \mathbb{C}f_1 \oplus \mathbb{C}f_w,$$

where $f_g \in I(\mu)^{(\mathbf{J}',\xi)}$ is supported on $\tilde{P}g\mathbf{J}'$ and is given by

$$f_g([n(b)m(a), z][g, 1]_L\mathbf{k}) = z\mu(a)|a|\xi(\mathbf{k}).$$

(ii) When $\psi^+ = 1$, $\psi^- \neq 1$, and $\chi = \chi_2\mu$ with $\chi(5) = \psi^-(4)$,

$$I(\mu)^{(\mathbf{J}',\xi)} = \mathbb{C}f_1.$$

(iii) When $\psi^- = 1$, $\psi^+ \neq 1$, and $\chi = \chi_2\mu$ with $\chi(5) = \psi^+(4)$,

$$I(\mu)^{(\mathbf{J}',\xi)} = \mathbb{C}f_w.$$

(iv) In all other cases, $I(\mu)^{(\mathbf{J}',\xi)} = 0$.

Proof. Since

$$G' = P'\mathbf{J}' \ \cup \ P'[n_-(2), 1]_{\mathrm{L}}\mathbf{J}' \ \cup \ P'[w, 1]_{\mathrm{L}}\mathbf{J}',$$

we have

$$I(\mu)^{(\mathbf{J}',\xi)} = \mathbb{C}f_1 \oplus \mathbb{C}f_{n_-(2)} \oplus \mathbb{C}f_w. \tag{8.4.25}$$

Here it is understood that if f_g, as given in the proposition, is not well defined, it is deleted from the the right-hand side of (8.4.25). Now, f_g is well defined if and only if the obvious compatibility condition of characters on $P' \cap g\mathbf{J}'g^{-1}$ is satisfied.

First, for a given ξ, f_1 is well defined if and only if

$$\mu(a) = f_1([n(b)m(a), 1]_L) = \chi(a)\chi_2(a)\psi^+(b), \quad a \in \mathbb{Z}_2^\times, b \in \mathbb{Z}_2.$$

This is equivalent to $\psi^+ = 1$ and $\chi = \mu\chi_2$ on $\mathbb{Z}_2^\times$. The fact that $\psi^-(4) = \chi(5)$ comes from the previous proposition.

Next, assume that f_w is well defined. Note that

$$wn(b)m(a)n_-(c)w^{-1} = n_-(-b)m(a^{-1})n(-c) \in \underline{P}' \quad \Longleftrightarrow \quad b = 0.$$

Now

$$[w, 1]_{\mathrm{L}}[m(a), \chi_2(a)]_{\mathrm{L}} = [m(a^{-1}), \chi_2(a)]_{\mathrm{L}}[w, 1]_{\mathrm{L}}$$

so that we must have

$$\chi(a) = \mu(a)^{-1}\chi_2(a), \quad a \in \mathbb{Z}_2^\times.$$

On the other hand, it is easy to check that

$$(8.4.26) \qquad [w,1]_L[n_-(c),\lambda(n_-(c))]_L = [n(-c),1]_L[w,1]_L,$$

(see Lemma 8.5.13), and this implies that $\psi^-(c) = 1$ for all $c \in \mathbb{Z}_2$. This proves that, if f_w is well defined, then $\psi^- = 1$ and $\chi = \chi_2\mu$ on $\mathbb{Z}_2^\times$. The converse is also true and left to the reader to check.

Finally, we prove that $f_{n_-(2)}$ is never well defined. This will finish the proof of the proposition. Note that

$$n_-(2)n(b)m(a)n_-(c) = \frac{1}{a}\begin{pmatrix} a^2+bc & b \\ 2a^2+(1+2b)c & (1+2b)\end{pmatrix}$$

and that

$$n_-(2)n(b)m(a)n_-(c)n_-(2)^{-1} = \frac{1}{a}\begin{pmatrix} a^2+b(c-2) & b \\ 2a^2+(1+2b)(c-2) & (1+2b)\end{pmatrix}.$$

Writing $c = 4c_0$, we see that this last element lies in $\underline{P}'$ if and only if

$$(8.4.27) \qquad a^2 = (1+2b)(1-2c_0),$$

and that, when (8.4.27) holds, then

$$\begin{aligned} n_-(2)n(b)m(a)n_-(c)n_-(2)^{-1} &= \begin{pmatrix} a^{-1}(1-2c_0) & a^{-1}b \\ 0 & a^{-1}(1+2b)\end{pmatrix} \\ &= n(b(1+2b)^{-1})\, m(a^{-1}(1-2c_0)). \end{aligned}$$

Then, for $k = n(b)m(a)n_-(c) \in \underline{J}'$, we have

$$\begin{aligned} &[n_-(2),1]_L[n(b)m(a)n_-(c),\lambda_2(k)]_L \\ &\qquad = [n_-(2)n(b)m(a)n_-(c),*]_L \\ &\qquad = [n(b(1+2b)^{-1})\, m(a^{-1}(1-2c_0)),*]_L[n_-(2),1]_L, \end{aligned}$$

where, after a short calculation given in Lemma 8.5.13,

$$* = c_L(n_-(2),k)\cdot\lambda_2(k) = \gamma(1-2c_0,\psi).$$

Thus, if $f_{n_-(2)}$ is well defined, then

$$\chi(a)\psi^+(b)\psi^-(c_0) = \mu(\frac{1-2c_0}{a})\gamma(1-2c_0,\psi)$$

whenever

$$a^2 = (1+2b)(1-2c_0).$$

Setting $c_0 = 0$, we see that

$$\mu(a) = \chi(a)\psi^+(\frac{a^2-1}{2}) = \pm 1.$$

Setting $a = 1 = c_0$ and $b = -1$, we have

$$\psi^+(-1)\psi^-(1) = \mu(-1)\gamma(-1,\psi).$$

Squaring both sides, we find that

$$\psi^+(-2)\psi^-(2) = (-1,-1) = -1.$$

On the other hand, setting $a = 1$ and $-c_0 \equiv b \equiv 2 \mod 8$, we have

$$\psi^+(2)\psi^-(-2) = \mu(3)\gamma(3,\psi) = \pm i,$$

contradicting the previous equation. So $f_{n_-(2)}$ does not exist. □

Proposition 8.4.15 implies that, if $I(\mu)^{(\mathbf{J}',\xi)} \neq 0$ for some ξ, then $\mu = \mu_t = \chi_{2\alpha}|\ |^t$, so we assume from now on that μ has this form. The induced representation $I(\mu)$ is irreducible unless $t = \pm\frac{1}{2}$. When $t = \frac{1}{2}$, it has a unique irreducible subrepresentation, the special representation $\tilde{\sigma}(\chi_\alpha|\ |^{\frac{1}{2}})$, which is the kernel of the intertwining operator $M(t) : I(\mu) \to I(\mu^{-1})$ at the point $t = \frac{1}{2}$. When $t = -\frac{1}{2}$, then the unique irreducible subrepresentation of $I(\mu)$ is an even Weil representation. Recall that, for $\mathrm{Re}(t) > 1$, the intertwining operator is defined by

$$M(t)f(g) = \int_{\mathbb{Q}_2} f([w^{-1},1]_{\mathrm{L}}[n(b),1]_{\mathrm{L}}g)\,db, \tag{8.4.28}$$

where db is the self-dual measure with respect to ψ. Note that $M(t)$ carries the space $I(\mu_t)^{(\mathbf{J}',\xi)}$ to $I(\mu_t^{-1})^{(\mathbf{J}',\xi)}$. As usual, functions in $I(\mu)$ are determined by their restrictions to K', and the resulting space of functions on K' is independent of t. Thus the space of ξ-eigenfunctions for $\mathbf{J}'$ is also independent of t. For simplicity, we assume from now on that $\psi^\pm = 1$, i.e., that $\xi = \chi_\alpha$ is a character of $\mathbb{Z}_2^\times$ with $\alpha \in \mathbb{Z}_2^\times$.

Proposition 8.4.16. *Let $\mu = \chi_{2\alpha}|\ |^t$, and let $\xi = \chi_\alpha$ be a character of $\mathbf{J}'$ with $\alpha \in \mathbb{Z}_2^\times$. Then the matrix of $M(t)$ with respect to the basis f_1, f_w of*

$I(\mu)^{(\mathbf{J}',\chi_\alpha)}$ *is*

$$\begin{pmatrix} \dfrac{\mathfrak{g}_8(\alpha)}{2^{2t}(1-2^{-2t})} & \chi_\alpha(-1) \\ \frac{1}{4} & \dfrac{\mathfrak{g}_8(\alpha)}{1-2^{-2t}} \end{pmatrix},$$

where

$$\mathfrak{g}_8(\alpha) = \int_{\mathbb{Z}_2^\times} \chi_{2\alpha}(u)\,\psi(\frac{u}{8})\,du = \frac{1}{2\sqrt{2}\,\delta_\alpha}.$$

Here

$$\delta_\alpha = \begin{cases} 1 & \text{if } \alpha \equiv 1 \mod 4, \\ i & \text{if } \alpha \equiv -1 \mod 4. \end{cases}$$

In particular, when $t = \frac{1}{2}$, *the matrix directly above is*

$$\frac{1}{2\sqrt{2}\,\delta_\alpha} \begin{pmatrix} 1 & \dfrac{2\sqrt{2}}{\delta_\alpha} \\ \dfrac{\delta_\alpha}{\sqrt{2}} & 2 \end{pmatrix},$$

and its eigenfunctions are

$$\boldsymbol{f}_{\mathrm{sp}} = f_1 - \frac{\delta_\alpha}{2\sqrt{2}}\, f_w \in \tilde{\sigma}(\chi_\alpha |\,|^{\frac{1}{2}})$$

and

$$\boldsymbol{f}_{\mathrm{ev}} = f_1 + \frac{\delta_\alpha}{\sqrt{2}} f_w$$

with eigenvalues 0 *and* $\frac{3}{2\sqrt{2}\,\delta_\alpha}$, *respectively.*

Proof. It suffices to compute the values of $M(t)f_1$ and $M(t)f_w$ at the points 1 and $\mathbf{w}$. It is easy to check that

$$w^{-1}n(b) = \begin{pmatrix} & -1 \\ 1 & b \end{pmatrix} = \begin{cases} -wn(b) & \text{if } \mathrm{ord}(b) \geq 0, \\ n(-b^{-1})\,m(b^{-1})\,n_-(b^{-1}) & \text{if } \mathrm{ord}(b) < 0, \end{cases}$$

and thus, for $b \notin \mathbb{Z}_2$,

$$\mathbf{w}^{-1}\mathbf{n}(b) = \mathbf{n}(-b^{-1})\mathbf{m}(b^{-1})[n_-(b^{-1}), 1]_L.$$

Here, for convenience, we write $\mathbf{w}^{-1} = [w^{-1}, 1]_{\mathrm{L}}$, $\mathbf{n}(b) = [n(b), 1]_{\mathrm{L}}$, and $\mathbf{m}(a) = [m(a), 1]_L$. Then

$$f_1(\mathbf{w}^{-1}\mathbf{n}(b)) = \begin{cases} 0 & \text{if } b \in \frac{1}{2}\mathbb{Z}_2, \\ \gamma(\psi_{2b})\mu(b)^{-1}|b|^{-1} & \text{if } b \notin \frac{1}{2}\mathbb{Z}_2, \end{cases}$$

and

$$f_w(\mathbf{w}^{-1}\mathbf{n}(b)) = \begin{cases} \chi_\alpha(-1) & \text{if } b \in \mathbb{Z}_2, \\ 0 & \text{if } b \notin \mathbb{Z}_2. \end{cases}$$

Therefore

$$\begin{aligned} M(t)f_1(1) &= \int_{\mathrm{ord}(b)\le -2} f_1(\mathbf{w}^{-1}\mathbf{n}(b))\, db \\ &= \sum_{r=2}^{\infty} \int_{\mathrm{ord}(b)=-r} \chi_{2\alpha}(b^{-1})\, |b^{-1}|^{t+1}\, \gamma(\psi_{2b})\, db \\ &= \sum_{r=2}^{\infty} 2^{-rt}\, \chi_{2\alpha}(2^r) \int_{\mathbb{Z}_2^\times} \chi_{2\alpha}(u)\, \gamma(\psi_{2^{r-1}u})\, du. \end{aligned}$$

Recall that, by [19], Proposition A.12,

$$\gamma(\psi_{2^{r-1}b}) = \begin{cases} \psi(\frac{b}{8}) & \text{if } r \text{ is even}, \\ \psi(\frac{b}{8})\chi_2(b) & \text{if } r \text{ is odd}, \end{cases} \tag{8.4.29}$$

so that the integral becomes

$$\begin{aligned} &\int_{\mathbb{Z}_2^\times} \chi_{2\alpha}(u)\, \gamma(\psi_{2^{r-1}u})\, du \\ &= \begin{cases} \int_{\mathbb{Z}_2^\times} \chi_{2\alpha}(u)\psi(\frac{u}{8})\, du & \text{if } r \text{ is even}, \\ \int_{\mathbb{Z}_2^\times} \chi_{\alpha}(u)\psi(\frac{u}{8})\, du & \text{if } r \text{ is odd}, \end{cases} \\ &= \begin{cases} \mathfrak{g}_8(\alpha) & \text{if } r \text{ is even}, \\ 0 & \text{if } r \text{ is odd}. \end{cases} \end{aligned}$$

So we have

$$M(t)f_1(1) = \sum_{n=1}^{\infty} 2^{-2nt}\mathfrak{g}_8(\alpha) = \frac{2^{-2t}}{1 - 2^{-2t}} \cdot \mathfrak{g}_8(\alpha),$$

and

$$M(t)f_w(1) = \int_{\mathbb{Q}_2} f_w(\mathbf{w}^{-1}\mathbf{n}(b))\, db = \chi_\alpha(-1).$$

Next, we note that

$$\mathbf{w}^{-1}\mathbf{n}(b)\mathbf{w} = \begin{cases} [n_-(-b), \gamma(\psi_{2b})]_L & \text{if } b \in 2\mathbb{Z}_2, \\ \mathbf{n}(-b^{-1})\mathbf{m}(b^{-1})\mathbf{w}\mathbf{n}(-b^{-1}) & \text{if } b \notin 2\mathbb{Z}_2, \end{cases}$$

so that

$$f_1(\mathbf{w}^{-1}\mathbf{n}(b)\mathbf{w}) = \begin{cases} 1 & \text{if } b \in 4\mathbb{Z}_2, \\ 0 & \text{if } b \notin 4\mathbb{Z}_2, \end{cases}$$

and

$$f_w(\mathbf{w}^{-1}\mathbf{n}(b)\mathbf{w}) = \begin{cases} 0 & \text{if } b \in 2\mathbb{Z}_2, \\ \mu(b^{-1})|b|^{-1}\gamma(\psi_{2b}) & \text{if } b \notin 2\mathbb{Z}_2. \end{cases}$$

This gives

$$M(t)f_1(\mathbf{w}) = \int_{\text{ord}(b)\geq 2} db = \frac{1}{4}.$$

Finally, the same calculation as in the beginning gives

$$\begin{aligned} M(t)f_w(\mathbf{w}) &= \sum_{r=0}^{\infty} \int_{\text{ord}(b)=-r} \chi_{2\alpha}(b^{-1})\, |b^{-1}|^{t+1}\, \gamma(\psi_{2b})\, db \\ &= \sum_{r=0}^{\infty} 2^{-rt}\chi_\alpha(2)^r \int_{\mathbb{Z}_2^\times} \chi_{2\alpha}(b)\gamma(\psi_{2^{r-1}b})\, db \\ &= \sum_{r\geq 0,\ \text{even}} 2^{-rt}\chi_\alpha(2)^r \mathfrak{g}_8(\alpha) \\ &= \frac{\mathfrak{g}_8(\alpha)}{1-2^{-2t}}. \end{aligned}$$

The values of $\mathfrak{g}_8(\alpha)$ are easy to calculate. □

It will be useful to record some information about the Whittaker functions of f_1 and f_w. With the same notation as in Proposition 8.4.16, recall that, for $m \in \mathbb{Q}_2$ and $\text{Re}(t) > 1$, the Whittaker function attached to $f \in I(\mu)$ is given by

$$W_m(f)(g) = \int_{\mathbb{Q}_2} f([w^{-1}, 1]_{\text{L}}[n(b), 1]_{\text{L}}g)\, \psi(-mb)\, db. \tag{8.4.30}$$

Notice that

$$W_m(f)(\mathbf{m}(a)g) = \mu(a)^{-1}|a|W_{a^2m}(f)(g).$$

We will only need the following values:

Proposition 8.4.17. *With the same notation as in Proposition 8.4.16, suppose that $m \in \mathbb{Z}_2 - 4\mathbb{Z}_2$. Then*
(i) $W_m(f_1)(\mathbf{w}) = \chi_\alpha(-1)$, $W_m(f_w)(1) = \frac{1}{4}$, *and*

$$W_m(f_w)(\mathbf{w}) = W_m(f_1)(1) + \frac{1}{2\sqrt{2}\delta_\alpha}.$$

(ii)

$$W_m(f_1)(1) = \begin{cases} -\frac{1}{2\sqrt{2}\delta_\alpha}\,\mu(2)^2 & \textit{if } m \equiv 2, -\alpha \mod 4, \\ \frac{1}{2\sqrt{2}\delta_\alpha}\,\mu(2)^2 + \frac{1}{2}\mu(2)^3\psi(\frac{1-m}{8}) & \textit{if } m \equiv \alpha \mod 4. \end{cases}$$

(iii) Let

$$\boldsymbol{f}^+ = f_1 + \frac{\sqrt{2}\mu(2)^2}{\delta_\alpha} f_w.$$

Then

$$W_m(\boldsymbol{f}^+)(1) = 0 \quad \textit{if } m \equiv 2, -\alpha \mod 4.$$

We omit the slightly more complicated values in general; these can be computed by the same methods. We remark that part (iii) of the proposition means that $\boldsymbol{f}^+$ is in Kohnen's plus space locally. Additional discussion of this point is given in Section 9.3. Notice that

$$W_m(\boldsymbol{f}^+)(\mathbf{w}) \neq 0 \quad \text{if } m \equiv 2, -\alpha \mod 4,$$

so that the Fourier coefficients at other cusps will not vanish.

Proof. The same calculation as in the proof of Proposition 8.4.16 gives

$$\begin{aligned} W_m(f_1)(1) &= \sum_{n\geq 2} \mu(2)^n I_m(n, \chi_{2\alpha}, \psi), \\ W_m(f_1)(\mathbf{w}) &= \chi_\alpha(-1)\text{char}(\mathbb{Z}_2)(m), \\ W_m(f_w)(1) &= \frac{1}{4}\text{char}(4\mathbb{Z}_2)(m), \end{aligned} \tag{8.4.31}$$

and

$$W_m(f_w)(\mathbf{w}) = \sum_{n\geq 0} \mu(2)^n I_m(n, \chi_{2\alpha}, \psi),$$

where

$$I_m(n, \chi_{2\alpha}, \psi) = \int_{\mathbb{Z}_2^\times} \gamma(\psi_{2^{1-n}b})\chi_{2\alpha}(b)\psi(-2^n mb)\, db. \tag{8.4.32}$$

These formulas give a complete description for the ψ-Whittaker functions of f_1 and f_w if one knows how to compute $I_m(n, \chi_{2\alpha}, \psi)$.

Lemma 8.4.18. *Let the notation be as above. Then*

$$I_m(n, \chi_{2\alpha}, \psi) = \begin{cases} \chi_{2\alpha}(1 - 2^{3-n}m)\frac{1}{2\sqrt{2}\delta_\alpha} & 2 \mid n,\ 2^{3-n}m \in 2\mathbb{Z}_2, \\ \frac{1}{2}\psi(\frac{1-2^{3-n}m}{8}) & 2 \nmid n,\ 2^{3-n}m \equiv \alpha \mod 4, \\ 0 & \textit{otherwise.} \end{cases}$$

Proof. First we assume that n is even. By (8.4.29), we have

$$\begin{aligned} I_m(n, \chi_{2\alpha}, \psi) &= \int_{\mathbb{Z}_2^\times} \chi_{2\alpha}(b)\psi(\frac{b}{8}(1 - 2^{3-n}m))\, db \\ &= \chi_\alpha(1 - 2^{3-n}m)\frac{1}{2\sqrt{2}\delta_\alpha}\text{char}(\mathbb{Z}_2^\times)(1 - 2^{3-n}m) \end{aligned}$$

as claimed. Now we assume that n is odd. Then (8.4.29) implies

$$\begin{aligned} &I_m(n, \chi_{2\alpha}, \psi) \\ &= \int_{\mathbb{Z}_2^\times} \chi_\alpha(b)\psi(\frac{b}{8}(1 - 2^{3-n}m))\, db \\ &= \begin{cases} \frac{1}{2}\text{char}(4\mathbb{Z}_2)(1 - 2^{3-n}m)\psi(\frac{1-2^{3-n}m}{8}) & \text{if } \alpha \equiv 1 \mod 4, \\ \frac{1}{2}\text{char}(2\mathbb{Z}_2^\times)(1 - 2^{3-n}m)\psi(\frac{1-2^{3-n}m}{8}) & \text{if } \alpha \equiv -1 \mod 4, \end{cases} \\ &= \begin{cases} \frac{1}{2}\psi(\frac{1-2^{3-n}m}{8}) & \text{if } 2^{3-n}m \equiv \alpha \mod 4, \\ 0 & \text{otherwise.} \end{cases} \end{aligned}$$

□

Now we can complete the proof of Proposition 8.4.17. The first two claims in (i) are parts of (8.4.31). The same formula also gives, together with Lemma 8.4.18,

$$\begin{aligned} W_m(f_w)(\mathbf{w}) &= W_m(f_1)(1) + I_m(0, \chi_{2\alpha}, \psi) + \mu(2)I_m(1, \chi_{2\alpha}, \psi) \\ &= W_m(f_1)(1) + \frac{1}{2\sqrt{2}\delta_\alpha}. \end{aligned}$$

When $m \equiv 2, -\alpha \mod 4$, one has

$$\text{ord}_2(2^{3-n}m) \begin{cases} \geq 1 & \text{if } n \leq 2, \\ \leq 0 & \text{if } n \geq 4, \end{cases}$$

and $2^{3-n}m \not\equiv \alpha \mod 4$. So Lemma 8.4.18 implies

$$\begin{aligned} W_m(f_1)(1) &= \mu(2)^2 I_m(2, \chi_{2\alpha}, \psi) \\ &= \frac{\mu(2)^2 \chi_{2\alpha}(1-2m)}{2\sqrt{2}\delta_\alpha} \\ &= -\frac{\mu(2)^2}{2\sqrt{2}\delta_\alpha} \end{aligned}$$

as claimed.

When $m \equiv \alpha \mod 4$, one has $2^{3-n}m \equiv \alpha \mod 4$ iff $n = 3$. So Lemma 8.4.18 implies that

$$\begin{aligned} W_m(f_1)(1) &= \mu(2)^2 I_m(2, \chi_{2\alpha}, \psi) + \mu(2)^3 I_m(3, \chi_{2\alpha}, \psi) \\ &= \frac{\mu(2)^2 \chi_{2\alpha}(1-2m)}{2\sqrt{2}\delta_\alpha} + \frac{1}{2}\mu(2)^3 \psi(\frac{1-m}{8}) \\ &= \frac{\mu(2)^2}{2\sqrt{2}\delta_\alpha} + \frac{1}{2}\mu(2)^3 \psi(\frac{1-m}{8}). \end{aligned}$$

This proves (ii). Claim (iii) follows from (i) and (ii) directly. □

We now return to the local zeta integrals. For $\kappa \in \mathbb{Z}_2^\times$, we define lattices L^0, L^1, and L^{ra} as in (8.4.12). By Proposition 8.5.14, when $L = L^0, L^1$, or L^{ra}, the standard section $\Phi_L(s) \in I(s, \chi_V)$ determined by

$$\Phi_L(0) = \lambda_V(\varphi_L \otimes \varphi_L)$$

is an eigenfunction of $i(\mathbf{J}' \times \mathbf{J}')$ with character χ_κ. Thus, $Z(s, \Phi_L, \cdot)$ maps $I(\mu)$ into $I(\mu)^{(\mathbf{J}', \chi_\kappa)}$. Since the latter space is nonzero only when $\mu = \chi_{2\alpha}|\,|^t$, for some α with $\alpha \equiv \kappa \mod 4$, we assume this condition from now on and are led to compute the zeta integrals, as in (8.4.10),

(8.4.33) $Z(s, \Phi_L, f_{w_j})(\mathbf{w}_i)$

$$= \gamma(V) \int_{\mathbb{Q}_2^\times} \chi_V \mu(a) |a|^{r+\frac{1}{2}} B(r, a; \omega_V(\mathbf{w}_i)\varphi_L^{(r)}, I(f_{w_j}, \varphi_L))\, d^\times a,$$

for $L = L^0, L^1, L^{ra}$, and $i, j = 0, 1$, where $\mathbf{w}_0 = 1$ and $\mathbf{w}_1 = \mathbf{w}$.

Lemma 8.4.19. *(i)*

$$\omega_V(\mathbf{w}_i)\varphi_L = \begin{cases} \varphi_L & \text{if } i = 0, \\ \gamma(V)[L^\# : L]^{-\frac{1}{2}} \cdot \varphi_{L^\#} & \text{if } i = 1. \end{cases}$$

(ii)

$$I(f_1, \varphi_L) = \mathrm{vol}(\mathbf{J}')\varphi_L.$$

(iii)

$$I(f_w, \varphi_L)(x) = \gamma(V)^{-1}\,[L^\# : L]^{-\frac{1}{2}}\,\mathrm{vol}(\mathbf{J}') \cdot 4\,\mathrm{char}_{\mathbb{Z}_2}(Q(x)) \cdot \varphi_{L^\#}(x)$$

Proof. Only (iii) needs proof. Since

$$\underline{K}' \cap \underline{P}' w \underline{J}' = \bigcup_{c\in\mathbb{Z}/4} n(c) w \underline{J}',$$

we have

$$\begin{aligned} I(f_w, \varphi_L) &= \int_{\underline{K}'} f_w(k')\,\overline{\omega_V(k')\varphi_L(x)}\,dk \\ &= \sum_{c\in\mathbb{Z}/4} \int_{\mathbf{J}'} \chi_\kappa(\mathbf{k}')\,\overline{\omega_V(\mathbf{wk}')\varphi_L(x)\psi(c\,Q(x))}\,d\mathbf{k}' \\ &= \gamma(V)^{-1}\,\mathrm{vol}(\mathbf{J}')\,[L^\# : L]^{-\frac{1}{2}} \cdot \varphi_{L^\#}(x) \sum_{c\in\mathbb{Z}/4} \psi(-c\,Q(x)) \\ &= \gamma(V)^{-1}\,\mathrm{vol}(\mathbf{J}')\,[L^\# : L]^{-\frac{1}{2}} \cdot \varphi_{L^\#}(x) \cdot 4\,\mathrm{char}_{\mathbb{Z}_2}(Q(x)). \quad \square \end{aligned}$$

Note that

$$\mathrm{char}_{\mathbb{Z}_2} \circ Q \cdot \varphi_{L^\#} = \sum_{\substack{y\in L^\sharp/L \\ Q(y)\in\mathbb{Z}_2}} \varphi_{y+L}.$$

Lemma 8.4.20. *For $y \in L^\# - L$ with $Q(y) \notin \mathbb{Z}_2$, and any lattice L',*

$$B(r, a; \varphi_{L'}, \varphi_{y+L}) = 0.$$

Proof. Indeed,

$$\begin{aligned} B(r, a; \varphi_{L'}, \varphi_{y+L}) &= \int_{\mathbb{Q}_2} \int_{V_r} \varphi_{L'}^{(r)}(x)\varphi_{y+L}^{(r)}(-ax)\,\psi(-b\,Q(x))\,dx\,db \\ &= \int_{L'_r} \varphi_{y+L}^{(r)}(-ax) \lim_{n\to\infty} \int_{2^{-n}\mathbb{Z}_2} \psi(-b\,Q(x))\,db\,dx. \end{aligned}$$

Now for $ax \in y + L_r$, one has $x \in a^{-1}y + a^{-1}L_r$, and thus, since $Q(y) \notin \mathbb{Z}_2$,

$$\mathrm{ord}(Q(x)) = -2\,\mathrm{ord}(a) + \mathrm{ord}(Q(y)) < -2\,\mathrm{ord}(a).$$

Thus,

$$\int_{2^{-n}\mathbb{Z}_2} \psi(-b\,Q(x))\,db = 0$$

when $n > -2\,\mathrm{ord}(a)$, and hence $B(r, a; \varphi_{L'}, \varphi_{y+L}) = 0$. □

Corollary 8.4.21. *(i) For $L = L^0$ or L^{ra},*

$$\big(B(r, a; \omega_V(\mathbf{w}_i)\varphi_L, I(f_{w_j}, \varphi_L))\big)_{i,j=0,1}$$
$$= \mathrm{vol}(\mathbf{J}')\, B(r, a; \varphi_L, \varphi_L) \cdot \begin{pmatrix} 1 & \frac{4\gamma(V)^{-1}}{\sqrt{[L^{\#}:L]}} \\ \frac{\gamma(V)}{\sqrt{[L^{\#}:L]}} & \frac{4}{[L^{\#}:L]} \end{pmatrix}.$$

(ii) For $L = L^1$, and $L' = L$ or $L^{\#}$,

$$B(r, a; \varphi_{L'}, I(f_w, \varphi_L)) = \mathrm{vol}(\mathbf{J}')\cdot B(r, a; \varphi_{L'}, \varphi_{L^{\#}})\cdot 4\,\gamma(V)^{-1}\,[L^{\#} : L]^{-\frac{1}{2}}.$$

Proof. Claim (ii) follows immediately from the previous two lemmas. To prove (i), by the same argument as above, it suffices to prove

$$B(r, a; \varphi_{L_1}, \varphi_{L_2}) = B(r, a; \varphi_L, \varphi_L),$$

for $L_i = L$ or $L^{\#}$, $i = 1, 2$. It is easy to check that for $y \in L^{\#}$, $L = L^0$, or L^{ra}, $Q(y) \in \mathbb{Z}_2$ if and only if $y \in L$. So one has, by the previous lemma and Lemma 8.4.4,

$$\begin{aligned} B(r, a; \varphi_{L_1}, \varphi_{L_2}) &= \sum_{y \in L_2/L} B(r, a; \varphi_{L_1}, \varphi_{y+L}) \\ &= B(r, a; \varphi_{L_1}, \varphi_L) \\ &= |a|^{-2r-1} B(r, a^{-1}; \varphi_L, \varphi_{L_1}) \\ &= |a|^{-2r-1} B(r, a^{-1}; \varphi_L, \varphi_L) \\ &= B(r, a; \varphi_L, \varphi_L). \end{aligned}$$

□

Lemma 8.4.22.

$$\gamma(V^{\pm}) = \pm\delta_\kappa,$$

where

$$\delta_\kappa = \psi(\frac{\kappa - 1}{8})\,\chi_2(\kappa) = \begin{cases} 1 & \text{if } \kappa \equiv 1 \mod 4, \\ i & \text{if } \kappa \equiv -1 \mod 4. \end{cases}$$

Proof. By Lemma 8.5.6,

$$\gamma(V) = \chi_V(-1)\gamma(\psi_{1/2})\left(\gamma(\det V, \psi_{1/2})\gamma(\psi_{1/2})^3\epsilon(V)\right)^{-1}.$$

Recall that $\det V = -2\kappa$, and that the matrix of $V^{\pm}$ is $2\kappa\,\mathrm{diag}(1, 1, -1)$ and $-2\kappa\,\mathrm{diag}(1, 1, 1)$ respectively. Thus

$$\epsilon(V) = \begin{cases} (2\kappa, -1) & \text{if } V = V^+, \\ (-2\kappa, -1) & \text{if } V = V^-, \end{cases}$$
$$= \chi_{2\kappa}(-1)\,\mathrm{inv}(B),$$

where $\mathrm{inv}(B) = \pm 1$ is the invariant of the quaternion B associated to V. Then, by a short calculation, $\mathrm{inv}(B)\,\gamma(V) = \delta_\kappa$. □

Proposition 8.4.23. *(i) For $L = L^0$,*

$$\left(Z(s, \Phi_L, f_{w_j})(\mathbf{w}_i) \right)_{i,j=0,1}$$
$$= \gamma(V)\,\frac{\mathrm{vol}(\mathbf{J}')}{\sqrt{2}}\,\frac{L(\frac{1}{2}+s, \chi_V\mu)L(\frac{1}{2}+s, \chi_V\mu^{-1})}{\zeta(2s+2)} \begin{pmatrix} 1 & \frac{2\sqrt{2}}{\delta_\kappa} \\ \frac{\delta_\kappa}{\sqrt{2}} & 2 \end{pmatrix}.$$

In particular, the eigenfunctions of $Z(s, \Phi_{L^0}, \cdot)$ in $I(\mu)^{(\mathbf{J}', \chi_\kappa)}$ are

$$\boldsymbol{f}_{\mathrm{ev}} = f_1 + \frac{\delta_\kappa}{\sqrt{2}} f_w \quad \text{and} \quad \boldsymbol{f}_{\mathrm{sp}} = f_1 - \frac{\delta_\kappa}{2\sqrt{2}} f_w$$

with eigenvalues

$$\gamma(V)\,\frac{3\,\mathrm{vol}(\mathbf{J}')}{\sqrt{2}}\,\frac{L(\frac{1}{2}+s, \chi_V\mu)L(\frac{1}{2}+s, \chi_V\mu^{-1})}{\zeta(2s+2)}$$

and 0, respectively.
(ii) When $L = L^{ra}$,

$$\left(Z(s, \Phi_L, f_{w_j})(\mathbf{w}_i) \right)_{i,j=0,1}$$
$$= \frac{1}{4}\gamma(V)\,\frac{\mathrm{vol}(\mathbf{J}')}{\sqrt{2}}\,\frac{L(\frac{1}{2}+s, \chi_V\mu)L(\frac{1}{2}+s, \chi_V\mu^{-1})}{\zeta(2s)} \begin{pmatrix} 2 & -\frac{2\sqrt{2}}{\delta_\kappa} \\ -\frac{\delta_\kappa}{\sqrt{2}} & 1 \end{pmatrix}.$$

In particular, its eigenfunctions in $I(\mu)^{(\mathbf{J}', \chi_\kappa)}$ are $\boldsymbol{f}_{\mathrm{ev}}$ and $\boldsymbol{f}_{\mathrm{sp}}$ with eigenvalues 0 and

$$\frac{1}{4}\gamma(V)\,\frac{3\,\mathrm{vol}(\mathbf{J}')}{\sqrt{2}}\,\frac{L(\frac{1}{2}+s, \chi_V\mu)L(\frac{1}{2}+s, \chi_V\mu^{-1})}{\zeta(2s)},$$

respectively.

Proof. By the previous corollary and (8.4.10), it suffices to prove that

$$\int_{\mathbb{Q}_2^\times} \chi_V\mu(a)|a|^{\frac{1}{2}+r} B(r,a;\varphi_L,\varphi_L)\, d^\times a$$

$$= \begin{cases} \dfrac{L(\frac{1}{2}+r,\chi_V\mu)L(\frac{1}{2}+r,\chi_V\mu^{-1})}{\sqrt{2}\,\zeta(2r+2)} & \text{if } L = L^0, \\ \dfrac{L(\frac{1}{2}+r,\chi_V\mu)L(\frac{1}{2}+r,\chi_V\mu^{-1})}{2\sqrt{2}\,\zeta(2r)} & \text{if } L = L^{\mathrm{ra}}. \end{cases}$$

By Lemma 8.4.4, the integral is equal to

$$\frac{L(\frac{1}{2}+r,\chi_V\mu)L(\frac{1}{2}+r,\chi_V\mu^{-1})}{\zeta(2r+1)} \cdot W_0(r,L).$$

On the other hand, [14] Proposition 13.4 gives (taking into account the different Haar measures used there)

$$W_0(r,L) = \begin{cases} \dfrac{1}{\sqrt{2}}\dfrac{\zeta(2r+1)}{\zeta(2r+2)} & \text{if } L = L^0, \\ \dfrac{1}{2\sqrt{2}}\dfrac{\zeta(2r+1)}{\zeta(2r)} & \text{if } L = L^{\mathrm{ra}}. \end{cases}$$

□

Next, we consider the case $L = L^1$.

Lemma 8.4.24.

$$W_0(r,L^1) = \frac{1}{2\sqrt{2}}\zeta(2r+1)$$

and

$$W_0(r,L^{1,\#}) = \frac{1}{2\sqrt{2}}(2+\zeta(2r+1)).$$

Proof. Recall that here the Haar measure on V is the self-dual Haar measure so that $\mathrm{vol}(L^1,dx) = [L^{1,\#} : L^1]^{-1/2} = \frac{1}{2\sqrt{2}}$. The same calculation as in

[30], Section 4 gives

$$
\begin{aligned}
&W_0(r, L^{1,\#}) \\
&= \int_{\mathbb{Q}_2} \int_{L_r^{1,\#}} \psi(-b\,Q(x))\,dx\,db \\
&= \frac{1}{2\sqrt{2}} \int_{\mathbb{Q}_2} \int_{\frac{1}{2}\mathbb{Z}_2} \psi(-\kappa b x_1^2)\,dx_1 \\
&\qquad \times \int_{(\frac{1}{2}\mathbb{Z}_2)^2} \psi(-2\kappa b x_2 x_3)\,dx_2\,dx_3 \left(\int_{\mathbb{Z}_2^2} \psi(-\kappa b y_1 y_2)\,dy_1\,dy_2 \right)^r db \\
&= 2\sqrt{2} \int_{\mathbb{Q}_2} \min(1, |b^{-1}|) \cdot \min(1, |\frac{2}{b}|) \cdot \int_{\mathbb{Z}_2} \psi(-\frac{\kappa b}{4} x^2)\,dx\,db \\
&= 2\sqrt{2} \left(\int_{4\mathbb{Z}_2} db + 2^{-\frac{3}{2}} \sum_{n \geq 0} 2^{-n(r+\frac{1}{2})} \int_{\mathbb{Z}_2^\times} \psi(\frac{b}{8}) \chi_2(b)^{-n+1}\,db \right) \\
&= 2\sqrt{2} \left(\frac{1}{4} + \frac{1}{8} \sum_{n \geq 0,\ \text{even}} 2^{-n(r+\frac{1}{2})} \right) \\
&= \frac{1}{2\sqrt{2}} \left(2 + \frac{1}{1 - 2^{-2r-1}} \right),
\end{aligned}
$$

as claimed. The case $W_0(r, L^1)$ is a special case of [30], Section 8. ☐

Proposition 8.4.25. *Let $L = L^1$ and $X = 2^{-s}$. Then*

$$
\begin{aligned}
&\big(Z(s, \Phi_{L^1}, f_{w_j})(\mathbf{w}_i) \big)_{i,j=0,1} \\
&= \gamma(V) \frac{\operatorname{vol}(\mathbf{J}')}{4\sqrt{2}} L(s + \frac{1}{2}, \chi_V \mu) L(s + \frac{1}{2}, \chi_V \mu^{-1}) \\
&\times \begin{pmatrix} 2 & \frac{2\sqrt{2}}{\delta_\kappa}(1 + \sqrt{2}\,\chi_V \mu^{-1}(2) X - X^2) \\ \frac{\delta_\kappa}{\sqrt{2}}(1 + \sqrt{2}\,\chi_V \mu(2) X - X^2) & 3 - X^2 \end{pmatrix}.
\end{aligned}
$$

Proof. For convenience, let $\chi = \chi_V \mu$. By (8.4.10) and the previous lemmas, the zeta integral matrix, for $s = r$, is equal to

$$
\gamma(V) \operatorname{vol}(\mathbf{J}') \begin{pmatrix} I(r, \chi; \varphi_L, \varphi_L) & \frac{\sqrt{2}}{\delta_\kappa} I(r, \chi; \varphi_L, \varphi_{L^\#}) \\ \frac{\delta_\kappa}{2\sqrt{2}} I(r, \chi; \varphi_{L^\#}, \varphi_L) & \frac{1}{2} I(r, \chi; \varphi_{L^\#}, \varphi_{L^\#}) \end{pmatrix},
$$

where

$$
I(r, \chi; \varphi_{L_1}, \varphi_{L_2}) = \int_{\mathbb{Q}_2^\times} \chi(a) |a|^{r+\frac{1}{2}} B(r, a; \varphi_{L_1}, \varphi_{L_2})\,d^\times a.
$$

Now Lemma 8.4.4 gives

$$I(r,\chi;\varphi_L,\varphi_L) = \frac{L(r+\frac{1}{2},\chi)L(r+\frac{1}{2},\chi^{-1})}{\zeta(2r+1)} \cdot W_0(r,L)$$
$$= \frac{1}{2\sqrt{2}} L(r+\frac{1}{2},\chi)L(r+\frac{1}{2},\chi^{-1}),$$

and

$$I(r,\chi;\varphi_{L^\#},\varphi_{L^\#}) = \frac{L(r+\frac{1}{2},\chi)L(r+\frac{1}{2},\chi^{-1})}{\zeta(2r+1)} \cdot W_0(r,L^\#)$$
$$= \frac{1}{2\sqrt{2}} L(r+\frac{1}{2},\chi)L(r+\frac{1}{2},\chi^{-1}) \cdot (3 - X^2).$$

Lemma 8.4.4 also gives

$$\begin{aligned} &I(r,\chi;\varphi_L,\varphi_{L^\#}) \\ &= L(r+\frac{1}{2},\chi) \cdot W_0(r,L) \\ &\qquad + \chi^{-1}(2)\, 2^{-r-\frac{1}{2}} L(r+\frac{1}{2},\chi^{-1}) \cdot (\frac{1}{\sqrt{2}} + W_0(r,L)) \\ &= \frac{1}{2\sqrt{2}} L(r+\frac{1}{2},\chi)L(r+\frac{1}{2},\chi^{-1}) + \chi^{-1}(2)\, 2^{-1} X L(r+\frac{1}{2},\chi^{-1}) \\ &= \frac{1}{2\sqrt{2}} L(r+\frac{1}{2},\chi)L(r+\frac{1}{2},\chi^{-1}) \cdot (1 + \sqrt{2}\chi(2)^{-1}X - X^2). \end{aligned}$$

Switching the roles of L and $L^\#$, and of χ and χ^{-1}, by Lemma 8.4.4, one sees that

$$I(r,\chi;\varphi_{L^\#},\varphi_L) = \frac{1}{2\sqrt{2}} L(r+\frac{1}{2},\chi)L(r+\frac{1}{2},\chi^{-1}) \cdot (1+\sqrt{2}\chi(2)X - X^2).$$

□

Note that the matrix for $Z(s,\Phi^1,\cdot)$ is not diagonal with respect to the basis $\boldsymbol{f}_{\text{ev}}$ and $\boldsymbol{f}_{\text{sp}}$ of $I(\mu)^{(\mathbf{J}',\chi_\kappa)}$.

Now we consider the case $\mu = \chi_{2\alpha}|\,|^{\frac{1}{2}}$ with $\alpha \equiv \kappa \mod 4$, where the special representation occurs as the unique irreducible submodule of $I(\mu)$. In this case, $\chi_V\mu = \chi_{\alpha\kappa}|\,|^{\frac{1}{2}}$ is unramified. Then matrix in the right hand side of the previous proposition is

$$\begin{pmatrix} 2 & \frac{2\sqrt{2}}{\delta_\kappa}(1 + 2\chi_{\alpha\kappa}(2)X - X^2) \\ \frac{\delta_\kappa}{\sqrt{2}}(1 + \chi_{\alpha\kappa}(2)X - X^2) & 3 - X^2 \end{pmatrix},$$

and the function

$$\boldsymbol{f}_{\rm sp} = f_1 - \frac{\delta_\kappa}{2\sqrt{2}} f_w$$

is an eigenfunction with eigenvalue $(1 - \chi_{\alpha\kappa}(2)\, X)^2$. Thus

$$\begin{aligned} &Z(s, \Phi^1, \boldsymbol{f}_{\rm sp}) \\ &= \gamma(V)\, \frac{\mathrm{vol}(\mathbf{J}')}{4\sqrt{2}}\, L(s+\frac{1}{2}, \chi_V \mu) L(s+\frac{1}{2}, \chi_V \mu^{-1}) \cdot (1 - \chi_{\alpha\kappa}(2)\, 2^{-s})^2 \cdot \boldsymbol{f}_{\rm sp}. \end{aligned}$$

As in Section 8.4.4 above, for $\sigma = \tilde{\sigma}(\mu\chi_2) = \tilde{\sigma}(\chi_\alpha |\ |^{\frac{1}{2}})$, we have

$$L(s + \frac{1}{2}, \chi_V \mu) = L(s+1, \chi_{\alpha\kappa}) = L(s + \frac{1}{2}, \mathrm{Wald}(\sigma, \psi_\kappa)).$$

On the other hand,

$$L(s + \frac{1}{2}, \chi_V \mu^{-1}) = (1 - \chi_{\alpha\kappa}(2)\, 2^{\frac{1}{2}}\, 2^{-s-\frac{1}{2}})^{-1} = (1 - \chi_{\alpha\kappa}(2)\, 2^{-s})^{-1},$$

and $\delta(\sigma, \psi_\kappa) = -\chi_{2\alpha\kappa}(2)$. Also note that, for $\mathrm{vol}(\mathrm{Sp}_1(\mathbb{Z}_2)) = 1$, we have $\mathrm{vol}(\mathbf{J}') = \frac{1}{6}$. Thus

$$Z(s, \Phi^1, \boldsymbol{f}_{\rm sp}) = \delta_\kappa\, \frac{1}{24\sqrt{2}}\, L(s + \frac{1}{2}, \mathrm{Wald}(\sigma, \psi_\kappa)) \cdot (1 + \delta(\sigma, \psi_\kappa)\, 2^{-s}) \cdot \boldsymbol{f}_{\rm sp}.$$

In summary, we have obtained the following results.

Theorem 8.4.26. *(i) For an irreducible principal series representation* $\sigma = I(\mu) = \tilde{\pi}(\mu\chi_2)$ *with* $\mu = \chi_{2\alpha} |\ |^t$, $t \neq \pm\frac{1}{2}$ *and* $\alpha \equiv \kappa \mod 4$,

$$Z(s, \Phi^0, \boldsymbol{f}_{\rm ev}) = \delta_\kappa\, \frac{1}{2\sqrt{2}}\, \frac{L(s + \frac{1}{2}, \mathrm{Wald}(\sigma, \psi_\kappa))}{\zeta(2s+2)} \cdot \boldsymbol{f}_{\rm ev},$$

and $Z(s, \Phi^0, \boldsymbol{f}_{\rm sp}) = 0$.

(ii) For a special representation $\sigma = \tilde{\sigma}(\mu\chi_2) = \tilde{\sigma}(\chi_\alpha |\ |^{\frac{1}{2}})$ *with* $\mu = \chi_{2\alpha} |\ |^{\frac{1}{2}}$ *and* $\alpha \equiv \kappa \mod 4$,

$$\boldsymbol{f}_{\rm sp} = f_1 - \frac{\delta_\alpha}{2\sqrt{2}} f_w \in \sigma.$$

Then, $Z(s, \Phi^0, \boldsymbol{f}_{\rm sp}) = 0$,

$$\begin{aligned} &Z(s, \Phi^1, \boldsymbol{f}_{\rm sp}) \\ &= \delta_\kappa\, \frac{1}{2\sqrt{2}} \cdot 2^{-2}\, \frac{1}{3}\, (1 + \delta(\sigma, \psi_\kappa)\, 2^{-s}) \cdot L(s + \frac{1}{2}, \mathrm{Wald}(\sigma, \psi_\kappa)) \cdot \boldsymbol{f}_{\rm sp}, \end{aligned}$$

and

$$Z(s,\Phi^{ra},\boldsymbol{f}_{\mathrm{sp}}) = -\delta_\kappa\,\frac{1}{2\sqrt{2}}\cdot 2^{-2}\,(1-\delta(\sigma,\psi_\kappa)\,2^{-s})\cdot L(s+\frac{1}{2},\mathrm{Wald}(\sigma,\psi_\kappa))\cdot \boldsymbol{f}_{\mathrm{sp}}.$$

Note that, except for the factor $\delta_\kappa\,\frac{1}{2\sqrt{2}}$, these formulas are the same as those given in Corollary 8.4.6 and Corollary 8.4.8 for p odd. As in Corollary 8.4.9, we have

Corollary 8.4.27. *For $\boldsymbol{f}_{\mathrm{sp}}$ as in (ii) of Theorem 8.4.26, and $\widetilde{\Phi}$ as in* (8.4.14),

$$Z(s,\widetilde{\Phi},\boldsymbol{f}_{\mathrm{sp}}) = -\delta_\kappa\,\frac{1}{2\sqrt{2}}\,2^{-2}\Big[(1-\delta(\sigma,\psi_\kappa)\,2^{-s})-\frac{1}{3}\,(1+\delta(\sigma,\psi_\kappa)\,2^{-s})\,B(s)\Big] \times L(s+\frac{1}{2},\mathrm{Wald}(\sigma,\psi_\kappa))\cdot \boldsymbol{f}_{\mathrm{sp}}.$$

Finally, we consider the zeta integral for the Kohnen plus space section

$$\boldsymbol{f}^{+} = (1+4\chi_\kappa(-1)\mu(2)^2)\,\boldsymbol{f}_{\mathrm{ev}} + (2-4\chi_\kappa(-1)\mu(2)^2)\,\boldsymbol{f}_{\mathrm{sp}}, \tag{8.4.34}$$

where $\alpha\equiv\kappa \mod 4$. By Theorem 8.4.26 (i) for $\mu=\chi_{2\alpha}|\ |^t$ with $t\neq\frac{1}{2}$, and $\alpha\equiv\kappa \mod 4$,

$$Z(s,\Phi^0,\boldsymbol{f}^{+}) = \frac{\delta_\kappa}{2\sqrt{2}}\,(1+4\chi_\kappa(-1)\mu(2)^2)\,\frac{L(s+\frac{1}{2},\mathrm{Wald}(\sigma,\psi_\kappa))}{\zeta(2s+2)}\cdot \boldsymbol{f}_{\mathrm{ev}}. \tag{8.4.35}$$

8.4.6 The archimedean local integrals

In this section, we deal with the case $F=\mathbb{R}$. We only consider the discrete series representations $\sigma=\tilde{\pi}_\ell^+$ of $G'=G'_{\mathbb{R}}$, where $\ell\in\frac{1}{2}+\mathbb{Z}_{>0}$. Recall from Table 2 in Section 8.2 that $\mathrm{Wald}(\sigma,\psi)=\mathrm{DS}_{2\ell-1}$, the discrete series representation of $\mathrm{PGL}_2(\mathbb{R})$ of weight $2\ell-1$. Since $\mathrm{DS}_{2\ell-1}\otimes(\mathrm{sgn})\simeq \mathrm{DS}_{2\ell-1}$, we see that $\mathrm{Wald}(\sigma,\psi_{-1})=\mathrm{DS}_{2\ell-1}$ as well. We fix the additive character $\psi(x)=e(x)=e^{2\pi i x}$.

Let (V,Q) be a positive definite quadratic space over $\mathbb{R}$ of dimension 2ℓ, and note that $\chi_V=((-1)^{\ell-\frac{1}{2}},\ )_{\mathbb{R}}=(\mathrm{sgn})^{\ell-\frac{1}{2}}$. The group G' acts on $S(V)$ via the Weil representation $\omega_V=\omega_{V,\psi}$, and the Gaussian $\varphi_0=e^{-2\pi Q(x)}\in S(V)$ of V is a eigenfunction of weight ℓ for the action of K', the inverse image of $\mathrm{SO}(2)\subset\mathrm{SL}_2(\mathbb{R})=\mathrm{Sp}_1(\mathbb{R})$, i.e.,

$$\omega_V(k')\varphi_0=\xi_\ell(k')\varphi_0.$$

Then $f_\infty := \lambda_V(\varphi_0) \in I(\ell-1,\chi_V)$ is the lowest weight vector in the unique irreducible submodule of $I(\ell-1,\chi_V)$, and this submodule is isomorphic to the discrete series representation $\tilde{\pi}_\ell^+$. In what follows, we identify $\sigma = \tilde{\pi}_\ell^+$ with this submodule. Note that $f_\infty(1)=1$.

Let $I(s,\chi_V)$ be the degenerate principal series representation of $G = G_{\mathbb{R}}$ and let $\Phi_\infty^\ell(s) \in I(s,\chi_V)$ be the unique standard section[12] of weight ℓ, normalized so that $\Phi_\infty^\ell(1,s)=1$. Note that, for k_1' and $k_2' \in K'$,

$$r(i(k_1',k_2'))\Phi_\infty^\ell(s) = \xi_\ell(k_1')\,\overline{\xi_\ell(k_2')}\cdot\Phi_\infty^\ell(s).$$

Thus, as explained in section 8.4.1, the local doubling integral defines a map

$$Z(s,\Phi_\infty^\ell,\cdot) : \sigma^{(K',\xi_\ell)} \longrightarrow \sigma^{(K',\xi_\ell)},$$

which is given by multiplication by the scalar

$$h_\infty(s,\ell) = Z(s,\Phi_\infty^\ell,f_\infty)(1) = \int_{\mathrm{Sp}_1(\mathbb{R})} \Phi_\infty^\ell(\delta i(1,g'),s)\,f_\infty(g')\,dg,$$

which we now compute using the same 'interpolation' method as in the nonarchimedean case.

Let $V_r = V + V_{r,r}$ be the direct sum of V and r copies of standard hyperbolic plane. Note that V_r has signature $(2\ell+r,r)$. Decompose V_r orthogonally as $V_r = V_r^+ \oplus V_r^-$ with $V_r^+ \supset V$ positive definite and V_r^- negative definite. Then the function $\varphi_0^{(r)} \in S(V_r)$ defined by

$$\varphi_0^{(r)}(x) = e^{-2\pi Q(x_r^+)+2\pi Q(x_r^-)},$$

where $x = x_r^+ + x_r^-$, with $x_r^\pm \in V_r^\pm$, is a Gaussian for V_r , and

$$\Phi_\infty^\ell(s_0+r) = \lambda_{V_r}(\varphi_0^{(r)} \otimes \varphi_0^{(r)}), \tag{8.4.36}$$

where $s_0 = \ell - \frac{3}{2}$.

By Lemma 8.4.3, writing $g' = \mathbf{n}(b)\mathbf{m}(a)k'$, we have

$$\begin{aligned}
&\Phi_\infty^\ell(\delta i(1,g'),s_0+r) \\
&= \gamma(V)\int_{V_r} \varphi_0^{(r)}(x)\,\overline{\omega_{V_r}(g')\varphi_0^{(r)}(-x)}\,dx \\
&= \gamma(V)\,\overline{\xi_\ell(k')}\,\chi_V(a)\,|a|^{r+\ell}\int_{V_r} \varphi_0^{(r)}(x)\,\varphi_0^{(r)}(-ax)\,\psi(-b\,Q(x))\,dx \\
&= \gamma(V)\,\overline{\xi_l(k')}\,\chi_V(a)\,|a|^{r+\ell}\int_{V_r} \varphi_0^{(r)}(\sqrt{a^2+1}\,x)\,\psi(-b\,Q(x))\,dx.
\end{aligned}$$

[12]As usual, we let K be the full inverse image in G of the standard maximal compact subgroup $U(2)$ of $\mathrm{Sp}_2(\mathbb{R})$. By weight ℓ we mean transforming by the character $\det^\ell$ under the right action of K. Also note that $\chi_V = \chi_{2\kappa}$ where $\kappa = (-1)^{\ell-\frac{1}{2}}$.

Therefore

$$
\begin{aligned}
&h_\infty(s_0+r,\ell)\\
&= \gamma(V)\,\mathrm{vol}(\mathrm{SO}(2)) \int_0^\infty |a|^{r+2\ell-2}\\
&\qquad\qquad \times \int_{-\infty}^{\infty}\int_{V_r} \varphi_0^{(r)}(\sqrt{a^2+1}\,x)\,\psi(-b\,Q(x))\,dx\,db\,d^\times a\\
&= \mathrm{vol}(\mathrm{SO}(2)) \int_0^\infty a^{r+2\ell-2}(a^2+1)^{1-r-\ell}d^\times a\\
&\qquad\qquad \times \gamma(V)\int_{-\infty}^{\infty}\int_{V_r} \varphi_0^{(r)}(x)\,\psi(-b\,Q(x))\,dx\,db.
\end{aligned}
$$

It is well known that

$$
\int_0^\infty a^{r+2\ell-2}(a^2+1)^{1-r-\ell}d^\times a = \frac{\Gamma(\ell+\frac{r}{2}-1)\,\Gamma(\frac{r}{2})}{2\,\Gamma(\ell+r-1)}.
$$

On the other hand,[13] by [14], Proposition 14.1(iii),

$$
\begin{aligned}
&\gamma(V)\int_{-\infty}^{\infty}\int_{V_r} \varphi_0^{(r)}(x)\,\psi(-b\,Q(x))\,dx\,db\\
&\qquad\qquad = 2\pi\, e(-\tfrac{1}{4}\,(\ell-\tfrac{1}{2}))\cdot \frac{2^{1-\ell-r}\,\Gamma(\ell+r-1)}{\Gamma(\ell+\frac{r}{2})\,\Gamma(\frac{r}{2})}.
\end{aligned}
$$

Thus,

$$
h_\infty(s_0+r,\ell) = \mathrm{vol}(\mathrm{SO}(2))\cdot 2\pi\, e(-\tfrac{1}{4}\,(\ell-\tfrac{1}{2}))\cdot \frac{2^{1-\ell-r}}{(r+2\ell-2)}.
$$

In summary,

Proposition 8.4.28. *Let the notation be as above. Then*

$$
Z(s,\Phi_\infty^\ell,f_\infty) = h_\infty(s,\ell)\cdot f_\infty,
$$

with

$$
h_\infty(s,\ell) = \mathrm{vol}(\mathrm{SO}(2))\cdot 2\pi\, e(-\tfrac{1}{4}\,(\ell-\tfrac{1}{2}))\cdot \frac{1}{2^{s+\frac{1}{2}}\,(s+\ell-\frac{1}{2})}.
$$

---

[13]Here we must be careful since we are working with $[w,1]_{\mathrm{L}}$ rather than $[w,1]_{\mathrm{R}}$, which was used in [14]. This means that we must multiply the value in [14] by $e(\frac{1}{8})$.

Recall that $\text{Wald}(\tilde{\pi}_\ell^+, \psi) = \text{Wald}(\tilde{\pi}_\ell^+, \psi_-)$ is the discrete series representation $DS_{2\ell-1}$ of $PGL_2(\mathbb{R})$ of weight $2\ell - 1$, so that

$$L(s + \frac{1}{2}, \text{Wald}(\tilde{\pi}_\ell^+, \psi_\infty)) = \Gamma_{\mathbb{C}}(s + \ell - \frac{1}{2}) = 2\,(2\pi)^{\frac{1}{2}-\ell-s}\,\Gamma(s + \ell - \frac{1}{2}).$$

Since

$$\zeta_\infty(2s + 2) = \pi^{-1-s}\,\Gamma(s + 1),$$

we have

$$(8.4.37)\quad h_\infty(s, \ell) = \text{vol}(SO(2)) \cdot 2^{\ell-1}\,\pi^{\ell-\frac{1}{2}}\,e(-\tfrac{1}{4}\,(\ell-\tfrac{1}{2})) \times \frac{L(s + \frac{1}{2}, \text{Wald}(\sigma, \psi_\infty))}{\zeta_\infty(2s + 2)} \cdot \prod_{j=1}^{\ell-\frac{1}{2}} (s + j)^{-1}.$$

Since we are using Tamagawa measure on $Sp_1(\mathbb{A}) = SL_2(\mathbb{A})$, we have $\text{vol}(SL_2(\mathbb{Q})\backslash SL_2(\mathbb{A})) = 1$. In the nonarchimedean calculations, we have normalized the measures on $SL_2(\mathbb{Q}_p)$ by the condition $\text{vol}(SL_2(\mathbb{Z}_p), dg_p) = 1$, for all p. The measure on $SL_2(\mathbb{R})$ is then determined.

Lemma 8.4.29. $\text{vol}(SO(2)) = \frac{12}{\pi}$.

Proof. Recall that we have used, for $g = n(b)\,m(a)\,k$, with $a > 0$,

$$dg = a^{-3}\,da\,db\,dk.$$

By the remark above, we have

$$\begin{aligned} 1 &= \text{vol}(SL_2(\mathbb{Z})\backslash SL_2(\mathbb{R})) \\ &= \text{vol}(SL_2(\mathbb{Z})\backslash\mathfrak{H}, \frac{1}{2}\,y^{-2}\,dx\,dy)\,\frac{1}{2}\,\text{vol}(SO(2)) \\ &= \frac{\pi}{12} \cdot \text{vol}(SO(2)), \end{aligned}$$

since

$$\text{vol}(SL_2(\mathbb{Z})\backslash\mathfrak{H}, \frac{1}{2\pi}\,y^{-2}\,dx\,dy) = \frac{1}{6}. \quad \square$$

For the case $\ell = \frac{3}{2}$, this gives:

Corollary 8.4.30. *If $\sigma = \tilde{\pi}_{\frac{3}{2}}^+$ and f_∞ is the lowest weight vector in σ, then, with the notation above,*

$$\begin{aligned} Z(s, \Phi_\infty^{\frac{3}{2}}, f_\infty) &= -12\,\frac{\sqrt{2}\,i}{s+1} \cdot \frac{L(s + \frac{1}{2}, \text{Wald}(\sigma, \psi_{-1}))}{\zeta_\infty(2s + 2)} \cdot f_\infty \\ &= -12\,\sqrt{2}\,i \cdot \frac{1}{2^s(s + 1)} \cdot f_\infty. \end{aligned}$$

8.5 APPENDIX: COORDINATES ON METAPLECTIC GROUPS

In this appendix, we collect some information about the metaplectic extension, cocycles, coordinates, Weil representations, etc., which is used in this and other chapters of this book. References for this material include [19], [8], and [9]. However, none of these sources contains all of the facts we need, so we have provided some proofs.

8.5.1 Cocycles and coordinates

Let F be a nonarchimedean local field of characteristic zero with ring of integers $\mathcal{O}$ and uniformizer ϖ, and let ψ be an unramified additive character of F.[14] We change notation slightly and write $G = \mathrm{Sp}_n(F) = \mathrm{Sp}(W)$, where $W = F^{2n}$ with standard basis $e_1, \dots, e_n, f_1, \dots, f_n$ and with symplectic form given by $< e_i, e_j >=< f_i, f_j >= 0$ and $< e_i, f_j >= \delta_{ij}$. Let $W = X + Y$ be the complete polarization of W where X (resp. Y) is the span of the first (resp. last) n standard basis vectors. Let $P = P_Y$ be the stabilizer of Y in G, and let $K = \mathrm{Sp}_n(\mathcal{O}) = \mathrm{Sp}_n(F) \cap \mathrm{GL}_{2n}(\mathcal{O})$. For $0 \leq j \leq n$, let

$$w_j = \begin{pmatrix} 1_{n-j} & & & \\ & 0 & & 1_j \\ & & 1_{n-j} & \\ & -1_j & & 0 \end{pmatrix} \in K, \tag{8.5.1}$$

and let $w = w_n$.

Let $\tilde{G}$ be the metaplectic extension

$$1 \longrightarrow \mathbb{C}^1 \longrightarrow \tilde{G} \longrightarrow G \longrightarrow 1,$$

defined by $\tilde{G} = (\tilde{G}^{(2)} \times \mathbb{C}^1)/\{\pm 1\}$, where $\tilde{G}^{(2)}$ is the unique nontrivial twofold topological cover of $\mathrm{Sp}(W)$.

For each $g = \begin{pmatrix} a & b \\ c & d \end{pmatrix} \in G$, there is an operator $r(g)$ on the Schwartz space $S(F^n)$ defined by[15]

$$r(g)\varphi(x) = \int_{F^n/\ker(c)} \psi\Big(\frac{1}{2}(xa, xb) + (xb, yc) + \frac{1}{2}(yc, yd)\Big)\, \varphi(xa + yc)\, d_g(y), \tag{8.5.2}$$

---

[14]This means that ψ is trivial on $\mathcal{O}$ and nontrivial on $\varpi^{-1}\mathcal{O}$.

[15]In the corresponding formula in [9], the notation $< \, , \, >$ denotes the symplectic form on W and $x \in X$ and $y \in Y$. If we identify these elements with row vectors $x = [x_0, 0]$ and $y = [0, y_0]$ with x_0 and $y_0 \in F^n$ and write $g \in Sp(W)$ as a matrix, as above, then, the term $< xb, yc >=< [0, x_0 b], [y_0 c, 0] >= -(x_0 b, y_0 c)$. This accounts for the change in sign in the formula here.

where, for $x, y \in F^n$ (row vectors), $(x, y) = x^t y$, and the measure $d_g(y)$ on $F^n/\ker(c)$ is normalized to make this operator unitary [19], [29]. These operators defined a projective representation of G on $S(F^n)$ and

$$r(g_1)r(g_2) = c_{\mathrm{L}}(g_1, g_2)\, r(g_1 g_2), \tag{8.5.3}$$

where c_{L} is the Leray cocycle, given by [19] and [18]:

$$c_{\mathrm{L}}(g_1, g_2) = \gamma(\psi \circ q(g_1, g_2)). \tag{8.5.4}$$

Here

$$q(g_1, g_2) = \mathrm{Leray}(Yg_1, Y, Yg_2^{-1}) \tag{8.5.5}$$

is the Leray invariant of the triple of isotropic subspaces (Yg_1, Y, Yg_2^{-1}) and$\gamma(\psi \circ q)$ is the Weil index of the character of second degree $\psi \circ q$. We then obtain an isomorphism

$$G \times \mathbb{C}^1 \xrightarrow{\sim} \tilde{G}, \qquad (g, z) \mapsto [g, z]_{\mathrm{L}},$$

with product

$$[g_1, z_1]_{\mathrm{L}} \cdot [g_2, z_2]_{\mathrm{L}} = [g_1 g_2, z_1 z_2\, c_{\mathrm{L}}(g_1, g_2)]_{\mathrm{L}}.$$

The cocycle c_{L} is trivial on $G \times P$ and $P \times G$ but is *not* trivial on $K \times K$. Thus it must be modified for use in a global situation.

Note that

$$r(m(a))\varphi(x) = |\det(a)|^{\frac{1}{2}} \varphi(xa), \tag{8.5.6}$$

$$r(n(b))\varphi(x) = \psi(\frac{1}{2}(x, xb))\, \varphi(x), \tag{8.5.7}$$

and

$$r(w)\varphi(x) = \int_{F^n} \psi((x, y))\, \varphi(y)\, dy. \tag{8.5.8}$$

Similarly, $r(w_j)$ is given by the partial Fourier transform with respect to the last j coordinates.

Example 8.5.1 *In the case of* $n = 1$, *if* $g_i = \begin{pmatrix} a_i & b_i \\ c_i & d_i \end{pmatrix} \in \mathrm{Sp}_1(F)$ *with* $g_1 g_2 = g_3$, *then*

$$c_{\mathrm{L}}(g_1, g_2) = \gamma(\psi \circ \frac{1}{2} c_1 c_2 c_3).$$

In particular, if $c_1 c_2 c_3 = 0$, *then* $c_{\mathrm{L}}(g_1, g_2) = 1$.

Remark 8.5.2. The dependence of the Leray coordinates on ψ is given explicitly as follows:

$$[g, z]_{\mathrm{L},\psi_\alpha} = [g, z\,\xi(g, \alpha, \psi)]_{\mathrm{L},\psi},$$

where

$$\xi(g, \alpha, \psi) = \chi_\alpha(x(g))\,\chi_\alpha(2)^{j(g)}\,\gamma_F(\alpha, \psi)^{j(g)},$$

for $\chi_\alpha(x) = (x, \alpha)_F$. This can be checked easily using Proposition 4.3 in [9].

Suppose that the residue characteristic of F is odd and let $\varphi^0 \in S(F^n)$ be the characteristic function of $\mathcal{O}^n \subset F^n$. Then the formulas (8.5.6), (8.5.7) and (8.5.8) imply immediately that

$$r(k)\varphi^0 = \varphi^0 \qquad \text{if } k = m(a), n(b) \text{ or } w_j.$$

Since these elements generate K, it follows that φ^0 is an eigenfunction of $r(k)$ for all $k \in K$, and we define a function λ on K by

$$r(k)\varphi^0 = \lambda(k)^{-1}\varphi^0. \tag{8.5.9}$$

Note that $\lambda(m(a)) = \lambda(n(b)) = \lambda(w_j) = 1$ and that

$$c_{\mathrm{L}}(k_1, k_2)\,\lambda(k_1)\,\lambda(k_2)\,\lambda(k_1k_2)^{-1} = 1.$$

Thus, for $p = n(b)m(a) \in P \cap K$, we have

$$\lambda(p) = \lambda(n(b))\,\lambda(m(a))\,c_{\mathrm{L}}(n(b), m(a)) = 1.$$

In fact, λ is bi-invariant under $P \cap K$. For example,

$$\lambda(pk) = \lambda(p)\,\lambda(k)\,c_{\mathrm{L}}(p, k) = \lambda(k).$$

Thus, for any $g \in G$, writing $g = pk$ with $p \in P$ and $k \in K$, we may define

$$\lambda(g) = \lambda(pk) := \lambda(k). \tag{8.5.10}$$

This allows us to renormalize coordinates and define an isomorphism

$$G \times \mathbb{C}^1 \xrightarrow{\sim} \tilde{G}, \qquad (g, z) \mapsto [g, z] := [g, z\,\lambda(g)]_{\mathrm{L}}, \tag{8.5.11}$$

with cocycle

$$c(g_1, g_2) = c_{\mathrm{L}}(g_1, g_2)\,\lambda(g_1)\,\lambda(g_2)\,\lambda(g_1g_2)^{-1}.$$

By construction, this cocycle is trivial on $K \times K$, $P \times P$, and $P \times K$, via

$$c(p, k) = c_{\mathrm{L}}(p, k)\,\lambda(p)\,\lambda(k)\,\lambda(pk)^{-1} = \lambda(k)\,\lambda(k)^{-1} = 1.$$

We will refer to the coordinates given by (8.5.11) as *normalized coordinates*. Note that, for $p \in P$, $[p,1] = [p,1]_{\mathrm{L}}$ and $[w_j,1] = [w_j,1]_{\mathrm{L}}$, for example. We write $\mathbf{K}$ for the image of $K = \mathrm{Sp}_n(\mathcal{O})$ in $\tilde{G}$ under the splitting homomorphism

$$K \longrightarrow \tilde{G}, \qquad k \mapsto \mathbf{k} = [k,1] = [k,\lambda(k)]_{\mathrm{L}}.$$

Remark 8.5.3. We have, for $n_-(c) \in K$, i.e., for $c \in \mathrm{Sym}_n(\mathcal{O})$,

$$\lambda(n_-(c)) = c_L(n_-(c),w)^{-1}. \tag{8.5.12}$$

In fact, since $n_-(c)w = wn(-c)$, we have $\lambda(n_-(c)w) = \lambda(wn(-c)) = \lambda(w) = 1$, whereas

$$\lambda(n_-(c)w) = \lambda(n_-(c))\lambda(w)\,c_L(n_-(c),w).$$

In the case $n=1$, this gives the useful formulas

$$\lambda(n_-(c)) = \gamma(\psi_{-\frac{1}{2}c}), \tag{8.5.13}$$

and

$$\lambda(\begin{pmatrix} a & b \\ c & d \end{pmatrix}) = \begin{cases} \gamma(\psi_{-2cd}) & \text{if } c \neq 0 \text{ and } \mathrm{ord}(c) > 0, \\ 1 & \text{otherwise.} \end{cases} \tag{8.5.14}$$

8.5.2 The lifts of some homomorphisms

We now establish several facts used in the doubling calculation. We continue to work in greater generality than will be needed in this chapter and will use the notation $G' = \mathrm{Sp}_n(F)$ and $G = \mathrm{Sp}_{2n}(F)$ for the linear groups and $\tilde{G}'$ and $\tilde{G}$ for their metaplectic extensions. This differs slightly from the conventions of the rest of the chapter. There is no restriction on the residue characteristic of F.

First, recall the embedding $\underline{i}_0 : G' \times G' \to G$ given by the formula (8.1.1). Since the Leray cocycle is compatible with this embedding — this follows from the fact that $Y = Y \cap W_1 + Y \cap W_2$ for the decomposition $W = W_1 + W_2$ used to define $\underline{i}_0$ — there is a lift i_0 of $\underline{i}_0$ to $\tilde{G}' \times \tilde{G}' \to \tilde{G}$, given in Leray coordinates by

$$i_0 : [g_1,z_1]_{\mathrm{L}} \times [g_2,z_2]_{\mathrm{L}} \mapsto [i_0(g_1,g_2), z_1z_2]_{\mathrm{L}}.$$

The operators $r(g)$ are also compatible with $\underline{i}_0$, i.e.,

$$r(\underline{i}_0(g_1,g_2)) = r(g_1) \otimes r(g_2)$$

on $S(F^{2n}) \simeq S(F^n) \otimes S(F^n)$. In the case of odd residue characteristic, since $\varphi^0_{2n} = \varphi^0_n \otimes \varphi^0_n$, this implies that

$$\lambda(\underline{i}_0(k_1, k_2)) = \lambda(k_1)\lambda(k_2),$$

and hence

$$\lambda(\underline{i}_0(g_1, g_2)) = \lambda(g_1)\lambda(g_2).$$

Thus, in normalized coordinates,

$$i_0 : [g_1, z_1] \times [g_2, z_2] \mapsto [\underline{i}_0(g_1, g_2), z_1 z_2],$$

so that i_0 is compatible with the splittings homomorphisms of $K' \times K'$ and K.

Next, we recall the homomorphism

$$^\vee : G' \longrightarrow G', \qquad g^\vee = \mathrm{Ad}\begin{pmatrix} 1 & \\ & -1 \end{pmatrix} g$$

and the twisted embedding

$$\underline{i}(g_1, g_2) = \underline{i}_0(g_1, g_2^\vee).$$

Lemma 8.5.4. *The homomorphism* $^\vee : G' \to G'$ *has a lift to a homomorphism* $^\vee : \tilde{G}' \to \tilde{G}'$, *given in Leray coordinates by*

$$^\vee : [g, z]_{\mathrm{L}} \longmapsto [g^\vee, z^{-1}]_{\mathrm{L}}.$$

Proof. We note that

(8.5.15)

$$\begin{aligned} r(g^\vee)\bar{\varphi}(x) &= \int_{F^n/\ker(-c)} \psi\big(-\frac{1}{2}(xa, xb) + (xb, yc) - \frac{1}{2}(yc, yd)\big)\, \bar{\varphi}(xa - yc)\, d_g(y) \\ &= \int_{F^n/\ker(c)} \psi\big(-\frac{1}{2}(xa, xb) - (xb, yc) - \frac{1}{2}(yc, yd)\big)\, \bar{\varphi}(xa + yc)\, d_g(y) \\ &= \overline{r(g)\varphi(x)}. \end{aligned}$$

This implies that

$$c_{\mathrm{L}}(g_1^\vee, g_2^\vee) = c_{\mathrm{L}}(g_1, g_2)^{-1}.$$

□

Thus, the homomorphism

$$i : \tilde{G}' \times \tilde{G}' \longrightarrow \tilde{G}, \qquad i(g_1, g_2) = i_0\,(g_1, g_2^{\vee})$$

lifts $\underline{i}$.

In the case of odd residue characteristic, taking $\varphi = \varphi^0$ in (8.5.15), we have $r(k^{\vee})\varphi^0 = \lambda(k)\,\varphi^0$, so that

$$\lambda(k^{\vee}) = \lambda(k)^{-1} \qquad \text{and} \qquad \lambda(g^{\vee}) = \lambda(g)^{-1},$$

and we obtain

$$^{\vee} : \tilde{G}' \longrightarrow \tilde{G}', \qquad [g, z] \longmapsto [g^{\vee}, z^{-1}],$$

in normalized coordinates as well. In particular, $^{\vee}$ and i are compatible with the splitting homomorphisms on K' and $K' \times K'$ respectively.

We next prove (i) of Lemma 8.4.1. Recall that

$$\underline{\delta}\,\underline{i}(g, g) = p(g)\,\underline{\delta}. \tag{8.5.16}$$

Lemma 8.5.5.

$$[\underline{\delta}, 1]_{\mathrm{L}}[i(g, g), 1]_{\mathrm{L}} = [p(g), 1]_{\mathrm{L}}[\underline{\delta}, 1]_{\mathrm{L}}.$$

Proof. On the right side,

$$[p(g), 1]_{\mathrm{L}}[\underline{\delta}, 1]_{\mathrm{L}} = [p(g)\underline{\delta}, 1]_{\mathrm{L}},$$

since the cocycle c_{L} is trivial on $P \times G$, whereas on the left side,

$$[\underline{\delta}, 1]_{\mathrm{L}}[i(g, g), 1]_{\mathrm{L}} = [\underline{\delta}\, i(g, g), c_{\mathrm{L}}(\underline{\delta}, i(g, g))]_{\mathrm{L}}.$$

Thus, we must show that the cocycle $c_{\mathrm{L}}(\underline{\delta}, i(g, g))$ is trivial. Note that, by (8.5.16),

$$\mathrm{Leray}(Y\underline{\delta}, Y, Y i(g, g)^{-1}) = \mathrm{Leray}(Y\underline{\delta}, Y i(g, g), Y),$$

so that, as a function of g the cocycle $c_{\mathrm{L}}(\underline{\delta}, i(g, g))$ is bi-invariant under P'. Thus it suffices to compute it for $g = w_j$. To compute the Leray invariant, note that the isotropic $2n$-planes $Y\underline{\delta}$, Y and $Y i(g, g)^{-1}$ are spanned by the $2n$ rows of

$$\begin{pmatrix} -1_n & 1_n & 0 & 0 \\ 0 & 0 & 1_n & 1_n \end{pmatrix},$$

$$\begin{pmatrix} 0 & & 1_n & \\ & 0 & & 1_n \end{pmatrix},$$

and

$$\begin{pmatrix} -{}^t c & & {}^t a & \\ & {}^t c & & {}^t a \end{pmatrix},$$

for

$$a = \begin{pmatrix} 1_{n-j} & \\ & 0 \end{pmatrix}, \qquad c = \begin{pmatrix} 0 & \\ & -1_j \end{pmatrix},$$

respectively. Thus $Y\underline{\delta} \cap Y$ is spanned by the rows of

$$\begin{pmatrix} 0 & 0 & 1_n & 1_n \end{pmatrix},$$

$Y\underline{\delta} \cap Yi(g,g)^{-1}$ is spanned by the rows of

$$\begin{pmatrix} -{}^t c & {}^t c & {}^t a & {}^t a \end{pmatrix},$$

and $Y \cap Yi(g,g)^{-1}$ is spanned by the rows of

$$\begin{pmatrix} 0 & 0 & {}^t a & 0 \\ 0 & 0 & 0 & {}^t a \end{pmatrix}.$$

Thus

$$R = Y\underline{\delta} \cap Y + Y \cap Yi(g,g)^{-1} + Y\underline{\delta} \cap Yi(g,g)^{-1}$$

has rank $2n$ and $q(\underline{\delta}, i(g,g)) = 0$, as required. □

Since any choice of δ in the metaplectic extension of $\mathrm{Sp}_{2n}(F)$ has the form $\delta = [\underline{\delta}, z]_{\mathrm{L}} = [\underline{\delta}, 1]_{\mathrm{L}} \cdot [1, z]_{\mathrm{L}}$ and since the element $[1, z]_{\mathrm{L}}$ is central, statement (i) of Lemma 8.4.1 is clear.

8.5.3 The Weil representation

We now return to the notation in Section 8.5.1, so that $\tilde{G}$ is the metaplectic extension of $\mathrm{Sp}_n(F)$, for example. We consider the Weil representation $(\omega_V, S(V^n))$ of $\tilde{G}$ associated to a quadratic space (V, Q). Here, as before, ψ has been fixed and $\eta = \psi_{\frac{1}{2}}$. In [9], Proposition 4.3, this representation is described in the *Rao coordinates* which are defined as follows:

$$G \times \mathbb{C}^1 \xrightarrow{\sim} \tilde{G}, \qquad (g, z) \mapsto [g, z]_{\mathrm{R}},$$

where

$$[g, z]_{\mathrm{R}} = [g, z\beta(g)]_{\mathrm{L}},$$

for

$$\beta(g) = \gamma(x(g), \eta)^{-1}\, \gamma(\eta)^{-j(g)}. \tag{8.5.17}$$

Here the notation is as in [19] and [8]. Multiplication is then

$$[g_1, z_1]_{\mathrm{R}} \cdot [g_2, z_2]_{\mathrm{R}} = [g_1 g_2, z_1 z_2\, c_{\mathrm{R}}(g_1, g_2)]_{\mathrm{R}}.$$

Note that, since the Rao cocycle c_{R} is valued in $\{\pm 1\}$, there is a character

$$\zeta : \tilde{G} \longrightarrow \mathbb{C}^1, \qquad [g, z]_{\mathrm{R}} \mapsto z^2. \tag{8.5.18}$$

In Leray coordinates, this character becomes

$$\zeta([g, z]_{\mathrm{L}}) = z^2 \beta(g)^{-2}. \tag{8.5.19}$$

A short calculation using Proposition 4.3 of [9] yields the following formula for the Weil representation in Leray coordinates.

Lemma 8.5.6. *Let* $m = \dim_F(V)$. *Then*

$$\omega_V([g, z]_{\mathrm{L}})\varphi(x) = \chi_V(x(g))\,(\, z\, \gamma(\eta)^{j(g)}\,)^{\bullet}\, \gamma(\eta \circ V)^{-j(g)} \cdot r_V(g)\varphi(x),$$

where

$$\bullet = \begin{cases} 1 & \textit{if } m \textit{ is odd,} \\ 0 & \textit{if } m \textit{ is even,} \end{cases}$$

and

$$r_V(g)\varphi(x) = \int_{V^n/\mathrm{ker}(c)} \psi(\mathrm{tr}(\,\tfrac{1}{2}(xa, xb) + (xb, yc) + \tfrac{1}{2}(yc, yd)\,))\, \varphi(xa + yc)\, d_g(y).$$

Here, for $x,\ y \in V^n$, $(x, y) = ((x_i, y_j))$, *and the measure* $d_g(y)$ *on the quotient* $V^n/\mathrm{ker}(c)$ *is normalized to make the operator* $r_V(g)$ *unitary. Also,*

$$\chi_V(x) = (x, (-1)^{\frac{m(m-1)}{2}} \det(V))_F,$$

and $\det(V)$ *is the determinant of the matrix of the bilinear form on* V. *In particular*

$$\omega_V([w, 1]_{\mathrm{L}})\varphi(x) = \gamma(V)^n \int_{V^n} \psi(\mathrm{tr}(x, y))\, \varphi(y)\, dy, \tag{8.5.20}$$

where

$$\gamma(V) = \chi_V(-1)\, \gamma(\eta)\, \gamma(\eta \circ V)^{-1}, \tag{8.5.21}$$

and

$$\gamma(\eta \circ V) = \gamma(\det V, \eta)\, \gamma(\eta)^m\, \epsilon(V).$$

Remark 8.5.7. In the formulas of Lemma 8.5.6, we should write $\omega_{V,\psi}$, $[g,z]_{\mathrm{L},\psi}$ and $r_{V,\psi}$ to indicate the dependence on the additive character ψ. The following useful relation is easy to check:

$$\omega_{V_\alpha,\psi_{\alpha^{-1}}}([g,z]_{\mathrm{L},\psi}) = \omega_{V,\psi}([g,z]_{\mathrm{L},\psi}). \tag{8.5.22}$$

Since, in general, we will work with a fixed additive character ψ, we will suppress it from the notation and write simply ω_V.

If $\dim(V)$ is odd, we have

$$\omega_V([n(b)m(a),z]_{\mathrm{L}})\varphi(0) = z\, \chi_V(\det a)\, |\det a|^{\frac{m}{2}}\, \varphi(0),$$

so that we obtain a $\tilde{G}$-intertwining map

$$\lambda_V : S(V^n) \longrightarrow I(s_0, \chi_V), \qquad \varphi \mapsto (g \mapsto \omega(g)\varphi(0),$$

where $s_0 = \frac{m}{2} - \frac{n+1}{2}$. We note several facts needed in this chapter.

Lemma 8.5.8.

$$\omega_V([g^\vee, z^{-1}]_{\mathrm{L}})\overline{\varphi}(x) = \overline{\omega_V([g,z]_{\mathrm{L}})\varphi(x)}.$$

Proof. Note that $x(g^\vee) = (-1)^{j(g)}x(g)$ and that

$$r_V(g^\vee)\overline{\varphi}(x) = \overline{r_V(g)\varphi(x)},$$

by the same argument as for r; see (8.5.15). The claimed formula then follows from the easily checked identity

$$\chi_V(-1)\, \gamma(\eta)^2\, \gamma(\eta \circ V)^{-2} = 1. \quad \square$$

Recall that the discriminant of V is given by

$$\mathrm{discr}(V) = (-1)^{\frac{m(m-1)}{2}} \det(V) \in F^\times/F^{\times,2}, \tag{8.5.23}$$

where $\det(V) = \det((v_i, v_j))$, for any basis $\{v_i\}$ of V. Suppose that $V = V_1 + V_2$ is an orthogonal direct sum of quadratic spaces V_i with $\dim_F(V_i) = m_i$. Then

$$\begin{aligned} r_V(g) &= r_{V_1}(g) \otimes r_{V_2}(g), \\ \gamma(\psi \circ V) &= \gamma(\psi \circ V_1) \cdot \gamma(\psi \circ V_2), \\ \mathrm{discr}(V) &= \mathrm{discr}(V_1) \cdot \mathrm{discr}(V_2) \cdot (-1)^{m_1 m_2}, \end{aligned}$$

and

$$\chi_V(x) = \chi_{V_1}(x) \cdot \chi_{V_2}(x) \cdot (x, -1)_F^{m_1 m_2}.$$

It follows that

Lemma 8.5.9. *The Weil representation ω_V of $\tilde{G}$ on $S(V^n) \simeq S(V_1^n) \otimes S(V_2^n)$ is given by*

$$\omega_V(g) = \omega_{V_1}(g) \otimes \omega_{V_2}(g) \cdot \begin{cases} \zeta(g)^{-1} & \text{if } m_1 m_2 \text{ is odd,} \\ 1 & \text{otherwise.} \end{cases}$$

The extra factor $\zeta(g)^{-1}$ in the case where both V_1 and V_2 are odd-dimensional is due to the fact that, in our definition of ω_V, [9], we have 'twisted' the naturally defined Weil representation by a power of ζ in order to obtain trivial central character in the even-dimensional case and central character $[1, z]_{\mathrm{L}} \mapsto z$ in the odd-dimensional case. Here we have used the identity

$$(x(g), -1)_F \, z^2 \, \gamma(\eta)^{2j(g)} = \zeta([g, z]_{\mathrm{L}}).$$

We finish with two examples.

First, suppose that $V = V_0 = F$ with quadratic form $Q(x) = \frac{1}{2}x^2$. Then the operator $r_V(g)$ reduces to the operator $r(g)$ on $S(F^n)$. On the other hand, $\det(V) = \mathrm{discr}(V) = 1$ and $\gamma(\eta \circ V) = \gamma(\eta)$, so that, in the odd residue characteristic case and in normalized coordinates,

$$\begin{aligned} \omega_V([g, z]) &= z\,\lambda(g)\,\gamma(\eta)^{j(g)}\gamma(\eta \circ V)^{-j(g)}\,r_V(g) \\ &= z\,\lambda(g)\,r(g). \end{aligned}$$

It follows that the function $\varphi^0 \in S(F^n)$ is invariant under $\mathbf{K}$, and so

$$g \mapsto \omega_V(g)\varphi^0(0)$$

is the $\mathbf{K}$-fixed vector in the induced representation $I(-\frac{n}{2}, \chi_1)$, where χ_1 is the trivial character.

Next, suppose that $V = V_{r,r}$, where the matrix for the bilinear form is

$$\begin{pmatrix} & 1_r \\ 1_r & \end{pmatrix}.$$

Then $\mathrm{discr}(V) = 1$, $\chi_V = \chi_0 = 1$, and $\gamma(\eta \circ V) = 1$. Thus

$$\omega_V([g, z]_{\mathrm{L}}) = r_V(g),$$

so the operators $r_V(g)$ give a representation of $\mathrm{Sp}_n(F)$. If φ_r^0 is the characteristic function of $M_{2r,n}(\mathcal{O}) \subset V^n$, then, for ψ unramified, φ_r^0 is invariant under the generators of K and hence under all of K. Thus, we again have

$$\Phi^0(g, r - \frac{n+1}{2}) = \omega_V(g)\varphi_r^0(0), \tag{8.5.24}$$

where $\Phi^0(g,s)$ is the K-fixed vector in $I(s,\chi_0)$, the induced representation of $\mathrm{Sp}_n(F)$. Note that even residue characteristic is allowed here.

Taking $V_r = V + V_{r,r}$ and using Lemma 8.5.9, we have

$$\omega_{V_r}(g)(\varphi \otimes \varphi_r^0)(0) = \Phi(g, s_0 + r), \tag{8.5.25}$$

where $\Phi(s)$ is the extension to a standard section of the image $\lambda_V(\varphi)$ of $\varphi \in S(V^n)$ in $I(s_0, \chi_V)$, with $s_0 = \frac{m}{2} - \frac{n+1}{2}$. For $V = V_0$, $\varphi = \varphi^0$, and odd residue characteristic, $\Phi(s) = \Phi^0(s)$ is the unique $\mathbf{K}$-fixed vector in $I(r - \frac{n}{2}, \chi_0)$ with $\Phi^0(1,s) = 1$. The following useful fact is easy to check. We refer to Lemma 8.5.14 for the case of even residue characteristic.

Lemma 8.5.10. *Assume that the residue characteristic of F is odd. Let $L \subset V$ be a lattice such that $\varpi^r L^\sharp \subset L \subset L^\sharp$, where*

$$L^\sharp = \{\, x \in V \mid (x,y) \in \mathcal{O},\ \forall y \in L \,\}$$

is the dual lattice and $r > 0$. Let

$$J_r = \{ \begin{pmatrix} a & b \\ c & d \end{pmatrix} \in K \mid c \equiv 0 \mod \varpi^r \mathcal{O} \},$$

and let $\mathbf{J}_r \subset \mathbf{K}$ be the corresponding subgroup. Then the characteristic function $\varphi_L \in S(V^n)$ of $L^n \subset V^n$ is an eigenfunction of $\mathbf{J}_r$ with

$$\omega_V(\mathbf{k})\,\varphi_L = \chi_V(\det(a))\,\varphi_L, \qquad \text{for all } \mathbf{k} \in \mathbf{J}_r.$$

Finally, for a quaternion algebra B over $\mathbb{Q}_p$, we take $V = \{\, x \in B \mid \mathrm{tr}(x) = 0 \}$ with $Q(x) = \nu(x) = -x^2$. Let ψ_p be the local component of the global unramified character ψ of $\mathbb{A}/\mathbb{Q}$ with $\psi_\infty(x) = e(x)$. Then $\chi_V(a) = (a, -2)_p$,

$$\gamma_p(\eta \circ V) = \mathrm{inv}_p(B) \cdot \begin{cases} e(\frac{1}{8}) & \text{if } p = 2, \\ e(-\frac{1}{8}) & \text{if } p = \infty, \\ 1 & \text{otherwise,} \end{cases}$$

and

$$\gamma_p(V) = \mathrm{inv}_p(B) \cdot \begin{cases} i & \text{if } p = 2, \\ -i & \text{if } p = \infty, \\ 1 & \text{otherwise.} \end{cases}$$

Note that $\gamma_p(\eta) = 1$ for $p \neq 2, \infty$, $\gamma_2(\eta) = e(-\frac{1}{8})$, and $\gamma_\infty(\eta) = e(\frac{1}{8})$. Recall that $\gamma(\eta \circ V)^{-1}$ is computed in Lemma 13.3 of [14].

8.5.4 The case $p = 2$

Almost everything in the previous sections is valid for any residue characteristic, except for the results involving the splitting homomorphism over K where p was required to be odd. In this section, we discuss the changes needed when $p = 2$. Let F be a 2-adic local field with ring of integers $\mathcal{O}$ and a uniformizer ϖ. Let ψ be a unramified additive character of F.

Let $G = \mathrm{Sp}_n(F)$, $K = \mathrm{Sp}_n(\mathcal{O})$, and $J = \{\gamma \in K \mid c \equiv 0 \mod (4)\}$. We can define a splitting $J \to \tilde{G}$ as follows. Let $V = F$ with $Q(x) = x^2$, so that $(x, y) = 2xy$. Then $\det(V) = 2$, $\chi_V(a) = (a, 2)_F = \chi_2(a)$, and

$$\gamma(\eta \circ V) = \gamma(2, \eta)\gamma(\eta).$$

For $L = \mathcal{O}$, the dual lattice is $L^\sharp = \frac{1}{2} L$, and $|L^\sharp : L| = |\mathcal{O}/2\mathcal{O}|$. Then the function $\varphi_L = \mathrm{char}(L)^n \in S(F^n)$ is an eigenvector for the operators $\omega_V([m(a), 1]_L)$, $\omega_V([n(b), 1]_L)$, and $\omega_V([n_-(c), 1]_L)$, for $c \equiv 0 \mod (4)$. Explicitly,

Lemma 8.5.11. *For an integer $0 \leq r \leq n$, let w_r be as in (8.5.1). Then*

$$\omega_V([m(a), 1]_L)\varphi_L = \chi_2(\det a)\, \varphi_L,$$
$$\omega_V([n(b), 1]_L)\varphi_L = \varphi_L,$$
$$\omega_V([w_r, 1]_L)\varphi_L = |2|^{\frac{r}{2}}\, \gamma(2, \eta)^{-r} \cdot \varphi_{L_r},$$
$$\omega_V([w_r, 1]_L)\varphi_{L_r} = |2|^{-\frac{r}{2}}\, \gamma(2, \eta)^{-r} \cdot \varphi_L,$$

and

$$\omega_V([n_-(c), 1]_L)\varphi_L = c_L(n_-(c), w) \cdot \varphi_L.$$

Proof. Notice that $w = w_n$ and $\varphi_{L^\sharp} = \varphi_{L_n}$.

We prove the last formula. Note that $n_-(c)w = wn(-c)$. Setting

$$* = |2|^{\frac{n}{2}}\, \gamma(2, \eta)^n,$$

we have

$$\begin{aligned}
\omega_V([n_-(c), 1]_L)\varphi_L &= (*) \cdot \omega_V([n_-(c), 1]_L)\, \omega_V([w, 1]_L)\varphi_{L_n} \\
&= (*)\, c_L(n_-(c), w) \cdot \omega_V([n_-(c)\, w, 1]_L)\varphi_{L_n} \\
&= (*)\, c_L(n_-(c), w) \cdot \omega_V([w, 1]_L)\, \omega_V([n(-c), 1]_{\mathrm{L}})\varphi_{L_n} \\
&= c_L(n_-(c), w) \cdot \varphi_L.
\end{aligned}$$

Here, in the last step, we have used the fact that, for $x \in L^\sharp$, $Q(x) \in \frac{1}{4}\,\mathcal{O}$, so that $\omega_V([n(-c), 1]_{\mathrm{L}})\varphi_{L^\sharp} = \varphi_{L^\sharp}$, since $c \equiv 0 \mod (4)$. $\square$

Every element of J can be written uniquely in the form

$$k = n(b)m(a)n_-(c), \tag{8.5.26}$$

for $a \in \mathrm{GL}_n(\mathcal{O})$, $b \in \mathrm{Sym}_n(\mathcal{O})$, $c \in 4\,\mathrm{Sym}_n(\mathcal{O})$, and

$$\omega_V([k,1]_{\mathrm{L}})\varphi_L = \lambda_2(k)^{-1}\,\varphi_L,$$

with

$$\lambda_2(k) = \chi_2(\det a)\, c_L(n_-(c), w)^{-1}. \tag{8.5.27}$$

Thus, we obtain a splitting homomorphism

$$J \to \tilde{G}, \quad k \mapsto \mathbf{k} = [k, \lambda_2(k)]_L, \tag{8.5.28}$$

where $k \in J$. Let $\mathbf{J} \subset \tilde{G}$ be the image of J under this homomorphism. By construction, φ_L is invariant under $\mathbf{J}$.

Remark 8.5.12. In the case $n = 1$,

$$\lambda_2(\begin{pmatrix} a & b \\ c & d \end{pmatrix}) = \chi_2(a)\,\gamma(\psi_{-2cd}).$$

In Section 4.5.2, we will need some additional cocycle information. For example, note that

$$wn(b)m(a)n_-(c)w^{-1} = n_-(-b)m(a^{-1})n(-c) \in \tilde{P} \quad \iff \quad b = 0,$$

and that $[w,1]_{\mathrm{L}}[m(a), \chi_2(a)]_{\mathrm{L}} = [m(a), \chi_2(a)]_{\mathrm{L}}[w,1]_{\mathrm{L}}$.

Lemma 8.5.13. *(i)*

$$[w,1]_L[n_-(c), \lambda_2(n_-(c))]_L = [n(-c),1]_L[w,1]_L.$$

(ii) For $k = n(b)m(a)n_-(c) \in J$ with $a^2 = (1+2b)(1-2c_0)$,

$$\begin{aligned}&[n_-(2),1]_L[n(b)m(a)n_-(c), \lambda_2(k)]_L \\ &\qquad = [n(b(1+2b)^{-1})\, m(a^{-1}(1-2c_0)), *]_L[n_-(2),1]_L,\end{aligned}$$

where

$$* = c_L(n_-(2), k) \cdot \lambda_2(k) = \gamma(1-2c_0, \psi).$$

Proof. First consider (i). Since the Leray cocycle satisfies $c_L(P, G) = 1 = c_L(G, P)$, we have

$$\begin{aligned}[w,1]_L[n_-(c),\lambda_2(n_-(c))]_L &= [wn_-(c), c_L(w,n_-(c))\lambda_2(n_-(c))]_L \\ &= [n(-c)w, c_L(w,n_-(c))\lambda_2(n_-(c))]_L \\ &= [n(-c), c_L(w,n_-(c))\lambda_2(n_-(c))]_L[w,1]_L \\ &= [n(-c),1]_L[w,1]_L,\end{aligned}$$

since $c_L(w, n_-(c))\lambda_2(n_-(c)) = 1$ by definition of λ_2

For (ii), we take $k = n(b)m(a)n_-(c) \in J$ with

$$a^2 = (1+2b)(1-2c_0). \tag{8.5.29}$$

Then

$$\begin{aligned}&[n_-(2),1]_L[n(b)m(a)n_-(c),\lambda_2(k)]_L \\ &\qquad = [n(b(1+2b)^{-1})\,m(a^{-1}(1-2c_0)),*]_L[n_-(2),1]_L,\end{aligned}$$

where $* = c_L(n_-(2), k) \cdot \lambda_2(k)$. When equation (8.5.29) holds, then

$$2a + a^{-1}(1+2b)c = 2a^{-1}(1+2b),$$

so that

$$2 \cdot a^{-1}c \cdot 2a^{-1}(1+2b) = 4a^{-2}c(1+2b)$$

and

$$c_L(n_-(2), n(b)m(a)n_-(c)) = \gamma(\psi_{2c(1+2b)}),$$

where we use the fact that $\gamma(\psi_{\alpha^2\beta}) = \gamma(\psi_\beta)$. Now

$$\begin{aligned}* &= \gamma(\psi_{2c(1+2b)}) \cdot \gamma(\psi_{-2c}) \\ &= \gamma(\psi_{-2c})\gamma(2c(1+2b),\psi)\gamma(\psi) \\ &= \gamma(\psi_{-2c})\gamma(2c,\psi)\gamma(\psi)(2c,1+2b)\gamma(1+2b,\psi) \\ &= (2c_0, 1-2c_0)\gamma(1+2b,\psi) \\ &= \gamma(1-2c_0,\psi),\end{aligned}$$

where (8.5.29) was used in the last two steps. □

Finally, we consider the Weil representation and record the analogue of Lemma 8.5.10 in the $p = 2$ case.

Proposition 8.5.14. *Supppose that (V, Q) is a quadratic space over F and that $L \subset V$ is a $\mathcal{O}$-lattice such that, for some $r \geq 0$,*

$$2\varpi^r L^\sharp \subset L \subset L^\sharp,$$

where $L^\sharp$ is the dual lattice. Let

$$J_r = \{ k \in K = \mathrm{Sp}_n(\mathcal{O}) \mid c \equiv 0 \mod (4\varpi^r) \},$$

and let $\mathbf{J}_r$ be the corresponding subgroup of $\mathbf{J}$. Let $\varphi_L \in S(V^n)$ be the characteristic function of $L^n \subset V^n$. Then φ_L is an eigenfunction for $\mathbf{J}_r$ with eigencharacter $\chi_V\chi_2$, i.e.,

$$\omega_V(\mathbf{k})\varphi_L = \chi_V\chi_2(\det a)\cdot\varphi_L, \qquad \mathbf{k} = [k, \lambda_2(k)], \quad k = \begin{pmatrix} a & b \\ c & d \end{pmatrix} \in J_r.$$

Proof. It is enough to check on the generators, (8.5.26), of $\mathbf{J}_r$. We have

$$\omega_V([m(a), \chi_2(\det a)]_{\mathrm{L}}]\varphi_L = \chi_V\chi_2(\det a)\,\varphi_L,$$

and

$$\omega_V([n(b), 1]_{\mathrm{L}})\varphi_L = \varphi_L.$$

Since, for $x \in L^\sharp$,

$$4\varpi^r Q(x) = (2\varpi^r x, x) \in \mathcal{O},$$

we have

$$\omega_V([n(-c), 1]_L)\varphi_{L^\sharp} = \varphi_{L^\sharp}$$

for $c \equiv 0 \mod (4\varpi^r)$. The same argument as before then yields

$$\omega_V([n_-(c), \lambda(n_-(c))]_L)\varphi_L = \lambda(n_-(c))\,c_L(n_-(c), w)\varphi_L = \varphi_L. \qquad \square$$

8.5.5 The global metaplectic group

We now turn to the cocycle for the global metaplectic group. Recall that ψ is our fixed unramified character of $\mathbb{A}/\mathbb{Q}$ with $\psi_\infty(x) = e(x)$. Let $\tilde{G}_{\mathbb{A}}$ be the metaplectic cover of $G(\mathbb{A}) = \mathrm{Sp}_n(\mathbb{A})$. For p odd, we have the normalized coordinate system for $\tilde{G}_p$, given by (8.5.11),

$$(8.5.30) \qquad G_p \times \mathbb{C}^1 \to \tilde{G}_p, \quad (g, z) \mapsto [g, z]_p = [g, z\lambda_p(g)]_L,$$

and the associated cocycle $c_p^0(\cdot,\cdot)$ is trivial on $K_p \times K_p$. For $g_1, g_2 \in G(\mathbb{A})$, the global cocycle

$$(8.5.31) \qquad c(g_1, g_2) = c_L(g_{1,\infty}, g_{2,\infty})\,c_L(g_{1,2}, g_{2,2}) \prod_{p \neq 2,\infty} c_p^0(g_{1,p}, g_{2,p})$$

is then well defined and gives a normalized coordinate system for $\tilde{G}_{\mathbb{A}}$,[16]
(8.5.32)
$$G(\mathbb{A}) \times \mathbb{C}^1 \to \tilde{G}_{\mathbb{A}}, \quad (g, z) \mapsto [g, z] = [g_\infty, z]_{\mathrm{L},\infty}[g_2, 1]_{\mathrm{L},2} \prod_{p \neq 2, \infty} [g_p, 1]_p.$$

Let

$$K_0(4) = \{g = \begin{pmatrix} a & b \\ c & d \end{pmatrix} \in \mathrm{Sp}_n(\hat{\mathbb{Z}}) \mid c \equiv 0 \mod 4\} = K_0(4)_2 \times \prod_{p \neq 2} K_p.$$

Then there is a splitting

$$K_0(4) \to \tilde{G}_{\mathbb{A}}, \qquad k \mapsto [k, \lambda_2(k)], \tag{8.5.33}$$

where $\lambda_2(k)$ is given by (8.5.27). Let $\mathbf{K}_0(4)$ be the image of $K_0(4)$ in $\tilde{G}_{\mathbb{A}}$. On the other hand, there is a unique splitting homomorphism of $G(\mathbb{Q})$ into $\tilde{G}_{\mathbb{A}}$, whose image we denote by $G_{\mathbb{Q}}$. In terms of the normalized coordinates, this splitting has the following description.

Lemma 8.5.15. *(i) For $g \in G(\mathbb{Q})$, $\lambda_p(g) = 1$ for almost all p. In particular,*

$$\lambda(g) := \prod_{p \neq 2, \infty} \lambda_p(g)$$

is well defined. Moreover,

$$c_p^0(g_1, g_2) = c_{\mathrm{L},p}(g_1, g_2) = 1$$

for almost all p.
(ii) For $g_1, g_2 \in G(\mathbb{Q})$, there is a product formula

$$\prod_{p \leq \infty} c_{\mathrm{L},p}(g_1, g_2) = 1.$$

(iii) The splitting homomorphism $s : G(\mathbb{Q}) \to \tilde{G}_{\mathbb{A}}$ is given by

$$s(g) = [g, \lambda(g)^{-1}].$$

Proof. To prove (i), we write $g = p_1 w_r p_2$ with $p_1, p_2 \in P_{\mathbb{Q}}$, the Siegel parabolic subgroup of $G_{\mathbb{Q}}$. We have p_1, $p_2 \in K_p \cap P_{\mathbb{Q}}$, for almost all primes p. Since λ_p is bi-invariant under $K_p \cap P_p$, we then have

$$\lambda_p(g) = \lambda_p(p_1 w_r p_2) = \lambda_p(w_r) = 1.$$

If $\lambda_p(g_1) = \lambda_p(g_2) = 1$, then $c_p^0(g_1, g_2) = c_{\mathrm{L},p}(g_1, g_2)$. On the other hand, $c_p^0(g_1, g_2) = 1$ whenever g_1 and g_2 are both in K_p and $p \neq 2$. This give the

[16]Here, in the last expression, we are using the fact that $\tilde{G}_p$ maps to $\tilde{G}_{\mathbb{A}}$ for every $p \leq \infty$.

second statement in (i). To prove (ii), note that the product is well defined by the second statement in (i). Now, for $g_1, g_2 \in G_{\mathbb{Q}}$, there is a global quadratic form

$$Q = \mathrm{Leray}(Yg_1, Y, Yg_2^{-1}),$$

and the product formula

$$\prod_{p \le \infty} \gamma_p(\psi \circ Q) = 1$$

for the Weil indices implies (ii).
To prove (iii), it suffices to check

$$c(g_1, g_2) = \lambda(g_1)\lambda(g_2)\lambda(g_1 g_2)^{-1}.$$

By (i), (ii), and the definitions, we have

$$\begin{aligned} c(g_1, g_2) &= \prod_{p=2,\,\infty} c_{\mathrm{L},p}(g_1, g_2) \prod_{p \ne 2,\,\infty} c_{\mathrm{L},p}(g_1, g_2)\lambda_p(g_1)\lambda_p(g_2)\lambda_p(g_1 g_2)^{-1} \\ &= \lambda(g_1)\lambda(g_2)\lambda(g_1 g_2)^{-1} \prod_p c_{\mathrm{L},p}(g_p, g_p) \\ &= \lambda(g_1)\lambda(g_2)\lambda(g_1 g_2)^{-1}. \end{aligned}$$

□

8.5.6 The multiplier system for Siegel modular forms of half integral weight

In this subsection, we describe the transformation laws of Siegel modular forms of half integral weight in terms of the metaplectic cocycle.

Let $K_\infty \cong U(n)$ be the standard maximal compact subgroup of $G(\mathbb{R}) = \mathrm{Sp}_n(\mathbb{R})$ and let $\tilde{K}_\infty \cong \tilde{U}(n)$ be its preimage in $\tilde{G}_{\mathbb{R}}$. Then for each half integer $\ell \in \frac{1}{2} + \mathbb{Z}$, there is a unique genuine character ξ_ℓ of $\tilde{K}_\infty$ such that, for $g = \begin{pmatrix} A & B \\ -B & A \end{pmatrix} \in K_\infty$ with $A + iB \in U(n)$,

$$\xi_\ell([1, z]_{\mathrm{R}}) = z, \qquad \xi_\ell^2([k, z]_{\mathrm{R}}) = z^2 \det(A + iB)^{2\ell}. \tag{8.5.34}$$

Here $[g, z]_{\mathrm{R}}$ are Rao coordinates. Equivalently, in Leray coordinates, we have

$$\xi_\ell^2([k, z]_{\mathrm{L}}) = z^2 \beta(k)^{-2} \det(A + iB)^{2\ell}, \tag{8.5.35}$$

where β is given by (8.5.17).

Next we define the automorphy factor. For $\tau = u + iv \in \mathfrak{H}_n$, let

$$g_\tau = n(u)m(v^{\frac{1}{2}}) \qquad \text{and} \qquad g'_\tau = [g_\tau, 1]_{\mathrm{L}}.$$

Then, for $\gamma = \left(\begin{smallmatrix} a & b \\ c & d \end{smallmatrix}\right) \in G(\mathbb{R})$, write

$$\gamma \cdot g_\tau = g_{\gamma(\tau)} \cdot k(\gamma, \tau)^{-1},$$

with

$$k(\gamma, \tau) = \left(\begin{smallmatrix} A & B \\ -B & A \end{smallmatrix}\right) \in K_\infty, \qquad \det(A + iB) = \frac{c\tau + d}{|c\tau + d|}.$$

This implies that

$$\gamma' \cdot g'_\tau = g'_{\gamma(\tau)} \cdot k'(\gamma, \tau)^{-1},$$

where $g' = [g, 1]_L \in \tilde{G}_{\mathbb{R}}$.

Definition 8.5.16. For $\gamma \in \Gamma_0(4) = G(\mathbb{Q}) \cap K_0(4)$, and for $\ell \in \frac{1}{2} + \mathbb{Z}$, let

$$j_\ell(\gamma, \tau) := \lambda(\gamma)\, \lambda_2(\gamma) \cdot \xi_\ell(k'(\gamma, \tau)) \cdot |\det(c\tau + d)|^\ell,$$

where $\lambda(\gamma)$ and $\lambda_2(\gamma)$ are defined in the previous section.

The first part of the next lemma is immediate from the fact that $\lambda(\gamma)$, $\lambda_2(\gamma)$, and $\beta(k)$ are all 8-th roots of unity. The second part is a standard classical result.

Lemma 8.5.17. *(i) For* $\gamma \in \Gamma_0(4)$

$$j_\ell(\gamma, \tau)^8 = \det(c\tau + d)^{8\ell}.$$

In particular, $\eta_\ell(\gamma) = j_\ell(\gamma, \tau) j_{\frac{1}{2}}(\gamma, \tau)^{-2\ell}$ *is independent of* τ *and is a character of* $\Gamma_0(4)$ *of order dividing* 8.
(ii) When $n = 1$,

$$j_\ell(\gamma, \tau) = \epsilon_d^{-1} \left(\frac{c}{d}\right) \cdot (c\tau + d)^\ell,$$

and

$$\eta_\ell(\gamma) = \left(\frac{-1}{d}\right)^{\ell - \frac{1}{2}}.$$

Here $\epsilon_d = 1$ *or* i *depending on whether* $d \equiv 1$ *or* $-1 \mod 4$, *and*

$$z^{\frac{1}{2}} = \sqrt{r} e^{i\theta/2}, \quad \textit{if } z = re^{i\theta}, \quad -\pi < \theta \le \pi.$$

Remark 8.5.18. In the case of classical modular forms of half-integral weight, Shimura [21] uses the automorphy factor

$$(j_{\frac{1}{2}}(\gamma,\tau))^{\ell} = \left(\frac{-1}{d}\right)^{\ell-\frac{1}{2}} \cdot j_{\ell}(\gamma,\tau).$$

This is more natural when taking products of such forms.

Lemma 8.5.19. *For* $\gamma = \left(\begin{smallmatrix} a & b \\ 0 & d \end{smallmatrix}\right) \in \Gamma_0(4)$,

$$j_{\ell}(\gamma,\tau) = \operatorname{sgn}(\det d)^{\ell-\frac{1}{2}}.$$

Proof. Since, for $\gamma = \left(\begin{smallmatrix} a & b \\ 0 & d \end{smallmatrix}\right) \in \Gamma_0(4)$, $d \in \mathrm{GL}_n(\mathbb{Z})$ and $\det a = \det d = \pm 1$, (8.5.27) gives $\lambda_2(\gamma) = \chi_2(\det d) = 1$ and (8.5.9) gives $\lambda_p(\gamma) = 1$ for $p \neq 2$. To compute $\xi_l(k'(\gamma,\tau))$, let $\epsilon_l = (\mathrm{sgn})^{l-\frac{1}{2}}$, and let Φ^l_∞ be the eigenfunction of $\tilde{K}_\infty$ in the induced representation $I_\infty(\epsilon_l)$ with eigencharacter ξ_l. By evaluating Φ^ℓ_∞ on the element

$$\gamma' g'_\tau = g_{\gamma(\tau)} k'(\gamma,\tau)^{-1},$$

we find that

$$\xi_l(k'(\gamma,\tau)) = \epsilon_l(\det a),$$

as claimed. □

Let N be a positive integer and let χ be a character of $\Gamma_0(4N)$. We say that a function f on $\mathfrak{H}_n$ is of weight l, level $4N$, and character χ if

$$f(\gamma\tau) = j_l(\gamma,\tau)\chi(\gamma)f(\tau) \tag{8.5.36}$$

for every $\gamma \in \Gamma_0(4N)$. We say that a function ϕ on $G_{\mathbb{Q}}\backslash\tilde{G}_{\mathbb{A}}$ is of weight l, level $4N$, and character $\boldsymbol{\chi}$, where $\boldsymbol{\chi}$ is a character of $\mathbf{K}_0(4N)$, if

$$\phi(gk) = \boldsymbol{\chi}(k^{-1}) \cdot \phi(g) \quad \text{for } k \in \mathbf{K}_0(4N), \tag{8.5.37}$$

and

$$\phi(gk_\infty) = \xi_l(k_\infty) \cdot \phi \quad \text{for } k_\infty \in \tilde{K}_\infty. \tag{8.5.38}$$

Proposition 8.5.20. *If the function* ϕ *on* $G_{\mathbb{Q}}\backslash\tilde{G}_{\mathbb{A}}$ *is of weight* l*, level* $4N$*, and character* $\boldsymbol{\chi}$*, then the function*

$$f(\tau) := (\det v)^{-\frac{l}{2}} \cdot \phi(g'_\tau), \qquad \tau = u + iv \in \mathfrak{H}_n,$$

is of weight l*, level* $4N$*, and character* χ*, where* χ *is the pullback of* $\boldsymbol{\chi}$ *to* $\Gamma_0(4N)$*.*

Proof. The argument is just an adelic version of that given by Shintani [22]. We have

$$
\begin{aligned}
f(\gamma(\tau)) &= \big(\det v(\gamma(\tau))\big)^{-\frac{1}{2}\ell} \cdot \phi(g_{\gamma(\tau)}) \\
&= (\det v)^{-\frac{1}{2}\ell} \cdot \det|c\tau+d|^{\ell} \cdot \phi(\gamma' g'_\tau k'(\gamma,\tau)) \qquad (8.5.39)\\
&= (\det v)^{-\frac{1}{2}\ell} \cdot \det|c\tau+d|^{\ell} \cdot \xi_\ell(k'(\gamma,\tau)) \cdot \phi(\gamma' g'_\tau).
\end{aligned}
$$

We write

$$
\begin{aligned}
\gamma' &= [\gamma, 1]_{\mathrm{L},\infty} \\
&= s(\gamma)\cdot[1,\lambda(\gamma)]\cdot[\gamma,1]_{\mathrm{L},2}^{-1} \prod_{p\neq 2,\infty} [\gamma,1]_p^{-1} \qquad (8.5.40)\\
&= s(\gamma)\cdot[1,\lambda(\gamma)\,\lambda_2(\gamma)]\cdot[\gamma,\lambda_2(\gamma)]_{\mathrm{L},2}^{-1} \prod_{p\neq 2,\infty} [\gamma,1]_p^{-1} \\
&= s(\gamma)\cdot[1,\lambda(\gamma)\,\lambda_2(\gamma)]\cdot\boldsymbol{\gamma}^{-1},
\end{aligned}
$$

where $\boldsymbol{\gamma} \in \mathbf{K}_0(4N)$ is the image of γ under the splitting homomorphism. Then, in the last line of (8.5.39),

$$\phi(\gamma' g'_\tau) = \lambda(\gamma)\,\lambda_2(\gamma)\,\boldsymbol{\chi}(\boldsymbol{\gamma})\cdot\phi(g'_\tau),$$

as required. □

This proposition yields the transformation law of the genus two Eisenstein series $\mathcal{E}_2(\tau, s; B)$ studied in Chapter 5.

Proposition 8.5.21.

$$\mathcal{E}_2(\gamma(\tau), s; B) = \mathrm{sgn}(\det d)\cdot j_{\frac{3}{2}}(\gamma,\tau)\cdot\mathcal{E}_2(\tau, s; B),$$

for all $\gamma = \left(\begin{smallmatrix} a & b\\ c & d\end{smallmatrix}\right) \in \Gamma_0(4D(B)_o)$, *where* $D(B)_o$ *is the odd part of* $D(B)$.

Note that $\mathrm{sgn}(\det d) = (-1, \det d)_2$. If $c = 0$, we have $j_{\frac{3}{2}}(\gamma,\tau) = \mathrm{sgn}(\det d)$, so that

$$\mathcal{E}_2(\gamma(\tau), s; B) = \mathcal{E}_2(\tau, s; B), \qquad (8.5.41)$$

in this case.

Proposition 8.5.21 follows from the eigenproperties of the sections used to define $\mathcal{E}_2(\tau, s; B)$.

Proposition 8.5.22. *Let* $\chi = \chi_{2\kappa}$, *with* $\kappa = \pm 1$, *and let* $\Phi^\bullet(s) \in I_n(s, \chi)$, *with* $\bullet = 0,\ 1$, *or* ra, *be the standard sections defined in (8.4.13) and Section 5.1. Let*

$$J_r = \{\begin{pmatrix} a & b \\ c & d \end{pmatrix} \in K_p \mid c \equiv 0 \mod 4p^r\},$$

and let $\mathbf{J}_r \subset \mathbf{K}_0(4)$ *be the image of* $J_r \subset K_0(4)$ *in* $\tilde{G}_p$ *under the splitting homomorphism.*
(i) For $p \neq 2$,

$$r(\mathbf{k})\Phi^\bullet(s) = \Phi^\bullet(s), \qquad \text{for all } \mathbf{k} \in \mathbf{J}_r, \text{ where } r = \begin{cases} 0 & \text{if } \bullet = 0, \\ 1 & \text{if } \bullet = 1, \text{ra}. \end{cases}$$

(ii) For $p = 2$,

$$r(\mathbf{k})\Phi^\bullet(s) = \chi_\kappa(\det d) \cdot \Phi^\bullet(s), \qquad \text{for all } \mathbf{k} \in \mathbf{J}_0,$$

where $\mathbf{k} = (k, \lambda_2(k))$ *with* $k = \left(\begin{smallmatrix} a & b \\ c & d \end{smallmatrix}\right)$.

Proof. The map $\lambda_V : S(V^n) \to I_n(s_0, \chi_V)$ is $\tilde{G}_p$ intertwining, and the passage from an element of $I_n(s_0, \chi_V)$ to the associated standard section is compatible with the action of $\tilde{K}_p$. Thus, by (8.4.13), it suffices to observe the following facts, which are a special case of Lemma 8.5.10 and Proposition 8.5.14.

Lemma 8.5.23. *Let* $V = V^\pm$, *and let* L^0, L^1 *in* V^+ *and* L^{ra} *in* V^- *be the lattices defined in (8.4.12).*
(i) For $p \neq 2$,

$$\omega_{V,\psi_p}(\mathbf{k})\, \text{char}(L^\bullet) = \text{char}(L^\bullet), \qquad \text{for all } \mathbf{k} \in \mathbf{J}_r,$$

where $r = 0$ *if* $\bullet = 0$ *and* $r = 1$ *if* $\bullet = 1$ *or* ra.
(ii) For $p = 2$,

$$\omega_{V,\psi_p}(\mathbf{k})\, \text{char}(L^\bullet) = \chi_\kappa(\det d) \cdot \text{char}(L^\bullet), \qquad \text{for all } \mathbf{k} \in \mathbf{J}_0.$$

This finishes the proof of Proposition 8.5.22. □

8.5.7 The level of the vertical component of $\widehat{\phi}_1(\tau)$

In this section, we determine the level of the theta functions associated to the vertical component of the genus one generating function $\widehat{\phi}_1(\tau)$ considered in Section 4.3; see Remark 4.3.5. This will complete the proof of Theorem A. We fix a prime $p \mid D(B)$, and we determine the compact open subgroup of the metaplectic group $G' = G'_p$ which fixes the Schwartz function $\mu_{[\Lambda]}$ defined in Lemma 4.3.3. We slightly simplify the notation of that section by letting

$$V = V'(\mathbb{Q}_p) = \{\, x \in M_2(\mathbb{Q}_p) \mid \mathrm{tr}(x) = 0 \,\},$$

and $\Lambda = \mathbb{Z}_p^2 \subset \mathbb{Q}_p^2$. Then

$$\varphi_{[\Lambda]} = \varphi^0 := \mathrm{char}(L), \qquad L = V \cap M_2(\mathbb{Z}_p),$$

and we have

$$\varphi(x) := \mu_{[\Lambda]}(x) = \varphi^0(x) - p\,\varphi^0(p^{-1}x) - \varphi^{\sim}(x),$$

where the function $\varphi^{\sim}$ is defined as follows. If $p \neq 2$,

$$\varphi^{\sim}(x) = \chi(-\det(x)) \cdot \mathrm{char}(V \cap \mathrm{GL}_2(\mathbb{Z}_p))(x),$$

where $\chi(\epsilon) = (\epsilon, p)_p$. If $p = 2$,

$$\varphi^{\sim}(x) = p \cdot \mathrm{char}(V \cap (1 + p\,M_2(\mathbb{Z}_p)))(x).$$

Note that $\mathrm{supp}(\varphi) \subset L$ and that φ is constant on pL cosets. In addition, φ is invariant under conjugation by $\mathrm{GL}_2(\mathbb{Z}_p)$ and under scaling by elements of $\mathbb{Z}_p^{\times}$.

Let

$$K_r = \{\, k = \begin{pmatrix} a & b \\ c & d \end{pmatrix} \in \mathrm{SL}_2(\mathbb{Z}_p) \mid \mathrm{ord}(c) \geq r \,\}.$$

There is a splitting homomorphism

$$K_r \xrightarrow{\sim} \mathbf{K}_r \subset G', \qquad k \mapsto \mathbf{k} = [k, \lambda(k)]_{\mathrm{L}},$$

where $r = 0$, if $p \neq 2$, and $r = 2$, if $p = 2$.

Proposition 8.5.24. *(i) For $p \neq 2$, the function φ is invariant under the subgroup $\mathbf{K}_1$.*
(ii) For $p = 2$, the function φ is an eigenfunction for the subgroup $\mathbf{K}_2$ with character

$$\chi_{-1} : \begin{pmatrix} a & b \\ c & d \end{pmatrix} \mapsto \chi_{-1}(a) = (a, -1)_2.$$

Proof. If $r \geq 1$, any element $k \in K_r$ can be written uniquely in the form $k = n_-(c)m(a)n(b)$, where, as usual,

$$n_-(c) = \begin{pmatrix} 1 & \\ c & 1 \end{pmatrix}, \quad m(a) = \begin{pmatrix} a & \\ & a^{-1} \end{pmatrix}, \quad n(b) = \begin{pmatrix} 1 & b \\ & 1 \end{pmatrix}.$$

We write $\mathbf{n}_-(c)$, $\mathbf{m}(a)$, and $\mathbf{n}(b)$ for the images of these elements under the splitting homomorphism, where, if $p = 2$, we suppose that $r \geq 2$.

Note that $\chi_V = \chi_{-2}$. Then, for $a \in \mathbb{Z}_p^\times$, we have

$$\omega(\mathbf{m}(a))\varphi(x) = \chi_V\chi_2(a)\,\varphi(ax) = \chi_{-1}(a)\,\varphi(x),$$

since φ is invariant under scaling by units, and

$$\omega(\mathbf{n}(b))\varphi(x) = \psi(b\,Q(x))\,\varphi(x) = \varphi(x),$$

since $\text{supp}(\varphi) \subset L$, Q is $\mathbb{Z}_p$-valued on L and ψ is unramified.

Next note that we have the relation

$$\mathbf{n}_-(c) = [w, 1]_{\mathrm{L}}^{-1}\,\mathbf{n}(-c)\,[w, 1]_{\mathrm{L}},$$

by Lemma 8.5.13, when $p = 2$ and from the fact that $[w, 1]_{\mathrm{L}} = \mathbf{w} \in \mathbf{K}_0$, in the case $p \neq 2$. By (8.5.20), we have

$$\omega_V([w, 1]_{\mathrm{L}})\varphi(x) = \gamma(V)\int_V \psi(\text{tr}(x, y))\,\varphi(y)\,dy = \gamma(V)\cdot\hat{\varphi}(x).$$

Thus, it suffices to prove the following result about the support of the Fourier transform $\hat{\varphi}$.

Lemma 8.5.25. *If $\hat{\varphi}(x) \neq 0$, then*

$$\text{ord}(Q(x)) \geq \begin{cases} -1 & \text{if } p \neq 2, \\ -2 & \text{if } p = 2. \end{cases}$$

Proof. Note that $\text{supp}(\hat{\varphi}) \subset p^{-1}L^\sharp$ and that $\hat{\varphi}$ is constant on $L^\sharp$ cosets. Thus, we may view $\hat{\varphi}$ as a function on the vector space $p^{-1}L^\sharp/L^\sharp$ of dimension 3 over $\mathbb{F}_p$. In addition, this function is invariant under scaling by $\mathbb{F}_p^\times$ and under the action of $\text{GL}_2(\mathbb{F}_p)$ induced by the conjugation action of $\text{GL}_2(\mathbb{Z}_p)$ on $L^\sharp$.

First consider the case $p \neq 2$, so that $L^\sharp = L$ and we can identify $p^{-1}L/L \simeq L/pL$ with the set $V(\mathbb{F}_p)$ of trace zero 2×2 matrices over $\mathbb{F}_p$. The conjugation action of $\text{GL}_2(\mathbb{F}_p)$ gives the action of $\text{SO}(V(\mathbb{F}_p))$ and there are three orbits of $\text{GL}_2(\mathbb{F}_p) \times \mathbb{F}_p^\times$ with orbit representatives

$$0, \quad \begin{pmatrix} & 1 \\ 0 & \end{pmatrix}, \quad \begin{pmatrix} 1 & \\ & -1 \end{pmatrix}, \quad \begin{pmatrix} & 1 \\ \beta & \end{pmatrix},$$

where $\chi(\beta) = -1$. We must show that the function determined by $\hat{\varphi}$ vanishes on the last two representatives. Of course, setting

$$\varphi'(x) = \varphi^0(p^{-1}x) = \text{char}(pL)(x)$$

and

$$\varphi''(x) = \varphi^0(px) = \mathrm{char}(p^{-1}L)(x),$$

we have

$$\widehat{\varphi^0} - p\,\widehat{\varphi'} = \varphi^0 - p^{-2}\,\varphi'',$$

so that it remains to calculate the values of $\widehat{\varphi^\sim}$ on the last two representatives. For the first of these, we have

$$\begin{aligned}\widehat{\varphi^\sim}(p^{-1}\begin{pmatrix}1 & \\ & -1\end{pmatrix}) &= \int_{V\cap \mathrm{GL}_2(\mathbb{Z}_p)} \psi(-p^{-1}2y_0)\,\chi(y_0^2+y_1y_2)\,dy \\ &= p^{-3}\sum_{\substack{y\in V(\mathbb{F}_p)\\ \det y\neq 0}} \psi(-p^{-1}2y_0)\,\chi(y_0^2+y_1y_2).\end{aligned}$$

Now the contribution of the set of y where $y_0 = 0$ vanishes, since the character χ of $\mathbb{F}_p^\times$ is nontrivial. This leaves the quantity

$$\begin{aligned}p^{-3}&\sum_{y_0\in\mathbb{F}_p^\times}\sum_{\substack{y_1,y_2\in\mathbb{F}_p\\ y_0^2+y_1y_2\neq 0}} \psi(-p^{-1}2y_0)\,\chi(y_0^2+y_1y_2) \\ &= p^{-3}\sum_{y_0\in\mathbb{F}_p^\times}\sum_{\substack{y_1,y_2\in\mathbb{F}_p\\ y_1y_2\neq -1}} \psi(-p^{-1}2y_0)\,\chi(1+y_1y_2) \\ &= -p^{-3}\sum_{\substack{y_1,y_2\in\mathbb{F}_p\\ y_1y_2\neq -1}} \chi(1+y_1y_2) \\ &= -p^{-3}\sum_{u\in\mathbb{F}_p^\times}\chi(u)\sum_{\substack{y_1,y_2\in\mathbb{F}_p\\ y_1y_2=u-1}} 1 \\ &= -p^{-3}(p-1)\sum_{u\in\mathbb{F}_p^\times}\chi(u) - p^{-2} \quad = -p^{-2}.\end{aligned}$$

Here, in the last two steps, we note that the number of solutions of $y_1y_2 = u-1$ is $p-1$, if $u\neq 1$, and $2p-1$, if $u=1$. It follows that

$$\widehat{\varphi}(p^{-1}\begin{pmatrix}1 & \\ & -1\end{pmatrix}) = 0,$$

as claimed. Next we calculate

$$\widehat{\varphi^{\sim}}(p^{-1}\begin{pmatrix} & 1\\ \beta & \end{pmatrix}) = \int_{V\cap \mathrm{GL}_2(\mathbb{Z}_p)} \psi(-p^{-1}(y_2+\beta y_1))\,\chi(y_0^2+y_1y_2)\,dy$$

$$= p^{-3} \sum_{\substack{y\in V(\mathbb{F}_p)\\ \det y\neq 0}} \psi(-p^{-1}(y_2+\beta y_1))\,\chi(y_0^2+y_1y_2).$$

The terms where $y_0 = 0$ give

$$p^{-3}\,\chi(\beta) \sum_{y_1,y_2} \psi(p^{-1}(y_1+y_2))\,\chi(y_1y_2)$$

$$= p^{-3}\,\chi(\beta) \sum_{y_1,y_2} \psi(p^{-1}y_1(1+y_2))\,\chi(y_2) \quad = -p^{-2}\,\chi(-1),$$

where, in the last step, we note that the sum on y_1 gives $-\chi(y_2)$, if $y_2 \neq -1$, and $(p-1)\,\chi(y_2)$, if $y_2 = -1$. Next, the terms with $y_0 \neq 0$ give

$$p^{-3} \sum_{\substack{y_1,y_2\\ y_1y_2\neq -1}} \chi(1+y_1y_2) \sum_{y_0\neq 0} \psi(-p^{-1}y_0(y_2+\beta y_1))$$

$$= p^{-3} \sum_{u\in\mathbb{F}_p^\times} \chi(u) \sum_{y_1y_2=u-1}\ \sum_{y_0\neq 0} \psi(-p^{-1}y_0(y_2+\beta y_1)).$$

The inner sum is -1, if $y_2+\beta y_1 \neq 0$, and $p-1$, if $y_2+\beta y_1 = 0$. When $u = 1$, the sum on y_1 and y_2 then gives

$$-2(p-1)+p-1 = -(p-1).$$

If $u \neq 1$, then $y_2 = (u-1)y_1^{-1}$, and we note that the equation

$$(u-1)y_1^{-1} + \beta y_1 = 0$$

has two solutions if $\chi(1-u) = -1$ and no solutions if $\chi(1-u) = 1$. It follows that, for a fixed u, the sum on y_1 and y_2 is equal $-(p-1)$, if $\chi(1-u) = 1$, and $p+1$, if $\chi(1-u) = -1$. This may be written as

$$1 - p\,\chi(1-u),$$

so that the whole expression becomes

$$p^{-3}\sum_{u\neq 1}\chi(u) - p^{-2}\sum_{u\neq 1}\chi(u)\,\chi(1-u) - p^{-3}(p-1). \tag{8.5.42}$$

It is easy to check that

$$p^{-2}\sum_{u\neq 1}\chi(u)\,\chi(1-u) = -\chi(-1)\,p^{-2},$$

so that (8.5.42) becomes

$$p^{-2}(\chi(-1)-1).$$

Thus,

$$\widehat{\varphi^{\sim}}(p^{-1}\begin{pmatrix} & 1\\ \beta & \end{pmatrix}) = -p^{-2},$$

and

$$\widehat{\varphi}(p^{-1}\begin{pmatrix} & 1\\ \beta & \end{pmatrix}) = 0,$$

as claimed.

Remark 8.5.26. For $p \neq 2$, similar calculations yield the values

$$\widehat{\varphi^{\sim}}(0) = p^{-2}(p-1) \qquad \text{and} \qquad \widehat{\varphi^{\sim}}(p^{-1}\begin{pmatrix} & 1\\ 0 & \end{pmatrix}) = 0,$$

and hence

$$\widehat{\varphi}(0) = p^{-1}(p-1) \qquad \text{and} \qquad \widehat{\varphi}(p^{-1}\begin{pmatrix} & 1\\ 0 & \end{pmatrix}) = p^{-2}(p-1).$$

Next suppose that $p=2$. In this case, we can take the following elements as orbit representatives for the action of $\mathrm{GL}_2(\mathbb{F}_2)$ on the $\mathbb{F}_2$-vector space $L^\sharp/2L^\sharp$:

$$0, \quad \begin{pmatrix} & 1\\ 0 & \end{pmatrix}, \quad \begin{pmatrix} \frac{1}{2} & \\ & -\frac{1}{2} \end{pmatrix}, \quad \begin{pmatrix} \frac{1}{2} & 1\\ 1 & -\frac{1}{2} \end{pmatrix}.$$

Once again, we need to show that the function on $L^\sharp/2L^\sharp$ determined by $\widehat{\varphi}$ vanishes on the last two elements. Note that the self-dual measure on V gives $\mathrm{vol}(L) = 1/\sqrt{2}$. Setting $\varphi' = \mathrm{char}(2L)$, we have

$$\widehat{\varphi^0} - 2\,\widehat{\varphi'} = \frac{1}{\sqrt{2}}\,\mathrm{char}(L^\sharp) - \frac{1}{4\sqrt{2}}\,\mathrm{char}(\frac{1}{2}\,L^\sharp).$$

Thus, we must determine

$$\widehat{\varphi^{\sim}}(x) = 2\int_{V\cap(1_2+2M_2(\mathbb{Z}_2))} \psi((x,y))\,dy.$$

If we write

$$y = \begin{pmatrix} y_0 & 2y_1\\ 2y_2 & -y_0 \end{pmatrix},$$

with $y_0 \in \mathbb{Z}_2^\times$, we have $dy = \frac{1}{4\sqrt{2}}\, dy_0\, dy_1\, dy_2$, and

$$(x, y) = \begin{cases} -\frac{1}{2}\, y_0 & \text{for } x = \begin{pmatrix} \frac{1}{4} & \\ & -\frac{1}{4} \end{pmatrix}, \\ -\frac{1}{2}(y_0 + 2y_1 + 2y_2) & \text{for } x = \begin{pmatrix} \frac{1}{4} & \frac{1}{2} \\ \frac{1}{2} & -\frac{1}{4} \end{pmatrix}. \end{cases}$$

In either case, we get

$$\begin{aligned} \widehat{\varphi^\sim}(x) &= \frac{1}{2\sqrt{2}} \int_{\mathbb{Z}_2^\times} \int_{(\mathbb{Z}_2)^2} \psi(-\frac{1}{2}\, y_0)\, dy_0\, dy_1\, dy_2 \\ &= \frac{1}{4\sqrt{2}}\, \psi(-\frac{1}{2}) \quad = -\frac{1}{4\sqrt{2}}. \end{aligned}$$

Thus,

$$\hat{\varphi}(x) = 0,$$

as claimed in the lemma. □

This finishes the proof of Proposition 8.5.24. □

Remark 8.5.27. Again, we record the other values

$$\widehat{\varphi^\sim}(0) = \widehat{\varphi^\sim}(\begin{pmatrix} & \frac{1}{2} \\ 0 & \end{pmatrix}) = \frac{1}{4\sqrt{2}},$$

so that

$$\hat{\varphi}(0) = \hat{\varphi}(\begin{pmatrix} & \frac{1}{2} \\ 0 & \end{pmatrix}) = \frac{1}{2\sqrt{2}}.$$

Bibliography

[1] E. M. Baruch and Z.Y. Mao, *Central value of automorphic L-functions*, preprint, 2003.

[2] S. Böcherer, *Über die Funktionalgleichung automorpher L-Funktionen zur Siegelschen Modulgruppe*, J. reine angew. Math., **362** (1985), 146–168.

[3] P. Garrett, *Pullbacks of Eisenstein series; applications*, in Automorphic Forms of Several Variables (Taniguchi Symposium, Katata, 1983), Progr. Math., **46**, 114–137, Birkhäuser, Boston, MA, 1984.

[4] S. Gelbart, Weil's Representation and the Spectrum of the Metaplectic Group, Lecture Notes in Math., **530**, Springer-Verlag, Berlin, 1976.

[5] S. Gelbart, I. Piatetski-Shapiro, and S. Rallis, Explicit Constructions of Automorphic L-functions, Lecture Notes in Math., **1254**, Springer-Verlag, Berlin, 1987.

[6] R. Howe, *θ–series and invariant theory*, Proc. Symp. Pure Math., **33** (1979), 275–285.

[7] R. Howe and I. I. Piatetski-Shapiro, *Some examples of automorphic forms on Sp_4*, Duke Math. J., **50** (1983), 55–106.

[8] S. Kudla, *Splitting metaplectic covers of dual reductive pairs*, Israel J. Math. **87** (1994), 361–401.

[9] ______, *Notes on the local theta correspondence*, Lecture Notes of the European School on Group Theory, Schloß Hirschberg, September 1996.

[10] ______, *Central derivatives of Eisenstein series and height pairings*, Annals of Math., **146** (1997), 545–646.

[11] ______, *Modular forms and arithmetic geometry*, in Current Developments in Mathematics, 2002, 135–179, International Press, Somerville, MA, 2003.

[12] S. Kudla and S. Rallis, *On first occurrence in the local theta correspondence*, forthcoming.

[13] S. Kudla and M. Rapoport, *Height pairings on Shimura curves and p-adic uniformization*, Invent. math., **142** (2000), 153–223.

[14] S. Kudla, M. Rapoport, and T. Yang, *Derivatives of Eisenstein series and Faltings heights*, Compositio Math., **140** (2004), 887–951.

[15] J.-S. Li, *Nonvanishing theorems for the cohomology of certain arithmetic quotients*, J. reine angew. Math., **428** (1992), 177–217.

[16] D. Manderscheid, *Waldpurger's involution and types*, J. London Math. Soc. (2), **70** (2004), 567–585.

[17] C. Moeglin, M.-F. Vigneras, and J.-L. Waldspurger, Correspondances de Howe sur un Corps p-adique, Lecture Notes in Math., **1291**, Springer-Verlag, Berlin, 1987.

[18] P. Perrin, *Représentations de Schrödinger, indice de Maslov et groupe métaplectique*, in Noncommutative Harmonic Analysis and Lie Groups (Marseille, 1980), Lecture Notes in Math., **880**, 370–407, Springer, Berlin-New York, 1981.

[19] R. Ranga Rao, *On some explicit formulas in the theory of Weil representation*, Pacific J. Math., **157** (1993), 335–370.

[20] S. Rallis and G. Schiffmann, *Représentations supercuspidales du groupe métaplectique*, J. Math. Kyoto Univ., **17** (1977), 567–603.

[21] G. Shimura, *On modular forms of half integral weight*, Annals of Math., **97** (1973), 440–481.

[22] T. Shintani, *On construction of holomorphic cusp forms of half integral weight*, Nagoya Math. J., **58** (1975), 83–126.

[23] J.-L. Waldspurger, *Correspondance de Shimura*, J. Math. Pures Appl., **59** (1980), 1–132.

[24] ______, *Sur les coefficients de Fourier des formes modulaires de poids demi-entier*, J. Math. Pures Appl., **60** (1981), 375–484.

[25] ______, *Correspondance de Shimura*, in Séminare de Théorie des Nombres (Paris 1979–80), Progr. Math., **12**, 357–369, Birkhäuser, Boston, 1981.

[26] ______, *Sur les valeurs de certaines fonctions L automorphes en leur centre de symétrie*, Compositio Math., **54** (1985), 173–242.

[27] ______, *Démonstration d'une conjecture de dualité de Howe dans le cas p-adique, $p \neq 2$*, in Festschrift in Honor of I. I. Piatetski-Shapiro, part 2, Israel Math. Conf. Proc., **2-3**, 267–234, Weizmann Science Press, Jerusalem, 1990.

[28] ______, *Correspondances de Shimura et quaternions*, Forum Math., **3** (1991), 219–307.

[29] A. Weil, *Sur certains groupes d'opérateurs unitaires*, Acta Math., **111** (1964), 143–211.

[30] T. H. Yang, *An explicit formula for local densities of quadratic forms*, J. Number Theory, **72** (1998), 309–356.

[31] ______, *Local densities of 2-adic quadratic forms*, J. Number Theory, **108** (2004), 287–345.

[32] D. Zagier, *Nombres de classes et formes modulaires de poids 3/2*, C. R. Acad. Sci. Paris, **281** (1975), 883–886.

Chapter Nine

Central derivatives of L-functions

In this chapter, we first use the Borcherds generating function, $\phi_{\text{Bor}}(\tau, \varphi)$, to define an arithmetic analogue of the classical Shimura-Waldspurger lift described in Section 8.2. We show that this lift, whose target is the Mordell-Weil group of a Shimura curve over $\mathbb{Q}$, is compatible with the local theta correspondence, and hence there are local obstructions to nonvanishing, just as in the classical case. We then formulate a conjectural analogue of the result of Waldspurger, Theorem 8.2.5, and characterize the nonvanishing of the arithmetic theta lift in terms of theta dichotomy (local obstructions) and the nonvanishing of the central derivative $L'(\frac{1}{2}, \text{Wald}(\sigma, \psi))$. In Section 9.2, we prove this conjecture in certain cases by means of an arithmetic version of Rallis's inner product formula, obtained by combining the arithmetic inner product formula of Chapter 7, the identity of Chapter 6 relating the genus two generating function and the central derivative $\mathcal{E}_2'(\tau, 0, B)$, and the explicit doubling formula of Chapter 8. In Section 9.3, we explain how the input for our arithmetic theta lift can be described in the classical language of normalized newforms of weight 2.

9.1 THE ARITHMETIC THETA LIFT

We begin by defining the arithmetic theta lift.

For an indefinite quaternion algebra B over $\mathbb{Q}$ and for any compact open subgroup $K \subset H^B(\mathbb{A}_f)$, where $H^B = B^\times \simeq \text{GSpin}(V^B)$, let

$$\text{MW}(M_K^B) = \text{Jac}(M_K^B)(\mathbb{Q}) \otimes_{\mathbb{Z}} \mathbb{C} \tag{9.1.1}$$

be the Mordell-Weil space of the associated Shimura curve M_K^B over $\mathbb{Q}$. There is an exact sequence

$$0 \longrightarrow \text{MW}(M_K^B) \longrightarrow \text{CH}^1(M_K^B) \longrightarrow H^2(M_K^B) \longrightarrow 0, \tag{9.1.2}$$

where $H^2(M_K^B)$ is the Betti cohomology with complex coefficients. We can pass to the limit over K and let

$$\text{MW}(M^B) = \varinjlim_{K} \text{MW}(M_K^B). \tag{9.1.3}$$

The spaces $\mathrm{CH}^1(M^B)$ and $H^2(M^B)$ are defined analogously and there is an exact sequence

$$0 \longrightarrow \mathrm{MW}(M^B) \longrightarrow \mathrm{CH}^1(M^B) \longrightarrow H^2(M^B) \longrightarrow 0. \tag{9.1.4}$$

of admissible $H^B(\mathbb{A}_f)$-modules, where, for example,

$$\mathrm{MW}(M^B)^K = \mathrm{MW}(M^B_K). \tag{9.1.5}$$

For any $\varphi \in S(V^B(\mathbb{A}_f))$, the associated arithmetic theta function $\widetilde{\phi}_{\mathrm{Bor}}(\cdot, \varphi)$, defined in Section 4.7, is a 'holomorphic' 'weight $\frac{3}{2}$' automorphic form on $G'_{\mathbb{A}}$, valued in $\mathrm{CH}^1(M^B)$. Recall that this is the lift to $G'_{\mathbb{A}}$ of the generating function

$$\phi_{\mathrm{Bor}}(\tau, \varphi) = \sum_{t \geq 0} Z(t, \varphi)\, q^t$$

for weighted 0-cycles. If $\varphi \in S(V^B(\mathbb{A}_f))^K$ for some compact open subgroup K of $H^B(\mathbb{A}_f)$, then $\widetilde{\phi}_{\mathrm{Bor}}(\cdot, \varphi)$ takes values in the finite dimensional space $\mathrm{CH}^1(M^B_K)$. Also recall, from Section 4.7, that if $g'_0 \in G'_{\mathbb{A}_f}$, then

$$\widetilde{\phi}_{\mathrm{Bor}}(g' g'_0, \varphi) = \widetilde{\phi}_{\mathrm{Bor}}(g', \omega(g'_0)\varphi), \tag{9.1.6}$$

just as for the classical theta function.

For any cusp form $f \in \mathcal{A}_{00}(G')$, the arithmetic theta lift of f is the class

$$\theta^{\mathrm{ar}}(f, \varphi) := \int_{\mathrm{Sp}_1(\mathbb{Q}) \backslash \mathrm{Sp}_1(\mathbb{A})} f(g')\, \overline{\widetilde{\phi}_{\mathrm{Bor}}(g', \varphi)}\, dg, \tag{9.1.7}$$

in $\mathrm{CH}^1(M^B)$. In fact, since f is orthogonal to all Eisenstein series and unary theta series, (4.4.29) and Proposition 4.4.7 imply that $\theta^{\mathrm{ar}}(f, \varphi)$ maps to zero in $H^2(M^B)$ in the sequence (9.1.4) and hence $\theta^{\mathrm{ar}}(f, \varphi) \in \mathrm{MW}(M^B)$.

Suppose that $\sigma \simeq \sigma_\infty \otimes \sigma_0$ is a genuine cuspidal automorphic representation of $G'_{\mathbb{A}}$ with

$$\sigma_\infty \simeq \tilde{\pi}^+_{\frac{3}{2}} = \mathrm{HDS}_{\frac{3}{2}}, \tag{9.1.8}$$

the holomorphic discrete series representation of $G'_{\mathbb{R}}$ of weight $\frac{3}{2}$. We write $\mathcal{V}(\sigma) \subset \mathcal{A}_{00}(G')$ for the space of σ, and we write $\mathcal{V}(\sigma)_{\frac{3}{2},\mathrm{hol}} \subset \mathcal{A}_{00}(G')_{\frac{3}{2},\mathrm{hol}}$ for the subspace of lowest weight vectors of weight $\frac{3}{2}$ for K'_∞, the full inverse image of $\mathrm{SO}(2)$ in $G'_{\mathbb{R}}$. Note that $\mathcal{V}(\sigma)_{\frac{3}{2},\mathrm{hol}} \simeq \sigma_0$. The arithmetic theta lift of σ,

$$\theta^{\mathrm{ar}}(\sigma, M^B) \subset \mathrm{MW}(M^B), \tag{9.1.9}$$

is the subspace spanned by the elements $\theta^{\text{ar}}(f,\varphi)$ as $f \in \sigma$ and $\varphi \in S(V^B(\mathbb{A}_f))$ vary. Note that $\theta^{\text{ar}}(\sigma, M^B)$ is an $H^B(\mathbb{A}_f)$-invariant subspace.

In analogy with (8.2.33), the theory of the local theta correspondence, in Howe's formulation [5], yields the following information about the space $\theta^{\text{ar}}(\sigma, M^B)$. Write $\sigma_0 \simeq \otimes_p \sigma_p$ and recall that, for each p, the maximal σ_p-isotypic quotient $S(\sigma_p, V_p^B)$ of $S(V_p^B)$ is either zero or

$$S(\sigma_p, V_p^B) \simeq \sigma_p \otimes \Theta(\sigma_p, V_p^B),$$

where $\Theta(\sigma_p, V_p^B)$ is a smooth representation of H_p^B which has a unique irreducible quotient $\theta_\psi(\sigma_p, V_p^B)$; see [10]. Note that, to be consistent with the notation of Chapter 8, we have written $\theta_\psi(\sigma_p, V_p^B) = \theta(\sigma_p, V_p^B)$, even though the additive character ψ is fixed throughout this chapter.

Proposition 9.1.1. *As a representation of* $H^B(\mathbb{A}_f)$,

$$\theta^{\text{ar}}(\sigma, M^B) \simeq \begin{cases} \otimes_{p<\infty}\theta_\psi(\sigma_p, V_p^B) \\ 0. \end{cases}$$

Proof. The nondegenerate $G'_{\mathbb{A}_f}$-invariant inner product

$$(9.1.10) \qquad \langle f_1, f_2 \rangle_{\text{Pet}} = \int_{G'_{\mathbb{Q}}\backslash G'_{\mathbb{A}}} f_1(g')\,\overline{f_2(g')}\,dg'$$

on $\mathcal{A}_0(G')_{\frac{3}{2},\text{hol}}$ allows us to identify $\mathcal{V}(\sigma)_{\frac{3}{2},\text{hol}}$ with its $\mathbb{C}$-antilinear admissible dual via the map

$$(9.1.11) \qquad f \longmapsto \langle f, \cdot \rangle_{\text{Pet}}.$$

Thus, we obtain a $\mathbb{C}$-linear map

$$(9.1.12)\ \ \theta_\sigma^{\text{ar}} : S(V^B(\mathbb{A}_f)) \longrightarrow \mathcal{V}(\sigma)_{\frac{3}{2},\text{hol}} \otimes \text{MW}(M^B), \qquad \varphi \mapsto \theta_\sigma^{\text{ar}}(\varphi),$$

where, for all $f \in \mathcal{V}(\sigma)_{\frac{3}{2},\text{hol}}$,

$$(9.1.13) \qquad \theta^{\text{ar}}(f,\varphi) = \langle f, \theta_\sigma^{\text{ar}}(\varphi) \rangle_{\text{Pet}}.$$

Note that $\theta_\sigma^{\text{ar}}(\varphi)$ is just the σ-isotypic part of $\widetilde{\phi}_{\text{Bor}}(g',\varphi)$.

The map $\theta_\sigma^{\text{ar}}$ is equivariant for the action of $G'_{\mathbb{A}_f} \times H^B(\mathbb{A}_f)$, by (9.1.6), and independent of the choice of the invariant measure dg'. In particular, $\theta_\sigma^{\text{ar}}$ factors through the maximal σ_0-isotypic quotient $S(\sigma_0, V^B)$ of $S(V^B(\mathbb{A}_f))$.

This quotient can be described in terms of the local theta correspondence.[1] As a representation of $G'_{\mathbb{A}_f} \times H^B(\mathbb{A}_f)$, $S(\sigma_0, V^B)$ is either zero or

$$S(\sigma_0, V^B) \simeq \sigma_0 \otimes \Theta(\sigma_0, V^B), \tag{9.1.14}$$

for a uniquely determined representation $\Theta(\sigma_0, V^B)$ of $H^B(\mathbb{A}_f)$. If σ_0 does not occur as a quotient of $S(V^B(\mathbb{A}_f))$ we set $\Theta(\sigma_0, V^B) = 0$. By the compatibility of the local and global theta correspondences and the results recalled above,

$$\Theta(\sigma_0, V^B) \simeq \otimes_p \Theta(\sigma_p, V^B)$$

is either zero or has finite length and has a unique irreducible quotient

$$\theta(\sigma_0, V^B) \simeq \otimes_p \theta(\sigma_p, V^B).$$

Since $\mathrm{MW}(M^B)$ is completely reducible as a representation of $H^B(\mathbb{A}_f)$,[2] the map $\theta^{\mathrm{ar}}_\sigma$ factors through the unique irreducible quotient $\sigma_0 \otimes \theta(\sigma_0, V^B)$ of $S(\sigma_0, V^B)$, and so we obtain
(9.1.15)

$$\theta^{\mathrm{ar}}_\sigma : S(V^B(\mathbb{A}_f)) \xrightarrow{\mathrm{pr}^B_\sigma} \sigma_0 \otimes \theta(\sigma_0, V^B) \xrightarrow{1 \otimes j^B_\sigma} \mathcal{V}(\sigma)_{\frac{3}{2},\mathrm{hol}} \otimes \mathrm{MW}(M^B),$$

for a unique determined $H^B(\mathbb{A}_f)$-equivariant map

$$j^B_\sigma : \theta(\sigma_0, V^B) \longrightarrow \theta^{\mathrm{ar}}(\sigma, M^B) \subset \mathrm{MW}(M^B) \tag{9.1.16}$$

with image $\theta^{\mathrm{ar}}(\sigma, M^B)$. Thus,

$$\theta^{\mathrm{ar}}(\sigma, M^B) \simeq \begin{cases} \otimes_{p<\infty} \theta_\psi(\sigma_p, V^B_p) \\ 0, \end{cases} \tag{9.1.17}$$

as claimed. □

By the previous proposition, we have

$$\theta^{\mathrm{ar}}(\sigma, M^B) \neq 0 \quad \Longrightarrow \quad \mathrm{inv}_p(B) = \delta_p(\sigma_p, \psi^-_p), \quad \forall p < \infty, \tag{9.1.18}$$

---

[1] Here we are following the original formalism introduced by Howe [5]. The particular dual pair $(G', O(V^B))$ was studied in detail by Waldspurger [12], [15] from a slightly different point of view. In our application to the arithmetic theta correspondence, which is modeled on [6], we need Howe's abstract formulation. The additional facts we need, some of which are special to this particular dual pair, can be found in [10].

[2] This follows from invariance under $H^B(\mathbb{A}_f)$ of the hermitian form on $\mathrm{MW}(M^B)$ determined by the Néron-Tate pairing.

where $\delta_p(\sigma_p, \psi_p^-)$ is the local dichotomy sign defined in (8.2.8). Note that the occurrence here of the additive ψ_p^-, where $\psi_p^-(x) = \psi_p(-x)$, is explained in Corollary 8.2.6. Since B is indefinite, this implies that

$$\prod_{p<\infty} \delta(\sigma_p, \psi_p^-) = 1, \tag{9.1.19}$$

and hence, by (8.2.29), that

$$\epsilon(\frac{1}{2}, \mathrm{Wald}(\sigma, \psi_{-1})) = \delta(\sigma_\infty, \psi_\infty^-). \tag{9.1.20}$$

The archimedean dichotomy invariant $\delta(\sigma_\infty, \psi_\infty^-)$ is given by Table 2 in Chapter 8.

Lemma 9.1.2. *Let* $\sigma_\infty = \mathrm{HDS}_{\frac{3}{2}}$ *and* $\psi_\infty(x) = e(x)$. *Then*
(i) $\mathrm{Wald}(\sigma_\infty, \psi_\infty^-) = \mathrm{DS}_2$, *the weight 2 discrete series representation of* $\mathrm{PGL}_2(\mathbb{R})$.
(ii) $\theta(\sigma_\infty, V_\infty^-) = \mathbb{1} \neq 0$ *and* $\theta(\sigma_\infty, V_\infty^+) = 0$. *In particular,*

$$\delta_\infty(\mathrm{HDS}_{\frac{3}{2}}, \psi_\infty^-) = -1.$$

Combining these facts, we obtain the following results concerning the vanishing of the arithmetic theta lift.

Proposition 9.1.3. *If* $\epsilon(\frac{1}{2}, \mathrm{Wald}(\sigma, \psi_-)) = +1$, *then* $\theta^{\mathrm{ar}}(\sigma, M^B) = 0$ *for all indefinite quaternion algebras* B *over* $\mathbb{Q}$.

This is the complement to the result of Waldspurger which says that all classical theta lifts $\theta(\sigma, V^B)$ vanish when $\epsilon(\frac{1}{2}, \mathrm{Wald}(\sigma, \psi_-)) = -1$. Of course, in the arithmetic case, we only consider σ's with $\sigma_\infty = \mathrm{HDS}_{\frac{3}{2}}$.

Corollary 9.1.4. *If* $\theta^{\mathrm{ar}}(\sigma, M^B) \neq 0$, *then* $L(\frac{1}{2}, \mathrm{Wald}(\sigma, \psi_-)) = 0$.

It is interesting to note that this assertion, in which the nonvanishing of a class in the Mordell-Weil group implies the vanishing of an L-function, is an immediate consequence of local Howe duality and theta dichotomy once the arithmetic theta lift has been constructed.[3]

Finally, in analogy with Waldspurger's result, part (ii) of Theorem 8.2.5, we have the following conjecture concerning the nonvanishing of the arithmetic theta lift $\theta^{\mathrm{ar}}(\sigma, M^B)$.

[3]Of course, we are only using the fact that $\Theta(\sigma_o, V^B)$ has a unique irreducible quotient.

Conjecture 9.1.5. *If* $\epsilon(\frac{1}{2}, \mathrm{Wald}(\sigma, \psi_{-1})) = -1$, *let* B *be the unique indefinite quaternion algebra over* $\mathbb{Q}$ *with* $\mathrm{inv}_p(B) = \delta_p(\sigma_p, \psi_p^-)$ *for all* $p < \infty$. *Then*

$$\theta^{\mathrm{ar}}(\sigma, M^B) \neq 0 \qquad \Longleftrightarrow \qquad L'(\frac{1}{2}, \mathrm{Wald}(\sigma, \psi_{-1})) \neq 0.$$

Remark 9.1.6. (i) In both Corollary 9.1.4 and Conjecture 9.1.5, we exclude, for the moment, the case of the modular curve M^B associated to $B = M_2(\mathbb{Q})$, where we have not defined the space $\theta^{\mathrm{ar}}(\sigma, M^B)$.
(ii) In [2], Gross formulated an arithmetic analogue of another result of Waldspurger [14]. In that case, the group involved is GU(2), and many cases of Gross's conjecture follow from the original results of Gross and Zagier [4] and from work of Shou-Wu Zhang [17]. A brief discussion of this work is given at the end of [9].

9.2 THE ARITHMETIC INNER PRODUCT FORMULA

In this section, we will prove certain cases of Conjecture 9.1.5 by using an arithmetic analogue of the Rallis inner product formula. The key step in doing this is to relate the arithmetic theta lift defined in section 1 using the Borcherds generating function $\phi^B_{\mathrm{Bor}}(\tau, \varphi)$ valued in $\mathrm{MW}(M^B)$ to another arithmetic theta lift defined using the $\widehat{\mathrm{CH}}^1(\mathcal{M}^B)$-valued generating function $\widehat{\phi}_1(\tau) = \widehat{\phi}^B_1(\tau)$. The lift defined in section 1 has the advantage that it depends only on the Shimura curve M^B over $\mathbb{Q}$ and hence can be defined for all levels, since it does not involve integral models. In contrast, the generating function $\widehat{\phi}^B_1(\tau)$ has only been constructed for the stack $\mathcal{M}^B$ over $\mathrm{Spec}\,\mathbb{Z}$, and so there is no level and no dependence on a variable Schwartz function $\varphi \in S(V^B(\mathbb{A}_f))$. On the other hand, the main results of Chapters 6 and 7 give a precise relation between $\widehat{\phi}_1(\tau)$ and the Eisenstein series $\mathcal{E}_2(\tau, s; B)$, so that the doubling formula provides the crucial link to the central derivative of the L-function.

We first define a variant of the arithmetic theta lift of the previous section. For an indefinite quaternion algebra B over $\mathbb{Q}$, recall that the generating function

$$\widehat{\phi}^B_1(\tau) = \sum_{t \in \mathbb{Z}} \widehat{\mathcal{Z}}(t, v)\, q^t$$

is a (nonholomorphic) modular form of weight $\frac{3}{2}$ for $\Gamma_0(4D(B)_o)$ valued in the arithmetic Chow group $\widehat{\mathrm{CH}}^1(\mathcal{M}^B)$, where

$$D(B)_o = \begin{cases} D(B) & \text{if } D(B) \text{ is odd,} \\ D(B)/2 & \text{if } D(B) \text{ is even.} \end{cases}$$

Slightly abusing notation, we also write $\widehat{\phi}_1^B$ for the left $G'_{\mathbb{Q}}$-invariant function on $G'_{\mathbb{A}}$ determined by $\widehat{\phi}_1^B$ as in Proposition 8.5.20. For any $f \in \mathcal{A}_{00}(G')$, we let

(9.2.1)

$$\begin{aligned}\widehat{\theta}^B(f) = \langle f, \widehat{\phi}_1 \rangle_{\mathrm{Pet}} &= \int_{\mathrm{Sp}_1(\mathbb{Q})\backslash \mathrm{Sp}_1(\mathbb{A})} f(g')\, \overline{\widehat{\phi}_1^B(g')}\, dg \\ &= \int_{\mathrm{Sp}_1(\mathbb{Q})\backslash \mathrm{Sp}_1(\mathbb{A})} f(g')\, \widehat{\phi}_1^B((g')^\vee)\, dg \quad \in \widehat{\mathrm{CH}}^1(\mathcal{M}),\end{aligned}$$

where dg is Tamagawa measure and

$$(g')^\vee = \begin{pmatrix} 1 & \\ & -1 \end{pmatrix} g' \begin{pmatrix} 1 & \\ & -1 \end{pmatrix}^{-1}.$$

Of course, this integral vanishes unless $f \in \mathcal{A}_{00}(G')_{\frac{3}{2},\mathrm{hol}}$. The analysis of the components of the function $\widehat{\phi}_1(\tau)$ given in Chapter 4 implies the following:

Proposition 9.2.1. *For any* $f \in \mathcal{A}_{00}(G')$,

(i)

$$\deg_{\mathbb{Q}}(\widehat{\theta}^B(f)) = 2\,\langle \widehat{\theta}^B(f), 1\!\!1 \rangle = 0,$$

(ii)

$$\langle \widehat{\theta}^B(f), \hat{\omega} \rangle = 0, \qquad \textit{and,}$$

(iii) for all $\phi \in A^0(\mathcal{M}_{\mathbb{R}})_0$,

$$\langle \widehat{\theta}^B(f), a(\phi) \rangle = 0.$$

Proof. For example,

$$\begin{aligned}\langle \widehat{\theta}^B(f), 1\!\!1 \rangle &= \langle f, \langle \widehat{\phi}_1^B, 1\!\!1 \rangle \rangle_{\mathrm{Pet}} \\ &= \langle f, \frac{1}{2}\,\mathcal{E}_1(\frac{1}{2}, B) \rangle_{\mathrm{Pet}} = 0,\end{aligned} \tag{9.2.2}$$

and

$$(9.2.3)\qquad \langle\, \widehat{\theta}^B(f), \hat{\omega}\,\rangle = \langle\, f, \langle\, \widehat{\phi}_1^B, \hat{\omega}\,\rangle\,\rangle_{\text{Pet}}$$
$$= \langle\, f, \mathcal{E}_1'(\frac{1}{2}, B)\,\rangle_{\text{Pet}} = 0,$$

since f is cuspidal. Here we have used relations (4.1.14), (4.2.12) and (4.2.13). □

It follows that $\widehat{\theta}^B(f)$ lies in $\widetilde{\text{MW}}(\mathcal{M}^B) \oplus \text{Vert}^0$ where Vert^0 consists of the classes in Vert which are orthogonal to $\hat{\omega}$.

Next we apply Proposition 4.5.2 to obtain the key link between the arithmetic theta lift just defined and that introduced in Section 9.1.

Proposition 9.2.2. *Let* $\text{res}_{\mathbb{Q}} : \widehat{\text{CH}}^1(\mathcal{M}^B) \to \text{CH}^1(\mathcal{M}^B_{\mathbb{Q}})$ *be the restriction map. Let* $\varphi^0 \in S(V^B(\mathbb{A}_f))$ *be the characteristic function of the set* $\widehat{O}_B \cap V^B(\mathbb{A}_f)$, *where* O_B *is a maximal order in* B. *Then*

$$\text{res}_{\mathbb{Q}}(\widehat{\theta}^B(f)) = \theta^{\text{ar}}(f, \varphi^0).$$

For a genuine cuspidal automorphic representation σ with $\sigma_\infty \simeq \text{HDS}_{\frac{3}{2}}$, as in the previous section, let

$$\widehat{\theta}(\sigma, \mathcal{M}^B) \quad \subset \quad \widehat{\text{CH}}^1(\mathcal{M}^B)$$

be the subspace generated by the $\widehat{\theta}^B(f)$'s as f varies in $\mathcal{V}(\sigma)$.

Using Proposition 9.2.2 together with Proposition 9.1.3 and the results of Section 4.3 about the vertical components of $\widehat{\phi}_1^B(\tau)$, we obtain the following result.

Corollary 9.2.3. *(i) If* $\epsilon(\frac{1}{2}, \text{Wald}(\sigma, \psi_-)) = +1$, *then*

$$\text{res}_{\mathbb{Q}}(\widehat{\theta}(\sigma, \mathcal{M}^B)) = 0,$$

so that $\widehat{\theta}(\sigma, \mathcal{M}^B) \subset \text{Vert}^0$.
(ii) If $\epsilon(\frac{1}{2}, \text{Wald}(\sigma, \psi_-)) = -1$, *then*

$$\langle\, \widehat{\theta}(\sigma, \mathcal{M}^B), \text{Vert}\,\rangle = 0,$$

so that $\widehat{\theta}(\sigma, \mathcal{M}^B) \subset \widetilde{\text{MW}}(\mathcal{M}^B)$.

From now on, we consider only σ's such that $\epsilon(\frac{1}{2}, \text{Wald}(\sigma, \psi_-)) = -1$, so that the space $\widehat{\theta}(\sigma, \mathcal{M}^B)$ is isomorphic, via $\text{res}_{\mathbb{Q}}$, to the subspace of $\text{MW}(\mathcal{M}^B_{\mathbb{Q}})$ spanned by the $\theta^{\text{ar}}(f, \varphi^0)$'s for $f \in \sigma$. Since the function $\widetilde{\phi}_{\text{Bor}}(g', \varphi^0)$ is an eigenfunction under right multiplication by $\mathbf{K}_0(4D(B)_o)$ with character χ_κ, where $\kappa = -1$, it follows that $\widehat{\theta}(\sigma, \mathcal{M}^B)$ is zero unless

$$(9.2.4)\qquad \sigma^{\mathbf{K}_0(4D(B)_o), \chi_\kappa} \neq 0,$$

and

$$\delta_p(\sigma_p, \psi_p^-) = \mathrm{inv}_p(B) \tag{9.2.5}$$

for all finite p. For such a representation σ, we must have

$$\begin{aligned} &\text{for } p \nmid D(B), && \sigma_p \simeq \tilde{\pi}(\chi_\alpha |\ |^t), && \text{with } t \neq \frac{1}{2} \text{ and } \alpha \in \mathbb{Z}_p^\times, \\ &\text{for } p \mid D(B), && \sigma_p \simeq \tilde{\sigma}(\chi_\kappa |\ |^{\frac{1}{2}}). \end{aligned} \tag{9.2.6}$$

Combining the main results of Chapters 6, 7 and 8, we obtain an arithmetic analogue of the Rallis inner product formula.

Theorem 9.2.4. *For σ as above, let*

$$\boldsymbol{f} \simeq \boldsymbol{f}_{\frac{3}{2}} \otimes (\otimes_p \boldsymbol{f}_p)$$

be the good newvector in σ of weight $\frac{3}{2}$, as in Section 8.3, and let $\pi =$ $\mathrm{Wald}(\sigma, \psi_-)$. Then

$$\langle \widehat{\phi}_1(\tau_1), \widehat{\theta}(\boldsymbol{f}) \rangle = \mathcal{C} \cdot L'(\frac{1}{2}, \pi) \cdot \underline{f}(\tau_1),$$

where

$$\mathcal{C} = \frac{3}{2\pi^2} \cdot \prod_{p | D(B)} \frac{2}{p+1}$$

and $\underline{f}$ is the classical newform attached to $\boldsymbol{f}$.

Proof. We compute:[4]

$$\begin{aligned} \langle \widehat{\phi}_1(\tau_1), \widehat{\theta}(\boldsymbol{f}) \rangle &= \langle \underline{f}, \langle \widehat{\phi}_1(\tau_1), \widehat{\phi}_1 \rangle \rangle_{\mathrm{Pet}} \\ &= \langle \underline{f}, \widehat{\phi}_2(\begin{pmatrix} \tau_1 & \\ & \cdot \end{pmatrix}) \rangle_{\mathrm{Pet}} \\ &= \langle \underline{f}, \mathcal{E}_2'(\begin{pmatrix} \tau_1 & \\ & \cdot \end{pmatrix}, 0; B) \rangle_{\mathrm{Pet}} \qquad (9.2.7) \\ &= \frac{\partial}{\partial s}\Big\{ \langle \underline{f}, \mathcal{E}_2(\begin{pmatrix} \tau_1 & \\ & \cdot \end{pmatrix}, s; B) \rangle_{\mathrm{Pet}} \Big\}\Big|_{s=0} \end{aligned}$$

[4] In the intermediate steps here, we should really write integrals of $\underline{f}(\tau)$ against functions $\langle \widehat{\phi}_1(\tau_1), \widehat{\phi}_1(-\bar{\tau}) \rangle$, etc. rather than using the complex conjugate as in the definition of $\langle\ ,\ \rangle_{\mathrm{Pet}}$. Also note that, since the classes $\widehat{\mathcal{Z}}(t, v)$ are real, $\overline{\widehat{\phi}_1(\tau)} = \widehat{\phi}_1(-\bar{\tau})$.

$$= \frac{\partial}{\partial s}\left\{C(s)\cdot L(s+\frac{1}{2}, \text{Wald}(\sigma,\psi_{-1}))\right\}\Big|_{s=0}\cdot \underline{f}(\tau_1)$$

$$= C(0)\cdot L'(\frac{1}{2}, \text{Wald}(\sigma,\psi_{-1}))\cdot \underline{f}(\tau_1).$$

Here the function $C(s)$ is as in Theorem 8.3.3. □

Corollary 9.2.5. *With the same notation as in Theorem 9.2.4,*

$$\langle\, \widehat{\theta}(\boldsymbol{f}), \widehat{\theta}(\boldsymbol{f})\,\rangle = C\cdot L'(\frac{1}{2},\pi)\cdot \langle\, \boldsymbol{f},\boldsymbol{f}\,\rangle.$$

In particular,

$$\widehat{\theta}(\boldsymbol{f}) \neq 0 \qquad \Longleftrightarrow \qquad L'(\frac{1}{2},\pi)\neq 0.$$

In the next section we will prove the following result, related to that of Kohnen [7], [1].

Proposition 9.2.6. *Let $S_2^{\text{new}}(D(B))^{(-)}$ be the set of normalized newforms of weight 2 and for $\Gamma_0(D(B))$ for which all of the Atkin-Lehner eigenvalues for $p \mid D(B)$ are -1. Then there is a bijection between $S_2^{\text{new}}(D(B))^{(-)}$ and the set of genuine cuspidal automorphic representations σ with $\sigma_\infty = \text{HDS}_{\frac{3}{2}}$ and with local components σ_p satisfying (9.2.6). This bijection is given by*

$$\sigma \longmapsto \text{Wald}(\sigma,\psi_{-1}) = \pi(F),$$

where, for $F\in S_2^{\text{new}}(D(B))^{(-)}$, $\pi(F)$ is the corresponding cuspidal automorphic representation of PGL_2.

We will sometimes write $\sigma(F)$ for the representation of $G'_{\mathbb{A}}$ determined by F.

We thus obtain the following 'explicit' formula for the restriction of our generating function $\widehat{\phi}_1^B(\tau)$ to the generic fiber, and, in particular, for its Mordell-Weil component.

Corollary 9.2.7. *For each $F\in S_2^{\text{new}}(D(B))^{(-)}$, let $\boldsymbol{f}\in\sigma(F)$ be the good newvector and let $\underline{f}$ be the corresponding holomorphic form of weight $\frac{3}{2}$ and level $4D(B)_o$. Then*

$$\text{res}_{\mathbb{Q}}(\,\widehat{\phi}_1^B(\tau)\,) = \phi^B_{\text{Bor}}(\tau,\varphi^0)$$

$$= \mathcal{E}_1(\tau,\frac{1}{2};B)\cdot\frac{\omega_{\mathbb{Q}}}{\deg\omega_{\mathbb{Q}}} + \sum_{F\in S_2^{\text{new}}(D(B))^{(-)}} \frac{\underline{f}(\tau)\cdot\widehat{\theta}(\boldsymbol{f})}{\langle\,\boldsymbol{f},\boldsymbol{f}\,\rangle},$$

where $\omega_{\mathbb{Q}}$ is the restriction of the Hodge bundle to $\mathcal{M}_{\mathbb{Q}}$. The class $\widehat{\theta}(\boldsymbol{f})$ is nonzero if and only if $L'(1,F)\neq 0$.

Here we have identified $\widehat{\theta}(\boldsymbol{f})$ with its image under $\mathrm{res}_{\mathbb{Q}}$.

If $t \in \mathbb{Z}_{\geq 0}$, let $Z(t) = Z(t, \varphi^0) \in \mathrm{CH}^1(M^B)$ be the class defined in Section 4.5, and recall that

$$\phi_{\mathrm{Bor}}(\tau, \varphi^0) = \sum_{t \geq 0} Z(t)\, q^t.$$

For $F \in S_2^{\mathrm{new}}(D(B))^{(-)}$, let $\pi^B = \pi^B(F)$ be the automorphic representation of $H^B(\mathbb{A}) = B^\times(\mathbb{A})$, with trivial central character, determined by $\pi = \pi(F)$ via the Jacquet-Langlands correspondence. By Waldspurger's results, $\pi_0^B \simeq \theta_\psi(\sigma_0, V^B)$, where π_0^B is the finite component of π^B. Let

$$\mathrm{CH}^1(M^B)(\pi^B) = \mathrm{MW}(M^B)(\pi^B)$$

be the π_0^B-isotypic component of $\mathrm{MW}(M^B)$, and let $Z(t)(\pi^B)$ be the image of $Z(t)$ in $\mathrm{MW}(M^B)(\pi^B)$. Then passing to π_0^B-isotypic components on both sides of the identity in Corollary 9.2.7, we have

$$Z(t)(\pi^B) = \frac{a_t(\underline{\boldsymbol{f}}) \cdot \widehat{\theta}(\boldsymbol{f})}{\langle\, \boldsymbol{f}, \boldsymbol{f}\, \rangle}, \tag{9.2.8}$$

where

$$\underline{\boldsymbol{f}}(\tau) = \sum_{t>0} a_t(\underline{\boldsymbol{f}})\, q^t. \tag{9.2.9}$$

By Corollary 9.2.5, we obtain a formula for the height pairing of the classes $Z(t)(\pi^B)$.

Corollary 9.2.8. *For t_1 and $t_2 \in \mathbb{Z}_{>0}$,*

$$\langle\, Z(t_1)(\pi^B), Z(t_2)(\pi^B)\, \rangle = C \cdot L'(1, F) \cdot \frac{a_{t_1}(\underline{\boldsymbol{f}}) \cdot a_{t_2}(\underline{\boldsymbol{f}})}{\langle\, \boldsymbol{f}, \boldsymbol{f}\, \rangle}.$$

Remark 9.2.9. (i) Note that, in this identity, the quantity on the right side is invariant under scaling of $\boldsymbol{f}$, as it should be, since $\boldsymbol{f}$ is only well defined up to a scalar. Similarly, the constant C depends on the normalization of the Petersson inner product in the denominator. Note that, if $\underline{\boldsymbol{f}}$ is the classical modular form of weight $\frac{3}{2}$ associated to $\boldsymbol{f}$, then

$$\begin{aligned} \langle\, \boldsymbol{f}, \boldsymbol{f}\, \rangle_{\mathrm{Pet}} &= C \cdot \int_{\Gamma_0(4D(B)_o)\backslash \mathfrak{H}} \underline{\boldsymbol{f}}(\tau)\, \overline{\underline{\boldsymbol{f}}(\tau)}\, v^{\frac{3}{2}} \cdot v^{-2}\, du\, dv \\ &= C \cdot \langle\, \underline{\boldsymbol{f}}, \underline{\boldsymbol{f}}\, \rangle_{\mathrm{Pet}}, \end{aligned} \tag{9.2.10}$$

where $\langle \underline{f}, \underline{f} \rangle_{\text{Pet}}$ is the classical Petersson inner product. Here

$$C = \frac{1}{2\pi} \prod_{p|D(B)_o} (p+1)^{-1};$$

see Lemma 8.4.29. Thus, if $\langle \underline{f}, \underline{f} \rangle_{\text{Pet}}$ is used in the denominator in Corollary 9.2.8, then the constant of proportionality becomes

$$\frac{\mathcal{C}}{C} = \frac{3}{\pi} \cdot 2^{o(D(B))} \cdot \begin{cases} \frac{1}{3} & \text{if } 2 \mid D(B), \\ 1 & \text{if } 2 \nmid D(B), \end{cases}$$

where $o(D(B))$ is the number of prime factors of $D(B)$.
(ii) The results of Corollaries 9.2.5, 9.2.7, and 9.2.8 are closely related to those of Gross-Kohnen-Zagier [3] and Zagier [16] for the Heegner points on the modular curve. Here we have derived them as consequences of the modularity of $\phi_{\text{Bor}}(\tau, \varphi^0)$, the relation between this function and the generating function $\widehat{\phi}_1^B(\tau)$, and the arithmetic inner product formula. In [3], the analogue of Corollary 9.2.8 is proved first and modularity of the generating series is a consequence; see the discussion on pp. 502–503 of [3].
(iii) Examples of newforms F with $\epsilon(\frac{1}{2}, \pi(F)) = -1$ and with all Atkin-Lehner signs equal to -1 can be found in Stein's tables, [11]. A few such examples are listed in Table 3 at the end of this section.

Finally, we would like to relate the nonvanishing criterion given in Corollary 9.2.5 to Conjecture 9.1.5 above.

Theorem 9.2.10. *Suppose that σ is a genuine cuspidal automorphic representation satisfying the conditions of (9.2.6) and with $\sigma_\infty \simeq \text{HDS}_{\frac{3}{2}}$. Suppose that, for $\pi = \text{Wald}(\sigma, \psi_{-1})$, $\epsilon(\frac{1}{2}, \pi) = -1$. Let B be the indefinite quaternion algebra over $\mathbb{Q}$ with $\text{inv}_p(B) = \delta_p(\sigma_p, \psi_p^-)$ for all p. Assume that $D(B) > 1$. Then*

$$\theta^{\text{ar}}(\sigma, M^B) \neq 0 \qquad \Longleftrightarrow \qquad L'(\frac{1}{2}, \pi) \neq 0.$$

Remark 9.2.11. (i) Note that, when $L'(\frac{1}{2}, \pi) \neq 0$,

$$\pi_0^B \simeq \theta^{\text{ar}}(\sigma, M^B) \subset \text{MW}(M^B),$$

where π_0^B is the finite component of the representation $\pi^B \simeq \pi_\infty^B \otimes \pi_0^B$ of $H^B(\mathbb{A})$ corresponding to $\pi = \text{Wald}(\sigma, \psi_{-1})$ under the Jacquet-Langlands correspondence.
(ii) In effect, we suppose that σ has square free level and that $\delta_p(\sigma_p, \psi_p^-) =$

-1 for all primes dividing the level. In order to remove these conditions on the level of σ and on the $\delta_p(\sigma_p, \psi_p^-)$'s and to allow $D(B) = 1$, we would have to extend some of our geometric results about generating functions for arithmetic cycles on $\mathcal{M}$ to the case nontrivial level and to the modular curve. Of course, for the application to the central derivative of the L-function, it is possible to work 'modulo oldforms', as is done in [4] and [17], so that complete geometric information would not be needed. Some additional remarks are made at the end of Section 9.3.

Proof. Let $\boldsymbol{f} \in \sigma$ be the good newvector. By Corollary 9.2.5, we have

(9.2.11)

$$L'(\frac{1}{2}, \mathrm{Wald}(\sigma, \psi_{-1})) \neq 0 \iff \widehat{\theta}(\boldsymbol{f}) \neq 0$$
$$\iff \theta^{\mathrm{ar}}(\boldsymbol{f}, \varphi^0) \neq 0 \implies \theta^{\mathrm{ar}}(\sigma, M^B) \neq 0.$$

Thus, it remains to prove that $\boldsymbol{f}$ and φ^0 are 'good test vectors' in the following sense:

Proposition 9.2.12.

$$\theta^{\mathrm{ar}}(\boldsymbol{f}, \varphi^0) \neq 0 \iff \theta^{\mathrm{ar}}(\sigma, M^B) \neq 0.$$

Proof. Assume that $\theta^{\mathrm{ar}}(\sigma, M^B) \neq 0$, so that the map j_σ^B of (9.1.16) is an isomorphism. It will then suffice to show that the vector $\mathrm{pr}_\sigma^B(\varphi^0) \in \sigma_0 \otimes \theta(\sigma_0, V^B)$ has nonzero pairing with $\boldsymbol{f}_0$, the finite component of $\boldsymbol{f}$. Here pr_σ^B is the natural projection in (9.1.15). This is then a local question. It suffices to show that, for each p, the image of φ_p^0 in the quotient $S(\sigma_p, V_p^B) \simeq \sigma_p \otimes \Theta(\sigma_p, V_p^B)$ has nonzero pairing with the good newvector $\boldsymbol{f}_p$.

First suppose that $p \neq 2$. Then, if $p \nmid D(B)$, σ_p is an unramified principal series and $\boldsymbol{f}_p$ is the $\mathbf{K}'$-invariant vector. The vector $\varphi_p^0 \in S(V_p)$ is also $\mathbf{K}'$-invariant and, as is shown in [10], has nonzero image $\mathrm{pr}_\sigma^B(\varphi_p^0)$ in $S(\sigma_p, V_p^B) \simeq \sigma_p \otimes \theta(\sigma_p, V_p^B)$. Thus the pairing of $\mathrm{pr}_\sigma^B(\varphi_p^0)$ and $\boldsymbol{f}_p$ is nonzero, as required. If $p \mid D(B)$, then σ_p is an unramified special representation with $\delta_p(\sigma_p, \psi_p^-) = -1$ and $\boldsymbol{f}_p$ is the $\mathbf{J}'$-invariant vector. Moreover, the condition $\delta_p(\sigma_p, \psi_p^-) = -1$ implies that $\sigma_p \simeq \tilde{\sigma}(\chi_{-1}|\,|^{\frac{1}{2}})$.[5] The map

$$\mathrm{pr}_\sigma^B : S(V_p^B) \longrightarrow \sigma_p \otimes \theta(\sigma_p, V_p^B) \simeq \tilde{\sigma}(\chi_{-1}|\,|^{\frac{1}{2}}) \otimes 1\!\!1$$

is given by $\lambda_{V_p} : \varphi \mapsto (g \mapsto \omega(g)\varphi(0))$. The vector $\varphi_p^0 \in S(V_p^B)$ is $\mathbf{J}'$-invariant and hence its image $\mathrm{pr}_\sigma^B(\varphi_p^0)$, which is nonzero by [10], is also $\mathbf{J}'$-invariant and has a nonzero pairing with $\boldsymbol{f}_p$, as required.

[5] Here we use the ψ-parametrization in Table 2 in Chapter 8.

Next suppose that $p = 2$. If $p \mid D(B)$, then $\sigma_p = \tilde{\sigma}(\chi_{-1}|\ |^{\frac{1}{2}})$. By Proposition 8.4.16, the space of χ_{-1}-eigenvectors for $\mathbf{J}'$ in σ_p has dimension 1 and $\boldsymbol{f}_p = \boldsymbol{f}_{\text{sp}}$ is a basis. On the other hand, by Proposition 8.5.14, the vector φ_p^0 is also a χ_{-1}-eigenvector for $\mathbf{J}'$ and the same argument as for odd p gives the required nonvanishing.

Finally suppose that $p = 2$ with $p \nmid D(B)$, so that $\sigma_p = \tilde{\pi}(\chi_{2\alpha}|\ |^t)$ with $\alpha \in \mathbb{Z}_2^\times$ and $t \neq \pm\frac{1}{2}$. Note that the condition on the global central character, discussed in Section 8.3, implies that $\alpha \equiv \kappa \mod 4$. Here $\kappa = -1$. In any case, the character $\mu^{-1}\chi_{2\kappa}$ is unramified. By Proposition 8.4.15, the space $I(\mu)^{(\mathbf{J}',\chi_\alpha)}$ of χ_α-eigenvectors for $\mathbf{J}'$ has dimension 2 and is spanned by the functions f_1 and f_w with support in $P\mathbf{J}'$ and $Pw\mathbf{J}'$ respectively, with $f_1(1) = 1$ and $f_w([w,1]_{\mathrm{L}}) = 1$. The good newvector in σ_p is $\boldsymbol{f}_p = \boldsymbol{f}_{\text{ev}} = f_1 + \frac{i}{\sqrt{2}} f_w$. The pairing between $I(\mu^{-1}) = I(\mu)^\vee$ and $I(\mu)$ is given by

$$(f, f') = \int_{\underline{K}'} f(k') \cdot f'(k') \cdot \zeta(k')^{-2}\, dk,$$

where $k' \in G'$ is any element which projects to $k \in \underline{K}' = \mathrm{SL}_2(\mathbb{Z}_p)$, and the factor $\zeta(k')^{-2}$ makes the integrand independent of the choice of k'; see Lemma 8.5.9. Note that $I(\mu)$ and $I(\mu^{-1})$ can be identified with the same space of functions on K'. Using this identification, the restriction of the pairing to the space of χ_α-eigenvectors for $\mathbf{J}'$ is given by $(f_1, f_1) = \mathrm{vol}(\underline{J}')$, $(f_1, f_w) = 0$ and

$$(f_w, f_w) = 4\,\mathrm{vol}(\underline{J}') \cdot \zeta([w,1]_{\mathrm{L}})^{-2} = -4.$$

Here we use the fact that

$$\zeta([w,1]_{\mathrm{L}}) = (-1,-1)_2\, \gamma(\psi_{\frac{1}{2}})^2 = i,$$

by (8.5.19) and (8.4.29). It follows that

$$(\boldsymbol{f}_p, \boldsymbol{f}_p) = \mathrm{vol}(\underline{J}') \cdot 3 = \frac{1}{2}.$$

Using the explicit formula given in [10] for the projection from $S(V_p^B)$ to the maximal σ_p-isotypic quotient, it is easily checked that, when $\mu^{-1}\chi_V$ is unramified, as it is in our case,

$$\mathrm{pr}_\sigma^B(\varphi^0) = C \cdot \boldsymbol{f}_{\text{ev}}$$

for a nonzero constant C. It follows that the pairing of this vector with $\boldsymbol{f}_p = \boldsymbol{f}_{\text{ev}}$ is nonzero. □

This completes the proof of Theorem 9.2.10. □

Table 3. Example of Newforms F from Stein's Tables

$D(B)$	genus of $\mathcal{M}_{\mathbb{Q}}$	F	dim of factor
$7 * 13$	7	91B	1
$3 * 41$	13	123B	1
$7 * 19$	9	133C	2
$5 * 29$	13	145B	2
$5 * 31$	11	155C	1
$5 * 37$	13	185C	1
$11 * 17$	13	187D	2
$7 * 37$	19	259E	3
$2 * 173$	14	346B	1
$13 * 29$	29	377A	1
		377D	5
$11 * 37$	31	407B	4
$31 * 43$	105	1333A	23
$31 * 113$	281	3503A	56
$43 * 83$	287	3569F	61
$43 * 89$	309	3827A	1
		3827B	65
$3 * 5 * 7 * 11$	41	1155N	1

9.3 THE RELATION WITH CLASSICAL NEWFORMS

In this section, we prove a slight generalization of Proposition 9.2.6.

Let $l \in \frac{3}{2} + \mathbb{Z}_{\geq 0}$ be a half integer and let $\kappa = (-1)^{l-\frac{1}{2}}$. Let N be a square free integer and let $S_{2l-1}^{\text{new}}(N)$ be the set of normalized newforms of weight $2l-1$ and level N. Associated to each $F \in S_{2l-1}^{\text{new}}(N)$, there is a cuspidal automorphic representation $\pi = \pi(F) \simeq \otimes_{p \leq \infty} \pi_p$ of $\mathrm{PGL}_2(\mathbb{A})$. The following result is essentially that of Kohnen [7], [1], although we allow N to be even.

Proposition 9.3.1. *(i) For each $F \in S_{2l-1}^{\text{new}}(N)$, there is a unique genuine cuspidal automorphic representation $\sigma(F) = \sigma \simeq \otimes_{p\leq\infty}\sigma_p$ of $G'_{\mathbb{A}}$ with* $\text{Wald}(\sigma, \psi_\kappa) = \pi$ *such that $\sigma_\infty = \text{HDS}_l$ is a holomorphic discrete series and such that, for all $p \mid N$, σ_p is a special representation. Moreover, the dichotomy signs for the local components of σ are given by*

$$\delta(\sigma_p, \psi_p^\kappa) = \begin{cases} \epsilon_p(F) & \textit{if } p \mid N, \\ (-1)^{l-\frac{1}{2}} & \textit{if } p = \infty, \\ 1 & \textit{otherwise,} \end{cases}$$

where, for $p \mid N$, $\epsilon_p(F)$ is the eigenvalue of F for the Atkin-Lehner involution,

$$F \mid W_p = \epsilon_p(F)\, F.$$

The classical modular $\underline{f}$ corresponding to the good test vector $\boldsymbol{f}$ in σ, as defined in Chapter 8, has weight l and level $4N_o$, where $N_o = N/2$ if N is even and $N_o = N$ if N is odd.
(ii) Conversely, suppose that $\sigma \simeq \otimes_{p\leq\infty}\sigma_p$ is a genuine cuspidal automorphic representation of $G'_{\mathbb{A}}$ with $\sigma_\infty =$ HDS_l and with finite components σ_p satisfying the conditions

$$\textit{for } p \nmid N, \qquad \sigma_p \simeq \tilde{\pi}(\chi_\alpha |\ |^t), \qquad \textit{with } t = t_p \neq \frac{1}{2} \textit{ and } \alpha = \alpha_p \in \mathbb{Z}_p^\times,$$

$$\textit{for } p \mid N, \qquad \sigma_p \simeq \tilde{\sigma}(\chi_\alpha |\ |^{\frac{1}{2}}), \qquad \textit{with } \alpha = \alpha_p \in \mathbb{Z}_p^\times,$$

for some square free integer N, and with $\alpha_2 \equiv 1 \mod 4$. Then the normalized newform F determined by $\pi =$ $\text{Wald}(\sigma, \psi_\kappa)$ lies in $S_{2l-1}^{\text{new}}(N)$, and $\sigma(F) = \sigma$.

Proof. The representation $\pi = \pi(F) \simeq \otimes_{p\leq\infty}\pi_p$ has local components $\pi_\infty = \text{DS}_{2l-1}$ and π_p, for $p \nmid N$, determined by the Hecke eigenvalue $a_p(F)$, where $F \mid T_p = a_p(F)\, F$. For $p \mid N$, the local component π_p is an unramified special representation $\pi_p = \sigma(\chi_\alpha |\ |^{\frac{1}{2}}, \chi_\alpha |\ |^{-\frac{1}{2}})$, where $\alpha = \alpha_p \in \mathbb{Z}_p^\times$ and $\alpha \equiv 1 \mod 4$ if $p = 2$. The quadratic character χ_{α_p} is determined by the Atkin-Lehner sign $\epsilon_p(F)$ according to the rule

$$\chi_{\alpha_p}(p) = -\epsilon_p(F). \tag{9.3.1}$$

By a basic result of Waldspurger, the set of genuine cuspidal automorphic representations σ in $\mathcal{A}_{00}(G'_{\mathbb{A}})$ with $\text{Wald}(\sigma, \psi_\kappa) = \pi$ is nonempty and has cardinality 1 if $N = 1$ and $2^{o(N)}$ if the number of prime factors $o(N)$ of N is positive. The possibilities for σ can be described by using the parametrization of local components given in Table 2 in Chapter 8. If $\text{Wald}(\sigma, \psi_\kappa) = \pi$, then $\sigma_\infty = \tilde{\pi}_l^{\pm}$, and the local components σ_p for $p \nmid N$ are irreducible principal series which are uniquely determined by the corresponding local components $\pi_p = \mathrm{Wald}(\sigma_p, \psi_p^\kappa)$ of π. For $p \mid N$ there are two possibilities,

$$\sigma_p \in \{\, \tilde{\sigma}_p(\chi_{\alpha_p\kappa}|\ |^{\frac{1}{2}}), \theta(\mathrm{sgn}, U_{\alpha_p\kappa}) \,\},$$

related by the Waldspurger involution; see Proposition 8.2.4. Moreover, any choice of σ_p's is allowed, subject to the condition (8.2.28) on the product of the central signs. One possible choice of σ_p's is given by the following.

Lemma 9.3.2. *The representation $\sigma \simeq \otimes_{p\le\infty}\sigma_p$ with local components $\sigma_\infty = \tilde{\pi}_l^+$ and $\sigma_p = \tilde{\sigma}_p(\chi_{\alpha\kappa}|\ |^{\frac{1}{2}})$ for all $p \mid N$ is a genuine cuspidal automorphic representation of $G'_{\mathbb{A}}$ with* $\mathrm{Wald}(\sigma, \psi_\kappa) = \pi$.

Proof. By the last row of Table 2 in Chapter 8, the central sign of the archimedean component σ_∞ is $z(\tilde{\pi}_l^+, \psi_\kappa) = 1$. Note that the fact that $\kappa = (-1)^{l-\frac{1}{2}}$ is used here. Then, by the last column of the table, we have the product formula for central signs

$$\prod_{p\le\infty} z(\sigma_p, \psi_\kappa) = \prod_{p|N} \chi_{\alpha_p}(-1) = 1,$$

since all of the χ_{α_p}'s are unramified. □

The dichotomy signs for the local components of σ are then given in the fourth column of Table 2. By the local results of Chapter 8, the good test vector $\boldsymbol{f} \in \sigma$ is unique up to a scalar factor. This vector has weight l for K'_∞, the inverse image of $\mathrm{SO}(2) \subset \mathrm{SL}_2(\mathbb{R})$ in $G'_{\mathbb{R}}$, and is an eigenvector for the group $\mathbf{K}_0(4N_o) \subset G'_{\mathbb{A}_f}$ with character χ_κ. The corresponding classical modular form $\underline{f}$ has weight l and level $4N_o$; see Proposition 8.5.20. □

Corollary 9.3.3. *(i) The map $F \mapsto \sigma(F)$ determines a bijection between the set $S_{2l-1}^{\mathrm{new}}(N)$ and the set of genuine cuspidal automorphic representations $\sigma(F) = \sigma \simeq \otimes_{p\le\infty}\sigma_p$ of $G'_{\mathbb{A}}$ with $\sigma_\infty = \mathrm{HDS}_l$ and with finite components σ_p satisfying the conditions of (ii) of Proposition 9.3.1.*
(ii) The representations in the set $\{\, \sigma \mid \mathrm{Wald}(\sigma, \psi_\kappa) = \pi(F),\ \sigma_\infty = \mathrm{HDS}_l\}$ have the form $\sigma = \sigma^\Sigma$ where, for a set Σ of primes dividing N, with $|\Sigma|$ even,

$$\sigma_p^\Sigma = \begin{cases} \theta(\mathrm{sgn}, U_{\alpha_p\kappa}) & \text{if } p \in \Sigma, \\ \sigma(F)_p & \text{otherwise.} \end{cases}$$

Of course, the second part here is just a restatement of Waldspurger's result, Proposition 8.2.4.

Remark 9.3.4. If we take the classical modular form $\underline{f}$ associated to the good test vector $\boldsymbol{f}$ in each $\sigma = \sigma(F)$, we obtain a bijection $F \mapsto \underline{f}$ between $S_{2l-1}^{\text{new}}(N)$ and a set of modular forms of weight l for $\Gamma_0(4N_o)$. When N is square free and odd, such a bijection was defined by Kohnen [7] where the image is his plus-space, characterized by the support of the Fourier expansion at the cusp at infinity. These two bijections do not agree. The point is that when N is odd, the local component $\sigma(F)_2$ of the representation $\sigma(F)$ is an irreducible principal series representation with a 2-dimensional space $\tilde{\pi}(\chi_\alpha|\ |^t)^{\mathbf{K}_0(4N),\chi_\kappa}$ of newvectors. Our local component

$$\boldsymbol{f}_2 = \boldsymbol{f}_{\text{ev}} = f_1 + \frac{\delta_\kappa}{\sqrt{2}} f_w$$

is an eigenfunction for a local zeta operator analyzed in Chapter 8. On the other hand, according to Baruch and Mao [1] or the computation of Whittaker functions given in Proposition 8.4.17, the local component of a function in the Kohnen plus space is, in our notation,

$$\begin{aligned}\boldsymbol{f}^+ &= f_1 + \frac{\sqrt{2}\mu(2)^2}{\delta_\alpha} f_w \\ &= (1 + 4\chi_\kappa(-1)\mu(2)^2)\,\boldsymbol{f}_{\text{ev}} + (2 - 4\chi_\kappa(-1)\mu(2)^2)\,\boldsymbol{f}_{\text{sp}};\end{aligned}$$

see (8.4.34). Of course, the underlying bijection $\pi(F) \mapsto \sigma(F)$ on representations is the same, while the choice of local component at $p = 2$ can be made according to the particular application at hand.

Now take $l = \frac{3}{2}$, and let $F \in S_2^{\text{new}}(N)$ be a newform of weight 2 with $\epsilon(\frac{1}{2}, \pi(F)) = -1$. Since

$$-\prod_{p|N} \epsilon_p(F) = \epsilon(\frac{1}{2}, \pi(F)), \tag{9.3.2}$$

there is a unique indefinite quaternion algebra B over $\mathbb{Q}$ such that $\text{inv}_p(B) = \epsilon_p(F)$ for all $p \mid N$ and $\text{inv}_p(B) = 1$ otherwise. When $D(B) = N$, i.e., when all of the Atkin-Lehner signs of F are -1, we take $\sigma = \sigma(F)$, and the results of the previous section describe the nonvanishing of the arithmetic theta lift $\theta^{\text{ar}}(\alpha, M^B)$ and of $\widehat{\theta}(\boldsymbol{f})$, where $\boldsymbol{f} \in \sigma(F)$ is the good newvector.

In the case in which some Atkin-Lehner signs of F are $+1$, we can again take $\sigma = \sigma(F)$ and consider the arithmetic theta lift $\theta^{\text{ar}}(\sigma, M^B)$, as in Section 9.1. To extend the results of Section 9.2 to this case, we should define a generating function $\widehat{\phi}_1^{B,N_0}(\tau)$ for the arithmetic surface $\mathcal{M}_0^B(N_0)$, where

$N = D(B)N_0$, associated to an Eichler order of level N_0 in B. This will involve a definition of the special cycles $\mathcal{Z}(t)$ with level structure and an extension of the results of this book to the generating function $\widehat{\phi}_1^{B,N_0}(\tau)$.

More generally, let B' be an indefinite quaternion algebra ramified at a set of primes dividing N, and let

$$\Sigma = \{ \, p \mid \epsilon_p(F) = -\mathrm{inv}_p(B') \, \}.$$

Take $\sigma = \sigma^{\Sigma}(F)$. Then, the arithmetic theta lift $\theta^{\mathrm{ar}}(\sigma, M^{B'})$ should be nonzero if and only if $L'(1, F) \neq 0$. The local component of σ^{Σ} at $p \in \Sigma$ is the odd Weil representation $\theta(\mathrm{sgn}, U_{\alpha_p \kappa})$ where χ_{α_p} is unramified and $\chi_{\alpha_p}(p) = -\epsilon_p(F)$. In order to define the analogue of $\widehat{\phi}_1^B$ in this case, we need to define cycles $\mathcal{Z}(t)$ which are anti-invariant under the Atkin-Lehner involution at the primes $p \in \Sigma$. This can be done by using a weighted average of the oriented cycles mentioned in the introduction and described in Chapter 3. Again, it should be possible to extend the main results of this book to the generating functions constructed from such cycles.

Bibliography

[1] M. Baruch and Z.Y. Mao, *Central value of automorphic L-functions*, preprint, 2003.

[2] B. H. Gross,, *Heegner points and representation theory*, in Heegner Points and Rankin L-series, Math. Sci. Res. Inst. Publ., **49**, 37–65, Cambridge Univ. Press, Cambridge, 2004.

[3] B. H. Gross, W. Kohnen, and D. Zagier, *Heegner points and derivatives of L-functions. II*, Math. Annalen, **278** (1987), 497–562.

[4] B. H. Gross, and D. Zagier, *Heegner points and the derivatives of L-series*, Invent. math., **84** (1986), 225–320.

[5] R. Howe, *θ–series and invariant theory*, Proc. Symp. Pure Math., **33** (1979), 275–285.

[6] R. Howe and I. I. Piatetski-Shapiro, *Some examples of automorphic forms on* Sp_4, Duke Math. J., **50** (1983), 55–106.

[7] W. Kohnen, *Newforms of half-integral weight*, J. reine angew. Math., **333** (1982) 32–72.

[8] S. Kudla, *Special cycles and derivatives of Eisenstein series*, in Heegner points and Rankin L-series, Math. Sci. Res. Inst. Publ., **49**, 243–270, Cambridge Univ. Press, Cambridge, 2004.

[9] ______, *Modular forms and arithmetic geometry*, in Current Developments in Mathematics, 2002, 135–179, International Press, Somerville, MA, 2003.

[10] ______, *Notes on the local theta correspondence for* $(\widetilde{\mathrm{SL}}_2, \mathrm{O}(3))$, preprint, 2005.

[11] W. Stein, The Modular Form Database, http://modular.ucsd.edu

[12] J.-L. Waldspurger, *Correspondance de Shimura*, J. Math. Pures Appl., **59** (1980), 1–132.

[13] ______, *Sur les coefficients de Fourier des formes modulaires de poids demi-entier*, J. Math. Pures Appl., **60** (1981), 375–484.

[14] ______, *Sur les valeurs de certaines fonctions L automorphes en leur centre de symétrie*, Compositio Math., **54** (1985), 173–242.

[15] ______, *Correspondances de Shimura et quaternions*, Forum Math., **3** (1991), 219–307.

[16] D. Zagier, *Modular points, modular curves, modular surfaces and modular forms*, in Lecture Notes in Math. **1111**, 225–248, Springer-Verlag, Berlin, 1985.

[17] Shou-Wu Zhang, *Gross–Zagier formula for* GL_2, Asian J. of Math., **5** (2001), 183–290.

Index

www.ingramcontent.com/pod-product-compliance
Ingram Content Group UK Ltd.
Pitfield, Milton Keynes, MK11 3LW, UK
UKHW022242121224
452420UK00005B/341